T0207201

Introduction to Plasma Physics and Controlled Fusion

Francis F. Chen

Introduction to Plasma Physics and Controlled Fusion

Third Edition

 Springer

Francis F. Chen
Electrical Engineering
University of California at Los Angeles
Los Angeles, CA, USA

ISBN 978-3-319-79391-7   ISBN 978-3-319-22309-4   (eBook)
DOI 10.1007/978-3-319-22309-4

Springer Cham Heidelberg New York Dordrecht London
© Springer International Publishing Switzerland 2016, corrected publication 2018
Softcover re-print of the Hardcover 3rd edition 2016

Printed on acid-free paper

Springer International Publishing AG Switzerland is part of Springer Science+Business Media (www.springer.com)

Preface

It has been 30 years since the Second Edition. Plasma physics has grown so much that the temptation is to include all that's new, but I haven't done that. This is a book for those learning plasma physics for the first time and could care less about the fine points. The basics remain the same, but new areas have arisen: dusty plasmas and plasma accelerators, for instance, and these subjects had to be added. The semi-conductor industry has sprouted since the 1980s, affecting all our lives. Computer chips cannot be made without plasmas, but these are partially ionized, radiofrequency ones, which are new. I've spent 25 years helping to make these industrial plasmas into an interesting subject, one that was previously avoided because it was too messy. Meanwhile, Al Gore has warned us about global warming. The best solution to this, and to energy shortage, is of course hydrogen fusion. With the ITER (International Thermonuclear Experimental Reactor) project going swimmingly in southern France, we are on schedule to get a fusion reactor by 2050. In laser fusion, the NIF (National Ignition Facility) at Livermore can focus the entire energy output of the United States onto a pinpoint for a fraction of a nanosecond to get inertial fusion. It is a technical achievement without peer, but so far we don't have a laser that can pulse rapidly enough. Fusion plasma physics was planned to be Vol. 2 of this textbook; and, in answer to so many questions, it is still in future plans. The job will be easier now that I have written *An Indispensable Truth, How Fusion Power Can Save the Planet*, which includes a chapter on fusion physics.

On the personal side, you will remember from the second edition that the "poet" (my father, who used to recite Chinese poetry from memory) had passed away, and that the "eternal scholar" (my mother) had finally gotten her Ph.D. at 72. She lived to 99 and died with a smile on her face. I myself am now 85, happily married for 49 years to artist Ande (Edna backwards), and we have two daughters and a son: Sheryl, a Cistercian nun in Norway; Patricia, a dancer and English teacher in France; and Bob, a chemical oceanography professor and education expert at UMass Boston, as well as an active soccer coach. We have four granddaughters, the three Boston ones being third-generation soccer players, and the French one a four-language linguist and former equestrian.

Finally, I would like to thank all the instructors and students who have used the second edition and have pointed out errors and made suggestions for improvement.

Los Angeles, CA   Francis F. Chen
2015

Contents

The original version of this book was revised. An erratum to this book can be found at https://doi.
org/10.1007/978-3-319-22309-4_11

Chapter 1
Introduction

1.1 Occurrence of Plasmas in Nature

It is now believed that the universe is made of 69 % dark energy, 27 % dark matter, and 1 % normal matter. All that we can see in the sky is the part of normal matter that is in the plasma state, emitting radiation. Plasma in physics, not to be confused with blood plasma, is an "ionized" gas in which at least one of the electrons in an atom has been stripped free, leaving a positively charged nucleus, called an ion. Sometimes plasma is called the "fourth state of matter." When a solid is heated, it becomes a liquid. Heating a liquid turns it into a gas. Upon further heating, the gas is ionized into a plasma. Since a plasma is made of ions and electrons, which are charged, electric fields are rampant everywhere, and particles "collide" not just when they bump into one another, but even at a distance where they can feel their electric fields. Hydrodynamics, which describes the flow of water through pipes, say, or the flow around boats in yacht races, or the behavior of airplane wings, is already a complicated subject. Adding the electric fields of a plasma greatly expands the range of possible motions, especially in the presence of magnetic fields.

Plasma usually exists only in a vacuum. Otherwise, air will cool the plasma so that the ions and electrons will recombine into normal neutral atoms. In the laboratory, we need to pump the air out of a vacuum chamber. In the vacuum of space, however, much of the gas is in the plasma state, and we can see it. Stellar interiors and atmospheres, gaseous nebulas, and entire galaxies can be seen because they are in the plasma state. On earth, however, our atmosphere limits our experience with plasmas to a few examples: the flash of a lightning bolt, the soft glow of the Aurora Borealis, the light of a fluorescent tube, or the pixels of a plasma TV. We live in a small part of the universe where plasmas do not occur naturally; otherwise, we would not be alive.

© Springer International Publishing Switzerland 2016
F.F. Chen, *Introduction to Plasma Physics and Controlled Fusion*,
DOI 10.1007/978-3-319-22309-4_1

The reason for this can be seen from the Saha equation, which tells us the amount of ionization to be expected in a gas in thermal equilibrium:

$$\frac{n_i}{n_n} \approx 2.4 \times 10^{21} \frac{T^{3/2}}{n_i} e^{-U_i/KT} \tag{1.1}$$

Here n_i and n_n are, respectively, the density (number per m^3) of ionized atoms and of neutral atoms, T is the gas temperature in °K, K is Boltzmann's constant, and U_i is the ionization energy of the gas—that is, the number of joules required to remove the outermost electron from an atom. (The mks or International System of units will be used in this book.) For ordinary air at room temperature, we may take $n_n \approx 3 \times 10^{25}$ m^{-3} (see Problem 1.1), $T \approx 300°$ K, and $U_i = 14.5$ eV (for nitrogen), where 1 eV $= 1.6 \times 10^{-19}$ J. The fractional ionization $n_i/(n_n + n_i) \approx n_i/n_n$ predicted by Eq. (1.1) is ridiculously low:

$$\frac{n_i}{n_n} \approx 10^{-122}$$

As the temperature is raised, the degree of ionization remains low until U_i is only a few times KT. Then n_i/n_n rises abruptly, and the gas is in a plasma state. Further increase in temperature makes n_n less than n_i, and the plasma eventually becomes fully ionized. This is the reason plasmas exist in astronomical bodies with temperatures of millions of degrees, but not on the earth. Life could not easily coexist with a plasma— at least, plasma of the type we are talking about. The natural occurrence of plasmas at high temperatures is the reason for the designation "the fourth state of matter."

Although we do not intend to emphasize the Saha equation, we should point out its physical meaning. Atoms in a gas have a spread of thermal energies, and an atom is ionized when, by chance, it suffers a collision of high enough energy to knock out an electron. In a cold gas, such energetic collisions occur infrequently, since an atom must be accelerated to much higher than the average energy by a series of "favor-able" collisions. The exponential factor in Eq. (1.1) expresses the fact that the number of fast atoms falls exponentially with U_i /KT. Once an atom is ionized, it remains charged until it meets an electron; it then very likely recombines with the electron to become neutral again. The recombination rate clearly depends on the density of electrons, which we can take as equal to n_i. The equilibrium ion fraction, therefore, should decrease with n_i; and this is the reason for the factor n_i^{-1} on the right-hand side of Eq. (1.1). The plasma in the interstellar medium owes its existence to the low value of n_i (about 1 per cm^3), and hence the low recombination rate.

1.2   Definition of Plasma

Any ionized gas cannot be called a plasma, of course; there is always some small degree of ionization in any gas. A useful definition is as follows:

A plasma is a quasineutral gas of charged and neutral particles which exhibits collective behavior.

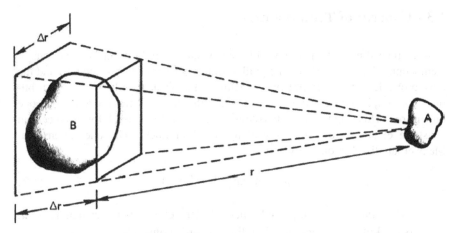

Fig. 1.1 Illustrating the long range of electrostatic forces in a plasma

We must now define "quasineutral" and "collective behavior." The meaning of quasineutrality will be made clear in Sect. 1.4. What is meant by "collective behavior" is as follows.

Consider the forces acting on a molecule of, say, ordinary air. Since the molecule is neutral, there is no net electromagnetic force on it, and the force of gravity is negligible. The molecule moves undisturbed until it makes a collision with another molecule, and these collisions control the particle's motion. A macroscopic force applied to a neutral gas, such as from a loudspeaker generating sound waves, is transmitted to the individual atoms by collisions. The situation is totally different in a plasma, which has *charged* particles. As these charges move around, they can generate local concentrations of positive or negative charge, which give rise to electric fields. Motion of charges also generates currents, and hence magnetic fields. These fields affect the motion of other charged particles far away.

Let us consider the effect on each other of two slightly charged regions of plasma separated by a distance r (Fig. 1.1). The Coulomb force between A and B diminishes as $1/r^2$. However, for a given solid angle (that is, $\Delta r/r = $ constant), the volume of plasma in B that can affect A increases as r^3. Therefore, elements of plasma exert a force on one another even at large distances. It is this long-ranged Coulomb force that gives the plasma a large repertoire of possible motions and enriches the field of study known as plasma physics. In fact, the most interesting results concern so-called "collisionless" plasmas, in which the long-range electro-magnetic forces are so much larger than the forces due to ordinary local collisions that the latter can be neglected altogether. By "collective behavior" we mean motions that depend not only on local conditions but on the state of the plasma in remote regions as well.

The word "plasma" seems to be a misnomer. It comes from the Greek πλάσμα, −ατος, τό, which means something molded or fabricated. Because of collective behavior, a plasma does not tend to conform to external influences; rather, it often behaves as if it had a mind of its own.

1.3   Concept of Temperature

Before proceeding further, it is well to review and extend our physical notions of "temperature." A gas in thermal equilibrium has particles of all velocities, and the most probable distribution of these velocities is known as the Maxwellian distribution. For simplicity, consider a gas in which the particles can move only in one dimension. (This is not entirely frivolous; a strong magnetic field, for instance, can constrain electrons to move only along the field lines.) The one-dimensional Maxwellian distribution is given by

$$f(u) = A \exp\left(-\tfrac{1}{2}mu^2/KT\right) \tag{1.2}$$

where $f\,du$ is the number of particles per m^3 with velocity between u and $u+du$, $\tfrac{1}{2}mu^2$ is the kinetic energy, and K is Boltzmann's constant,

$$K = 1.38 \times 10^{-23} \mathrm{J/^\circ K}$$

Note that a capital K is used here, since lower-case k is reserved for the propagation constant of waves. The density n, or number of particles per m^3, is given by (see Fig. 1.2)

$$n = \int_{-\infty}^{\infty} f(u)\,du \tag{1.3}$$

The constant A is related to the density n by (see Problem 1.2)

$$A = n\left(\frac{m}{2\pi KT}\right)^{1/2} \tag{1.4}$$

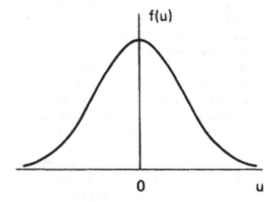

Fig. 1.2 A Maxwellian velocity distribution

The width of the distribution is characterized by the constant T, which we call the temperature. To see the exact meaning of T, we can compute the average kinetic energy of particles in this distribution:

$$E_{av} = \frac{\int_{-\infty}^{\infty} \frac{1}{2} m u^2 f(u) du}{\int_{-\infty}^{\infty} f(u) du} \tag{1.5}$$

Defining

$$v_{th} = (2KT/m)^{1/2} \quad \text{and} \quad y = u/v_{th} \tag{1.6}$$

we can write Eq. (1.2) as

$$f(u) = A \exp\left(-u^2/v_{th}^2\right)$$

and Eq. (1.5) as

$$E_{av} = \frac{\frac{1}{2} m A v_{th}^3 \int_{-\infty}^{\infty} \left[\exp\left(-y^2\right)\right] y^2 dy}{A v_{th} \int_{-\infty}^{\infty} \exp\left(-y^2\right) dy}$$

The integral in the numerator is integrable by parts:

$$\int_{-\infty}^{\infty} y \cdot \left[\exp\left(-y^2\right)\right] y \, dy = \left[-\frac{1}{2}[\exp(-y^2)]y\right]_{-\infty}^{\infty} - \int_{-\infty}^{\infty} -\frac{1}{2}\exp\left(-y^2\right) dy$$

$$= \frac{1}{2} \int_{-\infty}^{\infty} \exp\left(-y^2\right) dy$$

Canceling the integrals, we have

$$E_{av} = \frac{\frac{1}{2} m A v_{th}^3 \frac{1}{2}}{A v_{th}} = \frac{1}{4} m v_{th}^2 = \frac{1}{2} KT \tag{1.7}$$

Thus the average kinetic energy is $\frac{1}{2} KT$.

It is easy to extend this result to three dimensions. Maxwell's distribution is then

$$f(u, v, w) = A_3 \exp\left[-\frac{1}{2} m (u^2 + v^2 + w^2)/KT\right] \tag{1.8}$$

where

$$A_3 = n \left(\frac{m}{2\pi KT}\right)^{3/2} \tag{1.9}$$

The average kinetic energy is

$$
E_{av} = \frac{\iiint_{-\infty}^{\infty} A_3 \tfrac{1}{2} m(u^2 + v^2 + w^2) \exp\left[-\tfrac{1}{2} m(u^2 + v^2 + w^2)/KT\right] du\, dv\, dw}{\iiint_{-\infty}^{\infty} A_3 \exp\left[-\tfrac{1}{2} m(u^2 + v^2 + w^2)/KT\right] du\, dv\, dw}
$$

We note that this expression is symmetric in u, v, and w, since a Maxwellian distribution is isotropic. Consequently, each of the three terms in the numerator is the same as the others. We need only to evaluate the first term and multiply by three:

$$
E_{av} = \frac{3A_3 \int \tfrac{1}{2} mu^2 \exp\left(-\tfrac{1}{2} mu^2/KT\right) du \iint \exp\left[-\tfrac{1}{2} m(v^2 + w^2)/KT\right] dv\, dw}{A_3 \int \exp\left(-\tfrac{1}{2} mu^2/KT\right) du \iint \exp\left[-\tfrac{1}{2} m(v^2 + w^2)/KT\right] dv\, dw}
$$

Using our previous result, we have

$$
E_{av} = \tfrac{3}{2} KT \tag{1.10}
$$

The general result is that E_{ay} equals $\tfrac{1}{2} KT$ per degree of freedom.

Since T and E_{av} are so closely related, it is customary in plasma physics to give temperatures in units of energy. To avoid confusion on the number of dimensions involved, it is not E_{av} but the energy corresponding to KT that is used to denote the temperature. For $KT = 1$ eV $= 1.6 \times 10^{-19}$ J, we have

$$
T = \frac{1.6 \times 10^{-19}}{1.38 \times 10^{-23}} = 11,600
$$

Thus the conversion factor is

$$
1\,\text{eV} = 11,600\,^{\circ}\text{K} \tag{1.11}
$$

By a 2-eV plasma we mean that $KT = 2$ eV, or $E_{av} = 3$ eV in three dimensions.

It is interesting that a plasma can have several temperatures at the same time. It often happens that the ions and the electrons have separate Maxwellian distributions with different temperatures T_i and T_e. This can come about because the collision rate among ions or among electrons themselves is larger than the rate of collisions between an ion and an electron. Then each species can be in its own thermal equilibrium, but the plasma may not last long enough for the two temperatures to equalize. When there is a magnetic field $\mathbf{B}$, even a single species, say ions, can have two temperatures. This is because the forces acting on an ion along $\mathbf{B}$ are different from those acting perpendicular to $\mathbf{B}$ (due to the Lorentz force). The components of velocity perpendicular to $\mathbf{B}$ and parallel to $\mathbf{B}$ may then belong to different Maxwellian distributions with temperatures $T_\perp$ and $T_\parallel$.

Before leaving our review of the notion of temperature, we should dispel the popular misconception that high temperature necessarily means a lot of heat. People are usually amazed to learn that the electron temperature inside a fluorescent light bulb is about 20,000° K. "My, it doesn't feel that hot!" Of course, the heat capacity must also be taken into account. The density of electrons inside a fluorescent tube is much less than that of a gas at atmospheric pressure, and the total amount of heat transferred to the wall by electrons striking it at their thermal velocities is not that great. Everyone has had the experience of a cigarette ash dropped innocuously on his hand. Although the temperature is high enough to cause a burn, the total amount of heat involved is not. Many laboratory plasmas have temperatures of the order of 1,000,000° K (100 eV), but at densities of only 10^{18}–10^{19} per m^3, the heating of the walls is not a serious consideration.

Problems

1.1. Compute the density (in units of m^{-3}) of an ideal gas under the following conditions:

   (a) At 0 °C and 760 Torr pressure (1 Torr = 1 mmHg). This is called the Loschmidt number.
   (b) In a vacuum of 10^{-3} Torr at room temperature (20 °C). This number is a useful one for the experimentalist to know by heart (10^{-3} Torr = 1 μ).

1.2. Derive the constant A for a normalized one-dimensional Maxwellian distribution

$$\hat{f}(u) = A\exp(-mu^2/2KT)$$

such that

$$\int_{-\infty}^{\infty} \hat{f}(u)du = 1$$

Hint: To save writing, replace $(2KT/m)^{1/2}$ by v_{th} (Eq. 1.6).

1.2a. (Advanced problem). Find A for a two-dimensional distribution which integrates to unity. Extra credit for a solution in cylindrical coordinates.

$$\hat{f}(u,v) = A\exp[-m(u^2+v^2)/2KT]$$

1.4 Debye Shielding

A fundamental characteristic of the behavior of plasma is its ability to shield out electric potentials that are applied to it. Suppose we tried to put an electric field inside a plasma by inserting two charged balls connected to a battery (Fig. 1.3). The balls would attract particles of the opposite charge, and almost immediately a cloud of ions would surround the negative ball and a cloud of electrons would surround

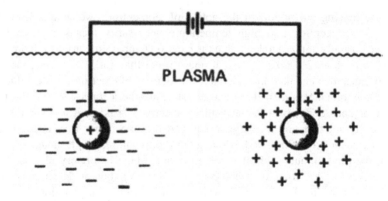

Fig. 1.3 Debye shielding

Fig. 1.4 Potential
distribution near a grid
in a plasma

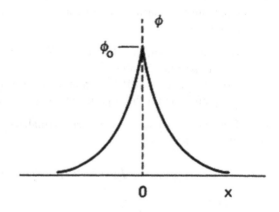

the positive ball. (We assume that a layer of dielectric keeps the plasma from actually recombining on the surface, or that the battery is large enough to maintain the potential in spite of this.) If the plasma were cold and there were no thermal motions, there would be just as many charges in the cloud as in the ball, the shielding would be perfect, and no electric field would be present in the body of the plasma outside of the clouds. On the other hand, if the temperature is finite, those particles that are at the edge of the cloud, where the electric field is weak, have enough thermal energy to escape from the electrostatic potential well. The "edge" of the cloud then occurs at the radius where the potential energy is approximately equal to the thermal energy KT of the particles, and the shielding is not complete. Potentials of the order of KT/e can leak into the plasma and cause finite electric fields to exist there.

Let us compute the approximate thickness of such a charge cloud. Imagine that the potential ϕ on the plane $x = 0$ is held at a value ϕ_0 by a perfectly transparent grid (Fig. 1.4). We wish to compute $\phi(x)$. For simplicity, we assume that the ion–electron mass ratio M/m is infinite, so that the ions do not move but form a uniform background of positive charge. To be more precise, we can say that M/m is large enough that the inertia of the ions prevents them from moving significantly on the time scale of the experiment. Poisson's equation in one dimension is

$$\varepsilon_0 \nabla^2 \phi = \varepsilon_0 \frac{d^2\phi}{dx^2} = -e(n_i - n_e) \quad (Z = 1) \tag{1.12}$$

If the density far away is n_∞, we have

$$n_i = n_\infty$$

In the presence of a potential energy $q\phi$, the electron distribution function is

$$f(u) = A \exp\left[-\left(\tfrac{1}{2}mu^2 + q\phi\right)/KT_e\right] \tag{1.13}$$

It would not be worthwhile to prove this here. What this equation says is intuitively obvious: There are fewer particles at places where the potential energy is large, since not all particles have enough energy to get there. Integrating $f(u)$ over u, setting $q = -e$, and noting that $n_e(\phi \to 0) = n_\infty$, we find

$$n_e = n_\infty \exp(e\phi/KT_e)$$

This equation will be derived with more physical insight in Sect. 3.5. Substituting for n_i and n_e in Eq. (1.12), we have

$$\varepsilon_0 \frac{d^2\phi}{dx^2} = en_\infty \left(e^{e\phi/KT_e} - 1\right)$$

In the region where $|e\phi/KT_e| \ll 1$, we can expand the exponential in a Taylor series:

$$\varepsilon_0 \frac{d^2\phi}{dx^2} = en_\infty \left[\frac{e\phi}{KT_e} + \tfrac{1}{2}\left(\frac{e\phi}{KT_e}\right)^2 + \cdots\right] \tag{1.14}$$

No simplification is possible for the region near the grid, where $|e\phi/KT_e|$ may be large. Fortunately, this region does not contribute much to the thickness of the cloud (called a sheath), because the potential falls very rapidly there. Keeping only the linear terms in Eq. (1.13), we have

$$\varepsilon_0 \frac{d^2\phi}{dx^2} = \frac{n_\infty e^2}{KT_e}\phi \tag{1.15}$$

Defining

$$\boxed{\lambda_D \equiv \left(\frac{\varepsilon_0 KT_e}{ne^2}\right)^{1/2}} \tag{1.16}$$

where n stands for n_∞, and KT_e is in joules. KT_e is often given in eV, in which case, we will write it also as T_{eV}.

We can write the solution of Eq. (1.14) as

$$\phi = \phi_0 \exp(-|x|/\lambda_D) \tag{1.17}$$

The quantity λ_D, called the Debye length, is a measure of the shielding distance or thickness of the sheath.

Note that as the density is increased, λ_D decreases, as one would expect, since each layer of plasma contains more electrons. Furthermore, λ_D increases with increasing KT_e. Without thermal agitation, the charge cloud would collapse to an infinitely thin layer. Finally, it is the *electron* temperature which is used in the definition of λ_D because the electrons, being more mobile than the ions, generally do the shielding by moving so as to create a surplus or deficit of negative charge. Only in special situations is this not true (see Problem 1.5).

The following are useful forms of Eq. (1.16):

$$\lambda_D = 69(T_e/n)^{1/2}\mathrm{m}, \qquad T_e \text{ in } {}^\circ\mathrm{K}$$
$$\lambda_D = 7430(KT_e/n)^{1/2}\mathrm{m}, \qquad KT_e \text{ in eV} \tag{1.18}$$

We are now in a position to define "quasineutrality." If the dimensions L of a system are much larger than λ_D, then whenever local concentrations of charge arise or external potentials are introduced into the system, these are shielded out in a distance short compared with L, leaving the bulk of the plasma free of large electric potentials or fields. Outside of the sheath on the wall or on an obstacle, $\nabla^2\phi$ is very small, and n_i is equal to n_e, typically to better than one part in 10^6. It takes only a small charge imbalance to give rise to potentials of the order of KT/e. The plasma is "quasineutral"; that is, neutral enough so that one can take $n_i \simeq n_e \simeq n$, where n is a common density called the *plasma density*, but not so neutral that all the interesting electromagnetic forces vanish.

A criterion for an ionized gas to be a plasma is that it be dense enough that λ_D is much smaller than L.

The phenomenon of Debye shielding also occurs—in modified form—in single-species systems, such as the electron streams in klystrons and magnetrons or the proton beam in a cyclotron. In such cases, any local bunching of particles causes a large unshielded electric field unless the density is extremely low (which it often is). An externally imposed potential—from a wire probe, for instance—would be shielded out by an adjustment of the density near the electrode. Single-species systems, or unneutralized plasmas, are not strictly plasmas; but the mathematical tools of plasma physics can be used to study such systems.

Debye shielding can be foiled if electrons are so fast that they do not collide with one another enough to maintain a thermal distribution. We shall see later that electron collisions are infrequent if the electrons are very hot. In that case, some electrons, attracted by the positive charge of the ion, come in at an angle so fast that they orbit the ion like a satellite around a planet. How this works will be clear in the discussion of Langmuir probes in a later chapter. Some like to call this effect *anti-shielding*.

1.5 The Plasma Parameter

The picture of Debye shielding that we have given above is valid only if there are enough particles in the charge cloud. Clearly, if there are only one or two particles in the sheath region, Debye shielding would not be a statistically valid concept. Using Eq. (1.17), we can compute the number N_D of particles in a "Debye sphere":

$$N_D = n\tfrac{4}{3}\pi\lambda_D^3 = 1.38 \times 10^6 T^{3/2}/n^{1/2} \quad (T \text{ in } {}^\circ\text{K}) \tag{1.19}$$

In addition to $\lambda_D \ll L$, "collective behavior" requires

$$N_D \ggg 1 \tag{1.20}$$

1.6 Criteria for Plasmas

We have given two conditions that an ionized gas must satisfy to be called a plasma. A third condition has to do with collisions. The weakly ionized gas in an airplane's jet exhaust, for example, does not qualify as a plasma because the charged particles collide so frequently with neutral atoms that their motion is controlled by ordinary hydrodynamic forces rather than by electromagnetic forces. If ω is the frequency of typical plasma oscillations and τ is the mean time between collisions with neutral atoms, we require $\omega\tau > 1$ for the gas to behave like a plasma rather than a neutral gas.

The three conditions a plasma must satisfy are therefore:

1. $\lambda_D \ll L$.
2. $N_D \ggg 1$.
3. $\omega\tau > 1$.

Problems

1.3. Calculate n vs. KT_e curves for five values of λ_D from 10^{-8} to 1, and three values of N_D from 10^3 to 10^9. On a log-log plot of n_e vs. KT_e with n_e from 10^6 to 10^{28} m^{-3} and KT_e from 10^{-2} to 10^5 eV, draw lines of constant λ_D (solid) and N_D (dashed). On this graph, place the following points (n in m^{-3}, KT in eV):

1. Typical fusion reactor: $n = 10^{20}$, $KT = 30{,}000$.
2. Typical fusion experiments: $n = 10^{19}$, $KT = 100$ (torus); $n = 10^{23}$, $KT = 1000$ (pinch).
3. Typical ionosphere: $n = 10^{11}$, $KT = 0.05$.
4. Typical radiofrequency plasma: $n = 10^{17}$, $KT = 1.5$
5. Typical flame: $n = 10^{14}$, $KT = 0.1$.
6. Typical laser plasma; $n = 10^{25}$, $KT = 100$.
7. Interplanetary space: $n = 10^6$, $KT = 0.01$.

Convince yourself that these are plasmas.

1.4. Compute the pressure, in atmospheres and in tons/ft^2, exerted by a thermonuclear plasma on its container. Assume $KT_e = KT_i = 20$ keV, $n = 10^{21}$ m^{-3}, and $p = nKT$, where $T = T_i + T_e$.

1.5. In a strictly steady state situation, both the ions and the electrons will follow the Boltzmann relation

$$n_j = n_0 \exp\left(-q_i\phi/KT_j\right)$$

For the case of an infinite, transparent grid charged to a potential ϕ, show that the shielding distance is then given approximately by

$$\lambda_D^{-2} = \frac{ne^2}{\epsilon_0}\left(\frac{1}{KT_e} + \frac{1}{KT_i}\right)$$

Show that λ_D is determined by the temperature of the colder species.

1.6. An alternative derivation of λ_D will give further insight to its meaning. Consider two infinite parallel plates at $x = \pm d$, set at potential $\phi = 0$. The space between them is uniformly filled by a gas of density n of particles of charge q.

   (a) Using Poisson's equation, show that the potential distribution between the plates is

$$\phi = \frac{nq}{2\varepsilon_0}\left(d^2 - x^2\right)$$

   (b) Show that for $d > \lambda_D$, the energy needed to transport a particle from a plate to the midplane is greater than the average kinetic energy of the particles.

1.7. Compute λ_D and N_D for the following cases:

   (a) A glow discharge, with $n = 10^{16}$ m^{-3}, $KT_e = 2$ eV.
   (b) The earth's ionosphere, with $n = 10^{12}$ m^{-3}, $KT_e = 0.1$ eV.
   (c) A θ-pinch, with $n = 10^{23}$ m^{-3}, $KT_e = 800$ eV.

1.7 Applications of Plasma Physics

Plasmas can be characterized by the two parameters n and KT_e. Plasma applications cover an extremely wide range of n and KT_e: n varies over 28 orders of magnitude from 10^6 to 10^{34} m^{-3}, and KT can vary over seven orders from 0.1 to 10^6 eV. Some of these applications are discussed very briefly below. The tremendous range of density can be appreciated when one realizes that air and water differ in density by only 10^3, while water and white dwarf stars are separated by only a factor of 10^5. Even neutron stars are only 10^{15} times denser than water. Yet gaseous plasmas in the entire density range of 10^{28} can be described by the same set of equations, since only the classical (non-quantum mechanical) laws of physics are needed.

1.7.1 Gas Discharges (Gaseous Electronics)

The earliest work with plasmas was that of Langmuir, Tonks, and their collaborators in the 1920s. This research was inspired by the need to develop vacuum tubes that could carry large currents, and therefore had to be filled with ionized gases. The research was done with weakly ionized glow discharges and positive columns typically with $KT_e \simeq 2$ eV and $10^{14} < n < 10^{18}$ m^{-3}. It was here that the shielding phenomenon was discovered; the sheath surrounding an electrode could be seen visually as a dark layer. Before semiconductors, gas discharges were encountered only in mercury rectifiers, hydrogen thyratrons, ignitrons, spark gaps, welding arcs, neon and fluorescent lights, and lightning discharges. The semiconductor industry's rapid growth in the last two decades has brought gas discharges from a small academic discipline to an economic giant. Chips for computers and the ubiquitous handheld devices cannot be made without plasmas. Usually driven by radiofrequency power, partially ionized plasmas (gas discharges) are used for etching and deposition in the manufacture of semiconductors.

1.7.2 Controlled Thermonuclear Fusion

Modern plasma physics had its beginnings around 1952, when it was proposed that the hydrogen bomb fusion reaction be controlled to make a reactor. A seminal conference was held in Geneva in 1958 at which each nation revealed its classified controlled fusion program for the first time. Fusion power requires holding a 30-keV plasma with a magnetic field for as long as one second. Research was carried out by each individual country until 2007, when the ITER project was started. ITER stands for International Thermonuclear Experimental Reactor, a large experiment being built in France and funded by seven countries. Serendipitously, ITER is a Latin word meaning path or road. It is a road that mankind must take to solve the problems of global warming and oil shortage by 2050. Of all the "magnetic bottles" presented at Geneva, the USSR's TOKAMAK has survived as the leading idea and is the configuration for ITER. First plasma in ITER is scheduled for about the year 2021.

The fuel is heavy hydrogen (deuterium), which exists naturally as one part in $\sim$6000 of water. The principal reactions, which involve deuterium (D) and tritium (T) atoms, are as follows:

$$D + D \rightarrow {}^3He + n + 3.2\,MeV$$
$$D + D \rightarrow T + p + 4.0\,MeV$$
$$D + T \rightarrow {}^4He + n + 17.6\,MeV$$

These cross sections are appreciable only for incident energies above 5 keV. Accelerated beams of deuterons bombarding a target will not work, because most of the deuterons will lose their energy by scattering before undergoing a fusion reaction. It is necessary to create a plasma with temperatures above 10 keV so that there are enough ions in the 40-keV range where the reaction cross section maximizes. The problem of heating and containing such a plasma is responsible for the rapid growth of the science of plasma physics since 1952.

1.7.3 Space Physics

Another important application of plasma physics is in the study of the earth's environment in space. A continuous stream of charged particles, called the solar wind, impinges on the earth's magnetosphere, which shields us from this radiation and is distorted by it in the process. Typical parameters in the solar wind are $n = 9 \times 10^6\,m^{-3}$, $KT_i = 10$ eV, $KT_e = 12$ eV, $B = 7 \times 10^{-9}$ T, and drift velocity 450 km/s. The ionosphere, extending from an altitude of 50 km to 10 earth radii, is populated by a weakly ionized plasma with density varying with altitude up to $n = 10^{12}$ m^{-3}. The temperature is only 10^{-1} eV. The solar wind blows the earth's magnetic field into a long tail on the night side of the earth. The magnetic field lines there can reconnect and accelerate ions in the process. This will be discussed in a later chapter.

The Van Allen radiation belts are two rings of charged particles above the equator trapped by the earth's magnetic field. Here we have $n \leq 10^9$ m^{-3}, $KT_e \leq 1$ keV, $KT_i \simeq 1$ eV, and $B \simeq 500 \times 10^{-9}$ T. In addition, there is a hot component with $n = 10^3$ m^{-3} and $KT_e = 40$ keV, and some ions have 100s of MeV.

Exploration of other planets have revealed the presence of plasmas. Though Mercury, Venus, and Mars have little plasma phenomena, the giant plants Jupiter and Saturn and their moons can have plasma created by lightning strikes. In 2013 The Voyager 1 satellite reached the boundary of the solar system. This was ascertained by detecting an *increase* in the plasma frequency there (!).

1.7.4 Modern Astrophysics

Stellar interiors and atmospheres are hot enough to be in the plasma state. The temperature at the core of the sun, for instance, is estimated to be 2 keV;

thermonuclear reactions occurring at this temperature are responsible for the sun's radiation. The solar corona is a tenuous plasma with temperatures up to 200 eV. The interstellar medium contains ionized hydrogen with $n \simeq 10^6 \, \mathrm{m}^{-3}$ (1 per cc). Various plasma theories have been used to explain the acceleration of cosmic rays. Although the stars in a galaxy are not charged, they behave like particles in a plasma; and plasma kinetic theory has been used to predict the development of galaxies. Radio astronomy has uncovered numerous sources of radiation that most likely originate from plasmas. The Crab nebula is a rich source of plasma phenomena because it is known to contain a magnetic field. It also contains a visual pulsar. Current theories of pulsars picture them as rapidly rotating neutron stars with plasmas emitting synchrotron radiation from the surface. Active galactic nuclei and black holes have come to the forefront. Astrophysics now requires an understanding of plasma physics.

1.7.5 MHD Energy Conversion and Ion Propulsion

Getting back down to earth, we come to two practical applications of plasma physics. Magnetohydrodynamic (MHD) energy conversion utilizes a dense plasma jet propelled across a magnetic field to generate electricity (Fig. 1.5). The Lorentz force $q\mathbf{v} \times \mathbf{B}$, where $\mathbf{v}$ is the jet velocity, causes the ions to drift upward and the electrons downward, charging the two electrodes to different potentials. Electrical current can then be drawn from the electrodes without the inefficiency of a heat cycle.

A more important application uses this principle in reverse to develop engines for interplanetary missions. In Fig. 1.6, a current is driven through a plasma by applying a voltage to the two electrodes. The $\mathbf{j} \times \mathbf{B}$ force shoots the plasma out of the rocket, and the ensuing reaction force accelerates the rocket. A more modern device called a Hall thruster uses a magnetic field to stop the electrons while a high voltage accelerates the ions out the back, giving their momenta to the spacecraft. A separate discharge ejects an equal number of warm electrons; otherwise, the space ship will charge to a high potential.

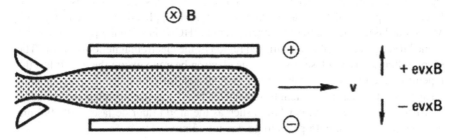

Fig. 1.5 Principle of the MHD generator

Fig. 1.6 Principle of
plasma-jet engine for
spacecraft propulsion

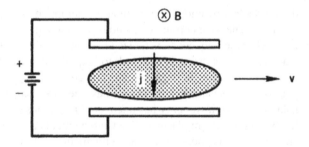

1.7.6 Solid State Plasmas

The free electrons and holes in semiconductors constitute a plasma exhibiting the same sort of oscillations and instabilities as a gaseous plasma. Plasmas injected into InSb have been particularly useful in studies of these phenomena. Because of the lattice effects, the effective collision frequency is much less than one would expect in a solid with $n \simeq 10^{29}$ m^{-3}. Furthermore, the holes in a semiconductor can have a very low effective mass—as little as 0.01 m_e—and therefore have high cyclotron frequencies even in moderate magnetic fields. If one were to calculate N_D for a solid state plasma, it would be less than unity because of the low temperature and high density. Quantum mechanical effects (uncertainty principle), however, give the plasma an effective temperature high enough to make N_D respectably large. Certain liquids, such as solutions of sodium in ammonia, have been found to behave like plasmas also.

1.7.7 Gas Lasers

The most common method to "pump" a gas laser—that is, to invert the population in the states that give rise to light amplification—is to use a gas discharge. This can be a low-pressure glow discharge for a dc laser or a high-pressure avalanche discharge in a pulsed laser. The He–Ne lasers commonly used for alignment and surveying and the Ar and Kr lasers used in light shows are examples of dc gas lasers. The powerful CO_2 laser has a commercial application as a cutting tool. Molecular lasers such as the hydrogen cyanide (HCN) laser make possible studies of the hitherto inaccessible far infrared region of the electromagnetic spectrum. The krypton fluoride (KrF) laser uses a large electron beam to excite the gas. It has the repetition rate but not the power for a laser-driven fusion reactor. In the semiconductor industry, short-wavelength ultraviolet lasers are used to etch ever smaller transistors on a chip. Excimer lasers such as the argon fluoride laser with 193 nm wavelength are used, with plans to go down to 5 nm.

1.7.8 Particle Accelerators

In high-energy particle research, linear accelerators are used to avoid synchrotron radiation on curves, especially for electrons. The 3-km long SLAC accelerator at Stanford produces 50-GeV electrons and positrons. Plasma accelerators generate plasma waves on which particles "surf" to gain energy. This technique has been applied at SLAC to double the energy from 42 to 84 GeV. A 31-km International Linear Collider is being built to generate 500 GeV colliding beams of electrons and positrons. In principle, plasma waves can add 1 GeV per cm.

1.7.9 Industrial Plasmas

Aside from their use in semiconductor production, partially ionized plasmas of high density have other industrial applications. Magnetrons are used in sputtering, a method of applying coatings of different materials. Eyeglasses can be coated with plasmas. Arcs, such as those in search lights, are plasmas. Instruments in medical research can be cleaned thoroughly with plasmas.

1.7.10 Atmospheric Plasmas

Most plasmas are created in vacuum systems, but it is also possible to produce plasmas at atmospheric pressure. For instance, a jet of argon and helium can be ionized with radiofrequency power. This makes possible small pencil-size devices for cauterizing skin. Industrial substrates can be processed by sweeping such a jet to cover a large area. Large atmospheric-pressure plasmas can also be produced for roll-to-roll processing.

Problems

1.8. In laser fusion, the core of a small pellet of DT is compressed to a density of 10^{33} m^{-3} at a temperature of 50,000,000 °K. Estimate the number of particles in a Debye sphere in this plasma.

1.9. A distant galaxy contains a cloud of protons and antiprotons, each with density $n = 10^6$ m^{-3} and temperature 100° K. What is the Debye length?

1.10. (Advanced problem) A spherical conductor of radius a is immersed in a uniform plasma and charged to a potential ϕ_0. The electrons remain Maxwellian and move to form a Debye shield, but the ions are stationary during the time frame of the experiment.

   (a) Assuming $e\phi/KT_e \ll 1$, write Poisson's equation for this problem in terms of λ_D.

   (b) Show that the equation is satisfied by a function of the form e^{-kr}/r. Determine k and derive an expression for $\phi(r)$ in terms of a, ϕ_0, and λ_D.

1.11. A field-effect transistor (FET) is basically an electron valve that operates on a finite-Debye-length effect. Conduction electrons flow from the source S to the drain D through a semiconducting material when a potential is applied between them. When a negative potential is applied to the insulated gate G, no current can flow through G, but the applied potential leaks into the semiconductor and repels electrons. The channel width is narrowed and the electron flow impeded in proportion to the gate potential. If the thickness of the device is too large, Debye shielding prevents the gate voltage from penetrating far enough. Estimate the maximum thickness of the conduction layer of an n-channel FET if it has doping level (plasma density) of $10^{22}\,\mathrm{m}^{-3}$, is at room temperature, and is to be no more than 10 Debye lengths thick. (See Fig. P1.11.)

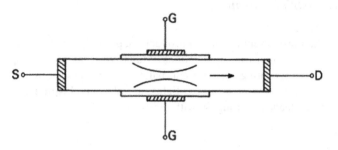

Fig. P1.11

1.12. (Advanced problem) Ionization is caused by electrons in the tail of a Maxwellian distribution which have energies exceeding the ionization potential. For instance, this potential is $E_{ioniz} = 15.8$ eV in argon. Consider a one-dimensional plasma with electron velocities u in the x direction only. What fraction of the electrons can ionize for given KT_e in argon? (Give an analytic answer in terms of error functions.)

Chapter 2
Single-Particle Motions

2.1 Introduction

What makes plasmas particularly difficult to analyze is the fact that the densities fall in an intermediate range. Fluids like water are so dense that the motions of individual molecules do not have to be considered. Collisions dominate, and the simple equations of ordinary fluid dynamics suffice. At the other extreme in very low-density devices like the alternating-gradient synchrotron, only single-particle trajectories need be considered; collective effects are often unimportant. Plasmas behave sometimes like fluids, and sometimes like a collection of individual particles. The first step in learning how to deal with this schizophrenic personality is to understand how single particles behave in electric and magnetic fields. This chapter differs from succeeding ones in that the **E** *and* **B** *fields are assumed to be prescribed* and not affected by the charged particles.

2.2 Uniform E and B Fields

2.2.1 *E = 0*

In this case, a charged particle has a simple cyclotron gyration. The equation of motion is

$$m\frac{dv}{dt} = q\mathbf{v} \times \mathbf{B} \tag{2.1}$$

Taking $\hat{z}$ to be the direction of **B** (**B** $= B\,\hat{z}$), we have

---

The original version of this chapter was revised. An erratum to this chapter can be found at
https://doi.org/10.1007/978-3-319-22309-4_11

© Springer International Publishing Switzerland 2016
F.F. Chen, *Introduction to Plasma Physics and Controlled Fusion*,
DOI 10.1007/978-3-319-22309-4_2

$$m\dot{v}_x = qBv_y \quad m\dot{v}_y = -qBv_x \quad m\dot{v}_z = 0$$

$$\ddot{v}_x = \frac{qB}{m}\dot{v}_y = -\left(\frac{qB}{m}\right)^2 v_x$$

$$\ddot{v}_y = -\frac{qB}{m}\dot{v}_x = -\left(\frac{qB}{m}\right)^2 v_y$$

(2.2)

This describes a simple harmonic oscillator at the *cyclotron frequency*, which we define to be

$$\boxed{\omega_c \equiv \frac{|q|B}{m}}$$

(2.3)

By the convention we have chosen, ω_c is always nonnegative. B is measured in tesla, or webers/m^2, a unit equal to 10^4 G. The solution of Eq. (2.2) is then

$$v_{x,y} = v_\perp \exp\left(\pm i\omega_c t + i\delta_{x,y}\right)$$

the $\pm$ denoting the sign of q. We may choose the phase δ so that

$$v_x = v_\perp e^{i\omega_c t} = \dot{x}$$

(2.4a)

where $v_\perp$ is a positive constant denoting the speed in the plane perpendicular to **B**. Then

$$v_y = \frac{m}{qB}\dot{v}_x = \pm\frac{1}{\omega_c}\dot{v}_x = \pm iv_\perp e^{i\omega_c t} = \dot{y}$$

(2.4b)

Integrating once again, we have

$$x - x_0 = -i\frac{v_\perp}{\omega_c}e^{i\omega_c t} \quad y - y_0 = \pm\frac{v_\perp}{\omega_c}e^{i\omega_c t}$$

(2.5)

We define the *Larmor radius* to be

$$\boxed{r_L \equiv \frac{v_\perp}{\omega_c} = \frac{mv_\perp}{|q|B}}$$

(2.6)

Taking the real part of Eq. (2.5), we have

$$x - x_0 = r_L \sin \omega_c t \quad y - y_0 = \pm r_L \cos \omega_c t$$

(2.7)

This describes a circular orbit around a *guiding center* (x_0, y_0) which is fixed (Fig. 2.1). The direction of the gyration is always such that the magnetic field

Fig. 2.1 Larmor orbits
in a magnetic field

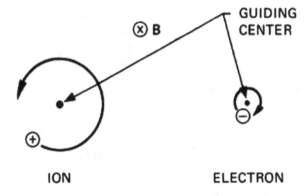

generated by the charged particle is opposite to the externally imposed field. Plasma particles, therefore, tend to *reduce* the magnetic field, and plasmas are *diamagnetic*. In Fig. 2.1, the right-hand rule with the thumb pointed in the **B** direction would give ions a clockwise gyration. Ions gyrate counterclockwise to generate an opposing **B**, thus lowering the energy of the system. In addition to this motion, there is an arbitrary velocity v_z along **B** which is not affected by **B**. The trajectory of a charged particle in space is, in general, a helix.

2.2.2   Finite E

If now we allow an electric field to be present, the motion will be found to be the sum of two motions: the usual circular Larmor gyration plus a drift of the guiding center. We may choose **E** to lie in the x–z plane so that $E_y = 0$. As before, the z component of velocity is unrelated to the transverse components and can be treated separately. The equation of motion is now

$$m\frac{d\mathbf{v}}{dt} = q(\mathbf{E} + \mathbf{v} \times \mathbf{B}) \tag{2.8}$$

whose z component is

$$\frac{dv_z}{dt} = \frac{q}{m}E_z$$

or

$$v_z = \frac{qE_z}{m}t + v_{z0} \tag{2.9}$$

This is a straightforward acceleration along **B**. The transverse components of Eq. (2.8) are

$$\frac{dv_x}{dt} = \frac{q}{m}E_x \pm \omega_c v_y$$

$$\frac{dv_y}{dt} = 0 \mp \omega_c v_x \qquad (2.10)$$

Differentiating, we have (for constant **E**)

$$\ddot{v}_x = -\omega_c^2 v_x$$

$$\ddot{v}_y = \mp\omega_c\left(\frac{q}{m}E_x \pm \omega_c v_y\right) = -\omega_c^2\left(\frac{E_x}{B} + v_y\right)$$

We can write this as

$$\frac{d^2}{dt^2}\left(v_y + \frac{E_x}{B}\right) = -\omega_c^2\left(v_y + \frac{E_x}{B}\right) \qquad (2.11)$$

so that Eq. (2.11) is reduced to the previous case (Eq. (2.2)) if we replace v_y by $v_y + (E_x/B)$. Equations (2.4a) and (2.4b) are therefore replaced by

$$v_x = v_\perp e^{i\omega_c t}$$

$$v_y = \pm i v_\perp e^{i\omega_c t} - \frac{E_x}{B} \qquad (2.12)$$

The Larmor motion is the same as before, but there is superimposed a drift $\mathbf{v}_{gc}$ of the guiding center in the $-y$ direction (for $E_x > 0$) (Fig. 2.2).

To obtain a general formula for $\mathbf{v}_{gc}$, we can solve Eq. (2.8) in vector form. We may omit the $m\, dv/dt$ term in Eq. (2.8), since this term gives only the circular motion at ω_c, which we already know about. Then Eq. (2.8) becomes

$$\mathbf{E} + \mathbf{v} \times \mathbf{B} = 0 \qquad (2.13)$$

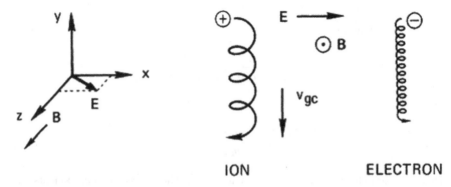

ION   ELECTRON

Fig. 2.2 Particle drifts in crossed electric and magnetic fields

Taking the cross product with **B**, we have

$$\mathbf{E} \times \mathbf{B} = \mathbf{B} \times (\mathbf{v} \times \mathbf{B}) = v B^2 - \mathbf{B}(\mathbf{v} \cdot \mathbf{B}) \tag{2.14}$$

The transverse components of this equation are

$$\mathbf{v}_{\perp gc} = \mathbf{E} \times \mathbf{B}/B^2 \equiv \mathbf{v}_E \tag{2.15}$$

We define this to be $\mathbf{v}_E$, the electric field drift of the guiding center. In magnitude, this drift is

$$\boxed{v_E = \frac{E(\text{V/m})}{B(\text{tesla})} \frac{\text{m}}{\text{sec}}} \tag{2.16}$$

It is important to note that $\mathbf{v}_E$ is independent of q, m, and $v_\perp$. The reason is obvious from the following physical picture. In the first half-cycle of the ion's orbit in Fig. 2.2, it gains energy from the electric field and increases in $v_\perp$ and, hence, in r_L. In the second half-cycle, it loses energy and decreases in r_L. This difference in r_L on the left and right sides of the orbit causes the drift v_E. A negative electron gyrates in the opposite direction but also gains energy in the opposite direction; it ends up drifting in the same direction as an ion. For particles of the same velocity but different mass, the lighter one will have smaller r_L and hence drift less per cycle. However, its gyration frequency is also larger, and the two effects exactly cancel. Two particles of the same mass but different energy would have the same ω_c. The slower one will have smaller r_L and hence gain less energy from **E** in a half-cycle. However, for less energetic particles the fractional change in r_L for a given change in energy is larger, and these two effects cancel (Problem 2.4).

The three-dimensional orbit in space is therefore a slanted helix with changing pitch (Fig. 2.3).

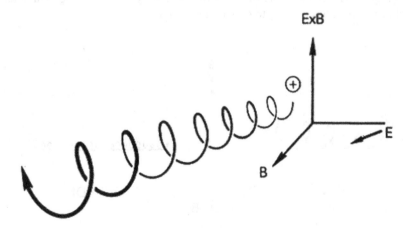

Fig. 2.3 The actual orbit of a gyrating particle in space

2.2.3 *Gravitational Field*

The foregoing result can be applied to other forces by replacing $q\mathbf{E}$ in the equation of motion (2.8) by a general force $\mathbf{F}$. The guiding center drift caused by $\mathbf{F}$ is then

$$\mathbf{v}_f = \frac{1}{q}\frac{\mathbf{F} \times \mathbf{B}}{B^2} \qquad (2.17)$$

In particular, if $\mathbf{F}$ is the force of gravity $m\mathbf{g}$, there is a drift

$$\mathbf{v}_g = \frac{m}{q}\frac{\mathbf{g} \times \mathbf{B}}{B^2} \qquad (2.18)$$

This is similar to the drift $\mathbf{v}_E$ in that it is perpendicular to both the force and $\mathbf{B}$, but it differs in one important respect. The drift $\mathbf{v}_g$ changes sign with the particle's charge. Under a gravitational force, ions and electrons drift in opposite directions, so there is a net current density in the plasma given by

$$\mathbf{j} = n(M+m)\frac{\mathbf{g} \times \mathbf{B}}{B^2} \qquad (2.19)$$

The physical reason for this drift (Fig. 2.4) is again the change in Larmor radius as the particle gains and loses energy in the gravitational field. Now the electrons gyrate in the opposite sense to the ions, but the force on them is in the same direction, so the drift is in the opposite direction. The magnitude of $\mathbf{v}_g$ is usually negligible (Problem 2.6), but when the lines of force are curved, there is an effective gravitational force due to centrifugal force. This force, which is *not* negligible, is independent of mass; this is why we did not stress the m dependence of Eq. (2.18). Centrifugal force is the basis of a plasma instability called the "gravitational" instability, which has nothing to do with real gravity.

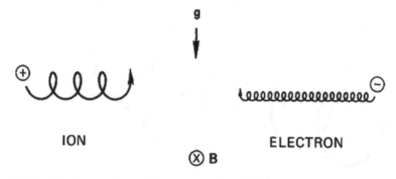

Fig. 2.4 The drift of a gyrating particle in a gravitational field

Problems

2.1. Compute r_L for the following cases if $v_\parallel$ is negligible:

   (a) A 10-keV electron in the earth's magnetic field of 5×10^{-5} T.
   (b) A solar wind proton with streaming velocity 300 km/s, $B = 5 \times 10^{-9}$ T.
   (c) A 1-keV He^+ ion in the solar atmosphere near a sunspot, where $B = 5 \times 10^{-2}$ T.
   (d) A 3.5-MeV He^{++} ash particle in an 8-T DT fusion reactor.

2.2. In the TFTR (Tokamak Fusion Test Reactor) at Princeton, the plasma was heated by injection of 200-keV neutral deuterium atoms, which, after entering the magnetic field, are converted to 200-keV D ions $(A = 2)$ by charge exchange. These ions are confined only if $r_L \ll a$, where $a = 0.6$ m is the minor radius of the toroidal plasma. Compute the maximum Larmor radius in a 5-T field to see if this is satisfied.

2.3. An ion engine (see Fig. 1.6) has a 1-T magnetic field, and a hydrogen plasma is to be shot out at an $E \times B$ velocity of 1000 km/s. How much internal electric field must be present in the plasma?

2.4. Show that v_E is the same for two ions of equal mass and charge but different energies, by using the following physical picture (see Fig. 2.2). Approximate the right half of the orbit by a semicircle corresponding to the ion energy after acceleration by the E field, and the left half by a semicircle corresponding to the energy after deceleration. You may assume that E is weak, so that the fractional change in $v_\perp$ is small.

2.5. Suppose electrons obey the Boltzmann relation of Problem 1.5 in a cylindrically symmetric plasma column in which $n(r)$ varies with a scale length λ; that is, $\partial n/\partial r \simeq -n/\lambda$.

   (a) Using $\mathbf{E} = -\nabla\phi$, find the radial electric field for given λ.
   (b) For electrons, show that finite Larmor radius effects are large if v_E is as large as v_{th}. Specifically, show that $r_L = 2\lambda$ if $v_E = v_{th}$.
   (c) Is (b) also true for ions?

   Hint: Do *not* use Poisson's equation.

2.6. Suppose that a so-called Q-machine has a uniform field of 0.2 T and a cylindrical plasma with $KT_e = KT_i = 0.2$ eV. The density profile is found experimentally to be of the form

$$n = n_0 \exp\left[\exp\left(-r^2/a^2\right) - 1\right]$$

Assume the density obeys the electron Boltzmann relation $n = n_0 \exp(e\phi/KT_e)$.

   (a) Calculate the maximum v_E if $a = 1$ cm.
   (b) Compare this with v_g due to the earth's gravitational field.
   (c) To what value can B be lowered before the ions of potassium $(A = 39, Z = 1)$ have a Larmor radius equal to a?

2.7. An unneutralized electron beam has density $n_e = 10^{14}\,\mathrm{m}^{-3}$ and radius $a = 1$ cm and flows along a 2-T magnetic field. If **B** is in the $+z$ direction and **E** is the electrostatic field due to the beam's charge, calculate the magnitude and direction of the $\mathbf{E} \times \mathbf{B}$ drift at $r = a$ (See Fig. P2.7).

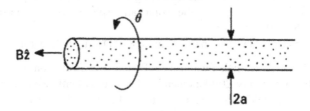

Fig. P2.7

2.3 Nonuniform B Field

Now that the concept of a guiding center drift is firmly established, we can discuss the motion of particles in inhomogeneous fields—**E** and **B** fields which vary in space or time. For uniform fields we were able to obtain exact expressions for the guiding center drifts. As soon as we introduce inhomogeneity, the problem becomes too complicated to solve exactly. To get an approximate answer, it is customary to expand in the small ratio r_L/L, where L is the scale length of the inhomogeneity. This type of theory, called *orbit theory*, can become extremely involved. We shall examine only the simplest cases, where only one inhomogeneity occurs at a time.

2.3.1 $\nabla B \perp B$: Grad-B Drift

Here the lines of force[1] are straight, but their density increases, say, in the y direction (Fig. 2.5). We can anticipate the result by using our simple physical picture. The gradient in $|B|$ causes the Larmor radius to be larger at the bottom of the orbit than at the top, and this should lead to a drift, in opposite directions for ions and electrons, perpendicular to both **B** and ∇B. The drift velocity should obviously be proportional to r_L/L and to $v_\perp$.

Consider the Lorentz force $\mathbf{F} = q\mathbf{v} \times \mathbf{B}$, averaged over a gyration. Clearly, $\overline{F}_x = 0$, since the particle spends as much time moving up as down. We wish to calculate $\overline{F}_y$, in an approximate fashion, by using the *undisturbed orbit* of the particle to find the average. The undisturbed orbit is given by Eqs. (2.4a),

---

[1] The magnetic field lines are often called "lines of force." They are not lines of force. The misnomer is perpetuated here to prepare the student for the treacheries of his profession.

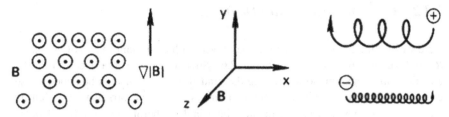

Fig. 2.5 The drift of a gyrating particle in a nonuniform magnetic field

(2.4b), and (2.7) for a uniform **B** field. Taking the real part of Eqs. (2.4a) and (2.4b), we have

$$F_y = -qv_x B_z(y) = -qv_\perp(\cos \omega_c t)\left[B_0 \pm r_L(\cos \omega_c t)\frac{\partial B}{\partial y}\right] \qquad (2.20)$$

where we have made a Taylor expansion of **B** field about the point $x_0 = 0$, $y_0 = 0$ and have used Eq. (2.7):

$$\mathbf{B} = \mathbf{B}_0 + (\mathbf{r} \cdot \nabla)\mathbf{B} + \cdots \qquad (2.21)$$
$$B_z = B_0 + y(\partial B_z/\partial y) + \cdots$$

This expansion of course requires $r_L/L \ll 1$, where L is the scale length of $\partial B_z/\partial y$. The first term of Eq. (2.20) averages to zero in a gyration, and the average of $\cos^2 \omega_c t$ is ½, so that

$$\overline{F}_y = \mp q v_\perp r_L \tfrac{1}{2}(\partial B/\partial y) \qquad (2.22)$$

The guiding center drift velocity is then

$$\mathbf{v}_{gc} = \frac{1}{q}\frac{\mathbf{F} \times \mathbf{B}}{B^2} = \frac{1}{q}\frac{\overline{F}_y}{|B|}\hat{\mathbf{x}} = \mp\frac{v_\perp r_L}{B}\frac{1}{2}\frac{\partial B}{\partial y}\hat{\mathbf{x}} \qquad (2.23)$$

where we have used Eq. (2.17). Since the choice of the y axis was arbitrary, this can be generalized to

$$\boxed{\mathbf{v}_{\nabla B} = \pm\frac{1}{2}v_\perp r_L\frac{\mathbf{B} \times \nabla B}{B^2}} \qquad (2.24)$$

This has all the dependences we expected from the physical picture; only the factor ½ (arising from the averaging) was not predicted. Note that the ± stands for the sign of the charge, and lightface B stands for $|B|$. The quantity $\mathbf{v}_{\nabla B}$ is called the *grad-B drift*; it is in opposite directions for ions and electrons and causes a current transverse to **B**. An exact calculation of $\mathbf{v}_{\nabla B}$ would require using the exact orbit, including the drift, in the averaging process.

2.3.2 Curved B: Curvature Drift

Here we assume the lines of force to be curved with a constant radius of curvature R_c, and we take $|B|$ to be constant (Fig. 2.6). Such a field does not obey Maxwell's equations in a vacuum, so in practice the grad-B drift will always be added to the effect derived here. A guiding center drift arises from the centrifugal force felt by the particles as they move along the field lines in their thermal motion. If $v_\parallel^2$ denotes the average square of the component of random velocity along $\mathbf{B}$, the average centrifugal force is

$$\mathbf{F}_{cf} = \frac{mv_\parallel^2}{R_c}\hat{\mathbf{r}} = mv_\parallel^2\frac{\mathbf{R}_c}{R_c^2} \tag{2.25}$$

According to Eq. (2.17), this gives rise to a drift

$$\boxed{\mathbf{v}_R = \frac{1}{q}\frac{\mathbf{F}_{cf} \times \mathbf{B}}{B^2} = \frac{mv_\parallel^2}{qB^2}\frac{\mathbf{R}_c \times \mathbf{B}}{R_c^2}} \tag{2.26}$$

The drift $\mathbf{v}_R$ is called the *curvature drift*.

We must now compute the grad-B drift which accompanies this when the decrease of $|B|$ with radius is taken into account. In a vacuum, we have $\nabla \times \mathbf{B} = 0$. In the cylindrical coordinates of Fig. 2.6, $\nabla \times \mathbf{B}$ has only a z component, since $\mathbf{B}$ has only a θ component and ∇B only an r component. We then have

Fig. 2.6 A curved magnetic field

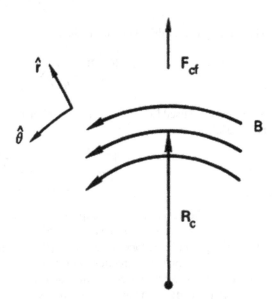

$$(\mathbf{\nabla} \times \mathbf{B})_z = \frac{1}{r}\frac{\partial}{\partial r}(rB_\theta) = 0 \qquad B_\theta \propto \frac{1}{r} \tag{2.27}$$

Thus

$$|B| \propto \frac{1}{R_c} \qquad \frac{\mathbf{\nabla}|B|}{|B|} = -\frac{R_c}{R_c^2} \tag{2.28}$$

Using Eq. (2.24), we have

$$\mathbf{v}_{\nabla B} = \mp\frac{1}{2}\frac{v_\perp r_L}{B^2}\mathbf{B} \times |B|\frac{R_c}{R_c^2} = \pm\frac{1}{2}\frac{v_\perp^2}{\omega_c}\frac{R_c \times \mathbf{B}}{R_c^2 B} = \frac{1}{2}\frac{m}{q}v_\perp^2\frac{R_c \times \mathbf{B}}{R_c^2 B^2} \tag{2.29}$$

Adding this to $\mathbf{v}_R$, we have the total drift in a curved vacuum field:

$$\boxed{\mathbf{v}_R + \mathbf{v}_{\nabla B} = \frac{m}{q}\frac{R_c \times \mathbf{B}}{R_c^2 B^2}\left(v_\parallel^2 + \frac{1}{2}v_\perp^2\right)} \tag{2.30}$$

It is unfortunate that these drifts add. This means that if one bends a magnetic field into a torus for the purpose of confining a thermonuclear plasma, the particles will drift out of the torus no matter how one juggles the temperatures and magnetic fields.

For a Maxwellian distribution, Eqs. (1.7) and (1.10) indicate that $\overline{v_\parallel^2}$ and $\frac{1}{2}\overline{v_\perp^2}$ are each equal to KT/m, since $v_\perp$ involves two degrees of freedom. Equations (2.3) and (1.6) then allow us to write the average curved-field drift as

$$\overline{\mathbf{v}}_{R+\nabla B} = \pm\frac{v_{th}^2}{R_c\omega_c}\hat{\mathbf{y}} = \pm\frac{\overline{r}_L}{R_c}v_{th}\hat{\mathbf{y}} \tag{2.30a}$$

where $\hat{\mathbf{y}}$ here is the direction of $R_c \times \mathbf{B}$. This shows that $\overline{\mathbf{v}}_{R+\nabla B}$ depends on the charge of the species but not on its mass.

2.3.3 $\nabla B \| B$: Magnetic Mirrors

Now we consider a magnetic field which is pointed primarily in the z direction and whose magnitude varies in the z direction. Let the field be axisymmetric, with $B_\theta = 0$ and $\partial/\partial\theta = 0$. Since the lines of force converge and diverge, there is necessarily a component B_r (Fig. 2.7). We wish to show that this gives rise to a force which can trap a particle in a magnetic field.

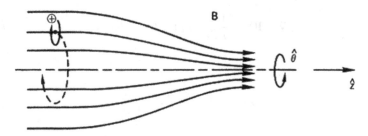

Fig. 2.7 Drift of a particle in a magnetic mirror field

We can obtain B_r from $\nabla \cdot \mathbf{B} = 0$:

$$\frac{1}{r}\frac{\partial}{\partial r}(rB_r) + \frac{\partial B_z}{\partial z} = 0 \tag{2.31}$$

If $\partial B_z/\partial z$ is given at $r = 0$ and does not vary much with r, we have approximately

$$rB_r = -\int_0^r r\frac{\partial B_z}{\partial z}dr \simeq -\frac{1}{2}r^2\left[\frac{\partial B_z}{\partial z}\right]_{r=0}$$

$$B_r = -\frac{1}{2}r\left[\frac{\partial B_z}{\partial z}\right]_{r=0} \tag{2.32}$$

The variation of $|B|$ with r causes a grad-B drift of guiding centers about the axis of symmetry, but there is no radial grad-B drift, because $\partial B/\partial \theta = 0$. The components of the Lorentz force are

$$F_r = q(\underset{①}{v_\theta B_z} - v_z \cancel{B_\theta})$$

$$F_\theta = q(\underset{②}{-v_r B_z} + \underset{③}{v_z B_r}) \tag{2.33}$$

$$F_z = q(v_r \cancel{B_\theta} - \underset{④}{v_\theta B_r})$$

Two terms vanish if $B_\theta = 0$, and terms 1 and 2 give rise to the usual Larmor gyration. Term 3 vanishes on the axis; when it does not vanish, this azimuthal force causes a drift in the radial direction. This drift merely makes the guiding centers follow the lines of force. Term 4 is the one we are interested in. Using Eq. (2.32), we obtain

$$F_z = \frac{1}{2}qv_\theta r(\partial B_z/\partial_z) \tag{2.34}$$

We must now average over one gyration. For simplicity, consider a particle whose guiding center lies on the axis. Then v_θ is a constant during a gyration; depending on the sign of q, v_θ is $\mp v_\perp$. Since $r = r_L$, the average force is

$$\overline{F}_z = \mp \frac{1}{2} q v_\perp r_L \frac{\partial B_z}{\partial z} = \mp \frac{1}{2} q \frac{v_\perp^2}{\omega_c} \frac{\partial B_z}{\partial z} = -\frac{1}{2} \frac{m v_\perp^2}{B} \frac{\partial B_z}{\partial z} \qquad (2.35)$$

We define the *magnetic moment* of the gyrating particle to be

$$\boxed{\mu \equiv \frac{1}{2} m v_\perp^2 / B} \qquad (2.36)$$

so that

$$\overline{F}_z = -\mu (\partial B_z / \partial z) \qquad (2.37)$$

This is a specific example of the force on a diamagnetic particle, which in general can be written

$$\mathbf{F}_\parallel = -\mu \partial B / \partial \mathbf{s} = -\mu \nabla_\parallel B \qquad (2.38)$$

where $d\mathbf{s}$ is a line element along $\mathbf{B}$. Note that the definition (2.36) is the same as the usual definition for the magnetic moment of a current loop with area A and current I: $\mu = IA$. In the case of a singly charged ion, I is generated by a charge e coming around $\omega_c / 2\pi$ times a second: $I = e \omega_c / 2\pi$. The area A is $\pi r_L^2 = \pi v_\perp^2 / \omega_c^2$. Thus

$$\mu = \frac{\pi v_\perp^2}{\omega_c^2} \frac{e \omega_c}{2\pi} = \frac{1}{2} \frac{v_\perp^2 e}{\omega_c} = \frac{1}{2} \frac{m v_\perp^2}{B}.$$

As the particle moves into regions of stronger or weaker $\mathbf{B}$, its Larmor radius changes, but μ *remains invariant*. To prove this, consider the component of the equation of motion along $\mathbf{B}$:

$$m \frac{dv_\parallel}{dt} = -\mu \frac{\partial B}{\partial s} \qquad (2.39)$$

Multiplying by $v_\parallel$ on the left and its equivalent ds/dt on the right, we have

$$m v_\parallel \frac{dv_\parallel}{dt} = \frac{d}{dt} \left(\frac{1}{2} m v_\parallel^2 \right) = -\mu \frac{\partial B}{\partial s} \frac{ds}{dt} = -\mu \frac{dB}{dt} \qquad (2.40)$$

Here dB/dt is the variation of B as seen by the particle; B itself is constant. The particle's energy must be conserved, so we have

$$\frac{d}{dt}\left(\frac{1}{2}mv_{\parallel}^2 + \frac{1}{2}mv_{\perp}^2\right) = \frac{d}{dt}\left(\frac{1}{2}mv_{\parallel}^2 + \mu B\right) = 0 \tag{2.41}$$

With Eq. (2.40) this becomes

$$-\mu\frac{dB}{dt} + \frac{d}{dt}(\mu B) = 0$$

so that

$$d\mu/dt = 0 \tag{2.42}$$

The invariance of μ is the basis for one of the primary schemes for plasma confinement: the *magnetic mirror*. As a particle moves from a weak-field region to a strong-field region in the course of its thermal motion, it sees an increasing B, and therefore its $v_{\perp}$ must increase in order to keep μ constant. Since its total energy must remain constant, $v_{\parallel}$ must necessarily decrease. If B is high enough in the "throat" of the mirror, $v_{\parallel}$ eventually becomes zero; and the particle is "reflected" back to the weak-field region. It is, of course, the force $\mathbf{F}_{\parallel}$ which causes the reflection. The nonuniform field of a simple pair of coils forms two magnetic mirrors between which a plasma can be trapped (Fig. 2.8). This effect works on both ions and electrons.

The trapping is not perfect, however. For instance, a particle with $v_{\perp} = 0$ will have no magnetic moment and will not feel any force along $\mathbf{B}$. A particle with small $v_{\perp}/v_{\parallel}$ at the midplane ($B = B_0$) will also escape if the maximum field B_m is not large enough. For given B_0 and B_m, which particles will escape? A particle with $v_{\perp} = v_{\perp 0}$ and $v_{\parallel} = v_{\parallel 0}$ at the midplane will have $v_{\perp} = v_{\perp}'$ and $v_{\parallel} = 0$ at its turning point. Let the field be B' there. Then the invariance of μ yields

$$\frac{1}{2}mv_{\perp 0}^2/B_0 = \frac{1}{2}mv_{\perp}'^2/B' \tag{2.43}$$

Conservation of energy requires

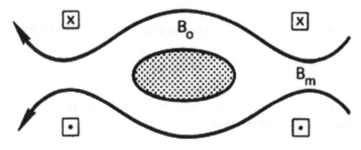

Fig. 2.8 A plasma trapped between magnetic mirrors

$$v_{\perp}'^2 = v_{\perp 0}^2 + v_{\parallel 0}^2 \equiv v_0^2 \tag{2.44}$$

Combining Eqs. (2.43) and (2.44), we find

$$\frac{B_0}{B'} = \frac{v_{\perp 0}^2}{v_{\perp}'^2} = \frac{v_{\perp 0}^2}{v_0^2} \equiv \sin^2 \theta \tag{2.45}$$

where θ is the pitch angle of the orbit in the weak-field region. Particles with smaller θ will mirror in regions of higher B. If θ is too small, B' exceeds B_m; and the particle does not mirror at all. Replacing B' by B_m in Eq. (2.45), we see that the smallest θ of a confined particle is given by

$$\sin^2 \theta_m = B_0/B_m \equiv 1/R_m \tag{2.46}$$

where R_m is the *mirror ratio*. Equation (2.46) defines the boundary of a region in velocity space in the shape of a cone, called a *loss cone* (Fig. 2.9). Particles lying within the loss cone are not confined. Consequently, a mirror-confined plasma is never isotropic. Note that the loss cone is independent of q or m. Without collisions, both ions and electrons are equally well confined. When collisions occur, particles are lost when they change their pitch angle in a collision and are scattered into the loss cone. Generally, electrons are lost more easily because they have a higher collision frequency.

The magnetic mirror was first proposed by Enrico Fermi as a mechanism for the acceleration of cosmic rays. Protons bouncing between magnetic mirrors approaching each other at high velocity could gain energy at each bounce (Fig. 2.10). How such mirrors could arise is another story. A further example of the mirror effect is the confinement of particles in the Van Allen belts. The magnetic field of the earth, being strong at the poles and weak at the equator, forms a natural mirror with rather large R_m.

Fig. 2.9 The loss cone

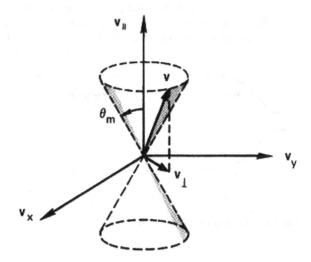

Problems

2.8. Suppose the earth's magnetic field is 3×10^{-5} T at the equator and falls off as $1/r^3$, as for a perfect dipole. Let there be an isotropic population of 1-eV protons and 30-keV electrons, each with density $n = 10^7$ m^{-3} at $r = 5$ earth radii in the equatorial plane.

  (a) Compute the ion and electron ∇B drift velocities.
  (b) Does an electron drift eastward or westward?
  (c) How long does it take an electron to encircle the earth?
  (d) Compute the ring current density in A/m^2.

   Note: The curvature drift is not negligible and will affect the numerical answer, but neglect it anyway.

2.9. An electron lies at rest in the magnetic field of an infinite straight wire carrying a current I. At $t = 0$, the wire is suddenly charged to a positive potential ϕ without affecting I. The electron gains energy from the electric field and begins to drift.

  (a) Draw a diagram showing the orbit of the electron and the relative directions of I, **B**, $\mathbf{v}_E$, $\mathbf{v}_{\nabla B}$, and $\mathbf{v}_R$.
  (b) Calculate the magnitudes of these drifts at a radius of 1 cm if $I = 500$ A, $\phi = 460$ V, and the radius of the wire is 1 mm. Assume that ϕ is held at 0 V on the vacuum chamber walls 10 cm away.

   Hint: A good intuitive picture of the motion is needed in addition to the formulas given in the text.

2.10. A 20-keV deuteron in a large mirror fusion device has a pitch angle θ of 45° at the midplane, where $B = 0.7$ T. Compute its Larmor radius.

2.11. A plasma with an isotropic velocity distribution is placed in a magnetic mirror trap with mirror ratio $R_m = 4$. There are no collisions, so the particles in the loss cone simply escape, and the rest remain trapped. What fraction is trapped?

2.12. A cosmic ray proton is trapped between two moving magnetic mirrors with $R_m = 5$ and initially has $W = 1$ keV and $v_\perp = v_\parallel$ at the midplane. Each mirror moves toward the midplane with a velocity $v_m = 10$ km/s (Fig. 2.10).

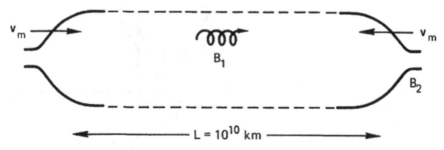

Fig. 2.10 Acceleration of cosmic rays

(a)   Using the loss cone formula and the invariance of μ, find the energy to
      which the proton will be accelerated before it escapes.
(b)   How long will it take to reach that energy?

    1. Treat the mirrors as flat pistons and show that the velocity gained at
       each bounce is $2v_m$.
    2. Compute the number of bounces necessary.
    3. Compute the time T it takes to traverse L that many times. Factor-of-
       two accuracy will suffice.

2.4   Nonuniform E Field

Now we let the magnetic field be uniform and the electric field be nonuniform. For
simplicity, we assume **E** to be in the x direction and to vary sinusoidally in the
x direction (Fig. 2.11):

$$\mathbf{E} \equiv E_0(\cos kx)\hat{\mathbf{x}} \tag{2.47}$$

This field distribution has a wavelength $\lambda = 2\pi/k$ and is the result of a sinusoidal
distribution of charges, which we need not specify. In practice, such a charge
distribution can arise in a plasma during a wave motion. The equation of motion is

$$m(d\mathbf{v}/dt) = q[\mathbf{E}(x)+\mathbf{v} \times \mathbf{B}] \tag{2.48}$$

whose transverse components are

$$\dot{v}_x = \frac{qB}{m}v_y + \frac{q}{m}E_x(x) \qquad \dot{v}_y = -\frac{qB}{m}v_x \tag{2.49}$$

$$\ddot{v}_x = -\omega_c^2 v_x \pm \omega_c \frac{\dot{E}_x}{B} \tag{2.50}$$

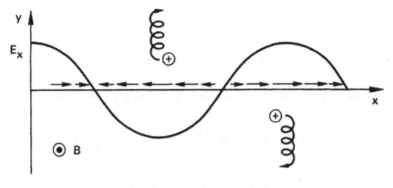

Fig. 2.11   Drift of a gyrating particle in a nonuniform electric field

$$\ddot{v}_y = -\omega_c^2 v_y - \omega_c^2 \frac{E_x(x)}{B} \qquad (2.51)$$

Here $E_x(x)$ is the electric field at the position of the particle. To evaluate this, we need to know the particle's orbit, which we are trying to solve for in the first place. If the electric field is weak, we may, as an approximation, use the *undisturbed orbit* to evaluate $E_x(x)$. The orbit in the absence of the E field was given in Eq. (2.7):

$$x = x_0 + r_L \sin \omega_c t \qquad (2.52)$$

From Eqs. (2.51) and (2.47), we now have

$$\ddot{v}_y = -\omega_c^2 v_y - \omega_c^2 \frac{E_0}{B} \cos k(x_0 + r_L \sin w_c t) \qquad (2.53)$$

Anticipating the result, we look for a solution which is the sum of a gyration at ω_c and a steady drift v_E. Since we are interested in finding an expression for v_E, we take out the gyratory motion by averaging over a cycle. Equation (2.50) then gives $\bar{v}_x = 0$. In Eq. (2.53), the oscillating term $\ddot{v}_y$ clearly averages to zero, and we have

$$\bar{\ddot{v}}_y = 0 = -\omega_c^2 \bar{v}_y - \omega_c^2 \frac{E_0}{B} \overline{\cos k(x_0 + r_L \sin \omega_c t)} \qquad (2.54)$$

Expanding the cosine, we have

$$\cos k(x_0 + r_L \sin \omega_c t) = \cos (kx_0) \cos (kr_L \sin \omega_c t)$$
$$- \sin (kx_0) \sin (kr_L \sin \omega_c t) \qquad (2.55)$$

It will suffice to treat the small Larmor radius case, $kr_L \ll 1$. The Taylor expansions

$$\cos \varepsilon = 1 - \frac{1}{2} \varepsilon^2 + \cdots$$
$$\sin \varepsilon = \varepsilon + \cdots \qquad (2.56)$$

allow us to write

$$\cos k(x_0 + r_L \sin \omega_c t) \approx (\cos kx_0)\left(1 - \frac{1}{2} k^2 r_L^2 \sin^2 \omega_c t\right) - (\sin kx_0) kr_L \sin \omega_c t$$

The last term vanishes upon averaging over time, and Eq. (2.54) gives

$$\bar{v}_y = -\frac{E_0}{B} (\cos kx_0)\left(1 - \frac{1}{4} k^2 r_L^2\right) = -\frac{E_x(x_0)}{B}\left(1 - \frac{1}{4} k^2 r_L^2\right) \qquad (2.57)$$

Thus the usual $\mathbf{E} \times \mathbf{B}$ drift is modified by the inhomogeneity to read

$$\mathbf{v}_E = \frac{\mathbf{E} \times \mathbf{B}}{B^2}\left(1 - \tfrac{1}{4}k^2 r_L^2\right) \tag{2.58}$$

The physical reason for this is easy to see. An ion with its guiding center at a maximum of $\mathbf{E}$ actually spends a good deal of its time in regions of weaker $\mathbf{E}$. Its average drift, therefore, is less than E/B evaluated at the guiding center. In a linearly varying $\mathbf{E}$ field, the ion would be in a stronger field on one side of the orbit and in a field weaker by the same amount on the other side; the correction to $\mathbf{v}_E$ then cancels out. From this it is clear that the correction term depends on the *second derivative* of $\mathbf{E}$. For the sinusoidal distribution we assumed, the second derivative is always negative with respect to $\mathbf{E}$. For an arbitrary variation of $\mathbf{E}$, we need only replace ik by ∇ and write Eq. (2.58) as

$$\boxed{\mathbf{v}_E = \left(1 + \tfrac{1}{4}r_L^2\nabla^2\right)\frac{\mathbf{E} \times \mathbf{B}}{B^2}} \tag{2.59}$$

The second term is called the *finite-Larmor-radius effect*. What is the significance of this correction? Since r_L is much larger for ions than for electrons, $\mathbf{v}_E$ is no longer independent of species. If a density clump occurs in a plasma, an electric field can cause the ions and electrons to separate, generating another electric field. If there is a feedback mechanism that causes the second electric field to enhance the first one, $\mathbf{E}$ grows indefinitely, and the plasma is unstable. Such an instability, called a *drift instability*, will be discussed in a later chapter. The grad-B drift, of course, is also a finite-Larmor-radius effect and also causes charges to separate. According to Eq. (2.24), however, $\mathbf{v}_{\nabla B}$ is proportional to $k r_L$, whereas the correction term in Eq. (2.58) is proportional to $k^2 r_L^2$. The nonuniform-E-field effect, therefore, is important at relatively large k, or small scale lengths of the inhomogeneity. For this reason, drift instabilities belong to a more general class called *microinstabilities*.

2.5   Time-Varying E Field

Let us now take $\mathbf{E}$ and $\mathbf{B}$ to be uniform in space but varying in time. First, consider the case in which $\mathbf{E}$ alone varies sinusoidally in time, and let it lie along the x axis:

$$\mathbf{E} = E_0 e^{i\omega t}\hat{\mathbf{x}} \tag{2.60}$$

Since $\dot{E}_x = i\omega E_x$, we can write Eq. (2.50) as

$$\ddot{v}_x = -\omega_c^2\left(v_x \mp \frac{i\omega}{\omega_c}\frac{\tilde{E}_x}{B}\right) \tag{2.61}$$

Let us define

$$\tilde{v}_p \equiv \pm \frac{i\omega \tilde{E}_x}{\omega_c B}$$

$$\tilde{v}_E \equiv \frac{\tilde{E}_x}{B} \tag{2.62}$$

where the tilde has been added merely to emphasize that the drift is oscillating. The upper (lower) sign, as usual, denotes positive (negative) q. Now Eqs. (2.50) and (2.51) become

$$\ddot{v}_x = -\omega_c^2 \left(v_x - \tilde{v}_p \right)$$

$$\ddot{v}_y = -\omega_c^2 \left(v_y - \tilde{v}_E \right) \tag{2.63}$$

By analogy with Eq. (2.12), we try a solution which is the sum of a drift and a gyratory motion:

$$v_x = v_\perp e^{i\omega_c t} + \tilde{v}_p$$

$$v_y = \pm i v_\perp e^{i\omega_c t} + \tilde{v}_E \tag{2.64}$$

If we now differentiate twice with respect to time, we find

$$\ddot{v}_x = -\omega_c^2 v_x + \left(\omega_c^2 - \omega^2 \right) \tilde{v}_p$$

$$\ddot{v}_y = -\omega_c^2 v_y + \left(\omega_c^2 - \omega^2 \right) \tilde{v}_E \tag{2.65}$$

This is not the same as Eq. (2.63) unless $\omega^2 \ll \omega_c^2$. If we now make the assumption that $\mathbf{E}$ varies slowly, so that $\omega^2 \ll \omega_c^2$, then Eq. (2.64) is the approximate solution to Eq. (2.63).

Equation (2.64) tells us that the guiding center motion has two components. The y component, perpendicular to $\mathbf{B}$ and $\mathbf{E}$, is the usual $\mathbf{E} \times \mathbf{B}$ drift, except that v_E now oscillates slowly at the frequency ω. The x component, a new drift *along the direction of* $\mathbf{E}$, is called the *polarization drift*. By replacing $i\omega$ by $\partial/\partial t$, we can generalize Eq. (2.62) and define the polarization drift as

$$\boxed{\mathbf{v}_p = \pm \frac{1}{\omega_c B} \frac{d\mathbf{E}}{dt}} \tag{2.66}$$

Since $\mathbf{v}_p$ is in opposite directions for ions and electrons, there is a *polarization current*; for $Z = 1$, this is

$$\mathbf{j}_p = ne \left(v_{ip} - v_{ep} \right) = \frac{ne}{eB^2} (M + m) \frac{d\mathbf{E}}{dt} = \frac{\rho}{B^2} \frac{d\mathbf{E}}{dt} \tag{2.67}$$

where ρ is the mass density.

Fig. 2.12 The polarization
drift

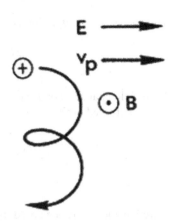

The physical reason for the polarization current is simple (Fig. 2.12). Consider an ion at rest in a magnetic field. If a field $\mathbf{E}$ is suddenly applied, the first thing the ion does is to move in the direction of $\mathbf{E}$. Only after picking up a velocity $\mathbf{v}$ does the ion feel a Lorentz force $e\mathbf{v} \times \mathbf{B}$ and begin to move downward in Fig. (2.12). If $\mathbf{E}$ is now kept constant, there is no further $\mathbf{v}_p$ drift but only a $\mathbf{v}_E$ drift. However, if $\mathbf{E}$ is reversed, there is again a momentary drift, this time to the left. Thus $\mathbf{v}_p$ is a startup drift due to inertia and occurs only in the first half-cycle of each gyration during which $\mathbf{E}$ changes. Consequently, $\mathbf{v}_p$ goes to zero with ω/ω_c.

The polarization effect in a plasma is similar to that in a solid dielectric, where $\mathbf{D} = \varepsilon_0\mathbf{E} + \mathbf{P}$. The dipoles in a plasma are ions and electrons separated by a distance r_L. But since ions and electrons can move around to preserve quasineutrality, the application of a steady $\mathbf{E}$ field does not result in a polarization field $\mathbf{P}$. However, if $\mathbf{E}$ oscillates, an oscillating current $\mathbf{j}_p$ results from the lag due to the ion inertia.

2.6   Time-Varying B Field

Finally, we allow the magnetic field to vary in time. Since the Lorentz force is always perpendicular to $\mathbf{v}$, a magnetic field itself cannot impart energy to a charged particle. However, associated with $\mathbf{B}$ is an electric field given by

$$\nabla \times \mathbf{E} = - \dot{\mathbf{B}} \qquad (2.68)$$

and this can accelerate the particles. We can no longer assume the fields to be completely uniform. Let $\mathbf{v}_\perp = d\mathbf{l}/dt$ be the transverse velocity, $\mathbf{l}$ being the element of path along a particle trajectory (with $v_\parallel$ neglected). Taking the scalar product of the equation of motion (2.8) with $\mathbf{v}_\perp$, we have

$$\frac{d}{dt}\left(\tfrac{1}{2}mv_\perp^2\right) = q\mathbf{E} \cdot \mathbf{v}_\perp = q\mathbf{E} \cdot \frac{d\mathbf{l}}{dt} \qquad (2.69)$$

The change in one gyration is obtained by integrating over one period:

$$\delta\left(\frac{1}{2}mv_\perp^2\right) = \int_0^{2\pi/\omega_c} q\mathbf{E}\cdot\frac{d\mathbf{l}}{dt}\,dt$$

If the field changes slowly, we can replace the time integral by a line integral over the unperturbed orbit:

$$\delta\left(\frac{1}{2}mv_\perp^2\right) = \oint q\mathbf{E}\cdot d\mathbf{l} = q\int_S (\nabla\times\mathbf{E})\cdot d\mathbf{S}$$

$$= -q\int_S \dot{\mathbf{B}}\cdot d\mathbf{S} \tag{2.70}$$

Here $\mathbf{S}$ is the surface enclosed by the Larmor orbit and has a direction given by the right-hand rule when the fingers point in the direction of $\mathbf{v}$. Since the plasma is diamagnetic, we have $\mathbf{B}\cdot d\mathbf{S} < 0$ for ions and >0 for electrons. Then Eq. (2.70) becomes

$$\delta\left(\frac{1}{2}mv_\perp^2\right) = \pm q\dot{B}\pi r_L^2 = \pm q\pi\dot{B}\frac{v_\perp^2}{\omega_c}\frac{m}{\pm qB} = \frac{\frac{1}{2}mv_\perp^2}{B}\cdot\frac{2\pi\dot{B}}{\omega_c} \tag{2.71}$$

The quantity $2\pi\dot{B}/\omega_c = \dot{B}/f_c$ is just the change δB during one period of gyration. Thus

$$\delta\left(\frac{1}{2}mv_\perp^2\right) = \mu\,\delta B \tag{2.72}$$

Since the left-hand side is $\delta(\mu B)$, we have the desired result

$$\delta\mu = 0 \tag{2.73}$$

The magnetic moment is invariant in slowly varying magnetic fields.

As the B field varies in strength, the Larmor orbits expand and contract, and the particles lose and gain transverse energy. This exchange of energy between the particles and the field is described very simply by Eq. (2.73). The invariance of μ allows us to prove easily the following well-known theorem:

The magnetic flux through a Larmor orbit is constant.

The flux Φ is given by BS, with $S = \pi r_L^2$ Thus

$$\Phi = B\pi\frac{v_\perp^2}{\omega_c^2} = B\pi\frac{v_\perp^2 m^2}{q^2 B^2} = \frac{2\pi m}{q^2}\frac{\frac{1}{2}mv_\perp^2}{B} = \frac{2\pi m}{q^2}\mu \tag{2.74}$$

Therefore, Φ is constant if μ is constant.

This property is used in a method of plasma heating known as *adiabatic compression*. Figure 2.13 shows a schematic of how this is done. A plasma is injected into the region between the mirrors A and B. Coils A and B are then pulsed

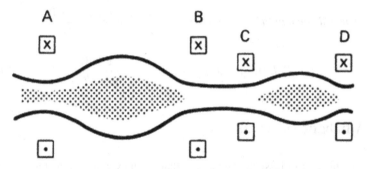

Fig. 2.13 Two-stage adiabatic compression of a plasma

to increase **B** and hence $v_\perp^2$. The heated plasma can then be transferred to the region C–D by a further pulse in A, increasing the mirror ratio there. The coils C and D are then pulsed to further compress and heat the plasma. Early magnetic mirror fusion devices employed this type of heating. Adiabatic compression has also been used successfully on toroidal plasmas and is an essential element of laser-driven fusion schemes using either magnetic or inertial confinement.

2.7 Summary of Guiding Center Drifts

$$\text{General force } F: \qquad \mathbf{v}_f = \frac{1}{q}\frac{\mathbf{F}\times\mathbf{B}}{B^2} \qquad (2.17)$$

$$\text{Electric field}: \qquad \mathbf{v}_E = \frac{\mathbf{E}\times\mathbf{B}}{B^2} \qquad (2.15)$$

$$\text{Gravitational field}: \qquad \mathbf{v}_g = \frac{m}{q}\frac{\mathbf{g}\times\mathbf{B}}{B^2} \qquad (2.18)$$

$$\text{Nonuniform } \mathbf{E}: \qquad \mathbf{v}_E = \left(1+\tfrac{1}{4}r_L^2\nabla^2\right)\frac{\mathbf{E}\times\mathbf{B}}{B^2} \qquad (2.59)$$

Nonuniform **B** *field*

$$\text{Grad} - B \text{ drift}: \qquad \mathbf{v}_{\nabla B} = \pm\frac{1}{2}v_\perp r_L\frac{\mathbf{B}\times\nabla B}{B^2} \qquad (2.24)$$

$$\text{Curvature drift}: \qquad \mathbf{v}_R = \frac{mv_\parallel^2}{q}\frac{\mathbf{R}_c\times\mathbf{B}}{R_c^2B^2} \qquad (2.26)$$

$$\text{Curved vacuum field:} \quad \mathbf{v}_R + \mathbf{v}_{\nabla B} = \frac{m}{q}\left(v_\parallel^2 + \frac{1}{2}v_\perp^2\right)\frac{\mathbf{R}_c \times \mathbf{B}}{R_c^2 B^2} \tag{2.30}$$

$$\text{Polarization drift:} \quad \mathbf{v}_p = \pm\frac{1}{\omega_c B}\frac{d\mathbf{E}}{dt} \tag{2.66}$$

2.8   Adiabatic Invariants

It is well known in classical mechanics that whenever a system has a periodic motion, the action integral $\oint p\,dq$ taken over a period is a constant of the motion. Here p and q are the generalized momentum and coordinate which repeat themselves in the motion. If a slow change is made in the system, so that the motion is not quite periodic, the constant of the motion does not change and is then called an *adiabatic invariant*. By slow here we mean slow compared with the period of motion, so that the integral $\oint p\,dq$ is well defined even though it is strictly no longer an integral over a closed path. Adiabatic invariants play an important role in plasma physics; they allow us to obtain simple answers in many instances involving complicated motions. There are three adiabatic invariants, each corresponding to a different type of periodic motion.

2.8.1   The First Adiabatic Invariant, μ

We have already met the quantity

$$\mu = mv_\perp^2/2B$$

and have proved its invariance in spatially and temporally varying $\mathbf{B}$ fields. The periodic motion involved, of course, is the Larmor gyration. If we take p to be angular momentum $mv_\perp r$ and dq to be the coordinate $d\theta$, the action integral becomes

$$\oint p\,dq = \oint mv_\perp r_L d\theta = 2\pi r_L mv_\perp = 2\pi \frac{mv_\perp^2}{\omega_c} = 4\pi \frac{m}{|q|}\mu \tag{2.75}$$

Thus μ is a constant of the motion as long as q/m is not changed. We have proved the invariance of μ only with the implicit assumption $\omega/\omega_c \ll 1$, where ω is a frequency characterizing the rate of change of $\mathbf{B}$ as seen by the particle. A proof exists, however, that μ is invariant even when $\omega \leq \omega_c$. In theorists' language, μ is invariant "to *all* orders in an expansion in ω/ω_c." What this means in practice is that μ remains much more nearly constant than $\mathbf{B}$ does during one period of gyration.

It is just as important to know when an adiabatic invariant does *not* exist as to know when it does. Adiabatic invariance of μ is violated when ω is not small compared with ω_c. We give three examples of this.

(A) *Magnetic Pumping*. If the strength of **B** in a mirror confinement system is varied sinusoidally, the particles' $v_\perp$ would oscillate; but there would be no gain of energy in the long run. However, if the particles make collisions, the invariance of μ is violated, and the plasma can be heated. In particular, a particle making a collision during the compression phase can transfer part of its gyration energy into $v_\parallel$ energy, and this is not taken out again in the expansion phase.

(B) *Cyclotron Heating*. Now imagine that the B field is oscillated at the frequency ω_c. The induced electric field will then rotate in phase with some of the particles and accelerate their Larmor motion continuously. The condition $\omega \ll \omega_c$ is violated, μ is not conserved, and the plasma can be heated.

(C) *Magnetic Cusps*. If the current in one of the coils in a simple magnetic mirror system is reversed, a magnetic cusp is formed (Fig. 2.14). This configuration has, in addition to the usual mirrors, a spindle-cusp mirror extending over 360° in azimuth. A plasma confined in a cusp device is supposed to have better stability properties than that in an ordinary mirror. Unfortunately, the loss-cone losses are larger because of the additional loss region; and the particle motion is nonadiabatic. Since the B field vanishes at the center of symmetry, ω_c is zero there; and μ is not preserved. The local Larmor radius near the center is larger than the device. Because of this, the adiabatic invariant μ does not guarantee that particles outside a loss cone will stay outside after passing through the nonadiabatic region. Fortunately, there is in this case another invariant: the canonical angular momentum $p_\theta = mrv_\theta - erA_\theta$. This ensures that there will be a population of particles trapped indefinitely until they make a collision.

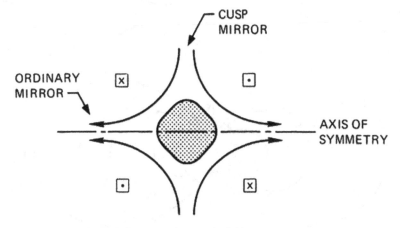

Fig. 2.14 Plasma confinement in a cusped magnetic field

2.8.2 The Second Adiabatic Invariant, J

Consider a particle trapped between two magnetic mirrors: It bounces between them and therefore has a periodic motion at the "bounce frequency." A constant of this motion is given by $\oint m v_{\|}\, ds$, where ds is an element of path length (of the guiding center) along a field line. However, since the guiding center drifts across field lines, the motion is not exactly periodic, and the constant of the motion becomes an adiabatic invariant. This is called the *longitudinal invariant J* and is defined for a half-cycle between the two turning points (Fig. 2.15):

$$\boxed{J = \int_a^b v_{\|}\, ds} \tag{2.76}$$

We shall prove that J is invariant in a static, nonuniform B field; the result is also true for a slowly time-varying B field.

Before embarking on this somewhat lengthy proof, let us consider an example of the type of problem in which a theorem on the invariance of J would be useful. As we have already seen, the earth's magnetic field mirror-traps charged particles, which slowly drift in longitude around the earth (Problem 2.8; see Fig. 2.16). If the magnetic field were perfectly symmetric, the particle would eventually drift back to the same line of force. However, the actual field is distorted by such effects as the solar wind. In that case, will a particle ever come back to the same line of force? Since the particle's energy is conserved and is equal to $\frac{1}{2} m v_{\perp}^2$ at the turning point, the invariance of μ indicates that $|B|$ remains the same at the turning point.

Fig. 2.15 A particle bouncing between turning points a and b in a magnetic field

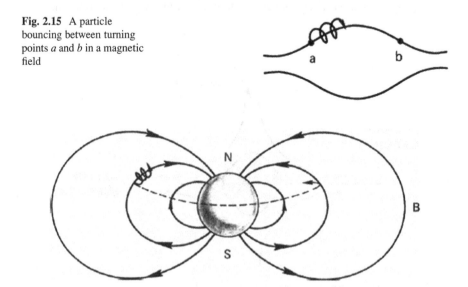

Fig. 2.16 Motion of a charged particle in the earth's magnetic field

Fig. 2.17 Proof of the
invariance of J

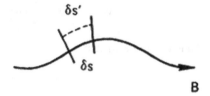

However, upon drifting back to the same longitude, a particle may find itself on
another line of force at a different altitude. This cannot happen if J is conserved. J
determines the length of the line of force between turning points, and no two lines
have the same length between points with the same $|B|$. Consequently, the particle
returns to the same line of force even in a slightly asymmetric field.

To prove the invariance of J, we first consider the invariance of $v_\| \delta s$, where δs is
a segment of the path along $\mathbf{B}$ (Fig. 2.17). Because of guiding center drifts, a
particle on s will find itself on another line of force $\delta s'$ after a time Δt. The length of
$\delta s'$ is defined by passing planes perpendicular to $\mathbf{B}$ through the end points of δs. The
length of δs is obviously proportional to the radius of curvature:

$$\frac{\delta s}{R_c} = \frac{\delta s'}{R'_c}$$

so that

$$\frac{\delta s' - \delta s}{\Delta t \delta s} = \frac{R'_c - R_c}{\Delta t R_c} \tag{2.77}$$

The "radial" component of $\mathbf{v}_{gc}$ is just

$$\mathbf{v}_{gc} \cdot \frac{\mathbf{R}_c}{R_c} = \frac{R'_c - R_c}{\Delta t} \tag{2.78}$$

From Eqs. (2.24) and (2.26), we have

$$\mathbf{v}_{gc} = \mathbf{v}_{\nabla B} + \mathbf{v}_R = \pm \frac{1}{2} v_\perp r_L \frac{\mathbf{B} \times \nabla B}{B^2} + \frac{m v_\|^2}{q} \frac{\mathbf{R}_c \times \mathbf{B}}{R_c^2 B^2} \tag{2.79}$$

The last term has no component along $\mathbf{R}_c$. Using Eqs. (2.78) and (2.79), we can
write Eq. (2.77) as

$$\frac{1}{\delta s} \frac{d}{dt} \delta s = \mathbf{v}_{gc} \cdot \frac{\mathbf{R}_c}{R_c^2} = \frac{1}{2} \frac{m v_\perp^2}{q B^3} (\mathbf{B} \times \nabla B) \cdot \frac{\mathbf{R}_c}{R_c^2} \tag{2.80}$$

This is the rate of change of δs as seen by the particle. We must now get the rate of
change of $v_\|$ as seen by the particle. The parallel and perpendicular energies are
defined by

$$W \equiv \frac{1}{2} m v_{\parallel}^2 + \frac{1}{2} m v_{\perp}^2 = \frac{1}{2} m v_{\parallel}^2 + \mu B \equiv W_{\parallel} + W_{\perp} \tag{2.81}$$

Thus $v_{\parallel}$ can be written

$$v_{\parallel} = [(2/m)(W - \mu B)]^{1/2} \tag{2.82}$$

Here W and μ are constant, and only B varies. Therefore,

$$\frac{\dot{v}_{\parallel}}{v_{\parallel}} = -\frac{1}{2} \frac{\mu \dot{B}}{W - \mu B} = -\frac{1}{2} \frac{\mu \dot{B}}{W_{\parallel}} = -\frac{\mu \dot{B}}{m v_{\parallel}^2} \tag{2.83}$$

Since B was assumed static, $\dot{B}$ is not zero only because of the guiding center motion:

$$\dot{B} = \frac{dB}{d\mathbf{r}} \cdot \frac{d\mathbf{r}}{dt} = \mathbf{v}_{gc} \cdot \nabla B = \frac{m v_{\parallel}^2}{q} \frac{\mathbf{R}_c \times \mathbf{B}}{R_c^2 B^2} \cdot \nabla B \tag{2.84}$$

Now we have

$$\frac{\dot{v}_{\parallel}}{v_{\parallel}} = -\frac{\mu}{q} \frac{(\mathbf{R}_c \times \mathbf{B}) \cdot \nabla B}{R_c^2 B^2} = -\frac{1}{2} \frac{m v_{\perp}^2}{q B} \frac{(\mathbf{B} \times \nabla B) \cdot \mathbf{R}_c}{R_c^2 B^2} \tag{2.85}$$

The fractional change in $v_{\parallel} \, \delta s$ is

$$\frac{1}{v_{\parallel} \delta s} \frac{d}{dt}(v_{\parallel} \delta s) = \frac{1}{\delta s} \frac{d \delta s}{dt} + \frac{1}{v_{\parallel}} \frac{d v_{\parallel}}{dt} \tag{2.86}$$

From Eqs. (2.80) and (2.85), we see that these two terms cancel, so that

$$v_{\parallel} \delta s = \text{constant} \tag{2.87}$$

This is not exactly the same as saying that J is constant, however. In taking the integral of $v_{\parallel} \delta s$ between the turning points, it may be that the turning points on $\delta s'$ do not coincide with the intersections of the perpendicular planes (Fig. 2.17). However, any error in J arising from such a discrepancy is negligible because near the turning points, $v_{\parallel}$ is nearly zero. Consequently, we have proved

$$J = \int_a^b v_{\parallel} ds = \text{constant} \tag{2.88}$$

An example of the violation of J invariance is given by a plasma heating scheme called *transit-time magnetic pumping*. Suppose an oscillating current is applied to the coils of a mirror system so that the mirrors alternately approach and withdraw from each other near the bounce frequency. Those particles that have the right bounce frequency will always see an approaching mirror and will therefore gain $v_\parallel$. J is not conserved in this case because the change of $\mathbf{B}$ occurs on a time scale not long compared with the bounce time.

2.8.3   The Third Adiabatic Invariant, Φ

Referring again to Fig. 2.16, we see that the slow drift of a guiding center around the earth constitutes a third type of periodic motion. The adiabatic invariant connected with this turns out to be the total magnetic flux Φ enclosed by the drift surface. It is almost obvious that, as $\mathbf{B}$ varies, the particle will stay on a surface such that the total number of lines of force enclosed remains constant. This invariant, Φ, has few applications because most fluctuations of $\mathbf{B}$ occur on a time scale short compared with the drift period. As an example of the violation of Φ invariance, we can cite some recent work on the excitation of hydromagnetic waves in the ionosphere. These waves have a long period comparable to the drift time of a particle around the earth. The particles can therefore encounter the wave in the same phase each time around. If the phase is right, the wave can be excited by the conversion of particle drift energy to wave energy.

Problems

2.13. Derive the result of Problem 2.12b directly by using the invariance of J.

(a)   Let $\int v_\parallel ds \simeq v_\parallel L$ and differentiate with respect to time

(b)   From this, get an expression for T in terms of dL/dt. Set $dL/dt = -2v_m$ to obtain the answer.

2.14. In plasma heating by adiabatic compression the invariance of μ requires that $KT_\perp$ increase as B increases. The magnetic field, however, cannot accelerate particles because the Lorentz force $q\mathbf{v} \times \mathbf{B}$ is always perpendicular to the velocity. How do the particles gain energy?

2.15. The polarization drift v_p can also be derived from energy conservation. If $\mathbf{E}$ is oscillating, the $\mathbf{E} \times \mathbf{B}$ drift also oscillates; and there is an energy $\frac{1}{2}mv_E^2$ associated with the guiding center motion. Since energy can be gained from an $\mathbf{E}$ field only by motion along $\mathbf{E}$, there must be a drift v_p in the $\mathbf{E}$ direction. By equating the rate of change of $\frac{1}{2}mv_E^2$ with the rate of energy gain from $\mathbf{v}_p \cdot \mathbf{E}$, find the required value of v_p.

2.16. A hydrogen plasma is heated by applying a radiofrequency wave with $\mathbf{E}$ perpendicular to $\mathbf{B}$ and with an angular frequency $\omega = 10^9$ rad/s. The confining magnetic field is 1 T. Is the motion of (a) the electrons and (b) the ions in response to this wave adiabatic?

2.17. A 1-keV proton with $v_\| = 0$ in a uniform magnetic field $B = 0.1$ T is accel-
erated as B is slowly increased to 1 T. It then makes an elastic collision with a
heavy particle and changes direction so that $v_\perp = v_\|$ The B-field is then
slowly decreased back to 0.1 T. What is the proton's energy now?

2.18. A collisionless hydrogen plasma is confined in a torus in which external
windings provide a magnetic field **B** lying almost entirely in the ϕ direction
(Fig. P2.18). The plasma is initially Maxwellian at $KT = 1$ keV. At $t = 0$, B is
gradually increased from 1 T to 3 T in 100 μs, and the plasma is compressed.

  (a) Show that the magnetic moment μ remains invariant for both ions and
       electrons.
  (b) Calculate the temperatures $T_\perp$ and $T_\|$ after compression.

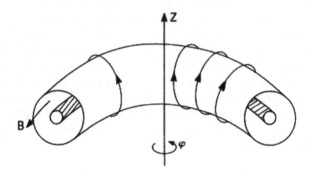

Fig. P2.18

2.19. A uniform plasma is created in a toroidal chamber with square cross section,
as shown. The magnetic field is provided by a current I along the axis of
symmetry. The dimensions are $a = 1$ cm, $R = 10$ cm. The plasma is Maxwellian at $KT = 100$ eV and has density $n = 10^{19}$ m^{-3}. There is no electric field.

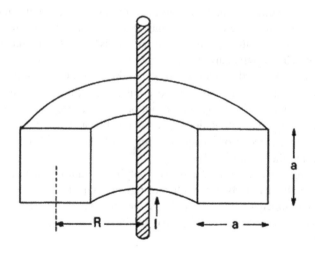

Fig. P2.19

(a) Draw typical orbits for ions and electrons with $v_\parallel = 0$ drifting in the nonuniform **B** field.

(b) Calculate the rate of charge accumulation (in coulombs per second) on the entire top plate of the chamber due to the combined $v_{\nabla B}$ and v_R drifts. The magnetic field at the center of the chamber is 1 T, and you may make a large aspect ratio ($R \gg a$) approximation where necessary.

2.20. Suppose the magnetic field along the axis of a magnetic mirror is given by $B_z = B_0(1 + \alpha^2 z^2)$.

(a) If an electron at $z = 0$ has a velocity given by $v^2 = 3v_\parallel^2 = 1.5v_\perp^2$, at what value of z is the electron reflected?

(b) Write the equation of motion of the guiding center for the direction parallel to the field.

(c) Show that the motion is sinusoidal, and calculate its frequency.

(d) Calculate the longitudinal invariant J corresponding to this motion.

2.21. An infinite straight wire carries a constant current I in the $+z$ direction. At $t = 0$, an electron of small gyroradius is at $z = 0$ and $r = r_0$ with $v_{\perp 0} = v_{\parallel 0}$. ($\perp$ and $\parallel$ refer to the direction relative to the magnetic field.)

(a) Calculate the magnitude and direction of the resulting guiding center drift velocity.

(b) Suppose that the current increases slowly in time in such a way that a constant electric field in the $\pm z$ direction is induced. Indicate on a diagram the relative directions of **I**, **B**, **E**, and v_E.

(c) Do $v_\perp$ and $v_\parallel$ increase, decrease, or remain the same as the current increases? Why?

Chapter 3
Plasmas as Fluids

3.1 Introduction

In a plasma the situation is much more complicated than that in the last chapter; the **E** and **B** fields are not prescribed but are determined by the positions and motions of the charges themselves. One must solve a self-consistent problem; that is, find a set of particle trajectories and field patterns such that the particles will generate the fields as they move along their orbits and the fields will cause the particles to move in those exact orbits. And this must be done in a time-varying situation. It sounds very hard, but it is not.

We have seen that a typical plasma density might be 10^{18} ion–electron pairs per m^3. If each of these particles follows a complicated trajectory and it is necessary to follow each of these, predicting the plasma's behavior would be a hopeless task. Fortunately, this is not usually necessary because, surprisingly, the majority—perhaps as much as 80 %—of plasma phenomena observed in real experiments can be explained by a rather crude model. This model is that used in fluid mechanics, in which the identity of the individual particle is neglected, and only the motion of fluid elements is taken into account. Of course, in the case of plasmas, the fluid contains electrical charges. In an ordinary fluid, frequent collisions between particles keep the particles in a fluid element moving together. It is surprising that such a model works for plasmas, which generally have infrequent collisions. But we shall see that there is a reason for this.

In the greater part of this book, we shall be concerned with what can be learned from the fluid theory of plasmas. A more refined treatment—the kinetic theory of plasmas—requires more mathematical calculation than is appropriate for an introductory course. An introduction to kinetic theory is given in Chap. 7.

The original version of this chapter was revised. An erratum to this chapter can be found at https://doi.org/10.1007/978-3-319-22309-4_11

© Springer International Publishing Switzerland 2016
F.F. Chen, *Introduction to Plasma Physics and Controlled Fusion*,
DOI 10.1007/978-3-319-22309-4_3

In some plasma problems, neither fluid theory nor kinetic theory is sufficient to describe the plasma's behavior. Then one has to fall back on the tedious process of following the individual trajectories. Modern computers can do this, although they have only enough memory to store the position and velocity components for about 10^6 particles if all three dimensions are involved. Nonetheless, computer simulation plays an important role in filling the gap between theory and experiment in those instances where even kinetic theory cannot come close to explaining what is observed.

3.2 Relation of Plasma Physics to Ordinary Electromagnetics

3.2.1 Maxwell's Equations

In vacuum:

$$\varepsilon_0 \mathbf{\nabla} \cdot \mathbf{E} = \sigma \tag{3.1}$$

$$\mathbf{\nabla} \cdot \mathbf{E} = -\dot{\mathbf{B}} \tag{3.2}$$

$$\mathbf{\nabla} \cdot \mathbf{B} = 0 \tag{3.3}$$

$$\mathbf{\nabla} \times \mathbf{B} = \mu_0 \left(\mathbf{j} + \varepsilon_0 \dot{\mathbf{E}} \right) \tag{3.4}$$

In a medium:

$$\mathbf{\nabla} \cdot \mathbf{D} = \sigma \tag{3.5}$$

$$\mathbf{\nabla} \times \mathbf{E} = -\dot{\mathbf{B}} \tag{3.6}$$

$$\mathbf{\nabla} \cdot \mathbf{B} = 0 \tag{3.7}$$

$$\mathbf{\nabla} \times \mathbf{H} = \mathbf{j} + \dot{\mathbf{D}} \tag{3.8}$$

$$\mathbf{D} = \varepsilon \mathbf{E} \tag{3.9}$$

$$\mathbf{B} = \mu \mathbf{H} \tag{3.10}$$

In Eqs. (3.5) and (3.8), σ and $\mathbf{j}$ stand for the "free" charge and current densities. The "bound" charge and current densities arising from polarization and magnetization of the medium are included in the definition of the quantities $\mathbf{D}$ and $\mathbf{H}$ in terms of ε and μ. In a plasma, the ions and electrons comprising the plasma are the equivalent of the "bound" charges and currents. Since these charges move in a complicated way, it is impractical to try to lump their effects into two constants ε and μ. Consequently, in plasma physics, one generally works with the vacuum equations (3.1)–(3.4), in which σ and $\mathbf{j}$ include *all* the charges and currents, both external and internal.

Note that we have used **E** and **B** in the vacuum equations rather than their counterparts **D** and **H**, which are related by the constants ε_0 and μ_0. This is because the forces $q\mathbf{E}$ and $\mathbf{j} \times \mathbf{B}$ depend on **E** and **B** rather than **D** and **H**, and it is not necessary to introduce the latter quantities as long as one is dealing with the vacuum equations.

3.2.2   Classical Treatment of Magnetic Materials

Since each gyrating particle has a magnetic moment, it would seem that the logical thing to do would be to consider a plasma as a magnetic material with a permeability μ_m. (We have put a subscript m on the permeability to distinguish it from the adiabatic invariant μ.) To see why this is *not* done in practice, let us review the way magnetic materials are usually treated.

The ferromagnetic domains, say, of a piece of iron have magnetic moments μ_i, giving rise to a bulk magnetization

$$\mathbf{M} = \frac{1}{V}\sum_i \boldsymbol{\mu}_i \tag{3.11}$$

per unit volume. This has the same effect as a bound current density equal to

$$\mathbf{j}_b = \nabla \times \mathbf{M} \tag{3.12}$$

In the vacuum equation (3.4), we must include in **j** both this current and the "free," or externally applied, current $\mathbf{j}_f$:

$$\mu_0^{-1}\nabla \times \mathbf{B} = \mathbf{j}_f + \mathbf{j}_b + \varepsilon_0\dot{\mathbf{E}} \tag{3.13}$$

We wish to write Eq. (3.13) in the simple form

$$\nabla \times \mathbf{H} = \mathbf{j}_f + \varepsilon_0\dot{\mathbf{E}} \tag{3.14}$$

by including $\mathbf{j}_b$ in the definition of **H**. This can be done if we let

$$\mathbf{H} = \mu_0^{-1}\mathbf{B} - \mathbf{M} \tag{3.15}$$

To get a simple relation between **B** and **H**, we assume **M** to be proportional to **B** or **H**:

$$\mathbf{M} = \chi_m\mathbf{H} \tag{3.16}$$

The constant χ_m is the magnetic susceptibility. We now have

$$\mathbf{B} = \mu_0(1 + \chi_m)\mathbf{H} \equiv \mu_m\mathbf{H} \tag{3.17}$$

This simple relation between $\mathbf{B}$ and $\mathbf{H}$ is possible because of the linear form of Eq. (3.16).

In a plasma with a magnetic field, each particle has a magnetic moment μ_α, and the quantity $\mathbf{M}$ is the sum of all these μ_α's in 1 m^3. But we now have

$$\mu_\alpha = \frac{mv_{\perp\alpha}^2}{2B} \propto \frac{1}{B} \qquad M \propto \frac{1}{B}$$

The relation between $\mathbf{M}$ and $\mathbf{H}$ (or $\mathbf{B}$) is no longer linear, and we cannot write $\mathbf{B} = \mu_m\mathbf{H}$ with μ_m constant. It is therefore not useful to consider a plasma as a magnetic medium.

3.2.3   Classical Treatment of Dielectrics

The polarization $\mathbf{P}$ per unit volume is the sum over all the individual moments $\mathbf{p}_i$ of the electric dipoles. This gives rise to a bound charge density

$$\sigma_b = -\nabla \cdot \mathbf{P} \tag{3.18}$$

In the vacuum equation (3.1), we must include both the bound charge and the free charge:

$$\varepsilon_0 \nabla \cdot \mathbf{E} = (\sigma_f + \sigma_b) \tag{3.19}$$

We wish to write this in the simple form

$$\nabla \cdot \mathbf{D} = \sigma_f \tag{3.20}$$

by including σ_b in the definition of $\mathbf{D}$. This can be done by letting

$$\mathbf{D} = \varepsilon_0\mathbf{E} + \mathbf{P} \equiv \varepsilon\,\mathbf{E} \tag{3.21}$$

If $\mathbf{P}$ is linearly proportional to $\mathbf{E}$,

$$\mathbf{P} = \varepsilon_0\chi_e\mathbf{E} \tag{3.22}$$

then ε is a constant given by

$$\varepsilon = (1 + \chi_e)\varepsilon_0 \tag{3.23}$$

There is no a priori reason why a relation like Eq. (3.22) cannot be valid in a plasma, so we may proceed to try to get an expression for ε in a plasma.

3.2.4   The Dielectric Constant of a Plasma

We have seen in Sect. 2.5 that a fluctuating $\mathbf{E}$ field gives rise to a polarization current $\mathbf{j}_p$. This leads, in turn, to a polarization charge given by the equation of continuity:

$$\frac{\partial \sigma_p}{\partial t} + \nabla \cdot \mathbf{j}_p = 0 \qquad (3.24)$$

This is the equivalent of Eq. (3.18), except that, as we noted before, a polarization effect does not arise in a plasma unless the electric field is time varying. Since we have an explicit expression for $\mathbf{j}_p$ but not for σ_p, it is easier to work with the fourth Maxwell equation, Eq. (3.4):

$$\nabla \times \mathbf{B} = \mu_0 \left(\mathbf{j}_f + \mathbf{j}_p + \varepsilon_0 \dot{\mathbf{E}} \right) \qquad (3.25)$$

We wish to write this in the form

$$\nabla \times \mathbf{B} = \mu_0 \left(\mathbf{j}_f + \varepsilon \dot{\mathbf{E}} \right) \qquad (3.26)$$

This can be done if we let

$$\varepsilon = \varepsilon_0 + \frac{j_p}{\dot{E}} \qquad (3.27)$$

From Eq. (2.67) for $\mathbf{j}_p$, we have

$$\varepsilon = \varepsilon_0 + \frac{\rho}{B^2} \qquad \text{or} \qquad \varepsilon_R \equiv \frac{\varepsilon}{\varepsilon_0} = 1 + \frac{\mu_0 \rho c^2}{B^2} \qquad (3.28)$$

This is the *low-frequency plasma dielectric constant for transverse motions*. The qualifications are necessary because our expression for $\mathbf{j}_p$ is valid only for $\omega^2 \ll \omega_c^2$ and for $\mathbf{E}$ perpendicular to $\mathbf{B}$. The general expression for ε, of course, is very complicated and hardly fits on one page.

Note that as $\rho \to 0$, ε_R approaches its vacuum value, unity, as it should. As $B \to \infty$, ε_R also approaches unity. This is because the polarization drift $\mathbf{v}_p$ then vanishes, and the particles do not move in response to the transverse electric field. In a usual laboratory plasma, the second term in Eq. (3.28) is large compared with unity. For instance, if $n = 10^{16}$ m^{-3} and $B = 0.1$ T we have (for hydrogen)

$$\frac{\mu_0 \rho c^2}{B^2} = \frac{\left(4\pi \times 10^{-7}\right)\left(10^{16}\right)\left(1.67 \times 10^{-27}\right)\left(9 \times 10^{16}\right)}{(0.1)^2} = 189$$

This means that the electric fields due to the particles in the plasma greatly alter the fields applied externally. A plasma with large ε shields out alternating fields, just as a plasma with small λ_D shields out dc fields.

Problems

3.1 Derive the uniform-plasma low-frequency dielectric constant, Eq. (3.28), by reconciling the time derivative of the equation $\nabla \cdot \mathbf{D} = \nabla \cdot (\varepsilon \mathbf{E}) = 0$ with that of the vacuum Poisson equation (3.1), with the help of equations (3.24) and (2.67).

3.2 If the ion cyclotron frequency is denoted by Ω_c and the ion plasma frequency is defined by

$$\Omega_p = \left(ne^2/\varepsilon_0 M\right)^{1/2}$$

where M is the ion mass, under what circumstances is the dielectric constant ε approximately equal to Ω_p^2/Ω_c^2?

3.3 The Fluid Equation of Motion

Maxwell's equations tell us what $\mathbf{E}$ and $\mathbf{B}$ are for a given state of the plasma. To solve the self-consistent problem, we must also have an equation giving the plasma's response to given $\mathbf{E}$ and $\mathbf{B}$. In the fluid approximation, we consider the plasma to be composed of two or more *interpenetrating fluids*, one for each species. In the simplest case, when there is only one species of ion, we shall need two equations of motion, one for the positively charged ion fluid and one for the negatively charged electron fluid. In a partially ionized gas, we shall also need an equation for the fluid of neutral atoms. The neutral fluid will interact with the ions and electrons only through collisions. The ion and electron fluids will interact with each other even in the absence of collisions, because of the $\mathbf{E}$ and $\mathbf{B}$ fields they generate.

3.3.1 The Convective Derivative

The equation of motion for a single particle is

$$m\frac{d\mathbf{v}}{dt} = q(\mathbf{E} + \mathbf{v} \times \mathbf{B}) \tag{3.29}$$

Assume first that there are no collisions and no thermal motions. Then all the particles in a fluid element move together, and the average velocity **u** of the particles in the element is the same as the individual particle velocity **v**. The fluid equation is obtained simply by multiplying Eq. (3.29) by the density n:

$$mn\frac{d\mathbf{u}}{dt} = qn(\mathbf{E} + \mathbf{u}\times\mathbf{B}) \tag{3.30}$$

This is, however, not a convenient form to use. In Eq. (3.29), the time derivative is to be taken *at the position of the particles*. On the other hand, we wish to have an equation for fluid elements *fixed in space*, because it would be impractical to do otherwise. Consider a drop of cream in a cup of coffee as a fluid element. As the coffee is stirred, the drop distorts into a filament and finally disperses all over the cup, losing its identity. A fluid element at a fixed spot in the cup, however, retains its identity although particles continually go in and out of it.

To make the transformation to variables in a fixed frame, consider $\mathbf{G}(x, t)$ to be any property of a fluid in one-dimensional x space. The change of **G** with time *in a frame moving with the fluid* is the sum of two terms:

$$\frac{d\mathbf{G}(x,t)}{dt} = \frac{\partial \mathbf{G}}{\partial t} + \frac{\partial \mathbf{G}}{\partial x}\frac{dx}{dt} = \frac{\partial \mathbf{G}}{\partial t} + u_x\frac{\partial \mathbf{G}}{\partial x} \tag{3.31}$$

The first term on the right represents the change of **G** at a fixed point in space, and the second term represents the change of **G** as the observer moves with the fluid into a region in which **G** is different. In three dimensions, Eq. (3.31) generalizes to

$$\frac{d\mathbf{G}}{dt} = \frac{\partial \mathbf{G}}{\partial t} + (\mathbf{u}\cdot\nabla)\mathbf{G} \tag{3.32}$$

This is called the *convective derivative* and is sometimes written DG/Dt. Note that $(\mathbf{u}\cdot\nabla)$ is a *scalar* differential operator. Since the sign of this term is sometimes a source of confusion, we give two simple examples.

Figure 3.1 shows an electric water heater in which the hot water has risen to the top and the cold water has sunk to the bottom. Let $G(x, t)$ be the temperature T; ∇G is then upward. Consider a fluid element near the edge of the tank. If the heater element is turned on, the fluid element is heated as it moves, and we have $dT/dt > 0$. If, in addition, a paddle wheel sets up a flow pattern as shown, the temperature in a *fixed* fluid element is lowered by the convection of cold water from the bottom. In this case, we have $\partial T/\partial x > 0$ and $u_x > 0$, so that $\mathbf{u}\cdot\nabla T > 0$. The temperature change in the fixed element, $\partial T/\partial t$, is given by a balance of these effects,

$$\frac{\partial T}{\partial t} = \frac{dT}{dt} - \mathbf{u}\cdot\nabla T \tag{3.33}$$

It is quite clear that $\partial T/\partial t$ can be made zero, at least for a short time.

Fig. 3.1 Motion of fluid
elements in a hot water
heater

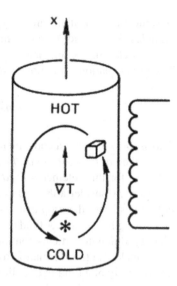

Fig. 3.2 Direction of the
salinity gradient at the
mouth of a river

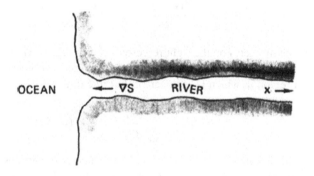

As a second example we may take G to be the salinity S of the water near the
mouth of a river (Fig. 3.2). If x is the upstream direction, there is normally a
gradient of S such that $\partial S/\partial x < 0$. When the tide comes in, the entire interface
between salt and fresh water moves upstream, and $u_x > 0$. Thus

$$\frac{\partial S}{\partial t} = -u_x \frac{\partial S}{\partial x} > 0 \qquad (3.34)$$

meaning that the salinity increases at any given point. Of course, if it rains, the
salinity decreases everywhere, and a negative term dS/dt is to be added to the
middle part of Eq. (3.34).

As a final example, take G to be the density of cars near a freeway entrance at
rush hour. A driver will see the density around him increasing as he approaches the

crowded freeway. This is the convective term $(\mathbf{u} \cdot \nabla)G$. At the same time, the local streets may be filling with cars that enter from driveways, so that the density will increase even if the observer does not move. This is the $\partial G/\partial t$ term. The total increase seen by the observer is the sum of these effects.

In the case of a plasma, we take $\mathbf{G}$ to be the fluid velocity $\mathbf{u}$ and write Eq. (3.30) as

$$mn\left[\frac{\partial \mathbf{u}}{\partial t} + (\mathbf{u} \cdot \nabla)\mathbf{u}\right] = qn(\mathbf{E} + \mathbf{u} \times \mathbf{B}) \tag{3.35}$$

where $\partial \mathbf{u}/\partial t$ is the time derivative in a fixed frame.

3.3.2 The Stress Tensor

When thermal motions are taken into account, a pressure force has to be added to the right-hand side of Eq. (3.35). This force arises from the random motion of particles in and out of a fluid element and does not appear in the equation for a single particle. Let a fluid element $\Delta x\, \Delta y\, \Delta z$ be centered at $(x_0, \frac{1}{2}\Delta y, \frac{1}{2}\Delta z)$ (Fig. 3.3). For simplicity, we shall consider only the x component of motion through the faces A and B. The number of particles per second passing through the face A with velocity v_x is

$$\Delta n_v v_x \Delta y \Delta z$$

where Δn_v is the number of particles per m^3 with velocity v_x:

$$\Delta n_v = \Delta v_x \iint f(v_x, v_y, v_z)dv_y dv_z$$

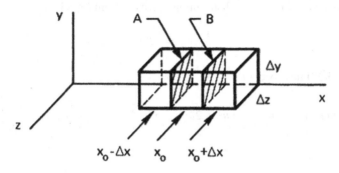

Fig. 3.3 Origin of the elements of the stress tensor

Each particle carries a momentum mv_x. The density n and temperature KT in each cube is assumed to have the value associated with the cube's center. The momentum P_{A+} carried into the element at x_0 through A is then

$$P_{A+} = \Sigma \, \Delta n_v \, m v_x^2 \, \Delta y \, \Delta z = \Delta y \, \Delta z \left[m \overline{v_x^2} \, \tfrac{1}{2} n \right]_{x_0 - \Delta x} \tag{3.36}$$

The sum over Δn_v results in the average $\overline{v_x^2}$ over the distribution. The factor $1/2$ comes from the fact that only half the particles in the cube at $x_0 - \Delta x$ are going *toward* face A. Similarly, the momentum carried out through face B is

$$P_{B+} = \Delta y \, \Delta z \left[m \overline{v_x^2} \, \tfrac{1}{2} n \right]_{x_0}$$

Thus the net gain in x momentum from right-moving particles is

$$
\begin{aligned}
P_{A+} - P_{B+} &= \Delta y \, \Delta z \tfrac{1}{2} m \left(\left[n \overline{v_x^2} \right]_{x_0 - \Delta x} - \left[n \overline{v_x^2} \right]_{x_0} \right) \\
&= \Delta y \, \Delta z \tfrac{1}{2} m (-\Delta x) \frac{\partial}{\partial x} \left(n \overline{v_x^2} \right)
\end{aligned}
\tag{3.37}
$$

This result will be just doubled by the contribution of left-moving particles, since they carry negative x momentum and also move in the opposite direction relative to the gradient of $n \overline{v_x^2}$. The total change of momentum of the fluid element at x_0 is therefore

$$\frac{\partial}{\partial t} (nmu_x) \Delta x \, \Delta y \, \Delta z = -m \frac{\partial}{\partial x} \left(n \overline{v_x^2} \right) \Delta x \, \Delta y \, \Delta z \tag{3.38}$$

Let the velocity v_x of a particle be decomposed into two parts,

$$v_x = u_x + v_{xr} \qquad u_x = \bar{v}_x$$

where u_x is the fluid velocity and v_{xr} is the random thermal velocity. For a one-dimensional Maxwellian distribution, we have from Eq. (1.7)

$$\tfrac{1}{2} m \overline{v_{xr}^2} = \tfrac{1}{2} KT \tag{3.39}$$

Equation (3.38) now becomes

$$\frac{\partial}{\partial t} (nmu_x) = -m \frac{\partial}{\partial x} \left[n \left(\overline{u_x^2} + 2\overline{uv}_{xr} + \overline{v_{xr}^2} \right) \right] = -m \frac{\partial}{\partial x} \left[n \left(u_x^2 + \frac{KT}{m} \right) \right]$$

We can cancel two terms by partial differentiation:

$$mn\frac{\partial u_x}{\partial t} + mu_x\frac{\partial n}{\partial t} = -mu_x\frac{\partial(nu_x)}{\partial x} - mnu_x\frac{\partial u_x}{\partial x} - \frac{\partial}{\partial x}(nKT) \qquad (3.40)$$

The equation of mass conservation[1]

$$\frac{\partial n}{\partial t} + \frac{\partial}{\partial x}(nu_x) = 0 \qquad (3.41)$$

allows us to cancel the terms nearest the equal sign in Eq. (3.40). Defining the pressure

$$\boxed{p \equiv nKT} \qquad (3.42)$$

we have finally

$$mn\left(\frac{\partial u_x}{\partial t} + u_x\frac{\partial u_x}{\partial x}\right) = -\frac{\partial p}{\partial x} \qquad (3.43)$$

This is the usual pressure-gradient force. Adding the electromagnetic forces and generalizing to three dimensions, we have the fluid equation

$$mn\left[\frac{\partial \mathbf{u}}{\partial t} + (\mathbf{u} \cdot \nabla)\mathbf{u}\right] = qn(\mathbf{E} + \mathbf{u}\times\mathbf{B}) - \nabla p \qquad (3.44)$$

What we have derived is only a special case: the transfer of x momentum by motion in the x direction; and we have assumed that the fluid is isotropic, so that the same result holds in the y and z directions. But it is also possible to transfer y momentum by motion in the x direction, for instance. Suppose, in Fig. 3.3, that u_y is zero in the cube at $x = x_0$ but is positive on both sides. Then as particles migrate across the faces A and B, they bring in more positive y momentum than they take out, and the fluid element gains momentum in the y direction. This *shear stress* cannot be represented by a scalar p but must be given by a tensor $\mathbf{P}$, the stress tensor, whose components $P_{ij} = mn\overline{v_i v_j}$ specify both the direction of motion and the component of momentum involved. In the general case the term $-\nabla p$ is replaced by $-\nabla \cdot \mathbf{P}$.

We shall not give the stress tensor here except for the two simplest cases. When the distribution function is an isotropic Maxwellian, $\mathbf{P}$ is written

---

[1] If the reader has not encountered this before, it is derived in Sect. 3.3.5.

$$\mathbf{P} = \begin{pmatrix} p & 0 & 0 \\ 0 & p & 0 \\ 0 & 0 & p \end{pmatrix} \tag{3.45}$$

$\nabla \cdot \mathbf{P}$ is just ∇p. In Sect. 1.3, we noted that a plasma could have two temperatures $T_\perp$ and $T_\parallel$ in the presence of a magnetic field. In that case, there would be two pressures $p_\perp = nKT_\perp$ and $p_\parallel = nKT_\parallel$. The stress tensor is then

$$\mathbf{P} = \begin{pmatrix} p_\perp & 0 & 0 \\ 0 & p_\perp & 0 \\ 0 & 0 & p_\parallel \end{pmatrix} \tag{3.46}$$

where the coordinate of the third row or column is the direction of $\mathbf{B}$. This is still diagonal and shows isotropy in a plane perpendicular to $\mathbf{B}$.

In an ordinary fluid, the off-diagonal elements of $\mathbf{P}$ are usually associated with viscosity. When particles make collisions, they come off with an average velocity in the direction of the fluid velocity $\mathbf{u}$ at the point where they made their last collision. This momentum is transferred to another fluid element upon the next collision. This tends to equalize $\mathbf{u}$ at different points, and the resulting resistance to shear flow is what we intuitively think of as viscosity. The longer the mean free path, the farther momentum is carried, and the larger is the viscosity. In a plasma there is a similar effect which occurs even in the absence of collisions. The Larmor gyration of particles (particularly ions) brings them into different parts of the plasma and tends to equalize the fluid velocities there. The Larmor radius rather than the mean free path sets the scale of this kind of collisionless viscosity. It is a finite-Larmor-radius effect which occurs in addition to collisional viscosity and is closely related to the $\mathbf{v}_E$ drift in a nonuniform $\mathbf{E}$ field (Eq. (2.58)).

3.3.3   Collisions

If there is a neutral gas, the charged fluid will exchange momentum with it through collisions. The momentum lost per collision will be proportional to the relative velocity $\mathbf{u} - \mathbf{u}_0$, where $\mathbf{u}_0$ is the velocity of the neutral fluid. If τ, the mean free time between collisions, is approximately constant, the resulting force term can be roughly written as $-mn(\mathbf{u} - \mathbf{u}_0)/\tau$. The equation of motion (3.44) can be generalized to include anisotropic pressure and neutral collisions as follows:

$$mn\left[\frac{\partial \mathbf{u}}{\partial t} + (\mathbf{u} \cdot \nabla)\mathbf{u}\right] = qn(\mathbf{E} + \mathbf{u} \times \mathbf{B}) - \nabla \cdot \mathbf{P} - \frac{mn(\mathbf{u} - \mathbf{u}_0)}{\tau} \tag{3.47}$$

Collisions between charged particles have not been included; these will be treated in Chap. 5.

3.3.4   Comparison with Ordinary Hydrodynamics

Ordinary fluids obey the Navier–Stokes equation

$$\rho\left[\frac{\partial \mathbf{u}}{\partial t} + (\mathbf{u} \cdot \nabla)\mathbf{u}\right] = -\nabla p + \rho\nu\nabla^2\mathbf{u} \tag{3.48}$$

This is the same as the plasma equation (3.47) except for the absence of electromagnetic forces and collisions between species (there being only one species). The viscosity term $\rho\nu\nabla^2\mathbf{u}$, where ν is the kinematic viscosity coefficient, is just the collisional part of $\nabla \cdot \mathbf{P} - \nabla p$ in the absence of magnetic fields. Equation (3.48) describes a fluid in which there are frequent collisions between particles. Equation (3.47), on the other hand, was derived without any explicit statement of the collision rate. Since the two equations are identical except for the E and B terms, can Eq. (3.47) really describe a plasma species? The answer is a guarded yes, and the reasons for this will tell us the limitations of the fluid theory.

In the derivation of Eq. (3.47), we did actually assume implicitly that there were many collisions. This assumption came in Eq. (3.39) when we took the velocity distribution to be Maxwellian. Such a distribution generally comes about as the result of frequent collisions. However, this assumption was used only to take the average of v_{xr}^2. Any other distribution with the same average would give us the same answer. The fluid theory, therefore, is not very sensitive to deviations from the Maxwellian distribution, but there are instances in which these deviations are important. Kinetic theory must then be used.

There is also an empirical observation by Irving Langmuir which helps the fluid theory. In working with the electrostatic probes which bear his name, Langmuir discovered that the electron distribution function was far more nearly Maxwellian than could be accounted for by the collision rate. This phenomenon, called *Langmuir's paradox*, has been attributed at times to high-frequency oscillations. There has been no satisfactory resolution of the paradox, but this seems to be one of the few instances in plasma physics where nature works in our favor.

Another reason the fluid model works for plasmas is that the magnetic field, when there is one, can play the role of collisions in a certain sense. When a particle is accelerated, say by an E field, it would continuously increase in velocity if it were allowed to free-stream. When there are frequent collisions, the particle comes to a limiting velocity proportional to E. The electrons in a copper wire, for instance, drift together with a velocity $\mathbf{v} = \mu\mathbf{E}$, where μ is the mobility. A magnetic field also limits free-streaming by forcing particles to gyrate in Larmor orbits. The electrons in a plasma also drift together with a velocity proportional to E, namely, $v_E = \mathbf{E} \times \mathbf{B}/B^2$. In this sense, a collisionless plasma behaves like a collisional fluid. Of course, particles do free-stream *along* the magnetic field, and the fluid picture is not particularly suitable for motions in that direction. *For motions perpendicular to* **B**, *the fluid theory is a good approximation.*

3.3.5   Equation of Continuity

The conservation of matter requires that the total number of particles N in a volume V can change only if there is a net flux of particles across the surface S bounding that volume. Since the particle flux density is $n\mathbf{u}$, we have, by the divergence theorem,

$$\frac{\partial N}{\partial t} = \int_V \frac{\partial n}{\partial t} dV = -\oint n\mathbf{u} \cdot d\mathbf{S} = -\int_V \boldsymbol{\nabla} \cdot (n\mathbf{u}) dV \qquad (3.49)$$

Since this must hold for any volume V, the integrands must be equal:

$$\frac{\partial n}{\partial t} + \boldsymbol{\nabla} \cdot (n\mathbf{u}) = 0 \qquad (3.50)$$

There is one such *equation of continuity* for each species. Any sources or sinks of particles are to be added to the right-hand side.

3.3.6   Equation of State

One more relation is needed to close the system of equations. For this, we can use the thermodynamic equation of state relating p to n:

$$p = C\rho^\gamma \qquad (3.51)$$

where C is a constant and γ is the ratio of specific heats C_p/C_v. The term $\boldsymbol{\nabla}p$ is therefore given by

$$\frac{\boldsymbol{\nabla}p}{p} = \gamma \frac{\boldsymbol{\nabla}n}{n} \qquad (3.52)$$

For isothermal compression, we have

$$\boldsymbol{\nabla}p = \boldsymbol{\nabla}(nKT) = KT\boldsymbol{\nabla}n$$

so that, clearly, $\gamma = 1$. For adiabatic compression, KT will also change, giving γ a value larger than one. If N is the number of degrees of freedom, γ is given by

$$\gamma = (2 + N)/N \qquad (3.53)$$

The validity of the equation of state requires that heat flow be negligible; that is, that thermal conductivity be low. Again, this is more likely to be true in directions perpendicular to $\mathbf{B}$ than parallel to it. Fortunately, most basic phenomena can be described adequately by the crude assumption of Eq. (3.51).

3.3.7   The Complete Set of Fluid Equations

For simplicity, let the plasma have only two species: ions and electrons; extension to more species is trivial. The charge and current densities are then given by

$$\sigma = n_i q_i + n_e q_e \tag{3.54}$$
$$\mathbf{j} = n_i q_i \mathbf{v}_i + n_e q_e \mathbf{v}_e$$

Since single-particle motions will no longer be considered, we may now use $\mathbf{v}$ instead of $\mathbf{u}$ for the fluid velocity. We shall neglect collisions and viscosity. Equations (3.1)–(3.4), (3.44), (3.50), and (3.51) form the following set:

$$\varepsilon_0 \nabla \cdot \mathbf{E} = n_i q_i + n_e q_e \tag{3.55}$$

$$\nabla \times \mathbf{E} = -\dot{\mathbf{B}} \tag{3.56}$$

$$\nabla \cdot \mathbf{B} = 0 \tag{3.57}$$

$$\mu_0^{-1} \nabla \times \mathbf{B} = n_i q_i \mathbf{v}_i + n_e q_e \mathbf{v}_e + \varepsilon_0 \dot{\mathbf{E}} \tag{3.58}$$

$$m_j n_j \left[\frac{\partial \mathbf{v}_j}{\partial t} + (\mathbf{v}_j \cdot \nabla)\mathbf{v}_j \right] = q_j n_j (\mathbf{E} + \mathbf{v}_j \times \mathbf{B}) - \nabla p_j \qquad j = i, e \tag{3.59}$$

$$\frac{\partial n_j}{\partial t} + \nabla \cdot (n_j \mathbf{v}_j) = 0 \qquad j = i, e \tag{3.60}$$

$$p_j = C_j n_j^{\gamma_j} \qquad j = i, e \tag{3.61}$$

There are 16 scalar unknowns: n_i, n_e, p_i, p_e, $\mathbf{v}_i$, $\mathbf{v}_e$, $\mathbf{E}$, and $\mathbf{B}$. There are apparently 18 scalar equations if we count each vector equation as three scalar equations. However, two of Maxwell's equations are superfluous, since Eqs. (3.55) and (3.57) can be recovered from the divergences of Eqs. (3.58) and (3.56) (Problem 3.3). The simultaneous solution of this set of 16 equations in 16 unknowns gives a self-consistent set of fields and motions in the fluid approximation.

3.4   Fluid Drifts Perpendicular to B

Since a fluid element is composed of many individual particles, one would expect the fluid to have drifts perpendicular to $\mathbf{B}$ if the individual guiding centers have such drifts. However, since the ∇p term appears only in the fluid equations, there is a drift associated with it which the fluid elements have but the particles do not have. For each species, we have an equation of motion

$$mn\left[\underset{①}{\frac{\partial \mathbf{v}}{\partial t}}+\underset{②}{(\mathbf{v}\cdot\nabla)\mathbf{v}}\right]=qn(\mathbf{E}+\underset{③}{\mathbf{v}\times\mathbf{B}})-\nabla p \qquad (3.62)$$

Consider the ratio of term ① to term ③:

$$\frac{①}{③}\approx\left|\frac{mni\omega v_{\perp}}{qnv_{\perp}B}\right|\approx\frac{\omega}{\omega_{c}}$$

Here we have taken $\partial/\partial t=i\omega$ and are concerned only with $\mathbf{v}_{\perp}$. For drifts slow compared with the time scale of ω_{c}, we may neglect term ①. We shall also neglect the $(\mathbf{v}\cdot\nabla)\mathbf{v}$ term and show *a posteriori* that this is all right. Let $\mathbf{E}$ and $\mathbf{B}$ be uniform, but let n and p have a gradient. This is the usual situation in a magnetically confined plasma column (Fig. 3.4). Taking the cross product of Eq. (3.62) with $\mathbf{B}$, we have (neglecting the left-hand side)

$$0 = qn[\mathbf{E}\times\mathbf{B}+(\mathbf{v}_{\perp}\times\mathbf{B})\times\mathbf{B}]-\nabla p\times\mathbf{B}$$
$$= qn[\mathbf{E}\times\mathbf{B}+\mathbf{B}(\mathbf{v}_{\perp}\cdot\mathbf{B})-\mathbf{v}_{\perp}B^{2}]-\nabla p\times\mathbf{B}$$

Therefore,

$$\mathbf{v}_{\perp}=\frac{\mathbf{E}\times\mathbf{B}}{B^{2}}-\frac{\nabla p\times\mathbf{B}}{qnB^{2}}\equiv \mathbf{v}_{E}+\mathbf{v}_{D} \qquad (3.63)$$

where

$$\boxed{\mathbf{v}_{E}\equiv\frac{\mathbf{E}\times\mathbf{B}}{B^{2}}}\qquad \mathbf{E}\times\mathbf{B}\quad\text{drift} \qquad (3.64)$$

$$\boxed{\mathbf{v}_{D}\equiv-\frac{\nabla p\times\mathbf{B}}{qnB^{2}}}\qquad \text{Diamagnetic drift} \qquad (3.65)$$

The drift $\mathbf{v}_{E}$ is the same as for guiding centers, but there is now a new drift $\mathbf{v}_{D}$, called the diamagnetic drift. Since $\mathbf{v}_{D}$ is perpendicular to the direction of the gradient, our neglect of $(\mathbf{v}\cdot\nabla)\mathbf{v}$ is justified if $\mathbf{E}=0$. If $\mathbf{E}=-\nabla\phi\neq 0$, $(\mathbf{v}\cdot\nabla)\mathbf{v}$ is still zero if $\nabla\phi$ and ∇p are in the same direction; otherwise, there could be a more complicated solution involving $(\mathbf{v}\cdot\nabla)\mathbf{v}$.

With the help of Eq. (3.52), we can write the diamagnetic drift as

$$\mathbf{v}_{D}=\pm\frac{\gamma KT}{eB}\frac{\hat{\mathbf{z}}\times\nabla n}{n} \qquad (3.66)$$

Fig. 3.4 Diamagnetic drifts
in a cylindrical plasma

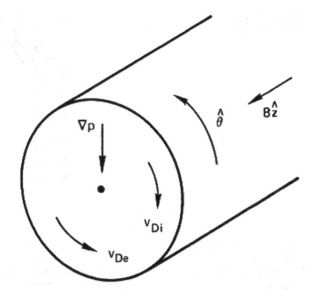

In particular, for an isothermal plasma in the geometry of Fig. 3.4, in which $\nabla n = n'\hat{\mathbf{r}}$, we have the following formulas familiar to experimentalists who have worked with Q-machines[2]:

$$\mathbf{v}_{Di} = \frac{KT_i}{eB}\frac{n'}{n}\hat{\boldsymbol{\theta}} \quad \left(n' \equiv \frac{\partial n}{\partial r} < 0\right) \tag{3.67}$$

$$\mathbf{v}_{De} = -\frac{KT_e}{eB}\frac{n'}{n}\hat{\boldsymbol{\theta}}$$

The magnitude of $\mathbf{v}_D$ is easily computed from the formula

$$\boxed{v_D = \frac{KT(eV)}{B(T)}\frac{1}{\Lambda}\frac{\text{m}}{\text{sec}}} \tag{3.68}$$

where Λ is the density scale length $|n'/n|$ in m.

The physical reason for this drift can be seen from Fig. 3.5. Here we have drawn the orbits of ions gyrating in a magnetic field. There is a density gradient toward the left, as indicated by the density of orbits. Through any fixed volume element there are more ions moving downward than upward, since the downward-moving ions come from a region of higher density. There is, therefore, a fluid drift perpendicular to ∇n and **B**, *even though the guiding centers are stationary*. The diamagnetic drift reverses sign with q because the direction of gyration reverses. The magnitude of $\mathbf{v}_D$

---

[2] A Q-machine produces a quiescent plasma by thermal ionization of Cs or K atoms impinging on hot tungsten plates. Diamagnetic drifts were first measured in Q-machines.

Fig. 3.5 Origin of the diamagnetic drift

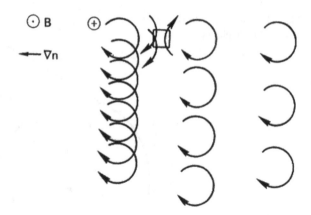

Fig. 3.6 Particle drifts in a bounded plasma, illustrating the relation to fluid drifts

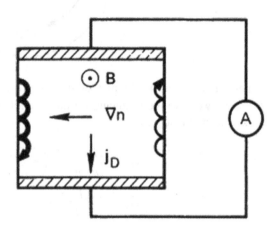

does not depend on mass because the $m^{-1/2}$ dependence of the velocity is cancelled by the $m^{1/2}$ dependence of the Larmor radius—less of the density gradient is sampled during a gyration if the mass is small.

Since ions and electrons drift in opposite directions, there is a diamagnetic current. For $\gamma = Z = 1$, this is given by

$$\mathbf{j}_D = ne(\mathbf{v}_{Di} - \mathbf{v}_{De}) = (KT_i + KT_e)\frac{\mathbf{B} \times \nabla n}{B^2} \tag{3.69}$$

In the particle picture, one would not expect to measure a current if the guiding centers do not drift. In the fluid picture, the current $\mathbf{j}_D$ flows wherever there is a pressure gradient. These two viewpoints can be reconciled if one considers that all experiments must be carried out in a finite-sized plasma. Suppose the plasma were in a rigid box (Fig. 3.6). If one were to calculate the current from the single-particle picture, one would have to take into account the particles at the edges which have

Fig. 3.7 Measuring the
diamagnetic current in an
inhomogeneous plasma

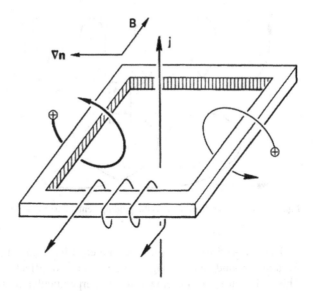

cycloidal paths. Since there are more particles on the left than on the right, there is a
net current downward, in agreement with the fluid picture.

The reader may not be satisfied with this explanation because it was necessary to
specify reflecting walls. If the walls were absorbing or if they were removed, one
would find that electric fields would develop because more of one species—the one
with larger Larmor radius—would collected than the other. Then the guiding
centers would drift, and the simplicity of the model would be lost. Alternatively,
one could imagine trying to measure the diamagnetic current with a current probe
(Fig. 3.7). This is just a transformer with a core of magnetic material. The primary
winding is the plasma current threading the core, and the secondary is a multiturn
winding all around the core. Let the whole thing be infinitesimally thin, so it does
not intercept any particles. It is clear from Fig. 3.7 that a net upward current would
be measured, there being higher density on the left than on the right, so that the
diamagnetic current is a real current. From this example, one can see that it can be
quite tricky to work with the single-particle picture. The fluid theory usually gives
the right results when applied straightforwardly, even though it contains "fictitious"
drifts like the diamagnetic drift.

What about the grad-B and curvature drifts which appeared in the single-particle
picture? The curvature drift also exists in the fluid picture, since the centrifugal
force is felt by all the particles in a fluid element as they move around a bend in the
magnetic field. A term $\overline{F}_{cf} = \overline{nmv_\parallel^2}/R_c = nKT_\parallel/R_c$ has to be added to the right-
hand side of the fluid equation of motion. This is equivalent to a gravitational force
Mng, with $g = KT_\parallel/MR_c$, and leads to a drift $v_g = (m/q)(g \times B)/B^2$, as in the
particle picture (Eq. (2.18)).

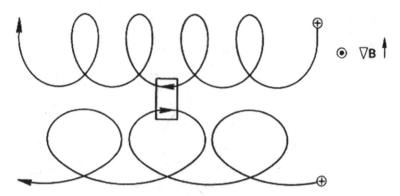

Fig. 3.8 In a nonuniform B field the guiding centers drift but the fluid elements do not

The grad-B drift, however, does not exist for fluids. It can be shown on thermodynamic grounds that a magnetic field does not affect a Maxwellian distribution. This is because the Lorentz force is perpendicular to $\mathbf{v}$ and cannot change the energy of any particle. The most probable distribution $f(\mathbf{v})$ in the absence of $\mathbf{B}$ is also the most probable distribution in the presence of $\mathbf{B}$. If $f(\mathbf{v})$ remains Maxwellian in a nonuniform $\mathbf{B}$ field, and there is no density gradient, then the net momentum carried into any fixed fluid element is zero. There is no fluid drift even though the individual guiding centers have drifts; the particle drifts in any fixed fluid element cancel out. To see this pictorially, consider the orbits of two particles moving through a fluid element in a nonuniform $\mathbf{B}$ field (Fig. 3.8). Since there is no $\mathbf{E}$ field, the Larmor radius changes only because of the gradient in $\mathbf{B}$; there is no acceleration, and the particle energy remains constant during the motion. If the two particles have the same energy, they will have the same velocity and Larmor radius while inside the fluid element. There is thus a perfect cancellation between particle pairs when their velocities are added to give the fluid velocity.

When there is a nonuniform $\mathbf{E}$ field, it is not easy to reconcile the fluid and particle pictures. Then the finite-Larmor-radius effect of Sect. 2.4 causes both a guiding center drift and a fluid drift, but these are not the same; in fact, they have opposite signs! The particle drift was calculated in Chap. 2, and the fluid drift can be calculated from the off-diagonal elements of $\mathbf{P}$. It is extremely difficult to explain how the finite-Larmor-radius effects differ. A simple picture like Fig. 3.6 will not work because one has to take into account subtle points like the following: In the presence of a density gradient, the density of guiding centers is not the same as the density of particles!

Problems

3.3 Show that Eqs. (3.55) and (3.57) are redundant in the set of Maxwell's equations.

3.4 Show that the expression for $\mathbf{j}_D$ on the right-hand side of Eq. (3.69) has the dimensions of a current density.

3.5 Show that if the current calculated from the particle picture (Fig. 3.6) agrees with that calculated from the diamagnetic drift for one width of the box, then it will agree for all widths.

3.6 An isothermal plasma is confined between the planes $x = \pm a$ in a magnetic field $\mathbf{B} = B_0 \hat{\mathbf{z}}$. The density distribution is

$$n = n_0 \left(1 - x^2/a^2 \right)$$

(a) Derive an expression for the electron diamagnetic drift velocity v_{De} as a function of x.

(b) Draw a diagram showing the density profile and the direction of v_{De} on both sides of the midplane if $\mathbf{B}$ is out of the paper.

(c) Evaluate v_{De} at $x = a/2$ if $B = 0.2$ T, $KT_e = 2$ eV, and $a = 4$ cm.

3.7 A cylindrically symmetric plasma column in a uniform $\mathbf{B}$ field has

$$n(r) = n_0 \exp\left(-r^2/r_0^2 \right) \quad \text{and} \quad n_i = n_e = n_0 \exp(e\phi/KT_e).$$

(The latter is the Boltzmann relation, Eq. (3.73).)

(a) Show that v_E and v_{De} are equal and opposite.

(b) Show that the plasma rotates as a solid body.

(c) In the frame which rotates with velocity v_E, some plasma waves (drift waves) propagate with a phase velocity $v_\phi = 0.5 v_{De}$. What is v_ϕ in the lab frame? On a diagram of the $r - \theta$ plane, draw arrows indicating the relative magnitudes and directions of v_E, v_{De}, and v_ϕ in the lab frame.

3.8 (a) For the plasma of Problem 3.7, find the diamagnetic current density j_D as a function of radius.

(b) Evaluate j_D in A/m^2 for $B = 0.4$ T, $n_0 = 10^{16}$ m^{-3}, $KT_e = KT_i = 0.25$ eV, $r = r_0 = 1$ cm.

(c) In the lab frame, is this current carried by ions or by electrons or by both?

3.9 In the preceding problem, by how much does the diamagnetic current reduce B on the axis? Hint: You may use Ampere's circuital law over an appropriate path.

3.10 In 2013, the Voyager 1 spacecraft left the heliosphere, the region dominated by solar winds, and entered outer space. The plasma frequency jumped from 2.2 to 2.6 kHz. What was the change in plasma density?

3.5 Fluid Drifts Parallel to B

The z component of the fluid equation of motion is

$$mn\left[\frac{\partial v_z}{\partial t} + (\mathbf{v} \cdot \nabla)v_z\right] = qnE_z - \frac{\partial p}{\partial z} \qquad (3.70)$$

The convective term can often be neglected because it is much smaller than the $\partial v_z/\partial t$ term. We shall avoid complicated arguments here and simply consider cases in which v_z is spatially uniform. Using Eq. (3.52), we have

$$\frac{\partial v_z}{\partial t} = \frac{q}{m}E_z - \frac{\gamma KT}{mn}\frac{\partial n}{\partial z} \qquad (3.71)$$

This shows that the fluid is accelerated along $\mathbf{B}$ under the combined electrostatic and pressure gradient forces. A particularly important result is obtained by applying Eq. (3.71) to massless electrons. Taking the limit $m \to 0$ and specifying $q = -e$ and $\mathbf{E} = -\nabla\phi$, we have[3]

$$qE_z = e\frac{\partial \phi}{\partial z} = \frac{\gamma KT_e}{n}\frac{\partial n}{\partial z} \qquad (3.72)$$

Electrons are so mobile that their heat conductivity is almost infinite. We may then assume isothermal electrons and take $\gamma = 1$. Integrating, we have

$$e\phi = KT_e\ln n + C$$

or

$$\boxed{n = n_0\exp(e\phi/KT_e)} \qquad (3.73)$$

This is just the *Boltzmann relation* for electrons.

What this means physically is that electrons, being light, are very mobile and would be accelerated to high energies very quickly if there were a net force on them. Since electrons cannot leave a region *en masse* without leaving behind a large ion charge, the electrostatic and pressure gradient forces on the electrons must be closely in balance. This condition leads to the Boltzmann relation. Note that Eq. (3.73) *applies to each line of force separately*. Different lines of force may be charged to different potentials arbitrarily unless a mechanism is provided for the electrons to move across $\mathbf{B}$. The conductors on which lines of force terminate can provide such a mechanism, and the experimentalist has to take these end effects into account carefully.

Figure 3.9 shows graphically what occurs when there is a local density clump in the plasma. Let the density gradient be toward the center of the diagram, and suppose KT is constant. There is then a pressure gradient toward the center. Since the plasma is quasineutral, the gradient exists for both the electron and ion fluids.

---

[3] Why can't $v_z \to \infty$ keeping mv_z constant? Consider the energy!

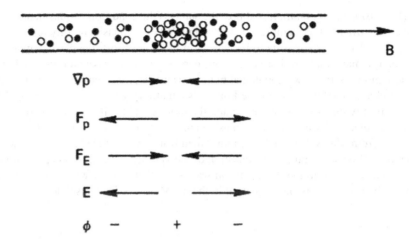

Fig. 3.9 Physical reason for the Boltzmann relation between density and potential

Consider the pressure gradient force $\mathbf{F}_p$ on the electron fluid. It drives the mobile electrons away from the center, leaving the ions behind. The resulting positive charge generates a field $\mathbf{E}$ whose force $\mathbf{F}_E$ on the electrons opposes $\mathbf{F}_p$. Only when $\mathbf{F}_E$ is equal and opposite to $\mathbf{F}_p$ is a steady state achieved. If $\mathbf{B}$ is constant, $\mathbf{E}$ is an electrostatic field $\mathbf{E} = -\nabla\phi$, and ϕ must be large at the center, where n is large. This is just what Eq. (3.73) tells us. The deviation from strict neutrality adjusts itself so that there is just enough charge to set up the $\mathbf{E}$ field required to balance the forces on the electrons.

3.6 The Plasma Approximation

The previous example reveals an important characteristic of plasmas that has wide application. We are used to solving for $\mathbf{E}$ from Poisson's equation when we are given the charge density σ. In a plasma, the opposite procedure is generally used. $\mathbf{E}$ is found from the equations of motion, and Poisson's equation is used only to find σ. The reason is that a plasma has an overriding tendency to remain neutral. If the ions move, the electrons will follow. $\mathbf{E}$ must adjust itself so that the orbits of the electrons and ions preserve neutrality. The charge density is of secondary importance; it will adjust itself so that Poisson's equation is satisfied. This is true, of course, only for low-frequency motions in which the electron inertia is not a factor.

In a plasma, it is usually possible to assume $n_i = n_e$ and $\nabla \cdot \mathbf{E} \neq 0$ at the same time. We shall call this the *plasma approximation*. It is a fundamental trait of plasmas, one which is difficult for the novice to understand. *Do not use Poisson's equation to obtain* $\mathbf{E}$ *unless it is unavoidable!* In the set of fluid equations (3.55)–(3.61), we may now eliminate Poisson's equation and also eliminate one of the unknowns by setting $n_i = n_e = n$.

The *plasma approximation* is almost the same as the condition of quasineutrality discussed earlier but has a more exact meaning. Whereas quasineutrality refers to a general tendency for a plasma to be neutral in its state of rest, the plasma approximation is a mathematical shortcut that one can use even for wave motions. As long as these motions are slow enough that both ions and electrons have time to move, it is a good approximation to replace Poisson's equation by the equation $n_i = n_e$. Of course, if only one species can move and the other cannot follow, such as in high-frequency electron waves, then the plasma approximation is not valid, and $\mathbf{E}$ must be found from Maxwell's equations rather than from the ion and electron equations of motion. We shall return to the question of the validity of the plasma approximation when we come to the theory of ion waves. At that time, it will become clear why we had to use Poisson's equation in the derivation of Debye shielding.

Chapter 4
Waves in Plasmas

4.1 Representation of Waves

Any periodic motion of a fluid can be decomposed by Fourier analysis into a superposition of sinusoidal oscillations with different frequencies ω and wavelengths λ. A simple wave is any one of these components. When the oscillation amplitude is small, the waveform is generally sinusoidal; and there is only one component. This is the situation we shall consider.

Any sinusoidally oscillating quantity—say, the density n—can be represented as follows:

$$n = \bar{n} \exp\left[i(\mathbf{k} \cdot \mathbf{r} - \omega t)\right] \tag{4.1}$$

where, in Cartesian coordinates,

$$\mathbf{k} \cdot \mathbf{r} = k_x x + k_y y + k_z z \tag{4.2}$$

Here $\bar{n}$ is a constant defining the amplitude of the wave, and $\mathbf{k}$ is called the propagation constant. If the wave propagates in the x direction, $\mathbf{k}$ has only an x component, and Eq. (4.1) becomes

$$n = \bar{n} e^{i(kx - \omega t)}$$

By convention, the exponential notation means that the real part of the expression is to be taken as the measurable quantity. Let us choose $\bar{n}$ to be real; we shall soon see that this corresponds to a choice of the origins of x and t. The real part of n is then

---

The original version of this chapter was revised. An erratum to this chapter can be found at https://doi.org/10.1007/978-3-319-22309-4_11

© Springer International Publishing Switzerland 2016
F.F. Chen, *Introduction to Plasma Physics and Controlled Fusion*,
DOI 10.1007/978-3-319-22309-4_4

$$\mathrm{Re}\,(n) = \bar{n}\cos{(kx - \omega t)} \tag{4.3}$$

A point of constant phase on the wave moves so that $(d/dt)(kx - \omega t) = 0$, or

$$\boxed{\frac{dx}{dt} = \frac{\omega}{k} \equiv v_\varphi} \tag{4.4}$$

This is called the *phase velocity*. If ω/k is positive, the wave moves to the right; that is, x increases as t increases, so as to keep $kx - \omega t$ constant. If ω/k is negative, the wave moves to the left. We could equally well have taken

$$n = \bar{n}\,e^{i(kx+\omega t)}$$

in which case positive ω/k would have meant negative phase velocity. This is a convention that is sometimes used, but we shall not adopt it. From Eq. (4.3), it is clear that reversing the sign of *both* ω and k makes no difference.

Consider now another oscillating quantity in the wave, say the electric field $\mathbf{E}$. Since we have already chosen the phase of n to be zero, we must allow $\mathbf{E}$ to have a different phase δ:

$$\mathbf{E} = \bar{\mathbf{E}}\cos{(kx - \omega t + \delta)} \quad \text{or} \quad \mathbf{E} = \bar{\mathbf{E}}\,e^{i(kx - \omega t + \delta)} \tag{4.5}$$

where $\bar{E}$ is a real, constant vector.

It is customary to incorporate the phase information into $\bar{E}$ by allowing $\bar{E}$ to be complex. We can write

$$\mathbf{E} = \bar{\mathbf{E}}\,e^{i\delta}e^{i(kx-\omega t)} \equiv \bar{\mathbf{E}}_c\,e^{i(kx-\omega t)}$$

where $\bar{E}_c$ is a complex amplitude. The phase δ can be recovered from $\bar{E}_c$, since Re $\left(\bar{\mathbf{E}}_c\right) = \bar{\mathbf{E}}\cos{\delta}$ and Im $\left(\bar{\mathbf{E}}_c\right) = \bar{\mathbf{E}}\sin{\delta}$, so that

$$\tan{\delta} = \frac{\mathrm{Im}\left(\bar{\mathbf{E}}_c\right)}{\mathrm{Re}\left(\bar{\mathbf{E}}_c\right)} \tag{4.6}$$

From now on, we shall assume that all amplitudes are complex and drop the subscript c. Any oscillating quantity $\mathbf{g}_1$ will be written

$$\mathbf{g}_1 = \mathbf{g}_1\exp{[i(\mathbf{k}\cdot\mathbf{r} - \omega t)]} \tag{4.7}$$

so that $\mathbf{g}_1$ can stand for either the complex amplitude or the entire expression Eq. (4.7). There can be no confusion, because in linear wave theory the same exponential factor will occur on both sides of any equation and can be cancelled out.

Problem

4.1 The oscillating density n_1 and potential ϕ_1 in a "drift wave" are related by

$$\frac{n_1}{n_0} = \frac{e\phi_1}{KT_e} \frac{\omega_* + ia}{\omega + ia}$$

where it is only necessary to know that all the other symbols (except i) stand for positive constants.

(a) Find an expression for the phase δ of ϕ_1 relative to n_1. (For simplicity, assume that n_1 is real.)

(b) If $\omega < \omega_*$, does ϕ_1 lead or lag n_1?

4.2 Group Velocity

The phase velocity of a wave in a plasma often exceeds the velocity of light c. This does not violate the theory of relativity, because an infinitely long wave train of constant amplitude cannot carry information. The carrier of a radio wave, for instance, carries no information until it is modulated. The modulation information does not travel at the phase velocity but at the group velocity, which is always less than c. To illustrate this, we may consider a modulated wave formed by adding ("beating") two waves of nearly equal frequencies. Let these waves be

$$E_1 = E_0 \cos\left[(k + \Delta k)x - (\omega + \Delta\omega)t\right] \qquad (4.8)$$
$$E_2 = E_0 \cos\left[(k - \Delta k)x - (\omega - \Delta\omega)t\right]$$

E_1 and E_2 differ in frequency by $2\Delta\omega$. Since each wave must have the phase velocity ω/k appropriate to the medium in which they propagate, one must allow for a difference $2\Delta k$ in propagation constant. Using the abbreviations

$$a = kx - \omega t$$
$$b = (\Delta k)x - (\Delta\omega)t$$

we have

$$E_1 + E_2 = E_0 \cos(a + b) + E_0 \cos(a - b)$$
$$= E_0(\cos a \cos b - \sin a \sin b + \cos a \cos b + \sin a \sin b)$$
$$= 2E_0 \cos a \cos b$$
$$E_1 + E_2 = 2E_0 \cos\left[(\Delta k)x - (\Delta\omega)t\right] \cos(kx - \omega t) \qquad (4.9)$$

Fig. 4.1 Spatial variation of the electric field of two waves with a frequency difference

This is a sinusoidally modulated wave (Fig. 4.1). The envelope of the wave, given by $\cos [(\Delta k)x - (\Delta\omega)t]$, is what carries information; it travels at velocity $\Delta\omega/\Delta k$. Taking the limit $\Delta\omega \to 0$, we define the *group velocity* to be

$$\boxed{v_g = d\omega/dk} \tag{4.10}$$

It is *this* quantity that cannot exceed c.

4.3   Plasma Oscillations

If the electrons in a plasma are displaced from a uniform background of ions, electric fields will be built up in such a direction as to restore the neutrality of the plasma by pulling the electrons back to their original positions. Because of their inertia, the electrons will overshoot and oscillate around their equilibrium positions with a characteristic frequency known as the *plasma frequency*. This oscillation is so fast that the massive ions do not have time to respond to the oscillating field and may be considered as fixed. In Fig. 4.2, the open rectangles represent typical elements of the ion fluid, and the darkened rectangles the alternately displaced elements of the electron fluid. The resulting charge bunching causes a spatially periodic **E** field, which tends to restore the electrons to their neutral positions.

We shall derive an expression for the plasma frequency ω_p in the simplest case, making the following assumptions: (1) There is no magnetic field; (2) there are no thermal motions ($KT = 0$); (3) the ions are fixed in space in a uniform distribution; (4) the plasma is infinite in extent; and (5) the electron motions occur only in the x direction. As a consequence of the last assumption, we have

$$\nabla = \hat{x}\,\partial/\partial x \quad \mathbf{E} = E\hat{x} \quad \nabla \times \mathbf{E} = 0 \quad \mathbf{E} = -\nabla\phi \tag{4.11}$$

There is, therefore, no fluctuating magnetic field; this is an electrostatic oscillation.

The electron equations of motion and continuity are

$$mn_e \left[\frac{\partial \mathbf{v}_e}{\partial t} + (\mathbf{v}_e \cdot \nabla)\mathbf{v}_e\right] = -en_e\mathbf{E} \tag{4.12}$$

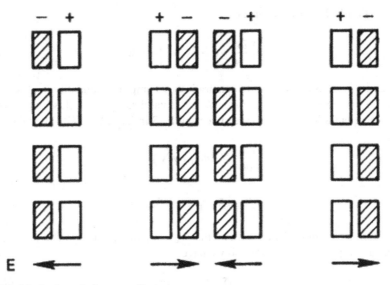

Fig. 4.2 Mechanism of plasma oscillations

$$\frac{\partial n_e}{\partial t} + \nabla \cdot (n_e v_e) = 0 \tag{4.13}$$

The only Maxwell equation we shall need is the one that does not involve **B**: Poisson's equation. This case is an exception to the general rule of Sect. 3.6 that Poisson's equation cannot be used to find **E**. This is a high-frequency oscillation; electron inertia is important, and the deviation from neutrality is the main effect in this particular case. Consequently, we write

$$\varepsilon_0 \nabla \cdot \mathbf{E} = \varepsilon_0 \partial \mathbf{E}/\partial x = e(n_i - n_e) \tag{4.14}$$

Equations (4.12)–(4.14) can easily be solved by the procedure of *linearization*. By this we mean that the amplitude of oscillation is small, and terms containing higher powers of amplitude factors can be neglected. We first separate the dependent variables into two parts: an "equilibrium" part indicated by a subscript 0, and a "perturbation" part indicated by a subscript 1:

$$n_e = n_0 + n_1 \quad v_e = v_0 + v_1 \quad \mathbf{E} = \mathbf{E}_0 + \mathbf{E}_1 \tag{4.15}$$

The equilibrium quantities express the state of the plasma in the absence of the oscillation. Since we have assumed a uniform neutral plasma at rest before the electrons are displaced, we have

$$\nabla n_0 = v_0 = \mathbf{E}_0 = 0 \tag{4.16}$$

$$\frac{\partial n_0}{\partial t} = \frac{\partial v_0}{\partial t} = \frac{\partial \mathbf{E}_0}{\partial t} = 0$$

Equation (4.12) now becomes

$$m\left[\frac{\partial \mathbf{v}_1}{\partial t} + (\mathbf{v}_1 \cdot \overset{0}{\nearrow}\nabla)\mathbf{v}_1\right] = -e\mathbf{E}_1 \tag{4.17}$$

The term $(\mathbf{v}_1 \cdot \nabla)\mathbf{v}_1$ is seen to be quadratic in an amplitude quantity, and we shall linearize by neglecting it. The *linear theory* is valid as long as $|\mathbf{v}_1|$ is small enough that such quadratic terms are indeed negligible. Similarly, Eq. (4.13) becomes

$$\frac{\partial n_1}{\partial t} + \nabla \cdot (n_0 \mathbf{v}_1 + n_1 \overset{0}{\nearrow}\mathbf{v}_1) = 0$$

$$\frac{\partial n_1}{\partial t} + n_0 \nabla \cdot \mathbf{v}_1 + \mathbf{v}_1 \cdot \nabla \overset{0}{\nearrow}n_0 = 0 \tag{4.18}$$

In Poisson's equation (4.14), we note that $n_{i0} = n_{e0}$ in equilibrium and that $n_{i1} = 0$ by the assumption of fixed ions, so we have

$$\varepsilon_0 \nabla \cdot \mathbf{E}_1 = -en_1 \tag{4.19}$$

The oscillating quantities are assumed to behave sinusoidally:

$$\mathbf{v}_1 = v_1 e^{i(kx-\omega t)}\hat{\mathbf{x}}$$

$$n_1 = n_1 e^{i(kx-\omega t)} \tag{4.20}$$

$$\mathbf{E} = E_1 e^{i(kx-\omega t)}\hat{\mathbf{x}}$$

The time derivative $\partial/\partial t$ can therefore be replaced by $-i\omega$, and the gradient ∇ by $ik\hat{\mathbf{x}}$. Equations (4.17)–(4.19) now become

$$-im\omega v_1 = -eE_1 \tag{4.21}$$

$$-i\omega n_1 = -n_0 ik v_1 \tag{4.22}$$

$$ik\varepsilon_0 E_1 = -en_1 \tag{4.23}$$

Eliminating n_1 and E_1, we have for Eq. (4.21)

$$-im\omega v_1 = -e\frac{-e}{ik\varepsilon_0}\frac{-n_0 ik v_1}{-i\omega} = -i\frac{n_0 e^2}{\varepsilon_0 \omega}v_1 \tag{4.24}$$

If v_1 does not vanish, we must have

$$\omega^2 = n_0 e^2/m\varepsilon_0$$

The *plasma frequency* is therefore

$$\omega_p = \left(\frac{n_0 e^2}{\varepsilon_0 m}\right)^{1/2} \quad \text{rad/sec} \tag{4.25}$$

Numerically, one can use the approximate formula

$$\omega_p/2\pi = f_p \approx 9\sqrt{n} \ \left(n \text{ in } m^{-3}\right) \tag{4.26}$$

This frequency, depending only on the plasma density, is one of the fundamental parameters of a plasma. Because of the smallness of m, the plasma frequency is usually very high. For instance, in a plasma of density $n = 10^{18}$ m^{-3}, we have

$$f_p \approx 9\left(10^{18}\right)^{1/2} = 9 \times 10^9 \text{ sec}^{-1} = 9\,\text{GHz}$$

Radiation at f_p normally lies in the microwave range. We can compare this with another electron frequency: ω_c. A useful numerical formula is

$$f_{ce} \simeq 28\,\text{GHz}\,/\,\text{Tesla} \tag{4.27}$$

Thus if $B \approx 0.32$ T and $n \approx 10^{18}$ m^{-3}, the cyclotron frequency is approximately equal to the plasma frequency for electrons.

Equation (4.25) tells us that if a plasma oscillation is to occur at all, it must have a frequency depending only on n. In particular, ω does not depend on k, so the group velocity $d\omega/dk$ is zero. The disturbance does not propagate. How this can happen can be made clear with a mechanical analogy (Fig. 4.3). Imagine a number of heavy balls suspended by springs equally spaced in a line. If all the springs are identical, each ball will oscillate vertically with the same frequency. If the balls are started in the proper phases relative to one another, they can be made to form a wave propagating in either direction. The frequency will be fixed by the springs, but the wavelength can be chosen arbitrarily. The two undisturbed balls at the ends will not be affected, and the initial disturbance does not propagate. Either traveling waves or standing waves can be created, as in the case of a stretched rope. Waves on a rope, however, must propagate because each segment is connected to neighboring segments.

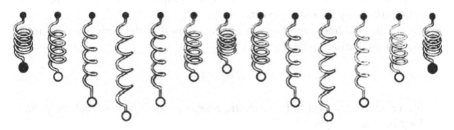

Fig. 4.3 Synthesis of a wave from an assembly of independent oscillators

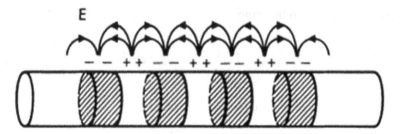

Fig. 4.4 Plasma oscillations propagate in a finite medium because of fringing fields

This analogy is not quite accurate, because plasma oscillations have motions in the direction of **k** rather than transverse to **k**. However, as long as electrons do not collide with ions or with each other, they can still be pictured as independent oscillators moving horizontally (in Fig. 4.3). But what about the electric field? Won't that extend past the region of initial disturbance and set neighboring layers of plasma into oscillation? In our simple example, it will not, because the electric field due to equal numbers of positive and negative infinite plane charge sheets is zero. In any finite system, however, plasma oscillations will propagate. In Fig. 4.4, the positive and negative (shaded) regions of a plane plasma oscillation are confined in a cylindrical tube. The fringing electric field causes a coupling of the disturbance to adjacent layers, and the oscillation does not stay localized.

Problems

4.2 The plasma density in the lower ionosphere has been measured during satellite re-entry to be about $10^{18}\,\mathrm{m}^{-3}$ at 50 km altitude, 10^{17} at 70 km, ad 10^{14} at 85 km. What are the plasma frequencies there?

4.3 Calculate the plasma frequency with the ion motions included, thus justifying our assumption that the ions are essentially fixed. (Hint: include the term n_{1i} in Poisson's equation and use the ion equations of motion and continuity.)

4.4 For a simple plasma oscillation with fixed ions and a space-time behavior $\exp[i\,(kx - \omega t)]$, calculate the phase δ for ϕ_1, E_1, and v_1 if the phase of n_1, is zero. Illustrate the relative phases by drawing sine waves representing n_1, ϕ_1, E_1, and v_1 (a) as a function of x at $t=0$, (b) as a function of t at $x=0$ for $\omega/k>0$, and (c) as a function of t at $x=0$ for $\omega/k<0$. Note that the time patterns can be obtained by translating the x patterns in the proper direction, as if the wave were passing by a fixed observer.

4.5 By writing the linearized Poisson's equation used in the derivation of simple plasma oscillations in the form

$$\nabla \cdot (\varepsilon \mathbf{E}) = 0$$

derive an expression for the dielectric constant ε applicable to high-frequency longitudinal motions.

4.4   Electron Plasma Waves

There is another effect that can cause plasma oscillations to propagate, and that is thermal motion. Electrons streaming into adjacent layers of plasma with their thermal velocities will carry information about what is happening in the oscillating region. The plasma *oscillation* can then properly be called a plasma *wave*. We can easily treat this effect by adding a term $-\nabla p_e$ to the equation of motion Eq. (4.12). In the one-dimensional problem, γ will be three, according to Eq. (3.53). Hence,

$$\nabla p_e = 3KT_e \nabla n_e = 3KT_e \nabla(n_0 + n_1) = 3KT_e \frac{\partial n_1}{\partial x}\hat{x}$$

and the linearized equation of motion is

$$mn_0 \frac{\partial v_1}{\partial t} = -en_0 E_1 - 3KT_e \frac{\partial n_1}{\partial x} \tag{4.28}$$

Note that in linearizing we have neglected the terms $n_1\, \partial v_1/\partial t$ and $n_1 E_1$ as well as the $(\mathbf{v}_1 \cdot \nabla)\mathbf{v}_1$ term. With Eq. (4.20), Eq. (4.28) becomes

$$-im\omega n_0 v_1 = -en_0 E_1 - 3KT_e ik n_1 \tag{4.29}$$

E_1 and n_1 are still given by Eqs. (4.23) and (4.22), and we have

$$im\omega n_0 v_1 = \left[en_0 \left(\frac{-e}{ik\varepsilon_0} \right) + 3KT_e ik \right] \frac{n_0 ik}{i\omega} v_1$$

$$\omega^2 v_1 = \left(\frac{n_0 e^2}{\varepsilon_0 m} + \frac{3KT_e}{m} k^2 \right) v_1$$

$$\boxed{\omega^2 = \omega_p^2 + \frac{3}{2}k^2 v_{th}^2} \tag{4.30}$$

where $v_{th}^2 \equiv 2KT_e/m$. The frequency now depends on k, and the group velocity is finite:

$$2\omega\, d\omega = \frac{3}{2}v_{th}^2 2k\, dk$$

$$v_g = \frac{d\omega}{dk} = \frac{3}{2}\frac{k}{\omega}v_{th}^2 = \frac{3}{2}\frac{v_{th}^2}{v_\phi} \tag{4.31}$$

That v_g is always less than c can easily be seen from a graph of Eq. (4.30). Figure 4.5 is a plot of the *dispersion relation* $\omega(k)$ as given by Eq. (4.30). At any point P on this curve, the slope of a line drawn from the origin gives the phase velocity ω/k.

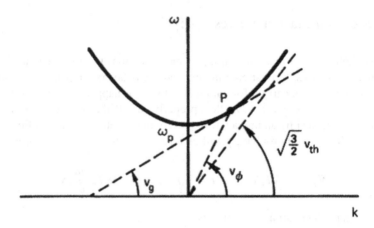

Fig. 4.5 Dispersion relation for electron plasma waves (Bohm–Gross waves)

The slope of the curve at P gives the group velocity. This is clearly always less than $(3/2)^{1/2} v_{th}$, which, in our nonrelativistic theory, is much less than c. Note that at large k (small λ), information travels essentially at the thermal velocity. At small k (large λ), information travels more slowly than v_{th} even though v_ϕ is greater than v_{th}. This is because the density gradient is small at large λ, and thermal motions carry very little net momentum into adjacent layers.

The existence of plasma oscillations has been known since the days of Langmuir in the 1920s. It was not until 1949 that Bohm and Gross worked out a detailed theory telling how the waves would propagate and how they could be excited. A simple way to excite plasma waves would be to apply an oscillating potential to a grid or a series of grids in a plasma; however, oscillators in the GHz range were not generally available in those days. Instead, one had to use an electron beam to excite plasma waves. If the electrons in the beam were bunched so that they passed by any fixed point at a frequency f_p, they would generate an electric field at that frequency and excite plasma oscillations. It is not necessary to form the electron bunches beforehand; once the plasma oscillations arise, they will bunch the electrons, and the oscillations will grow by a positive feedback mechanism. An experiment to test this theory was first performed by Looney and Brown in 1954. Their apparatus was entirely contained in a glass bulb about 10 cm in diameter (Fig. 4.6). A plasma filling the bulb was formed by an electrical discharge between the cathodes K and an anode ring A under a low pressure (3×10^{-3} Torr) of mercury vapor. An electron beam was created in a side arm containing a negatively biased filament. The emitted electrons were accelerated to 200 V and shot into the plasma through a small hole. A thin, movable probe wire connected to a radio receiver was used to pick up the oscillations. Figure 4.7 shows their experimental results for f^2 vs. discharge current, which is generally proportional to density. The points show a linear dependence, in rough agreement with Eq. (4.26). Deviations from the straight line could be attributed to the $k^2 v_{th}^2$ term in Eq. (4.30). However, not all frequencies were observed; k had to be such that an integral number of half wavelengths fit along

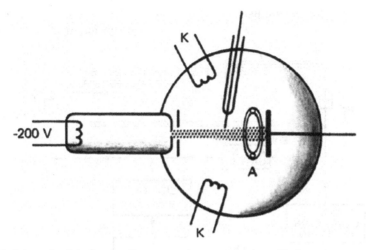

Fig. 4.6 Schematic of the Looney–Brown experiment on plasma oscillations

Fig. 4.7 Square of the observed frequency vs. plasma density, which is generally proportional to the discharge current. The *inset* shows the observed spatial distribution of oscillation intensity, indicating the existence of a different standing wave pattern for each of the groups of experimental points. [From D. H. Looney and S. C. Brown, *Phys. Rev.* **93**, 965 (1954).]

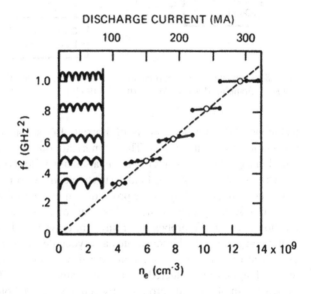

the plasma column. The standing wave patterns are shown at the left of Fig. 4.7. The predicted traveling plasma waves could not be seen in this experiment, probably because the beam was so thin that thermal motions carried electrons out of the beam, thus dissipating the oscillation energy. The electron bunching was accomplished not in the plasma but in the oscillating sheaths at the ends of the plasma column. In this early experiment, one learned that reproducing the conditions assumed in the uniform plasma theory requires considerable skill.

A number of recent experiments have verified the Bohm–Gross dispersion relation, Eq. (4.30), with precision. As an example of more modern experimental

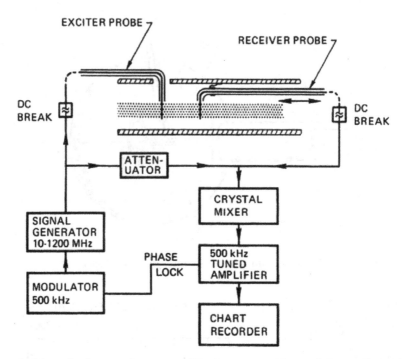

Fig. 4.8 Schematic of an experiment to measure plasma waves. [From P. J. Barrett, H. G. Jones, and R. N. Franklin, *Plasma Physics* **10**, 911 (1968).]

technique, we show the results of Barrett, Jones, and Franklin. Figure 4.8 is a schematic of their apparatus. The cylindrical column of quiescent plasma is produced in a Q-machine by thermal ionization of Cs atoms on hot tungsten plates (not shown). A strong magnetic field restricts electrons to motions along the column. The waves are excited by a wire probe driven by an oscillator and are detected by a second, movable probe. A metal shield surrounding the plasma prevents communication between the probes by ordinary microwave (electromagnetic wave) propagation, since the shield constitutes a waveguide beyond cutoff for the frequency used. The traveling waveforms are traced by interferometry: the transmitted and received signals are detected by a crystal which gives a large dc output when the signals are in phase and zero output when they are 90° out of phase. The resulting signal is shown in Fig. 4.9 as a function of position along the column. Synchronous detection is used to suppress the noise level. The excitation signal is chopped at 500 kHz, and the received signal should also be modulated at 500 kHz. By detecting only the 500-kHz component of the received signal, noise at other frequencies is eliminated. The traces of Fig. 4.9 give a measurement of k. When the oscillator frequency ω is varied, a plot of the dispersion curve $(\omega/\omega_p)^2$ vs. ka is obtained, where a is the radius of the column (Fig. 4.10). The various curves are labeled according to the value of $\omega_p a/v_{\text{th}}$. For $v_{\text{th}} = 0$, we have the curve labeled ∞, which corresponds to the dispersion relation $\omega = \omega_p$. For finite v_{th}, the curves correspond

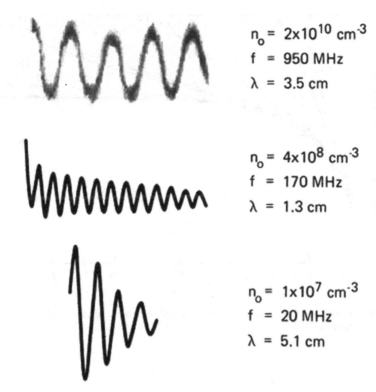

n_o = 2x10^{10} cm^{-3}
f = 950 MHz
λ = 3.5 cm

n_o = 4x10^8 cm^{-3}
f = 170 MHz
λ = 1.3 cm

n_o = 1x10^7 cm^{-3}
f = 20 MHz
λ = 5.1 cm

Fig. 4.9 Spatial variation of the perturbed density in a plasma wave, as indicated by an interfer-ometer, which multiplies the instantaneous density signals from two probes and takes the time average. The interferometer is tuned to the wave frequency, which varies with the density. The apparent damping at low densities is caused by noise in the plasma. [From Barrett, Jones, and Franklin, *loc. cit.*]

to that of Fig. 4.5. There is good agreement between the experimental points and the theoretical curves. The decrease of ω at small ka is the finite-geometry effect shown in Fig. 4.4. In this particular experiment, that effect can be explained another way. To satisfy the boundary condition imposed by the conducting shield, namely that $E = 0$ on the conductor, the plasma waves must travel at an angle to the magnetic field. Destructive interference between waves traveling with an outward radial component of k and those traveling inward enables the boundary condition to be satisfied. However, waves traveling at an angle to **B** have crests and troughs separated by a distance larger than $\lambda/2$ (Fig. 4.11). Since the electrons can move only along **B** (if B is very large), they are subject to less acceleration, and the frequency is lowered below ω_p.

Problems

4.6 Electron plasma waves are propagated in a uniform plasma with $KT_e = 100$ eV, $n = 10^{16}$ m^{-3}, and $B = 0$. If the frequency f is 1.1 GHz, what is the wavelength in cm?

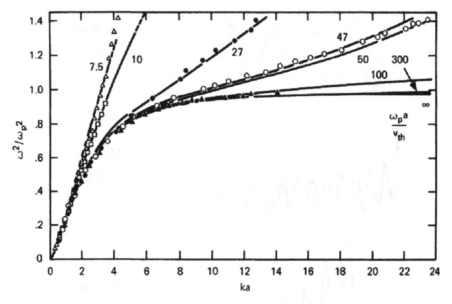

Fig. 4.10 Comparison of the measured and calculated dispersion curves for electron plasma waves in a cylinder of radius a. [From Barrett, Jones, and Franklin, *loc. cit.*]

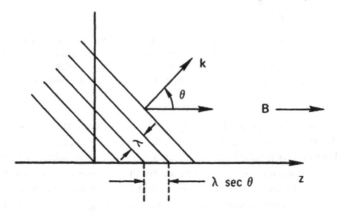

Fig. 4.11 Wavefronts traveling at an angle to the magnetic field are separated, in the field direction, by a distance larger than the wavelength λ

4.7 (a) Compute the effect of collisional damping on the propagation of Langmuir waves (plasma oscillations), by adding a term $-mn\nu\mathbf{v}$ to the electron equation of motion and rederiving the dispersion relation for $T_e = 0$. Here n is the electron collision frequency with ions and neutral atoms.

(b) Write an explicit expression for Im (ω) and show that its sign indicates that the wave is damped in time.

4.5   Sound Waves

As an introduction to ion waves, let us briefly review the theory of sound waves in ordinary air. Neglecting viscosity, we can write the Navier–Stokes equation (3.48), which describes these waves, as

$$\rho\left[\frac{\partial \mathbf{v}}{\partial t} + (\mathbf{v} \cdot \nabla)\mathbf{v}\right] = -\nabla p = -\frac{\gamma p}{\rho}\nabla\rho \qquad (4.32)$$

The equation of continuity is

$$\frac{\partial \rho}{\partial t} + \nabla \cdot (\rho\mathbf{v}) = 0 \qquad (4.33)$$

Linearizing about a stationary equilibrium with uniform p_0 and ρ_0, we have

$$-i\omega\rho_0\mathbf{v}_1 = -\frac{\gamma p_0}{\rho_0} i\mathbf{k}\rho_1 \qquad (4.34)$$

$$-i\omega\rho_1 + \rho_0 i\mathbf{k} \cdot \mathbf{v}_1 = 0 \qquad (4.35)$$

where we have again taken a wave dependence of the form

$$\exp[i(\mathbf{k} \cdot \mathbf{r} - \omega t)]$$

For a plane wave with $\mathbf{k} = k\hat{\mathbf{x}}$ and $\mathbf{v} = v\hat{\mathbf{x}}$, we find, upon eliminating ρ_1,

$$-i\omega\rho_0 v_1 = -\frac{\gamma p_0}{\rho_0} ik \frac{\rho_0 ik v_1}{i\omega}$$

$$\omega^2 v_1 = k^2 \frac{\gamma p_0}{\rho_0} v_1$$

or

$$\boxed{\frac{\omega}{k} = \left(\frac{\gamma p_0}{\rho_0}\right)^{1/2} = \left(\frac{\gamma KT}{M}\right)^{1/2} \equiv c_s} \qquad (4.36)$$

This is the expression for the velocity c_s of sound waves in a neutral gas. The waves are pressure waves propagating from one layer to the next by collisions among the air molecules. In a plasma with no neutrals and few collisions, an analogous phenomenon occurs. This is called an *ion acoustic wave*, or, simply, an *ion wave*.

4.6   Ion Waves

In the absence of collisions, ordinary sound waves would not occur. Ions can still transmit vibrations to each other because of their charge, however; and acoustic waves can occur through the intermediary of an electric field. Since the motion of massive ions will be involved, these will be low-frequency oscillations, and we can use the plasma approximation of Sect. 3.6. We therefore assume $n_i = n_e = n$ and do not use Poisson's equation. The ion fluid equation in the absence of a magnetic field is

$$Mn\left[\frac{\partial \mathbf{v}_i}{\partial t} + (\mathbf{v}_i \cdot \nabla)\mathbf{v}_i\right] = en\mathbf{E} - \nabla p = -en\nabla\phi - \gamma_i KT_i \nabla n \qquad (4.37)$$

We have assumed $\mathbf{E} = -\nabla \phi$ and used the equation of state. Linearizing and assuming plane waves, we have

$$-i\omega Mn_0 v_{i1} = -en_0 ik\phi_1 - \gamma_i KT_i ik n_1 \qquad (4.38)$$

As for the electrons, we may assume $m = 0$ and apply the argument of Sect. 3.5, regarding motions along $\mathbf{B}$, to the present case of $\mathbf{B} = 0$. The balance of forces on electrons, therefore, requires

$$n_e = n = n_0 \exp\left(\frac{e\phi_1}{KT_e}\right) = n_0\left(1 + \frac{e\phi_1}{KT_e} + \cdots\right)$$

The perturbation in density of electrons, and, therefore, of ions, is then

$$n_1 = n_0 \frac{e\phi_1}{KT_e} \qquad (4.39)$$

Here the n_0 of Boltzmann's relation also stands for the density in the equilibrium plasma, in which we can choose $\phi_0 = 0$ because we have assumed $\mathbf{E}_0 = 0$. In linearizing Eq. (4.39), we have dropped the higher-order terms in the Taylor expansion of the exponential.

The only other equation needed is the linearized ion equation of continuity. From Eq. (4.22), we have

$$i\omega n_1 = n_0 ik v_{i1} \qquad (4.40)$$

In Eq. (4.38), we may substitute for ϕ_1 and n_1 in terms of v_{i1} from Eqs. (4.39) and (4.40) and obtain

$$i\omega Mn_0 v_{i1} = \left(en_0 ik\frac{KT_e}{en_0} + \gamma_i KT_i ik\right)\frac{n_0 ik v_{i1}}{i\omega}$$

$$\omega^2 = k^2\left(\frac{KT_e}{M} + \frac{\gamma_i KT_i}{M}\right)$$

Fig. 4.12 Dispersion
relation for ion acoustic
waves in the limit of small
Debye length

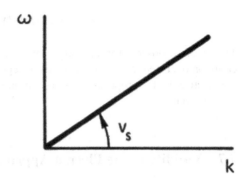

$$\frac{\omega}{k} = \left(\frac{KT_e + \gamma_i KT_i}{M}\right)^{1/2} \equiv v_s \qquad (4.41)$$

This is the dispersion relation for *ion acoustic waves*; v_s is the sound speed in a plasma. Since the ions suffer one-dimensional compressions in the plane waves we have assumed, we may set $\gamma_i = 3$ here. The electrons move so fast relative to these waves that they have time to equalize their temperature everywhere; therefore, the electrons are isothermal, and $\gamma_e = 1$. Otherwise, a factor γ_e would appear in front of KT_e in Eq. (4.41).

The dispersion curve for ion waves (Fig. 4.12) has a fundamentally different character from that for electron waves (Fig. 4.5). Plasma oscillations are basically *constant-frequency waves*, with a correction due to thermal motions. Ion waves are basically *constant-velocity* waves and exist *only* when there are thermal motions. For ion waves, the group velocity is equal to the phase velocity. The reasons for this difference can be seen from the following description of the physical mechanisms involved. In electron plasma oscillations, the other species (namely, ions) remains essentially fixed. In ion acoustic waves, the other species (namely, electrons) is far from fixed; in fact, electrons are pulled along with the ions and tend to shield out electric fields arising from the bunching of ions. However, this shielding is not perfect because, as we saw in Sect. 1.4, potentials of the order of KT_e/e can leak out because of electron thermal motions. What happens is as follows. The ions form regions of compression and rarefaction, just as in an ordinary sound wave. The compressed regions tend to expand into the rarefactions, for two reasons. First, the ion thermal motions spread out the ions; this effect gives rise to the second term in the square root of Eq. (4.41). Second, the ion bunches are positively charged and tend to disperse because of the resulting electric field. This field is largely shielded out by electrons, and only a fraction, proportional to KT_e, is available to act on the ion bunches. This effect gives rise to the first term in the square root of Eq. (4.41). The ions overshoot because of their inertia, and the compressions and rarefactions are regenerated to form a wave.

The second effect mentioned above leads to a curious phenomenon. When KT_i goes to zero, ion waves still exist. This does not happen in a neutral gas (Eq. (4.36)). The acoustic velocity is then given by

$$v_s = (KT_e/M)^{1/2} \tag{4.42}$$

This is often observed in laboratory plasmas, in which the condition $T_i \ll T_e$ is a common occurrence. The sound speed v_s depends on *electron* temperature (because the electric field is proportional to it) and on *ion* mass (because the fluid's inertia is proportional to it).

4.7 Validity of the Plasma Approximation

In deriving the velocity of ion waves, we used the neutrality condition $n_i = n_e$ while allowing $\mathbf{E}$ to be finite. To see what error was engendered in the process, we now allow n_i to differ from n_e and use the linearized Poisson equation:

$$\varepsilon_0 \nabla \cdot \mathbf{E}_1 = \varepsilon_0 k^2 \phi_1 = e(n_{i1} - n_{e1}) \tag{4.43}$$

The electron density is given by the linearized Boltzmann relation Eq. (4.39):

$$n_{e1} = \frac{e\phi_1}{KT_e} n_0 \tag{4.44}$$

Inserting this into Eq. (4.43), we have

$$\varepsilon_0 \phi_1 \left(k^2 + \frac{n_0 e^2}{\varepsilon_0 KT_e} \right) = e n_{i1} \tag{4.45}$$

$$\varepsilon_0 \phi_1 \left(k^2 \lambda_D^2 + 1 \right) = e n_{i1} \lambda_D^2$$

The ion density is given by the linearized ion continuity equation (4.40):

$$n_{i1} = \frac{k}{\omega} n_0 v_{i1} \tag{4.46}$$

Inserting Eqs. (4.45) and (4.46) into the ion equation of motion Eq. (4.38), we find

$$i\omega M n_0 v_{i1} = \left(\frac{e n_o i k}{\varepsilon_0} \frac{e \lambda_D^2}{1 + k^2 \lambda_D^2} + \gamma_i KT_i i k \right) \frac{k}{\omega} n_0 v_{i1} \tag{4.47}$$

$$\omega^2 = \frac{k^2}{M} \left(\frac{n_0 e^2 \varepsilon_0^{-1} \lambda_D^2}{1 + k^2 \lambda_D^2} + \gamma_i KT_i \right)$$

$$\frac{\omega}{k} = \left(\frac{KT_e}{M} \frac{1}{1 + k^2 \lambda_D^2} + \frac{\gamma_i KT_i}{M} \right)^{1/2} \tag{4.48}$$

This is the same as we obtained previously (Eq. (4.41)) except for the factor $1 + k^2$ λ_D^2. Our assumption $n_i = n_e$ has given rise to an error of order $k^2\lambda_D^2 = (2\pi\lambda_D/\lambda)^2$. Since λ_D is very small in most experiments, the plasma approximation is valid everywhere except in a thin layer, called a sheath (Chap. 8), a few λ_D's in thickness, next to a wall.

4.8   Comparison of Ion and Electron Waves

If we consider these short-wavelength waves by taking $k^2\lambda_D^2 \gg 1$, Eq. (4.47) becomes

$$\omega^2 = k^2 \frac{n_0 e^2}{\varepsilon_0 M k^2} = \frac{n_0 e^2}{\varepsilon_0 M} \equiv \Omega_p^2 \tag{4.49}$$

We have, for simplicity, also taken the limit $T_i \to 0$. Here Ω_p is the ion plasma frequency. For high frequencies (short wavelengths) the ion acoustic wave turns into a constant-frequency wave. There is thus a complementary behavior between electron plasma waves and ion acoustic waves: the former are basically constant frequency, but become constant velocity at large k; the latter are basically constant velocity, but become constant frequency at large k. This comparison is shown graphically in Fig. 4.13.

Experimental verification of the existence of ion waves was first accomplished by Wong, Motley, and D'Angelo. Figure 4.14 shows their apparatus, which was again a Q-machine. (It is no accident that we have referred to Q-machines so often; careful experimental checks of plasma theory were possible only after schemes to make quiescent plasmas were discovered.) Waves were launched and detected by grids inserted into the plasma. Figure 4.15 shows oscilloscope traces of the transmitted and received signals. From the phase shift, one can find the phase velocity (same as group velocity in this case). These phase shifts are plotted as

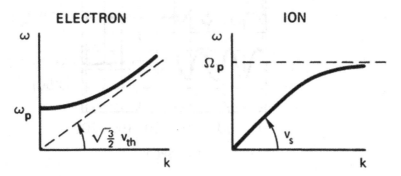

Fig. 4.13 Comparison of the dispersion curves for electron plasma waves and ion acoustic waves

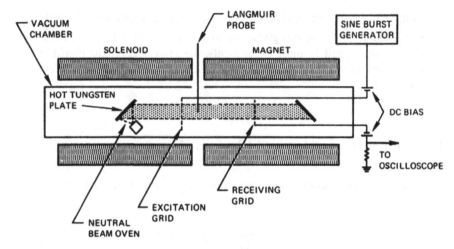

Fig. 4.14 Q-machine experiment to detect ion waves. [From N. Rynn and N. D'Angelo, *Rev. Sci. Instrum.* **31**, 1326 (1960).]

Fig. 4.15 Oscillograms of signals from the driver and receiver grids, separated by a distance *d*, showing the delay indicative of a traveling wave. [From A. Y. Wong, R. W. Motley, and N. D'Angelo, *Phys. Rev.* **133**, A436 (1964).]

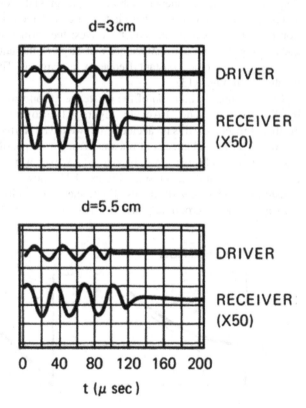

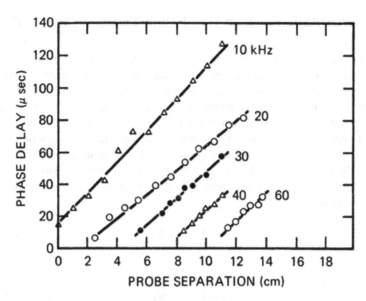

Fig. 4.16 Experimental measurements of delay vs. probe separation at various frequencies of the wave exciter. The slope of the lines gives the phase velocity. [From Wong, Motley, and D'Angelo, *loc. cit.*]

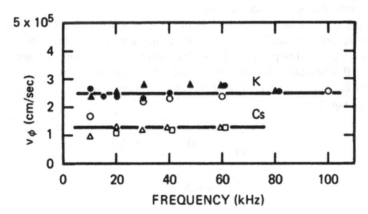

Fig. 4.17 Measured phase velocity of ion waves in potassium and cesium plasmas as a function of frequency. The different sets of points correspond to different plasma densities. [From Wong, Motley, and D'Angelo, *loc. cit.*]

functions of distance in Fig. 4.16 for a plasma density of 3×10^{17} m^{-3}. The slopes of such lines give the phase velocities plotted in Fig. 4.17 for the two masses and various plasma densities n_0. The constancy of v_s with ω and n_0 is demonstrated experimentally, and the two sets of points for K and Cs plasmas show the proper dependence on M.

4.9 Electrostatic Electron Oscillations Perpendicular to B

Up to now, we have assumed $\mathbf{B} = 0$. When a magnetic field exists, many more types of waves are possible. We shall examine only the simplest cases, starting with high-frequency, electrostatic, electron oscillations propagating at right angles to the magnetic field. First, we should define the terms perpendicular, parallel, longitudinal, transverse, electrostatic, and electromagnetic. *Parallel* and *perpendicular* will be used to denote the direction of $\mathbf{k}$ relative to the undisturbed magnetic field $\mathbf{B}_0$. *Longitudinal* and *transverse* refer to the direction of $\mathbf{k}$ relative to the *oscillating* electric field $\mathbf{E}_1$. If the *oscillating* magnetic field $\mathbf{B}_1$ is zero, the wave is *electrostatic*; otherwise, it is *electromagnetic*. The last two sets of terms are related by Maxwell's equation

$$\nabla \times \mathbf{E}_1 = -\dot{\mathbf{B}}_1 \tag{4.50}$$

or

$$\mathbf{k} \times \mathbf{E}_1 = \omega \mathbf{B}_1 \tag{4.51}$$

If a wave is longitudinal, $\mathbf{k} \times \mathbf{E}_1$ vanishes, and the wave is also electrostatic. If the wave is transverse, $\mathbf{B}_1$ is finite, and the wave is electromagnetic. It is of course possible for $\mathbf{k}$ to be at an arbitrary angle to $\mathbf{B}_0$ or $\mathbf{E}_1$; then one would have a mixture of the principal modes presented here.

Coming back to the electron oscillations perpendicular to $\mathbf{B}_0$, we shall assume that the ions are too massive to move at the frequencies involved and form a fixed, uniform background of positive charge. We shall also neglect thermal motions and set $KT_e = 0$. The equilibrium plasma, as usual, has constant and uniform n_0 and $\mathbf{B}_0$ and zero $\mathbf{E}_0$ and $\mathbf{v}_0$. The motion of electrons is then governed by the following linearized equations:

$$m \frac{\partial \mathbf{v}_{e1}}{\partial t} = -e(\mathbf{E}_1 + \mathbf{v}_{e1} \times \mathbf{B}_0) \tag{4.52}$$

$$\frac{\partial n_{e1}}{\partial t} + n_0 \nabla \cdot \mathbf{v}_{e1} = 0 \tag{4.53}$$

$$\varepsilon_0 \nabla \cdot \mathbf{E}_1 = -e n_{e1} \tag{4.54}$$

We shall consider only *longitudinal* waves with $\mathbf{k} \parallel \mathbf{E}_1$. Without loss of generality, we can choose the x axis to lie along $\mathbf{k}$ and $\mathbf{E}_1$, and the z axis to lie along $\mathbf{B}_0$ (Fig. 4.18). Thus $k_y = k_z = E_y = E_z = 0$, $\mathbf{k} = k\hat{\mathbf{x}}$, and $\mathbf{E} = E\hat{\mathbf{x}}$. Dropping the subscripts 1 and e and separating Eq. (4.52) into components, we have

$$-i\omega m v_x = -eE - e v_y B_0 \tag{4.55}$$

Fig. 4.18 Geometry of a longitudinal plane wave propagating at right angles to B_0

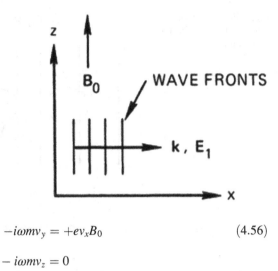

$$-i\omega m v_y = +e v_x B_0 \tag{4.56}$$

$$-i\omega m v_z = 0$$

Solving for v_y in Eq. (4.56) and substituting into Eq. (4.55), we have

$$i\omega m v_x = eE + eB_0 \frac{ieB_0}{m\omega} v_x$$

$$v_x = \frac{eE/im\omega}{1 - \omega_c^2/\omega^2} \tag{4.57}$$

Note that v_x becomes infinite at cyclotron resonance, $\omega = \omega_c$. This is to be expected, since the electric field changes sign with v_x and continuously accelerates the electrons. [Note that the fluid and single-particle equations are identical when the $(\mathbf{v} \cdot \nabla)\mathbf{v}$ and ∇p terms are both neglected, so that all the particles move together.] From the linearized form of Eq. (4.53), we have

$$n_1 = \frac{k}{\omega} n_0 v_x \tag{4.58}$$

Linearizing Eq. (4.54) and using the last two results, we have

$$ik\varepsilon_0 E = -e\frac{k}{\omega} n_0 \frac{eE}{im\omega} \left(1 - \frac{\omega_c^2}{\omega^2}\right)^{-1}$$

$$\left(1 - \frac{\omega_c^2}{\omega^2}\right) E = \frac{\omega_p^2}{\omega^2} E \tag{4.59}$$

The dispersion relation is therefore

$$\boxed{\omega^2 = \omega_p^2 + \omega_c^2 \equiv \omega_h^2} \tag{4.60}$$

Fig. 4.19 Motion of electrons in an upper hybrid oscillation

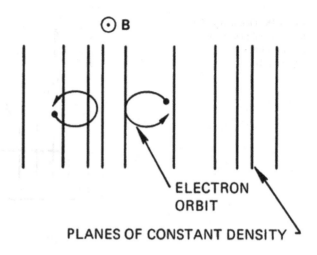

The frequency ω_h is called the *upper hybrid frequency*. Electrostatic electron waves *across* **B** have this frequency, while those *along* **B** are the usual plasma oscillations with $\omega = \omega_p$. The group velocity is again zero as long as thermal motions are neglected.

A physical picture of this oscillation is given in Fig. 4.19. Electrons in the plane wave form regions of compression and rarefaction, as in a plasma oscillation. However, there is now a **B** field perpendicular to the motion, and the Lorentz force turns the trajectories into ellipses. There are two restoring forces acting on the electrons: the electrostatic field and the Lorentz force. The increased restoring force makes the frequency larger than that of a plasma oscillation. As the magnetic field goes to zero, ω_c goes to zero in Eq. (4.60), and one recovers a plasma oscillation. As the plasma density goes to zero, ω_p goes to zero, and one has a simple Larmor gyration, since the electrostatic forces vanish with density.

The existence of the upper hybrid frequency has been verified experimentally by microwave transmission across a magnetic field. As the plasma density is varied, the transmission through the plasma takes a dip at the density that makes ω_h equal to the applied frequency. This is because the upper hybrid oscillations are excited, and energy is absorbed from the beam. From Eq. (4.60), we find a linear relationship between ω_c^2/ω^2 and the density:

$$\frac{\omega_c^2}{\omega^2} = 1 - \frac{\omega_p^2}{\omega^2} = 1 - \frac{ne^2}{\varepsilon_0 \, m\omega^2}$$

This linear relation is followed by the experimental points on Fig. 4.20, where ω_c^2/ω^2 is plotted against the discharge current, which is proportional to n.

If we now consider propagation at an angle θ to **B**, we will get two possible waves. One is like the plasma oscillation, and the other is like the upper hybrid oscillation, but both will be modified by the angle of propagation. The details of this

Fig. 4.20 Results of an experiment to detect the existence of the upper hybrid frequency by mapping the conditions for maximum absorption (minimum transmission) of microwave energy sent across a magnetic field. The field at which this occurs (expressed as ω_c^2/ω^2) is plotted against discharge current (proportional to plasma density). [From R. S. Harp, *Proceedings of the Seventh International Conference on Phenomena in Ionized Gases*, Belgrade, 1965, *II*, 294 (1966).]

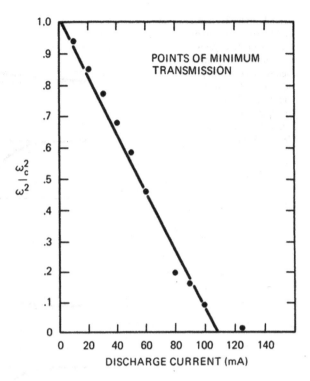

are left as an exercise (Problem 4.8). Figure 4.21 shows schematically the ω–k_z diagram for these two waves for fixed k_x, where $k_x/k_z = \tan\theta$. Because of the symmetry of Eq. (4.60), the case $\omega_c > \omega_p$ is the same as the case $\omega_p > \omega_c$ with the subscripts interchanged. For large k_z, the wave travels parallel to $\mathbf{B}_0$. One wave is the plasma oscillation at $\omega = \omega_p$; the other wave, at $\omega = \omega_c$, is a spurious root at $k_z \to \infty$. For small k_z, we have the situation of $\mathbf{k} \perp \mathbf{B}_0$ discussed in this section. The lower branch vanishes, while the upper branch approaches the hybrid oscillation at $\omega = \omega_h$. These curves were first calculated by Trivelpiece and Gould, who also verified them experimentally (Fig. 4.22). The Trivelpiece–Gould experiment was done in a cylindrical plasma column; it can be shown that varying k_z in this case is equivalent to propagating plane waves at various angles to $\mathbf{B}_0$.

Problems

4.8 For the upper hybrid oscillation, show that the elliptical orbits (Fig. 4.19) are always elongated in the direction of $\mathbf{k}$. (Hint: From the equation of motion, derive an expression for v_x/v_y in terms of ω/ω_c.)

4.9 Find the dispersion relation for electrostatic electron waves propagating at an arbitrary angle θ relative to $\mathbf{B}_0$. Hint: Choose the x axis so that $\mathbf{k}$ and $\mathbf{E}$ lie in the x–z plane (Fig. P4.9). Then

Fig. 4.21 The Trivelpiece–Gould dispersion curves for electrostatic electron waves in a conducting cylinder filled with a uniform plasma and a coaxial magnetic field. [From A. W. Trivelpiece and R. W. Gould, *J. Appl. Phys.* **30,** 1784 (1959).]

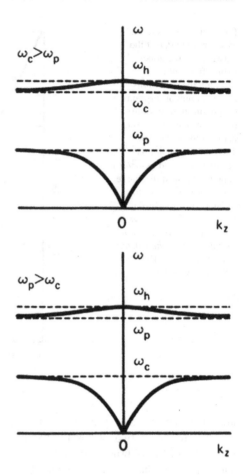

$$E_x = E_1 \sin\theta, \quad E_z = E_1 \cos\theta, \quad E_y = 0$$

and similarly for **k**. Solve the equations of motion and continuity and Poisson's equation in the usual way with n_0 uniform and $\mathbf{v_0} = \mathbf{E_0} = 0$.

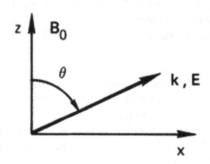

Fig. P4.9

Fig. 4.22 Experimental
verification of the
Trivelpiece –Gould curves,
showing the existence of
backward waves; that is,
waves whose group
velocity, as indicated by the
slope of the dispersion
curve, is opposite in
direction to the phase
velocity. [From Trivelpiece
and Gould, *loc. cit.*]

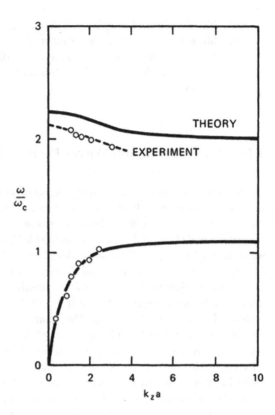

(a) Show that the answer is

$$\omega^2(\omega^2 - \omega_h^2) + \omega_c^2\omega_p^2 \cos^2\theta = 0$$

(b) Write out the two solutions of this quadratic for ω^2, and show that in the
limits $\theta \to 0$ and $\theta \to \pi/2$, our previous results are recovered. Show that in
these limits, one of the two solutions is a spurious root with no physical
meaning.

(c) By completing the square, show that the above equation is the equation of
an ellipse:

$$\frac{(y-1)^2}{1^2} + \frac{x^2}{a^2} = 1$$

where

$$x \equiv \cos\theta, \quad y \equiv 2\omega^2/\omega_h^2, \quad \text{and } a \equiv \omega_h^2/2\omega_c\omega_p.$$

(d)   Plot the ellipse for $\omega_p/\omega_c = 1, 2$, and ∞.

(e)   Show that if $\omega_c > \omega_p$, the lower root for ω is always less than ω_p for any $\theta > 0$ and the upper root always lies between ω_c and ω_h; and that if $\omega_p > \omega_c$, the lower root lies below ω_c while the upper root is between ω_p and ω_h.

4.10   Electrostatic Ion Waves Perpendicular to B

We next consider what happens to the ion acoustic wave when $\mathbf{k}$ is perpendicular to $\mathbf{B}_0$. It is tempting to set $\mathbf{k} \cdot \mathbf{B}_0$ exactly equal to zero, but this would lead to a result (Sect. 4.11) which, though mathematically correct, does not describe what usually happens in real plasmas. Instead, we shall let $\mathbf{k}$ be *almost* perpendicular to $\mathbf{B}_0$; what we mean by "almost" will be made clear later. We shall assume the usual infinite plasma in equilibrium, with n_0 and $\mathbf{B}_0$ constant and uniform and $\mathbf{v}_0 = \mathbf{E}_0 = 0$. For simplicity, we shall take $T_i = 0$; we shall not miss any important effects because we know that acoustic waves still exist if $T_i = 0$. We also assume electrostatic waves with $\mathbf{k} \times \mathbf{E} = 0$, so that $\mathbf{E} = -\nabla\phi$. The geometry is shown in Fig. 4.23. The angle $\frac{1}{2}\pi - \theta$ is taken to be so small that we may take $\mathbf{E} = E_1\hat{\mathbf{x}}$ and $\nabla = ik\hat{\mathbf{x}}$ as far as the *ion* motion is concerned. For the *electrons*, however, it makes a great deal of difference whether $\frac{1}{2}\pi - \theta$ is zero, or small but finite. The electrons have such small Larmor radii that they cannot move in the x direction to preserve charge neutrality; all that the $\mathbf{E}$ field does is make them drift back and forth in the y direction. If θ is not exactly $\pi/2$, however, the electrons can move along the dashed line (along $\mathbf{B}_0$) in Fig. 4.23 to carry charge from negative to positive regions in the wave and carry out Debye shielding. The ions cannot do this effectively because their inertia prevents them from moving such long distances in a wave period; this is why we can neglect k_z for ions. The critical angle $\chi = \frac{1}{2}\pi - \theta$ is proportional to the ratio of ion to electron parallel velocities: $\chi \simeq (m/M)^{1/2}$

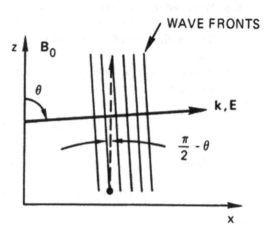

Fig. 4.23 Geometry of an electrostatic ion cyclotron wave propagating nearly at right angles to B_0

(in radians). For angles χ larger than this, the following treatment is valid. For angles χ smaller than this, the treatment of Sect. 4.11 is valid.

After this lengthy introduction, we proceed to the brief derivation of the result. For the ion equation of motion, we have

$$M\frac{\partial \mathbf{v}_{i1}}{\partial t} = -e\nabla\phi_1 + e\mathbf{v}_{i1} \times \mathbf{B}_0 \tag{4.61}$$

Assuming plane waves propagating in the x direction and separating into components, we have

$$-i\omega M v_{ix} = -eik\phi_1 + ev_{iy}B_0$$
$$-i\omega M v_{iy} = \qquad\quad -ev_{ix}B_0 \tag{4.62}$$

Solving as before, we find

$$v_{ix} = \frac{ek}{M\omega}\phi_1\left(1 - \frac{\Omega_c^2}{\omega^2}\right)^{-1} \tag{4.63}$$

where $\Omega_c = eB_0/M$ is the ion cyclotron frequency. The ion equation of continuity yields, as usual,

$$n_{i1} = n_0\frac{k}{\omega}v_{ix} \tag{4.64}$$

Assuming the electrons can move along $\mathbf{B}_0$ because of the finiteness of the angle χ, we can use the Boltzmann relation for electrons. In linearized form, this is

$$\frac{n_{e1}}{n_0} = \frac{e\phi_1}{KT_e} \tag{4.65}$$

The plasma approximation $n_i = n_e$ now closes the system of equations. With the help of Eqs. (4.64) and (4.65), we can write Eq. (4.63) as

$$\left(1 - \frac{\Omega_c^2}{\omega^2}\right)v_{ix} = \frac{ek}{M\omega}\frac{KT_e}{en_0}\frac{n_0k}{\omega}v_{ix}$$

$$\omega^2 - \Omega_c^2 = k^2\frac{KT_e}{M} \tag{4.66}$$

Since we have taken $KT_i = 0$, we can write this as

$$\boxed{\omega^2 = \Omega_c^2 + k^2 v_s^2} \tag{4.67}$$

This is the dispersion relation for *electrostatic ion cyclotron waves*.

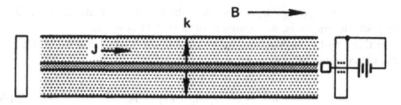

Fig. 4.24 Schematic of a Q-machine experiment on electrostatic ion cyclotron waves. [After R. W. Motley and N. D'Angelo, Phys. Fluids **6**, 296 (1963).]

Fig. 4.25 Measurements of frequency of electrostatic ion cyclotron waves vs. magnetic field. [From Motley and D'Angelo, *loc. cit.*]

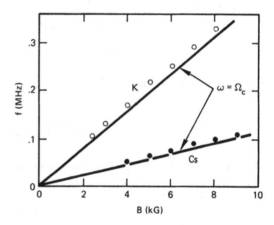

The physical explanation of these waves is very similar to that in Fig. 4.19 for upper hybrid waves. The ions undergo an acoustic-type oscillation, but the Lorentz force constitutes a new restoring force giving rise to the Ω_c^2 term in Eq. (4.67). The acoustic dispersion relation $\omega^2 = k^2 v_s^2$ is valid if the electrons provide Debye shielding. In this case, they do so by flowing long distances along $\mathbf{B}_0$.

Electrostatic ion cyclotron waves were first observed by Motley and D'Angelo, again in a *Q*-machine (Fig. 4.24). The waves propagated radially outward across the magnetic field and were excited by a current drawn along the axis to a small auxiliary electrode. The reason for excitation is rather complicated and will not be given here. Figure 4.25 gives their results for the wave frequency vs. magnetic field. In this experiment, the $k^2 v_s^2$ term was small compared to the Ω_c^2 term, and the measured frequencies lay only slightly above Ω_c.

4.11   The Lower Hybrid Frequency

We now consider what happens when θ is exactly $\pi/2$, and the electrons are not allowed to preserve charge neutrality by flowing along the lines of force. Instead of obeying Boltzmann's relation, they will obey the full equation of motion,

Eq. (3.62). If we keep the electron mass finite, this equation is nontrivial even if we assume $T_e = 0$ and drop the ∇p_e term: hence, we shall do so in the interest of simplicity. The ion equation of motion is unchanged from Eq. (4.63):

$$v_{ix} = \frac{ek}{M\omega}\phi_1\left(1 - \frac{\Omega_c^2}{\omega^2}\right)^{-1} \tag{4.68}$$

By changing e to $-e$, M to m, and Ω_c to $-\omega_c$ in Eq. (4.68), we can write down the result of solving Eq. (3.62) for electrons, with $T_e = 0$:

$$v_{ex} = -\frac{ek}{m\omega}\phi_1\left(1 - \frac{\omega_c^2}{\omega^2}\right)^{-1} \tag{4.69}$$

The equations of continuity give

$$n_{i1} = n_0\frac{k}{\omega}v_{i1} \qquad n_{e1} = n_0\frac{k}{\omega}v_{e1} \tag{4.70}$$

The plasma approximation $n_i = n_e$ then requires $v_{i1} = v_{e1}$. Setting Eqs. (4.68) and (4.69) equal to each other, we have

$$M\left(1 - \frac{\Omega_c^2}{\omega^2}\right) = -m\left(1 - \frac{\omega_c^2}{\omega^2}\right)$$

$$\omega^2(M + m) = m\omega_c^2 + M\Omega_c^2 = e^2B^2\left(\frac{1}{m} + \frac{1}{M}\right)$$

$$\omega^2 = \frac{e^2B^2}{Mm} = \Omega_c\omega_c$$

$$\boxed{\omega = (\Omega_c\omega_c)^{1/2} \equiv \omega_l} \tag{4.71}$$

This is called the *lower hybrid frequency*. These oscillations can be observed only if θ is very close to $\pi/2$. If we had used Poisson's equation instead of the plasma approximation, we would have obtained

$$\frac{1}{\omega_l^2} = \frac{1}{\omega_c\Omega_c} + \frac{1}{\Omega_p^2} \tag{4.71a}$$

In very low-density plasmas the latter term actually dominates because the plasma approximation is not valid when the Debye length is not negligibly small.

4.12   Electromagnetic Waves with $\mathbf{B}_0 = 0$

Next in the order of complexity come waves with $\mathbf{B}_1 \neq 0$. These are transverse electromagnetic waves—light waves or radio waves traveling through a plasma. We begin with a brief review of light waves in a vacuum. The relevant Maxwell equations are

$$\nabla \times \mathbf{E}_1 = -\dot{\mathbf{B}}_1 \qquad (4.72)$$

$$c^2 \nabla \times \mathbf{B}_1 = \dot{\mathbf{E}}_1 \qquad (4.73)$$

since in a vacuum $\mathbf{j} = 0$ and $\varepsilon_0 \mu_0 = c^{-2}$. Taking the curl of Eq. (4.73) and substituting into the time derivative of Eq. (4.72), we have

$$c^2 \nabla \times (\nabla \times \mathbf{B}_1) = \nabla \times \dot{\mathbf{E}}_1 = -\ddot{\mathbf{B}}_1 \qquad (4.74)$$

Again assuming planes waves varying as $\exp[i(kx - \omega t)]$, we have

$$\omega^2 \mathbf{B}_1 = -c^2 \mathbf{k} \times (\mathbf{k} \times \mathbf{B}_1) = -c^2 \left[\mathbf{k}(\mathbf{k} \cdot \mathbf{B}_1) - k^2 \mathbf{B}_1 \right] \qquad (4.75)$$

Since $\mathbf{k} \cdot \mathbf{B}_1 = -i\nabla \cdot \mathbf{B}_1 = 0$ by another of Maxwell's equations, the result is

$$\omega^2 = k^2 c^2 \qquad (4.76)$$

and c is the phase velocity ω/k of light waves.

In a plasma with $\mathbf{B}_0 = 0$, Eq. (4.72) is unchanged, but we must add a term $\mathbf{j}_1/\varepsilon_0$ to Eq. (4.73) to account for currents due to first-order charged particle motions:

$$c^2 \nabla \times \mathbf{B}_1 = \frac{\mathbf{j}_1}{\varepsilon_0} + \dot{\mathbf{E}}_1 \qquad (4.77)$$

The time derivative of this is

$$c^2 \nabla \times \dot{\mathbf{B}}_1 = \frac{1}{\varepsilon_0} \frac{\partial \mathbf{j}_1}{\partial t} + \ddot{\mathbf{E}}_1 \qquad (4.78)$$

while the curl of Eq. (4.72) is

$$\nabla \times (\nabla \times \mathbf{E}_1) = \nabla(\nabla \cdot \mathbf{E}_1) - \nabla^2 \mathbf{E}_1 = -\nabla \times \dot{\mathbf{B}}_1 \qquad (4.79)$$

Eliminating $\nabla \times \dot{\mathbf{B}}_1$ and assuming an $\exp[i(\mathbf{k} \cdot \mathbf{r} - \omega t)]$ dependence, we have

$$-\mathbf{k}(\mathbf{k} \cdot \mathbf{E}_1) + k^2 \mathbf{E}_1 = \frac{i\omega}{\varepsilon_0 c^2} \mathbf{j}_1 + \frac{\omega^2}{c^2} \mathbf{E}_1 \qquad (4.80)$$

By transverse waves we mean $\mathbf{k} \cdot \mathbf{E}_1 = 0$, so this becomes

$$\left(\omega^2 - c^2 k^2\right)\mathbf{E}_1 = -i\omega\,\mathbf{j}_1/\varepsilon_0 \tag{4.81}$$

If we consider light waves or microwaves, these will be of such high frequency that the ions can be considered as fixed. The current $\mathbf{j}_1$ then comes entirely from electron motion:

$$\mathbf{j}_1 = -n_0 e \mathbf{v}_{e1} \tag{4.82}$$

From the linearized electron equation of motion, we have (for $KT_e = 0$):

$$m\frac{\partial \mathbf{v}_{e1}}{\partial t} = -e\mathbf{E} \tag{4.83}$$

$$\mathbf{v}_{e1} = \frac{e\mathbf{E}_1}{im\omega}$$

Equation (4.81) now can be written

$$\left(\omega^2 - c^2 k^2\right)\mathbf{E}_1 = \frac{i\omega}{\varepsilon_0}n_0 e\,\frac{e\mathbf{E}_1}{im\omega} = \frac{n_0 e^2}{\varepsilon_0 m}\mathbf{E}_1 \tag{4.84}$$

The expression for ω_p^2 is recognizable on the right-hand side, and the result is

$$\omega^2 = \omega_p^2 + c^2 k^2 \tag{4.85}$$

This is the dispersion relation for *electromagnetic waves* propagating in a plasma with no dc magnetic field. We see that the vacuum relation Eq. (4.76) is modified by a term ω_p^2 reminiscent of plasma oscillations. The phase velocity of a light wave in a plasma is greater than the velocity of light:

$$v_\phi^2 = \frac{\omega^2}{k^2} = c^2 + \frac{\omega_p^2}{k^2} > c^2 \tag{4.86}$$

However, the group velocity cannot exceed the velocity of light. From Eq. (4.85), we find

$$\frac{d\omega}{dk} = v_g = \frac{c^2}{v_\phi} \tag{4.87}$$

so that v_g is less than c whenever v_ϕ is greater than c. The dispersion relation Eq. (4.85) is shown in Fig. 4.26. This diagram resembles that of Fig. 4.5 for plasma waves, but the dispersion relation is really quite different because the asymptotic velocity c in Fig. 4.26 is so much larger than the thermal velocity v_{th} in Fig. 4.5.

Fig. 4.26 Dispersion relation for electromagnetic waves in a plasma with no dc magnetic field

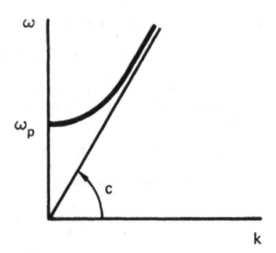

More importantly, there is a difference in damping of the waves. Plasma waves with large kv_{th} are highly damped, a result we shall obtain from kinetic theory in Chap. 7. Electromagnetic waves, on the other hand, become ordinary light waves at large kc and are not damped by the presence of the plasma in this limit.

A dispersion relation like Eq. (4.85) exhibits a phenomenon called *cutoff*. If one sends a microwave beam with a given frequency ω through a plasma, the wavelength $2\pi/k$ in the plasma will take on the value prescribed by Eq. (4.85). As the plasma density, and hence ω_p^2 is raised, k^2 will necessarily decrease; and the wavelength becomes longer and longer. Finally, a density will be reached such that k^2 is zero. For densities larger than this, Eq. (4.85) cannot be satisfied for any real k, and the wave cannot propagate. This cutoff condition occurs at a critical density n_c such that $\omega = \omega_p$; namely (from Eq. (4.25))

$$n_c = m\varepsilon_0 \omega^2 / e^2 \tag{4.88}$$

If n is too large or ω too small, an electromagnetic wave cannot pass through a plasma. When this happens, Eq. (4.85) tells us that k is imaginary:

$$ck = \left(\omega^2 - \omega_p^2\right)^{1/2} = i\left|\omega_p^2 - \omega^2\right|^{1/2} \tag{4.89}$$

Since the wave has a spatial dependence $\exp(ikx)$, it will be exponentially attenuated if k is imaginary. The skin depth δ is found as follows:

$$e^{ikx} = e^{-|k|x} = e^{-x/\delta} \quad \delta = |k|^{-1} = \frac{c}{\left(\omega_p^2 - \omega^2\right)^{1/2}} \tag{4.90}$$

For most laboratory plasmas, the cutoff frequency lies in the microwave range.

4.13 Experimental Applications

The phenomenon of cutoff suggests an easy way to measure plasma density. A beam of microwaves generated by a klystron is launched toward the plasma by a horn antenna (Fig. 4.27). The transmitted beam is collected by another horn and is detected by a crystal. As the frequency or the plasma density is varied, the detected signal will disappear whenever the condition Eq. (4.88) is satisfied somewhere in the plasma. This procedure gives the maximum density. It is not a convenient or versatile scheme because the range of frequencies generated by a single microwave generator is limited.

A widely used method of density measurement relies on the dispersion, or variation of index of refraction, predicted by Eq. (4.85). The index of refraction $\tilde{n}$ is defined as

$$\tilde{n} \equiv c/v_\phi = ck/\omega \tag{4.91}$$

This clearly varies with ω, and a plasma is a dispersive medium. A microwave interferometer employing the same physical principles as the Michelson interferometer is used to measure density (Fig. 4.28). The signal from a klystron is split into

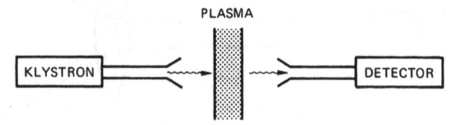

Fig. 4.27 Microwave measurement of plasma density by the cutoff of the transmitted signal

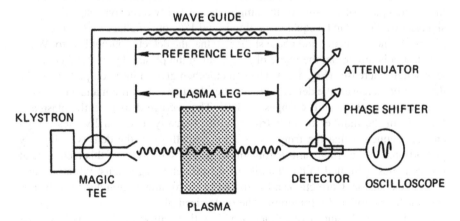

Fig. 4.28 A microwave interferometer for plasma density measurement

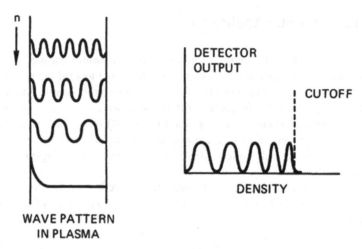

Fig. 4.29 The observed signal from an interferometer (*right*) as plasma density is increased, and the corresponding wave patterns in the plasma (*left*)

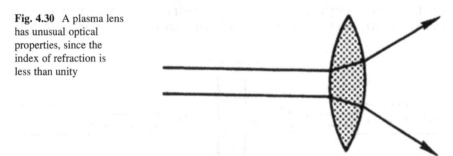

Fig. 4.30 A plasma lens has unusual optical properties, since the index of refraction is less than unity

two paths. Part of the signal goes to the detector through the "reference leg." The other part is sent through the plasma with horn antennas. The detector responds to the mean square of the sum of the amplitudes of the two received signals. These signals are adjusted to be equal in amplitude and 180° out of phase *in the absence of plasma* by the attenuator and phase shifter, so that the detector output is zero. When the plasma is turned on, the phase of the signal in the plasma leg is changed as the wavelength increases (Fig. 4.29). The detector then gives a finite output signal. As the density increases, the detector output goes through a maximum and a minimum every time the phase shift changes by 360°. The average density in the plasma is found from the number of such fringe shifts. Actually, one usually uses a high enough frequency that the fringe shift is kept small. Then the density is linearly proportional to the fringe shift (Problem 4.13). The sensitivity of this technique at low densities is limited to the stability of the reference leg against vibrations and thermal expansion. Corrections must also be made for attenuation due to collisions and for diffraction and refraction by the finite-sized plasma.

The fact that the index of refraction is less than unity for a plasma has some interesting consequences. A convex plasma lens (Fig. 4.30) is divergent rather

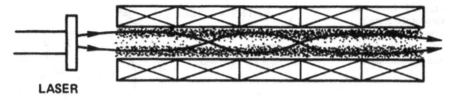

Fig. 4.31 A plasma confined in a long, linear solenoid will trap the CO_2 laser light used to heat it only if the plasma has a density minimum on axis. The vacuum chamber has been omitted for clarity

than convergent. This effect is important in the laser-solenoid proposal for a linear fusion reactor. A plasma hundreds of meters long is confined by a strong magnetic field and heated by absorption of CO_2 laser radiation (Fig. 4.31). If the plasma has a normal density profile (maximum on the axis), it behaves like a negative lens and causes the laser beam to diverge into the walls. If an inverted density profile (minimum on the axis) can be created, however, the lens effect becomes converging; and the radiation is focused and trapped by the plasma. The inverted profile can be produced by squeezing the plasma with a pulsed coil surrounding it, or it can be produced by the laser beam itself. As the beam heats the plasma, the latter expands, decreasing the density at the center of the beam. The CO_2 laser operates at $\lambda = 10.6$ μm, corresponding to a frequency

$$ f = \frac{c}{\lambda} = \frac{3 \times 10^8}{10.6 \times 10^{-6}} = 2.8 \times 10^{13} \text{Hz} $$

The critical density is, from Eq. (4.88),

$$ n_c = m\varepsilon_0 (2\pi f)^2 / e^2 = 10^{25} \text{m}^{-3} $$

However, because of the long path lengths involved, the refraction effects are important even at densities of 10^{22} m^{-3}. The focusing effect of a hollow plasma has been shown experimentally.

Perhaps the best known effect of the plasma cutoff is the application to short-wave radio communication. When a radio wave reaches an altitude in the ionosphere where the plasma density is sufficiently high, the wave is reflected (Fig. 4.32), making it possible to send signals around the earth. If we take the maximum density to be 10^{12} m^{-3}, the critical frequency is of the order of 10 MHz (*cf.* Eq. (4.26)). To communicate with space vehicles, it is necessary to use frequencies above this in order to penetrate the ionosphere. However, during reentry of a space vehicle, a plasma is generated by the intense heat of friction. This causes a plasma cutoff, resulting in a communications blackout during reentry (Fig. 4.32).

Fig. 4.32 Exaggerated
view of the earth's
ionosphere, illustrating the
effect of plasma on radio
communications

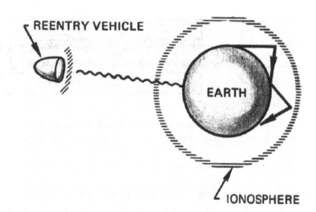

Problems

4.10. A space capsule making a reentry into the earth's atmosphere suffers a
communications blackout because a plasma is generated by the shock wave
in front of the capsule. If the radio operates at a frequency of 300 MHz, what
is the minimum plasma density during the blackout?

4.11. Hannes Alfvén, the first plasma physicist to be awarded the Nobel prize, has
suggested that perhaps the primordial universe was symmetric between
matter and antimatter. Suppose the universe was at one time a uniform
mixture of protons, antiprotons, electrons, and positrons, each species having
a density n_0.

(a) Work out the dispersion relation for high-frequency electromagnetic
waves in this plasma. You may neglect collisions, annihilations, and
thermal effects.

(b) Work out the dispersion relation for ion waves, using Poisson's equation. You may neglect T_i (but not T_e) and assume that all leptons follow
the Boltzmann relation.

4.12. For electromagnetic waves, show that the index of refraction is equal to the
square root of the appropriate plasma dielectric constant (*cf.* Problem 4.4).

4.13. In a potassium Q-machine plasma, a fraction κ of the electrons can be
replaced by negative Cl ions. The plasma then has n_0 K$^+$ ions, κn_0 Cl$^-$ ions,
and $(1 - \kappa)n_0$ electrons per m^3. Find the critical value of n_0 which will cut off
a 3-cm wavelength microwave beam if $\kappa = 0.6$.

4.14. An 8-mm microwave interferometer is used on an infinite plane-parallel
plasma slab 8 cm thick (Fig. P4.13).

(a) If the plasma density is uniform, and a phase shift of 1/10 fringe is
observed, what is the density? (Note: One fringe corresponds to a 360°
phase shift.)

(b) Show that if the phase shift is small, it is proportional to the density.

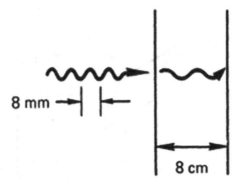

Fig. P4.13

4.14   Electromagnetic Waves Perpendicular to $\mathbf{B_0}$

We now consider the propagation of electromagnetic waves when a magnetic field is present. We treat first the case of perpendicular propagation, $\mathbf{k} \perp \mathbf{B_0}$. If we take transverse waves, with $\mathbf{k} \perp \mathbf{E_1}$, there are still two choices: $\mathbf{E_1}$ can be parallel to $\mathbf{B_0}$ or perpendicular to $\mathbf{B_0}$ (Fig. 4.33).

4.14.1   Ordinary Wave, $E_1 \parallel B_0$

If $\mathbf{E_1}$ is parallel to $\mathbf{B_0}$, we may take $\mathbf{B_0} = B_0\hat{\mathbf{z}}$, $\mathbf{E_1} = E_1\hat{\mathbf{z}}$, $\mathbf{k} = k\hat{\mathbf{x}}$. In a real experiment, this geometry is approximated by a beam of microwaves incident on a plasma column with the narrow dimension of the waveguide in line with $\mathbf{B_0}$ (Fig. 4.34).

Fig. 4.33 Geometry for
electromagnetic waves
propagating at right angles
to $\mathbf{B_0}$

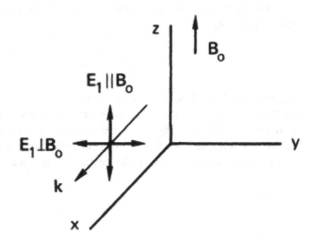

Fig. 4.34 An ordinary
wave launched from a
waveguide antenna toward
a magnetized plasma
column

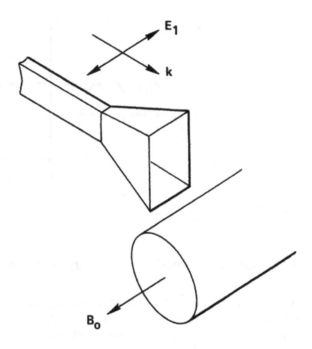

The wave equation for this case is still given by Eq. (4.81):

$$\left(\omega^2 - c^2k^2\right)\mathbf{E}_1 = -i\omega\mathbf{j}_1/\varepsilon_0 = in_0e\omega\mathbf{v}_{e1}/\varepsilon_0 \tag{4.92}$$

Since $\mathbf{E}_1 = E_1\hat{\mathbf{z}}$, we need only the component v_{ez}. This is given by the equation of motion

$$m\partial v_{ez}/\partial t = -eE_z \tag{4.93}$$

Since this is the same as the equation for $\mathbf{B}_0 = 0$, the result is the same as we had previously for $\mathbf{B}_0 = 0$:

$$\boxed{\omega^2 = \omega_p^2 + c^2k^2} \tag{4.94}$$

This wave, with $\mathbf{E}_1 \parallel \mathbf{B}_0$, is called the *ordinary* wave. The terminology "ordinary" and "extraordinary" is taken from crystal optics; however, the terms have been interchanged. In plasma physics, it makes more sense to let the "ordinary" wave be the one that is not affected by the magnetic field. Strict analogy with crystal optics would have required calling this the "extraordinary" wave.

Fig. 4.35 The E-vector of
an extraordinary wave is
elliptically polarized. The
components E_x and E_y
oscillate 90° out of phase,
so that the total electric field
vector E_1 has a tip that
moves in an ellipse once in
each wave period

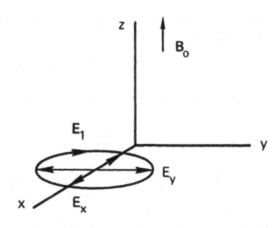

4.14.2 Extraordinary Wave, $E_1 \perp B_0$

If **E**1 is perpendicular to $\mathbf{B}_0$, the electron motion will be affected by $\mathbf{B}_0$, and the
dispersion relation will be changed. To treat this case, one would be tempted to take
$\mathbf{E}_1 = E_1\hat{\mathbf{y}}$ and $\mathbf{k} = k\hat{\mathbf{x}}$ (Fig. 4.33). However, it turns out that waves with $\mathbf{E}_1 \perp \mathbf{B}_0$
tend to be elliptically polarized instead of plane polarized. That is, as such a wave
propagates into a plasma, it develops a component E_x *along* **k**, thus becoming partly
longitudinal and partly transverse. To treat this mode properly, we must allow $\mathbf{E}_1$ to
have both x and y components (Fig. 4.35):

$$\mathbf{E}_1 = E_x\hat{\mathbf{x}} + E_y\hat{\mathbf{y}} \tag{4.95}$$

The linearized electron equation of motion (with $KT_e = 0$) is now

$$-im\omega\mathbf{v}_{e1} = -e(\mathbf{E} + \mathbf{v}_{e1} \times \mathbf{B}_0) \tag{4.96}$$

Only the x and y components are nontrivial; they are

$$
\begin{aligned}
v_x &= \frac{-ie}{m\omega}(E_x + v_yB_0) \\
v_y &= \frac{-ie}{m\omega}(E_y - v_xB_0)
\end{aligned}
\tag{4.97}
$$

The subscripts 1 and e have been suppressed. Solving for v_x and v_y as usual, we find

$$
\begin{aligned}
v_x &= \frac{e}{m\omega}\left(-iE_x - \frac{\omega_c}{\omega}E_y\right)\left(1 - \frac{\omega_c^2}{\omega^2}\right)^{-1} \\
v_y &= \frac{e}{m\omega}\left(-iE_y - \frac{\omega_c}{\omega}E_x\right)\left(1 - \frac{\omega_c^2}{\omega^2}\right)^{-1}
\end{aligned}
\tag{4.98}
$$

The wave equation is given by Eq. (4.80), where we must now keep the longitudinal term $\mathbf{k} \cdot \mathbf{E}_1 = kE_x$:

$$(\omega^2 - c^2k^2)\mathbf{E}_1 + c^2kE_x\mathbf{k} = -i\omega\mathbf{j}_1/\varepsilon_0 = in_0\omega ev_{e1}/\varepsilon_0 \qquad (4.99)$$

Separating this into x and y components and using Eq. (4.98), we have

$$\omega^2 E_x = -\frac{i\omega n_0 e}{\varepsilon_0}\frac{e}{m\omega}\left(iE_x + \frac{\omega_c}{\omega}E_y\right)\left(1 - \frac{\omega_c^2}{\omega^2}\right)^{-1}$$

$$(\omega^2 - c^2k^2)E_y = -\frac{i\omega n_0 e}{\varepsilon_0}\frac{e}{m\omega}\left(iE_y - \frac{\omega_c}{\omega}Ex\right)\left(1 - \frac{\omega_c^2}{\omega^2}\right)^{-1}$$

(4.100)

Introducing the definition of ω_p, we may write this set as

$$\underbrace{\left[\omega^2\left(1 - \frac{\omega_c^2}{\omega^2}\right) - \omega_p^2\right]}_{\text{Ⓐ}}E_x + \underbrace{i\frac{\omega_p^2\omega_c}{\omega}}_{\text{Ⓑ}}E_y = 0$$

(4.101)

$$\underbrace{\left[(\omega^2 - c^2k^2)\left(1 - \frac{\omega_c^2}{\omega^2}\right) - \omega_p^2\right]}_{\text{Ⓓ}}E_y - \underbrace{i\frac{\omega_p^2\omega_c}{\omega}}_{\text{Ⓒ}}E_x = 0$$

These are two simultaneous equations for E_x and E_y which are compatible only if the determinant of the coefficients vanishes:

$$\begin{Vmatrix} A & B \\ C & D \end{Vmatrix} = 0 \qquad (4.102)$$

Since the coefficient A is $\omega^2 - \omega_h^2$, where ω_h is the upper hybrid frequency defined by Eq. (4.60), the condition $AD = BC$ can be written

$$(\omega^2 - \omega_h^2)\left[\omega^2 - \omega_h^2 - c^2k^2\left(1 - \frac{\omega_c^2}{\omega^2}\right)\right] = \left(\frac{\omega_p^2\omega_c}{\omega}\right)^2$$

$$\frac{c^2k^2}{\omega^2} = \frac{\omega^2 - \omega_h^2 - \left[\left(\omega_p^2\omega_c/\omega\right)^2/(\omega^2 - \omega_h^2)\right]}{\omega^2 - \omega_c^2}$$

(4.103)

This can be simplified by a few algebraic manipulations. Replacing the first ω_h^2 on the right-hand side by $\omega_c^2 + \omega_p^2$ and multiplying through by $\omega^2 - \omega_h^2$, we have

$$\frac{c^2 k^2}{\omega^2} = 1 - \frac{\omega_p^2 \left(\omega^2 - \omega_h^2\right) + \left(\omega_p^4 \omega_c^2 / \omega\right)^2}{\left(\omega^2 - \omega_c^2\right)\left(\omega^2 - \omega_h^2\right)}$$

$$= 1 - \frac{\omega_p^2}{\omega^2} \frac{\omega^2 \left(\omega^2 - \omega_h^2\right) + \omega_p^2 \omega_c^2}{\left(\omega^2 - \omega_c^2\right)\left(\omega^2 - \omega_h^2\right)}$$

$$= 1 - \frac{\omega_p^2}{\omega^2} \frac{\omega^2 \left(\omega^2 - \omega_c^2\right) - \omega_p^2 \left(\omega^2 - \omega_c^2\right)}{\left(\omega^2 - \omega_c^2\right)\left(\omega^2 - \omega_h^2\right)}$$

$$\boxed{\frac{c^2 k^2}{\omega^2} = \frac{c^2}{v_\phi^2} = 1 - \frac{\omega_p^2}{\omega^2} \frac{\omega^2 - \omega_p^2}{\omega^2 - \omega_h^2}} \tag{4.104}$$

This is the dispersion relation for the *extraordinary* wave. It is an electromagnetic wave, partly transverse and partly longitudinal, which propagates perpendicular to $\mathbf{B}_0$ with $\mathbf{E}_1$ perpendicular to $\mathbf{B}_0$.

4.15 Cutoffs and Resonances

The dispersion relation for the extraordinary wave is considerably more complicated than any we have met up to now. To analyze what it means, it is useful to define the terms *cutoff* and *resonance*. A *cutoff* occurs in a plasma when the index of refraction goes to zero; that is, when the wavelength becomes infinite, since $\tilde{n} = ck/\omega$. A *resonance* occurs when the index of refraction becomes infinite; that is, when the wavelength becomes zero. As a wave propagates through a region in which ω_p and ω_c are changing, it may encounter cutoffs and resonances. A wave is generally reflected at a cutoff and absorbed at a resonance.

The resonance of the extraordinary wave is found by setting k equal to infinity in Eq. (4.104). For any finite ω, $k \to \infty$ implies $\omega \to \omega_h$, so that a resonance occurs at a point in the plasma where

$$\omega_h^2 = \omega_p^2 + \omega_c^2 = \omega^2 \tag{4.105}$$

This is easily recognized as the dispersion relation for electrostatic waves propagating across $\mathbf{B}_0$ (Eq. (4.60)). As a wave of given ω approaches the resonance point, both its phase velocity and its group velocity approach zero, and the wave energy is converted into upper hybrid oscillations. The extraordinary wave is partly electromagnetic and partly electrostatic; it can easily be shown (Problem 4.14) that at resonance this wave loses its electromagnetic character and becomes an electrostatic oscillation.

The cutoffs of the extraordinary wave are found by setting k equal to zero in Eq. (4.104). Dividing by $\omega^2 - \omega_p^2$, we can write the resulting equation for ω as follows:

$$1 = \frac{\omega_p^2}{\omega^2} \frac{1}{1 - \left[\omega_c^2/\left(\omega^2 - \omega_p^2\right)\right]} \tag{4.106}$$

A few tricky algebraic steps will yield a simple expression for ω:

$$1 - \frac{\omega_c^2}{\omega^2 - \omega_p^2} = \frac{\omega_p^2}{\omega^2}$$

$$1 - \frac{\omega_p^2}{\omega^2} = \frac{\omega_c^2/\omega^2}{1 - \left(\omega_p^2/\omega^2\right)}$$

$$\left(1 - \frac{\omega_p^2}{\omega^2}\right)^2 - \frac{\omega_c^2}{\omega^2}$$

$$1 - \frac{\omega_p^2}{\omega^2} = \pm \frac{\omega_c}{\omega}$$

$$\omega^2 \mp \omega\omega_c - \omega_p^2 = 0 \tag{4.107}$$

Each of the two signs will give a different cutoff frequency; we shall call these ω_R and ω_L. The roots of the two quadratics are

$$\omega_R = \frac{1}{2}\left[\omega_c + \left(\omega_c^2 + 4\omega_p^2\right)^{1/2}\right]$$

$$\omega_L = \frac{1}{2}\left[-\omega_c + \left(\omega_c^2 + 4\omega_p^2\right)^{1/2}\right] \tag{4.108}$$

We have taken the plus sign in front of the square root in each case because we are using the convention that ω is always positive; waves going in the $-x$ direction will be described by negative k. The cutoff frequencies ω_R and ω_L are called the *right-hand* and *left-hand cutoffs*, respectively, for reasons which will become clear in the next section.

The cutoff and resonance frequencies divide the dispersion diagram into regions of propagation and non propagation. Instead of the usual ω–k diagram, it is more enlightening to give a plot of phase velocity versus frequency; or, to be precise, a plot of $\omega^2/c^2k^2 = 1/\tilde{n}^2$ vs. ω (Fig. 4.36). To interpret this diagram, imagine that ω_c is fixed, and a wave with a fixed frequency ω is sent into a plasma from the outside. As the wave encounters regions of increasing density, the frequencies ω_L, ω_p, ω_h, and ω_R all increase, moving to the right in the diagram. This is the same as if the density

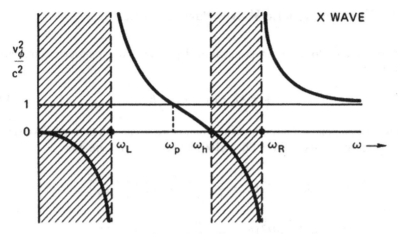

Fig. 4.36 The dispersion of the extraordinary wave, as seen from the behavior of the phase velocity with frequency. The wave does not propagate in the *shaded regions*

Fig. 4.37 A similar dispersion diagram for the ordinary wave

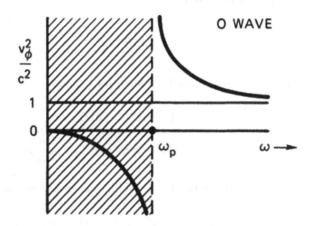

were fixed and the frequency ω were gradually being decreased. Taking the latter point of view, we see that at large ω (or low density), the phase velocity approaches the velocity of light. As the wave travels further, v_ϕ increases until the right-hand cutoff $\omega = \omega_R$ is encountered. There, v_ϕ becomes infinite. Between the $\omega = \omega_R$ and $\omega = \omega_h$ layers, ω^2/k^2 is negative, and there is no propagation possible. At $\omega = \omega_h$, there is a resonance, and v_ϕ goes to zero. Between $\omega = \omega_h$ and $\omega = \omega_L$, propagation is again possible. In this region, the wave travels either faster or slower than c depending on whether ω is smaller or larger than ω_p. From Eq. (4.104), it is clear that at $\omega = \omega_p$, the wave travels at the velocity c. For $\omega < \omega_L$, there is another region of nonpropagation. The extraordinary wave, therefore, has two regions of propagation separated by a stop band.

By way of comparison, we show in Fig. 4.37 the same sort of diagram for the ordinary wave. This dispersion relation has only one cutoff and no resonances.

4.16   Electromagnetic Waves Parallel to $\mathbf{B_0}$

Now we let $\mathbf{k}$ lie along the z axis and allow $\mathbf{E_1}$ to have both transverse components E_x and E_y:

$$\mathbf{k} = k\hat{\mathbf{z}} \quad \mathbf{E}_1 = E_x\hat{\mathbf{x}} + E_y\hat{\mathbf{y}} \tag{4.109}$$

The wave equation (4.99) for the extraordinary wave can still be used if we simply change $\mathbf{k}$ from $k\hat{\mathbf{x}}$ to $k\hat{\mathbf{z}}$. From Eq. (4.100), the components are now

$$
\begin{aligned}
(\omega^2 - c^2k^2)E_x &= \frac{\omega_p^2}{1 - \omega_c^2/\omega^2}\left(E_x - \frac{i\omega_c}{\omega}E_y\right) \\
(\omega^2 - c^2k^2)E_y &= \frac{\omega_p^2}{1 - \omega_c^2/\omega^2}\left(E_y + \frac{i\omega_c}{\omega}E_x\right)
\end{aligned}
\tag{4.110}
$$

Using the abbreviation

$$\alpha \equiv \frac{\omega_p^2}{1 - (\omega_c^2/\omega^2)} \tag{4.111}$$

we can write the coupled equations for E_x and E_y as

$$
\begin{aligned}
(\omega^2 - c^2k^2 - \alpha)E_x + i\alpha\frac{\omega_c}{\omega}E_y &= 0 \\
(\omega^2 - c^2k^2 - \alpha)E_y - i\alpha\frac{\omega_c}{\omega}E_x &= 0
\end{aligned}
\tag{4.112}
$$

Setting the determinant of the coefficients to zero, we have

$$\left(\omega^2 - c^2k^2 - \alpha\right)^2 = (\alpha\omega_c/\omega)^2 \tag{4.113}$$

$$\omega^2 - c^2k^2 - \alpha = \pm\alpha\omega_c/\omega \tag{4.114}$$

Thus

$$
\omega^2 - c^2k^2 = \alpha\left(1 \pm \frac{\omega_c}{\omega}\right) = \frac{\omega_p^2}{1 - (\omega_c^2/\omega^2)}\left(1 \pm \frac{\omega_c}{\omega}\right)
$$

$$
= \omega_p^2 \frac{1 \pm (\omega_c/\omega)}{[1 + (\omega_c/\omega)][1 - (\omega_c/\omega)]} = \frac{\omega_p^2}{1 \mp (\omega_c/\omega)}
\tag{4.115}
$$

The $\mp$ sign indicates that there are two possible solutions to Eq. (4.112) corresponding to two different waves that can propagate along $\mathbf{B}_0$. The dispersion relations are

$$\tilde{n}^2 = \frac{c^2 k^2}{\omega^2} = 1 - \frac{\omega_p^2/\omega^2}{1 - (\omega_c/\omega)} \quad (R\ \text{wave}) \qquad (4.116)$$

$$\tilde{n}^2 = \frac{c^2 k^2}{\omega^2} = 1 - \frac{\omega_p^2/\omega^2}{1 + (\omega_c/\omega)} \quad (L\ \text{wave}) \qquad (4.117)$$

The R and L waves turn out to be circularly polarized, the designations R and L meaning, respectively, *right-hand circular polarization* and *left-hand circular polarization* (Problem 4.17). The geometry is shown in Fig. 4.38. The electric field vector for the R wave rotates clockwise in time as viewed along the direction of $\mathbf{B}_0$, and vice versa for the L wave. Since Eqs. (4.116) and (4.117) depend only on k^2, the direction of rotation of the $\mathbf{E}$ vector is independent of the sign of k; the polarization is the same for waves propagating in the opposite direction. To summarize: The principal electromagnetic waves propagating *along* $\mathbf{B}_0$ are a right-hand (R) and a left-hand (L) circularly polarized wave; the principal waves propagating *across* $\mathbf{B}_0$ are a plane-polarized wave (O-wave) and an elliptically polarized wave (X-wave).

We next consider the cutoffs and resonances of the R and L waves. For the R wave, k becomes infinite at $\omega = \omega_c$; the wave is then in resonance with the cyclotron motion of the electrons. The direction of rotation of the plane of polarization is the same as the direction of gyration of electrons; the wave loses its energy in continuously accelerating the electrons, and it cannot propagate. The L wave, on the other hand, does not have a cyclotron resonance with the electrons because it rotates in the opposite sense. As is easily seen from Eq. (4.117), the L wave does not have a resonance for positive ω. If we had included ion motions in our treatment,

Fig. 4.38 Geometry of right- and left-handed circularly polarized waves propagating along B_0

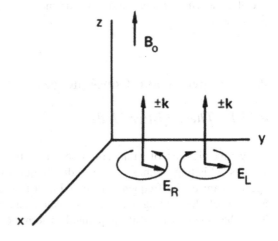

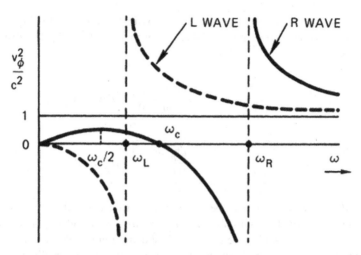

Fig. 4.39 The v_ϕ^2/c^2 vs. ω diagrams for the L and R waves. The regions of nonpropagation $\left(v_\phi^2/c^2 < 0\right)$ have not been shaded, since they are different for the two waves

the L wave would have been found to have a resonance at $\omega = \Omega_c$, since it would then rotate with the ion gyration.

The cutoffs are obtained by setting $k=0$ in Eqs. (4.116) and (4.117). We then obtain the same equations as we had for the cutoffs of the X wave (Eq. (4.107)). Thus the cutoff frequencies are the same as before. The R wave, with the minus sign in Eqs. (4.116) and (4.107), has the higher cutoff frequency ω_R given by Eq. (4.108); the L wave, with the plus sign, has the lower cutoff frequency ω_L. This is the reason for the notation ω_R, ω_L chosen previously. The dispersion diagram for the R and L waves is shown in Fig. 4.39. The L wave has a stop band at low frequencies; it behaves like the O wave except that the cutoff occurs at ω_L instead of ω_p. The R wave has a stop band between ω_R and ω_c, but there is a second band of propagation, with $v_\phi < c$, below ω_c. The wave in this low-frequency region is called the *whistler mode* and is of extreme importance in the study of ionospheric phenomena.

4.17   Experimental Consequences

4.17.1   The Whistler Mode

Early investigators of radio emissions from the ionosphere were rewarded by various whistling sounds in the audio frequency range. Figure 4.40 shows a spectrogram of the frequency received as a function of time. There is typically a series of descending glide tones, which can be heard over a loudspeaker. This phenomenon is easily explained in terms of the dispersion characteristics

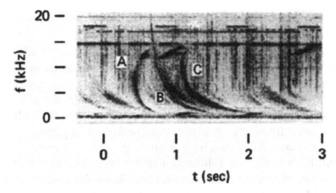

Fig. 4.40 Actual spectrograms of whistler signals, showing the curvature caused by the low-frequency branch of the R-wave dispersion relation (Fig. 4.39). At each time t, the receiver rapidly scans the frequency range between 0 and 20 kHz, tracing a vertical line. The recorder makes a spot whose darkness is proportional to the intensity of the signal at each frequency. The downward motion of the dark spot with time then indicates a descending glide tone. [Courtesy of D. L. Carpenter, *J. Geophys. Res.* **71**, 693 (1966).]

Fig. 4.41 Diagram showing how whistlers are created. The channels A, B, and C refer to the signals so marked in Fig. 4.40

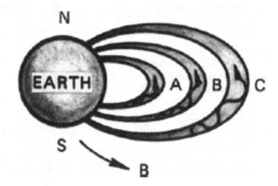

of the R wave. When a lightning flash occurs in the Southern Hemisphere, radio noise of all frequencies is generated. Among the waves generated in the plasma of the ionosphere and magnetosphere are R waves traveling along the earth's magnetic field. These waves are guided by the field lines and are detected by observers in Canada. However, different frequencies arrive at different times. From Fig. 4.39, it can be seen that for $\omega < \omega_c/2$, the phase velocity increases with frequency (Problem 4.19). It can also be shown (Problem 4.20) that the group velocity increases with frequency. Thus the low frequencies arrive later, giving rise to the descending tone. several whistles can be produced by a single lightning flash because of propagation along different tubes of force A, B, C (Fig. 4.41). Since the waves have $\omega < \omega_c$, they must have frequencies lower than the lowest gyrofrequency along the tube of force, about 100 kHz. Either the whistles lie directly in the audio range or they can easily be converted into audio signals by heterodyning.

4.17.2 Faraday Rotation

A plane-polarized wave sent along a magnetic field in a plasma will suffer a rotation
of its plane of polarization (Fig. 4.42). This can be understood in terms of the
difference in phase velocity of the R and L waves. From Fig. 4.39, it is clear that for
large ω the R wave travels faster than the L wave. Consider the plane-polarized
wave to be the sum of an R wave and an L wave (Fig. 4.43). Both waves are, of
course, at the same frequency. After N cycles, the E_L and E_R vectors will return to
their initial positions. After traversing a given distance d, however, the R and
L waves will have undergone a different number of cycles, since they require a
different amount of time to cover the distance. Since the L wave travels more
slowly, it will have undergone $N + \epsilon$ cycles at the position where the R wave has
undergone N cycles. The vectors then have the positions shown in Fig. 4.44. The
plane of polarization is seen to have rotated. A measurement of this rotation by
means of a microwave horn can be used to give a value of ω_p^2 and, hence, of the
density (Problem 4.22). The effect of Faraday rotation has been verified experimentally, but it is not as useful a method of density measurement as microwave
interferometry, because access at the ends of a plasma column is usually difficult,

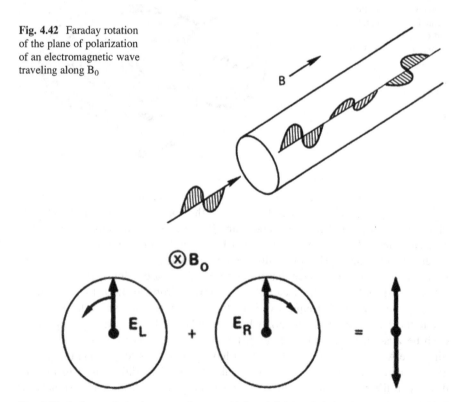

Fig. 4.42 Faraday rotation
of the plane of polarization
of an electromagnetic wave
traveling along B_0

Fig. 4.43 A plane-polarized wave as the sum of left and righthanded circularly polarized waves

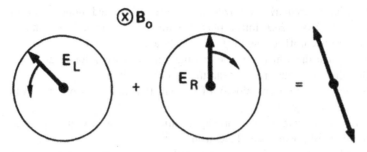

Fig. 4.44 After traversing the plasma, the L wave is advanced in phase relative to the R wave, and the plane of polarization is rotated

and because the effect is small unless the density is so high that refraction becomes a problem.

When powerful pulsed lasers are used to produce a dense plasma by vaporizing a solid target, magnetic fields of megagauss intensities are sometimes spontaneously generated. These have been detected by Faraday rotation using laser light of higher frequency than the main beam. In interstellar space, the path lengths are so long that Faraday rotation is important even at very low densities. This effect has been used to explain the polarization of microwave radiation generated by maser action in clouds of OH or H_2O molecules during the formation of new stars.

Problems

4.14. Prove that the extraordinary wave is purely electrostatic at resonance. Hint: Express the ratio E_y/E_x as a function of ω and set ω equal to ω_h.

4.15. Prove that the critical points on Fig. 4.36 are correctly ordered; namely, that
$$\omega_L < \omega_p < \omega_h < \omega_R.$$

4.16. Show that the X-wave group velocity vanishes at cutoffs and resonances. You may neglect ion motions.

4.17. Prove that the R and L waves are right- and left-circularly polarized as follows:

  (a) Show that the simultaneous equations for E_x and E_y can be written in the form

$$F(\omega)(E_x - iE_y) = 0, \quad G(\omega)(E_x + iE_y) = 0$$

  where $F(\omega) = 0$ for the R wave and $G(\omega) = 0$ for the L wave.

  (b) For the R wave, $G(\omega) \neq 0$; and therefore $E_x = -iE_y$. Recalling the exponential time dependence of $\mathbf{E}$, show that $\mathbf{E}$ then rotates in the electron gyration direction. Confirm that $\mathbf{E}$ rotates in the opposite direction for the L wave.

  (c) For the R wave, draw the helices traced by the tip of the $\mathbf{E}$ vector in space at a given time for (i) $k_z > 0$ and (ii) $k_z < 0$. Note that the rotation of $\mathbf{E}$ is in the same direction in both instances if one stays at a fixed position and watches the helix pass by.

4.18. Left-hand circularly polarized waves are propagated along a uniform magnetic field $\mathbf{B} = B_0\hat{\mathbf{z}}$ into a plasma with density increasing with z. At what density is cutoff reached if $f = 2.8$ GHz and $B_0 = 0.3$ T?

4.19. Show that the whistler mode has maximum phase velocity at $\omega = \omega_c/2$ and that this maximum is less than the velocity of light.

4.20. Show that the group velocity of the whistler mode is proportional to $\omega^{1/2}$ if $\omega \ll \omega_c$ and $\epsilon \gg 1$.

4.21. Show that there is no Faraday rotation in a positronium plasma (equal numbers of positrons and electrons).

4.22. Faraday rotation of an 8-mm-wavelength microwave beam in a uniform plasma in a 0.1-T magnetic field is measured. The plane of polarization is found to be rotated 90° after traversing 1 m of plasma. What is the density?

4.23. Show that the Faraday rotation angle, in degrees, of a linearly polarized transverse wave propagating along B_0 is given by

$$\theta = 1.5 \times 10^{-11}\lambda_0^2 \int_0^L B(z)n_e(z)dz$$

where λ_0 is the free-space wavelength and L the path length in the plasma. Assume $\omega^2 \gg \omega_p^2, \omega_c^2$.

4.24. In some laser-fusion experiments in which a plasma is created by a pulse of 1.06-μm light impinging on a solid target, very large magnetic fields are generated by thermoelectric currents. These fields can be measured by Faraday rotation of frequency-doubled light ($\lambda_0 = 0.53$ μm) derived from the same laser. If $B = 100$ T, $n = 10^{27}$ m^{-3}, and the path length in the plasma is 30 μm, what is the Faraday rotation angle in degrees? (Assume $\mathbf{k} \parallel \mathbf{B}$.)

4.25. A microwave interferometer employing the ordinary wave cannot be used above the cutoff density n_c. To measure higher densities, one can use the extraordinary wave.

(a) Write an expression for the cutoff density n_{cx} for the X wave.

(b) On a v_ϕ^2/c^2 vs. ω diagram, show the branch of the X-wave dispersion relation on which such an interferometer would work.

4.18 Hydromagnetic Waves

The last part of our survey of fundamental plasma waves concerns low-frequency ion oscillations in the presence of a magnetic field. Of the many modes possible, we shall treat only two: the hydromagnetic wave along $\mathbf{B}_0$, or *Alfvén wave*, and the magnetosonic wave. The Alfvén wave in plane geometry has $\mathbf{k}$ along $\mathbf{B}_0$; $\mathbf{E}_1$ and $\mathbf{j}_1$ perpendicular to $\mathbf{B}_0$; and $\mathbf{B}_1$ and $\mathbf{v}_1$ perpendicular to both $\mathbf{B}_0$ and $\mathbf{E}_1$ (Fig. 4.45). From Maxwell's equation we have, as usual (Eq. (4.80)),

$$\nabla \times \nabla \times \mathbf{E}_1 = -\mathbf{k}(\mathbf{k} \cdot \mathbf{E}_1) + k^2\mathbf{E}_1 = \frac{\omega^2}{c^2}\mathbf{E}_1 + \frac{i\omega}{\varepsilon_0}\frac{1}{c^2}\mathbf{j}_1 \qquad (4.118)$$

Fig. 4.45 Geometry of an
Alfvén wave propagating
along B_0

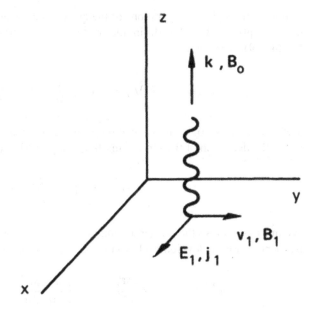

Since $\mathbf{k} = k\hat{\mathbf{z}}$ and $\mathbf{E}_1 = E_1\hat{\mathbf{x}}$ by assumption, only the x component of this equation is nontrivial. The current $\mathbf{j}_1$ now has contributions from both ions and electrons, since we are considering low frequencies. The x component of Eq. (4.118) becomes

$$\varepsilon_0\left(\omega^2 - c^2k^2\right)E_1 = -i\omega n_0 e(v_{ix} - v_{ex}) \tag{4.119}$$

Thermal motions are not important for this wave; we may therefore use the solution of the ion equation of motion with $T_i = 0$ obtained previously in Eq. (4.63). For completeness, we include here the component v_{iy}, which was not written explicitly before:

$$v_{ix} = \frac{ie}{M\omega}\left(1 - \frac{\Omega_c^2}{\omega^2}\right)^{-1} E_1$$

$$v_{iy} = \frac{e}{M\omega}\frac{\Omega_c}{\omega}\left(1 - \frac{\Omega_c^2}{\omega^2}\right)^{-1} E_1 \tag{4.120}$$

The corresponding solution to the electron equation of motion is found by letting $M \to m$, $e \to -e$, $\Omega_c \to -\omega_c$ and then taking the limit $\omega_c^2 \gg \omega^2$:

$$v_{ex} = \frac{ie}{m\omega}\frac{\omega^2}{\omega_c^2}E_1 \to 0$$

$$v_{ey} = -\frac{e}{m\omega}\frac{\omega_c}{\omega^2}\frac{\omega^2}{\omega_c^2}E_1 = -\frac{E_1}{B_0} \tag{4.121}$$

In this limit, the Larmor gyrations of the electrons are neglected, and the electrons have simply an $\mathbf{E} \times \mathbf{B}$ drift in the y direction. Inserting these solutions into Eq. (4.119), we obtain

$$\varepsilon_0 \left(\omega^2 - c^2 k^2 \right) E_1 = -i\omega n_0 e \frac{ie}{M\omega} \left(1 - \frac{\Omega_c^2}{\omega^2} \right)^{-1} E_1 \qquad (4.122)$$

The y components of $\mathbf{v}_1$ are needed only for the physical picture to be given later. Using the definition of the ion plasma frequency Ω_p (Eq. (4.49)), we have

$$\omega^2 - c^2 k^2 = \Omega_p^2 \left(1 - \frac{\Omega_c^2}{\omega^2} \right)^{-1} \qquad (4.123)$$

We must now make the further assumption $\omega^2 \ll \Omega_c^2$; hydromagnetic waves have frequencies well below ion cyclotron resonance. In this limit, Eq. (4.123) becomes

$$\omega^2 - c^2 k^2 = -\omega^2 \frac{\Omega_p^2}{\Omega_c^2} = -\omega^2 \frac{n_0 e^2}{\varepsilon_0 M} \frac{M^2}{e^2 B_0^2} = -\omega^2 \frac{\rho}{\varepsilon_0 B_0^2}$$

$$\frac{\omega^2}{k^2} = \frac{c^2}{1 + \left(\rho / \varepsilon_0 B_0^2 \right)} = \frac{c^2}{1 + \left(\rho \mu_0 / B_0^2 \right) c^2} \qquad (4.124)$$

where ρ is the mass density $n_0 M$. This answer is no surprise, since the denominator can be recognized as the relative dielectric constant for low-frequency perpendicular motions (Eq. (3.28)). Equation (4.124) simply gives the phase velocity for an electromagnetic wave in a dielectric medium:

$$\frac{\omega}{k} = \frac{c}{(\varepsilon_R \mu_R)^{1/2}} = \frac{c}{\varepsilon_R^{1/2}} \quad \text{for} \quad \mu_R = 1$$

As we have seen previously, ε is much larger than unity for most laboratory plasmas, and Eq. (4.124) can be written approximately as

$$\frac{\omega}{k} = v_\phi = \frac{B_0}{(\mu_0 \rho)^{1/2}} \qquad (4.125)$$

These hydromagnetic waves travel along $\mathbf{B}_0$ at a constant velocity v_A, called the *Alfvén velocity*:

$$\boxed{v_A \equiv B / (\mu_0 \rho)^{1/2}} \qquad (4.126)$$

This is a characteristic velocity at which perturbations of the lines of force travel. The dielectric constant of Eq. (3.28) can now be written

$$\epsilon_R = \epsilon/\epsilon_0 = 1 + \left(c^2/v_A^2\right) \qquad (4.127)$$

Note that v_A is small for well-developed plasmas with appreciable density, and therefore ϵ_R is large.

To understand what happens physically in an Alfvén wave, recall that this is an electromagnetic wave with a fluctuating magnetic field $\mathbf{B}_1$ given by

$$\nabla \times \mathbf{E}_1 = -\dot{\mathbf{B}}_1 \quad E_x = (\omega/k)B_y \qquad (4.128)$$

The small component B_y, when added to $\mathbf{B}_0$, gives the lines of force a sinusoidal ripple, shown exaggerated in Fig. 4.46. At the point shown, B_y is in the positive y direction, so, according to Eq. (4.128), E_x is in the positive x direction if ω/k is in the z direction. The electric field E_x gives the plasma an $\mathbf{E}_1 \times \mathbf{B}_0$ drift in the negative y direction. Since we have taken the limit $\omega^2 \ll \Omega_c^2$, both ions and electrons will have the same drift v_y, according to Eqs. (4.120) and (4.121). Thus, the fluid moves up and down in the y direction, as previously indicated in Fig. 4.45. The magnitude of this velocity is $|E_x/B_0|$. Since the ripple in the field is moving by at the phase velocity ω/k, the line of force is also moving downward at the point indicated in Fig. 4.46. The downward velocity of the line of force is $(\omega/k)|B_y/B_0|$, which, according to Eq. (4.128), is just equal to the fluid velocity $|E_x/B_0|$. Thus, the fluid and the field lines oscillate together *as if the particles were stuck to the lines*. The lines of force act as if they were mass-loaded strings under tension, and an Alfvén wave can be regarded as the propagating disturbance occurring when the strings are plucked. This concept of plasma frozen to lines of force and moving with them is a useful one for understanding many low-frequency plasma phenomena. It can be shown that this notion is an accurate one as long as there is no electric field *along* $\mathbf{B}$.

It remains for us to see what sustains the electric field E_x which we presupposed was there. As $\mathbf{E}_1$ fluctuates, the ions' inertia causes them to lag behind the electrons, and there is a polarization drift $\mathbf{v}_p$ in the direction of $\mathbf{E}_1$. This drift v_{ix} is given by Eq. (4.120) and causes a current $\mathbf{j}_1$ to flow in the x direction. The resulting $\mathbf{j}_1 \times \mathbf{B}_0$ force on the fluid is in the y direction and is 90° out of phase with the velocity $\mathbf{v}_1$.

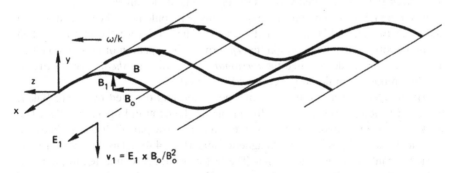

Fig. 4.46 Relation among the oscillating quantities in an Alfvén wave and the (exaggerated) distortion of the lines of force

Fig. 4.47 Geometry of a
shear Alfvén wave in a
cylindrical column

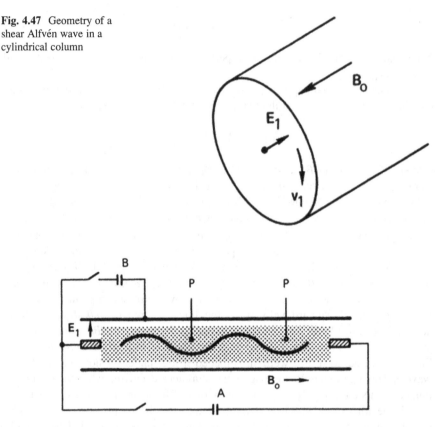

Fig. 4.48 Schematic of an experiment to detect Alfvén waves. [From J. M. Wilcox, F. I. Boley, and A. W. DeSilva, *Phys. Fluids* **3**, 15 (1960).]

This force perpetuates the oscillation in the same way as in any oscillator where the force is out of phase with the velocity. It is, of course, always the ion inertia that causes an overshoot and a sustained oscillation, but in a plasma the momentum is transferred in a complicated way via the electromagnetic forces.

In a more realistic geometry for experiments, E_1 would be in the radial direction and v_1 in the azimuthal direction (Fig. 4.47). The motion of the plasma is then incompressible. This is the reason the ∇p term in the equation of motion could be neglected. This mode is called the *torsional Alfvén wave* or *shear Alfvén wave*. It was first produced in liquid mercury by B. Lehnert.

Alfvén waves in a plasma were first generated and detected by Allen, Baker, Pyle, and Wilcox at Berkeley, California, and by Jephcott in England in 1959. The work was done in a hydrogen plasma created in a "slow pinch" discharge between two electrodes aligned along a magnetic field (Fig. 4.48). Discharge of a slow capacitor bank *A* created the plasma. The fast capacitor *B*, connected to the metal wall, was then fired to create an electric field E_1 perpendicular to B_0. The ringing of the capacitor generated a wave, which was detected, with an appropriate time delay,

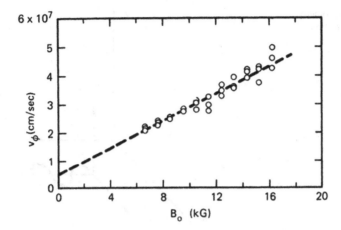

Fig. 4.49 Measured phase velocity of Alfvén waves vs. magnetic field. [From Wilcox, Boley, and DeSilva, *loc. cit.*]

by probes P. Figure 4.49 shows measurements of phase velocity vs. magnetic field, demonstrating the linear dependence predicted by Eq. (4.126).

This experiment was a difficult one, because a large magnetic field of 1 T was needed to overcome damping. With large $\mathbf{B}_0$, v_A (and hence the wavelength), become uncomfortably large unless the density is high. In the experiment of Wilcox *et al.*, a density of 6×10^{21} m^{-3} was used to achieve a low Alfvén speed of 2.8×10^5 m/s. Note that it is not possible to increase ρ by using a heavier atom. The frequency $\omega = k v_A$ is proportional to $M^{-1/2}$, while the cyclotron frequency Ω_c is proportional to M^{-1}. Therefore, the ratio ω/Ω_c is proportional to $M^{1/2}$. With heavier atoms it is not possible to satisfy the condition $\omega^2 \ll \Omega_c^2$.

4.19 Magnetosonic Waves

Finally, we consider low-frequency electromagnetic waves propagating *across* $\mathbf{B}_0$. Again we may take $\mathbf{B}_0 = B_0 \hat{\mathbf{z}}$ and $\mathbf{E}_1 = E_1 \hat{\mathbf{x}}$, but we now let $\mathbf{k} = k \hat{\mathbf{y}}$ (Fig. 4.50). Now we see that the $\mathbf{E}_1 \times \mathbf{B}_0$ drifts lie along $\mathbf{k}$, so that the plasma will be compressed and released in the course of the oscillation. It is necessary, therefore, to keep the ∇p term in the equation of motion. For the ions, we have

$$M n_0 \frac{\partial \mathbf{v}_{i1}}{\partial t} = e n_0 (\mathbf{E}_1 + \mathbf{v}_{i1} \times \mathbf{B}_0) - \gamma_i K T_i \nabla n_1 \tag{4.129}$$

With our choice of $\mathbf{E}_1$ and $\mathbf{k}$, this becomes

$$v_{ix} = \frac{ie}{M\omega} (E_x + v_{iy} B_0) c_s \tag{4.130}$$

Fig. 4.50 Geometry
of a magnetosonic
wave propagating
at right angles to B_0

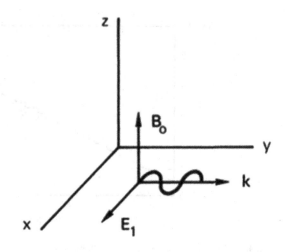

$$v_{iy} = \frac{ie}{M\omega}(-v_{ix}B_0) + \frac{k}{\omega}\frac{\gamma_i KT_i}{M}\frac{n_1}{n_0} \tag{4.131}$$

The equation of continuity yields

$$\frac{n_1}{n_0} = \frac{k}{\omega}v_{iy} \tag{4.132}$$

so that Eq. (4.131) becomes

$$v_{iy} = -\frac{ie}{M\omega}v_{ix}B_0 + \frac{k^2}{\omega^2}\frac{\gamma_i KT_i}{M}v_{iy} \tag{4.133}$$

With the abbreviation

$$A \equiv \frac{k^2}{\omega^2}\frac{\gamma_i KT_i}{M}$$

this becomes

$$v_{iy}(1 - A) = -\frac{i\Omega_c}{\omega}v_{ix} \tag{4.134}$$

Combining this with Eq. (4.130), we have

$$v_{ix} = \frac{ie}{M\omega}E_x + \frac{i\Omega_c}{\omega}\left(-\frac{i\Omega_c}{\omega}\right)(1 - A)^{-1}v_{ix}$$

$$v_{ix}\left(1 - \frac{\Omega_c^2/\omega^2}{1 - A}\right) = \frac{ie}{M\omega}E_x \tag{4.135}$$

This is the only component of v_{i1} we shall need, since the only nontrivial component of the wave equation (4.81) is

$$\varepsilon_0 \left(\omega^2 - c^2 k^2\right) E_x = -i\omega n_0 e (v_{ix} - v_{ex}) \tag{4.136}$$

To obtain v_{ex}, we need only to make the appropriate changes in Eq. (4.135) and take the limit of small electron mass, so that $\omega^2 \ll \omega_c^2$ and $\omega^2 \ll k^2 v_{the}^2$:

$$v_{ex} = \frac{ie}{m\omega} \frac{\omega^2}{\omega_c^2} \left(1 - \frac{k^2}{\omega^2} \frac{\gamma_e KT_e}{m}\right) E_x \rightarrow -\frac{ik^2}{\omega B_0^2} \frac{\gamma_e KT_e}{e} E_x. \tag{4.137}$$

Putting the last three equations together we have

$$\varepsilon_0 \left(\omega^2 - c^2 k^2\right) E_x = -i\omega n_0 e \left[\frac{ie}{M\omega} E_x \left(\frac{1-A}{1-A-(\Omega_c^2/\omega^2)}\right)\right.$$
$$\left. + \frac{ik^2 M}{\omega B_0^2} \frac{\gamma_e KT_e}{eM} E_x\right] \tag{4.138}$$

We shall again assume $\omega^2 \ll \Omega_c^2$, so that $1 - A$ can be neglected relative to Ω_c^2/ω^2. With the help of the definitions of Ω_p and v_A, we have

$$\left(\omega^2 - c^2 k^2\right) = -\frac{\Omega_p^2}{\Omega_c^2} \omega^2 (1 - A) + \frac{k^2 c^2}{v_A^2} \frac{\gamma_e KT_e}{M}$$
$$\omega^2 - c^2 k^2 \left(1 + \frac{\gamma_e KT_e}{M v_A^2}\right) + \frac{\Omega_p^2}{\Omega_c^2} \left(\omega^2 - k^2 \frac{\gamma_i KT_i}{M}\right) = 0 \tag{4.139}$$

Since

$$\Omega_p^2 / \Omega_c^2 = c^2 / v_A^2 \tag{4.140}$$

Eq. (4.139) becomes

$$\omega^2 \left(1 + \frac{c^2}{v_A^2}\right) = c^2 k^2 \left(1 + \frac{\gamma_e KT_e + \gamma_i KT_i}{M v_A^2}\right) = c^2 k^2 \left(1 + \frac{v_s^2}{v_A^2}\right) \tag{4.141}$$

where v_s is the acoustic speed. Finally, we have

$$\boxed{\frac{\omega^2}{k^2} = c^2 \frac{v_s^2 + v_A^2}{c^2 + v_A^2}} \tag{4.142}$$

This is the dispersion relation for the *magnetosonic wave* propagating perpendicular to $\mathbf{B}_0$. It is an acoustic wave in which the compressions and rarefactions are produced not by motions along $\mathbf{E}$, but by $\mathbf{E} \times \mathbf{B}$ drifts across $\mathbf{E}$. In the limit $\mathbf{B}_0 \to 0$, $v_A \to 0$, the magnetosonic wave turns into an ordinary ion acoustic wave. In the limit $KT \to 0$, $v_s \to 0$, the pressure gradient forces vanish, and the wave becomes a modified Alfvén wave. The phase velocity of the magnetosonic mode is almost always larger than v_A; for this reason, it is often called simply the "fast" hydromagnetic wave.

4.20 Summary of Elementary Plasma Waves

Electron waves (*electrostatic*)

$$\mathbf{B}_0 = 0 \ \text{ or } \mathbf{k} \parallel \mathbf{B}_0 : \quad \omega^2 = \omega_p^2 + \frac{3}{2} k^2 v_{\text{th}}^2 \qquad \text{(Plasma oscillations)} \qquad (4.143)$$

$$\mathbf{k} \perp \mathbf{B}_0 : \qquad \omega^2 = \omega_p^2 + \omega_c^2 = \omega_h^2 \qquad \begin{array}{l}\text{(Upper hybrid} \\ \text{oscillations)}\end{array} \qquad (4.144)$$

Ion waves (*electrostatic*)

$$\mathbf{B}_0 = 0 \text{ or } \mathbf{k} \parallel \mathbf{B}_0 : \quad \omega^2 = k^2 v_s^2$$
$$= k^2 \frac{\gamma_e K T_e + \gamma_i K T_i}{M} \ \text{(Acoustic waves)} \qquad (4.145)$$

$$\mathbf{k} \perp \mathbf{B}_0 : \qquad \omega^2 = \Omega_c^2 + k^2 v_s^2 \qquad \begin{array}{l}\text{(Electrostatic ion} \\ \text{cyclotron waves)}\end{array} \qquad (4.146)$$

or

$$\omega^2 = \omega_l^2 = \Omega_c \omega_c \qquad \begin{array}{l}\text{(Lower hybrid} \\ \text{oscillations)}\end{array} \qquad (4.147)$$

Electron waves (*electromagnetic*)

$$\mathbf{B}_0 = 0 : \qquad \omega^2 = \omega_p^2 + k^2 c^2 \qquad \text{(Light waves)} \qquad (4.148)$$

$$\mathbf{k} \perp \mathbf{B}_0, \ \mathbf{E}_1 \parallel \mathbf{B}_0 : \quad \frac{c^2 k^2}{\omega^2} = 1 - \frac{\omega_p^2}{\omega^2} \qquad \text{(O wave)} \qquad (4.149)$$

$$\mathbf{k} \perp \mathbf{B}_0, \ \mathbf{E}_1 \perp \mathbf{B}_0 : \quad \frac{c^2 k^2}{\omega^2} = 1 - \frac{\omega_p^2}{\omega^2} \frac{\omega^2 - \omega_p^2}{\omega^2 - \omega_h^2} \quad \text{(X wave)} \qquad (4.150)$$

$$\mathbf{k} \parallel \mathbf{B}_0 : \qquad \frac{c^2 k^2}{\omega^2} = 1 - \frac{\omega_p^2/\omega^2}{1 - (\omega_c/\omega)} \qquad \begin{array}{l} (R \text{ wave}) \\ (\text{whistler mode}) \end{array} \qquad (4.151)$$

$$\frac{c^2 k^2}{\omega^2} = 1 - \frac{\omega_p^2/\omega^2}{1 + (\omega_c/\omega)} \qquad (L \text{ wave}) \qquad (4.152)$$

Ion waves (electromagnetic)

$\mathbf{B}_0 = 0 :$ None

$$\mathbf{k} \parallel \mathbf{B}_0 : \qquad \omega^2 = k^2 v_A^2 \qquad (\text{Alfvén wave}) \qquad (4.153)$$

$$\mathbf{k} \perp \mathbf{B}_0 : \qquad \frac{\omega^2}{k^2} = c^2 \frac{v_s^2 + v_A^2}{c^2 + v_A^2} \qquad (\text{Magnetosonic wave}) \qquad (4.154)$$

This set of dispersion relations is a greatly simplified one covering only the principal directions of propagation. Nonetheless, it is a very useful set of equations to have in mind as a frame of reference for discussing more complicated wave motions. It is often possible to understand a complex situation as a modification or superposition of these basic modes of oscillation.

4.21 The CMA Diagram

When propagation occurs at an angle to the magnetic field, the phase velocities change with angle. Some of the modes listed above with $\mathbf{k} \parallel \mathbf{B}_0$ and $\mathbf{k} \perp \mathbf{B}_0$ change continuously into each other; other modes simply disappear at a critical angle. This complicated state of affairs is greatly clarified by the Clemmow–Mullaly–Allis (CMA) diagram, so named for its co-inventors by T. H. Stix. Such a diagram is shown in Fig. 4.51. The CMA diagram is valid, however, only for cold plasmas, with $T_i = T_e = 0$. Extension to finite temperatures introduces so much complexity that the diagram is no longer useful.

Figure 4.51 is a plot of ω_c/ω vs. ω_p^2/ω^2 or, equivalently, a plot of magnetic field vs. density. For a given frequency ω, any experimental situation characterized by ω_p and ω_c is denoted by a point on the graph. The total space is divided into sections by the various cutoffs and resonances we have encountered. For instance, the extraordinary wave cutoff at $\omega^2 = \omega_c^2 + \omega_p^2$ is a quadratic relation between ω_c/ω and ω_p^2/ω^2; the resulting parabola can be recognized on Fig. 4.51 as the curve labeled "upper hybrid resonance." These cutoff and resonance curves separate regions of propagation and nonpropagation for the various waves. The sets of waves that can exist in the different regions will therefore be different.

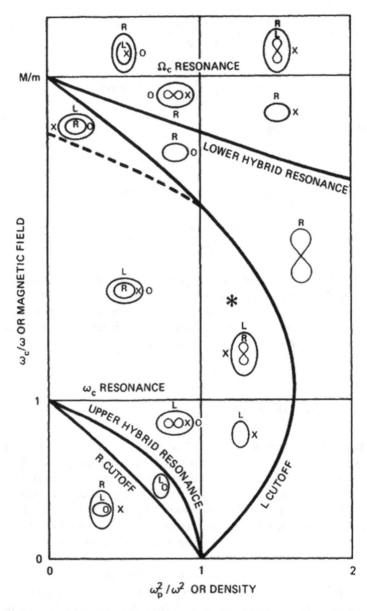

Fig. 4.51 A Clemmow–Mullaly–Allis diagram for classification of waves in a cold plasma

The small diagram in each region indicates not only which waves are present but also how the phase velocity varies qualitatively with angle. The magnetic field is imagined to be vertical on the diagram. The distance from the center to any point on an ellipse or figure-eight at an angle θ to the vertical is proportional to the phase velocity at that angle with respect to the magnetic field. For instance, in the triangular region marked with an * on Fig. 4.51, we see that the L wave becomes the X wave as θ varies from zero to $\pi/2$. The R wave has a velocity smaller than the L wave, and it disappears as θ varies from zero to $\pi/2$. It does not turn into the O wave, because $\omega^2 < \omega_p^2$ in that region, and the O wave does not exist.

The upper regions of the CMA diagram correspond to $\omega \ll \omega_c$. The low-frequency ion waves are found here. Since thermal velocities have been neglected on this diagram, the electrostatic ion waves do not appear; they propagate only in warm plasmas. One can regard the CMA diagram as a "plasma pond": A pebble dropped in each region will send out ripples with shapes like the ones shown.

Problems

4.26. A hydrogen discharge in a 1-T field produces a density of 10^{16} m^{-3}.

   (a) What is the Alfvén speed v_A?
   (b) Suppose v_A had come out greater than c. Does this mean that Alfvén waves travel faster than the speed of light?

4.27. Calculate the Alfvén speed in a region of the magnetosphere where $B = 10^{-8}$ T, $n = 10^8$ m^{-3}, and $M = M_H = 1.67 \times 10^{-27}$ kg.

4.28. Suppose you have created a laboratory plasma with $n = 10^{15}$ m^{-3} and $B = 10^{-2}$ T. You connect a 160-MHz signal generator to a probe inserted into the plasma.

   (a) Draw a CMA diagram and indicate the region in which the experiment is located.
   (b) What electromagnetic waves might be excited and propagated in the plasma?

4.29. Suppose you wish to design an experiment in which standing torsional Alfvén waves are generated in a cylindrical plasma column, so that the standing wave has maximum amplitude at the midplane and nodes at the ends. To satisfy the condition $\omega \ll \Omega_c$, you make $\omega = 0.1\Omega_c$.

   (a) If you could create a hydrogen plasma with $n = 10^{19}$ m^{-3} and $B = 1$ T, how long does the column have to be?
   (b) If you tried to do this with a 0.3 T Q-machine, in which the singly charged Cs ions have an atomic weight 133 and a density $n = 10^{18}$ m^{-3}, how long would the plasma have to be? Hint: Figure out the scaling factors and use the result of part (a).

4.30. A pulsar emits a broad spectrum of electromagnetic radiation, which is detected with a receiver tuned to the neighborhood of $f = 80$ MHz. Because of the dispersion in group velocity caused by the interstellar plasma, the observed frequency during each pulse drifts at a rate given by $df/dt = -5$ MHz/s.

   (a) If the interstellar magnetic field is negligible and $\omega^2 \ll \omega_p^2$, show that

$$\frac{df}{dt} \approx -\frac{c}{x}\frac{f^3}{f_p^2}$$

where f_p is the plasma frequency and x is the distance of the pulsar.

(b) If the average electron density in space is 2×10^5 m^{-3}, how far away is the pulsar? (1 parsec $= 3 \times 10^{16}$ m.)

4.31. A three-component plasma has a density n_0 of electrons, $(1 - \varepsilon)n_0$ of ions of mass M_1, and εn_0 of ions of mass M_2. Let $T_{i1} = T_{i2} = 0$, $T_e \neq 0$.

(a) Derive a dispersion relation for electrostatic ion cyclotron waves.
(b) Find a simple expression for ω^2 when ε is small.
(c) Evaluate the wave frequencies for a case when ε is not small: a 50–50 % D–T mixture at $KT_e = 10$ keV, $B_0 = 5$ T, and $k = 1$ cm^{-1}.

4.32. For a Langmuir plasma oscillation, show that the time-averaged electron kinetic energy per m^3 is equal to the electric field energy density $1/2\ \epsilon_0 \langle E^2 \rangle$.

4.33. For an Alfven wave, show that the time-averaged ion kinetic energy per m^3 is equal to the magnetic wave energy $\langle B_1^2 \rangle / 2\mu_0$.

4.34. Figure P4.34 shows a far-infrared laser operating at $\lambda = 337$ μm. When $B_0 = 0$, this radiation easily penetrates the plasma whenever ω_p is less than ω, or $n < n_c = 10^{22}$ m^{-3}. However, because of the long path length, the defocusing effect of the plasma (cf. Fig. 4.30) spoils the optical cavity, and the density is limited by the conditions $\omega_p^2 < \varepsilon\omega^2$, where $\varepsilon \ll 1$. In the interest of increasing the limiting density, and hence the laser output power, a magnetic field B_0 is added.

(a) If ε is unchanged, show that the limiting density can be increased if left-hand circularly polarized waves are propagated.
(b) If n is to be doubled, how large does B_0 have to be?

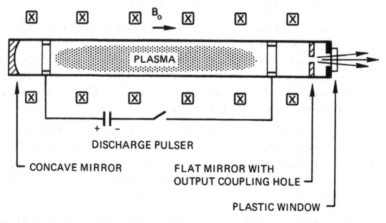

Fig. P4.34 Schematic of a pulsed HCN laser

(c) Show that the plasma is a focusing lens for the whistler mode.

(d) Can one use the whistler mode and therefore go to much higher densities?

4.35. Use Maxwell's equations and the electron equation of motion to derive the dispersion relation for light waves propagating through a uniform, unmagnetized, collisionless, isothermal plasma with density n and finite electron temperature T_e. (Ignore ion motions.)

4.36. Prove that transverse waves are unaffected by the ∇p term whenever $\mathbf{k} \times \mathbf{B}_0 = 0$, even if ion motion is included.

4.37. Consider the damping of an ordinary wave caused by a constant collision frequency v between electrons and ions.

(a) Show that the dispersion relation is

$$\frac{c^2 k^2}{\omega^2} = 1 - \frac{\omega_p^2}{\omega(\omega + iv)}$$

(b) For waves damped in time (k real) when $v/\omega \ll 1$, show that the damping rate $\gamma \equiv -\mathrm{Im}\,(\omega)$ is approximately

$$\gamma = \frac{\omega_p^2}{\omega^2} \frac{v}{2}$$

(c) For waves damped in space (ω real) when $v/\omega \ll 1$, show that the attenuation distance $\delta \equiv (\mathrm{Im}\ k)^{-1}$ is approximately

$$\delta = \frac{2c}{v} \frac{\omega^2}{\omega_p^2} \left(1 - \frac{\omega_p^2}{\omega^2}\right)^{1/2}$$

4.38. It has been proposed to build a solar power station in space with huge panels of solar cells collecting sunlight 24 h a day. The power is transmitted to earth in a 30-cm-wavelength microwave beam. We wish to estimate how much of the power is lost in heating up the ionosphere. Treating the latter as a weakly ionized gas with constant electron-neutral collision frequency, what fraction of the beam power is lost in traversing 100 km of plasma with $n_e = 10^{11}$ m^{-3}, $n_n = 10^{16}$ m^{-3}, and $\overline{\sigma v} = 10^{-14}$m^3/ sec ?

4.39. The Appleton–Hartree dispersion relation for high-frequency electromagnetic waves propagating at an angle θ to the magnetic field is

$$\frac{c^2 k^2}{\omega^2} = 1 - \frac{2\omega_p^2\left(1 - \omega_p^2/\omega^2\right)}{2\omega^2\left(1 - \omega_p^2/\omega^2\right) - \omega_c^2 \sin^2\theta \pm \omega_c \left[\omega_c^2 \sin^4\theta + 4\omega^2\left(1 - \omega_p^2/\omega^2\right)^2 \cos^2\theta\right]^{1/2}}$$

Discuss the cutoffs and resonances of this equation. Which are independent of θ?

4.40. Microwaves with free-space wavelength λ_0 equal to 1 cm are sent through a plasma slab 10 cm thick in which the density and magnetic field are uniform and given by $n_0 = 2.8 \times 10^{18}$ m^{-3} and $B_0 = 1.07$ T. Calculate the number of wavelengths *inside the slab* if (see Fig. P4.40)

    (a) the waveguide is oriented so that $\mathbf{E}_1$ is in the $\hat{z}$ direction;
    (b) the waveguide is oriented so that $\mathbf{E}_1$ is in the $\hat{y}$ direction.

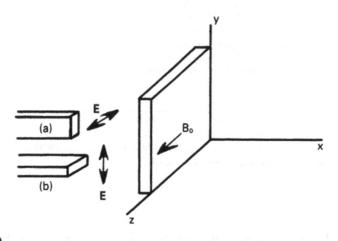

Fig. P4.40

4.41. A cold plasma is composed of positive ions of charge Ze and mass M_+ and negative ions of charge $-e$ and mass M_-. In the equilibrium state, there is no magnetic or electric field and no velocity; and the respective densities are n_{0+} and $n_{0-} = Zn_{0+}$. Derive the dispersion relation for plane electromagnetic waves.

4.42. Ion waves are generated in a gas-discharge plasma in a mixture of argon and helium gases. The plasma has the following constituents:

    (a) Electrons of density n_0 and temperature KT_e;
    (b) Argon ions of density n_A, mass M_A, charge $+Ze$, and temperature 0; and
    (c) He ions of density n_H, mass M_H, charge $+e$, and temperature 0.

    Derive an expression for the phase velocity of the waves using a linearized, one-dimensional theory with the plasma approximation and the Boltzmann relation for electrons.

4.43. In a remote part of the universe, there exists a plasma consisting of positrons and fully stripped antifermium nuclei of charge $-Ze$, where $Z = 100$. From the equations of motion, continuity, and Poisson, derive a dispersion relation for plasma oscillations in this plasma, including ion motions. Define the plasma frequencies. You may assume $KT = 0$, $B_0 = 0$, and all other simplifying initial conditions.

4.44. Intelligent life on a planet in the Crab nebula tries to communicate with us primitive creatures on the earth. We receive radio signals in the 10^8–10^9 Hz range, but the spectrum stops abruptly at 120 MHz. From optical measurements, it is possible to place an upper limit of 36 G on the magnetic field in the vicinity of the parent star. If the star is located in an HII region (one which contains ionized hydrogen), and if the radio signals are affected by some sort of cutoff in the plasma there, what is a reasonable lower limit to the plasma density? ($1 \text{ G} = 10^{-4}$ T.)

4.45. A space ship is moving through the ionosphere of Jupiter at a speed of 100 km/s, parallel to the 10^{-5}-T magnetic field. If the motion is supersonic ($v > v_s$), ion acoustic shock waves would be generated. If, in addition, the motion is super-Alfvénic ($v > v_A$), magnetic shock waves would also be excited. Instruments on board indicate the former but not the latter. Find limits to the plasma density and electron temperature and indicate whether these are upper or lower limits. Assume that the atmosphere of Jupiter contains cold, singly charged molecular ions of H_2, He, CH_4, CO_2, and NH_4 with an average atomic weight of 10.

4.46. An extraordinary wave with frequency ω is incident on a plasma from the outside. The variation of the right-hand cutoff frequency ω_R and the upper hybrid resonance frequency ω_h with radius are as shown. There is an evanescent layer in which the wave cannot propagate. If the density gradient at the point where $\omega \simeq \omega_h$ is given by $|\partial n/\partial r| \simeq n/r_0$, show that the distance d between the $\omega = \omega_R$ and ω_h points is approximately $d = (\omega_c/\omega)r_0$.

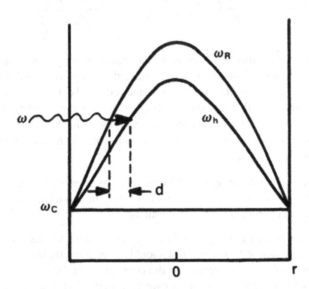

Fig. P4.46

4.47. By introducing a gradient in B_0, it is possible to make the upper hybrid resonance accessible to an X wave sent in from the outside of the plasma (cf. preceding problem).

   (a) Draw on an ω_c/ω vs. ω_p^2/ω^2 diagram the path taken by the wave, showing how the ω_R cutoff is avoided.
   (b) Show that the required change in B_0 between the plasma surface and the upper hybrid layer is

$$\Delta B_0 = B_0 \omega_p^2 / 2\omega_c^2$$

4.48. A certain plasma wave has the dispersion relation

$$\frac{c^2 k^2}{\omega^2} = 1 - \frac{\overline{\omega}_p^2}{\omega^2 - \omega_c \Omega_c + \frac{\omega^2 (\omega_c - \Omega_c)^2}{\overline{\omega}_p^2 - \omega^2 + \omega_c \Omega_c}}$$

where $\overline{\omega}^2 \equiv \omega_p^2 + \Omega_p^2$. Write explicit expressions for the resonance and cutoff frequencies (or for the squares thereof), when $\varepsilon \equiv m/M \ll 1$.

4.49. The extraordinary wave with ion motions included has the following dispersion relation:

$$\frac{c^2 k^2}{\omega^2} = 1 - \frac{\omega_p^2}{\omega^2 - \omega_c^2} - \frac{\Omega_p^2}{\omega^2 - \Omega_c^2} - \frac{\left(\frac{\omega_c}{\omega} \frac{\omega_p^2}{\omega^2 - \omega_c^2} - \frac{\Omega_c}{\omega} \frac{\Omega_p^2}{\omega^2 - \Omega_c^2} \right)^2}{1 - \frac{\omega_p^2}{\omega^2 - \omega_c^2} - \frac{\Omega_p^2}{\omega^2 - \Omega_c^2}}$$

   (a) Show that this is identical to the equation in the previous problem. (Warning: this problem may be hazardous to your mental health.)
   (b) If ω_l and ω_L are the lower hybrid and left-hand cutoff frequencies of this wave, show that the ordering $\Omega_c \leq \omega_l \leq \omega_L$ is always obeyed.
   (c) Using these results and the known phase velocity in the $\omega \to 0$ limit, draw a qualitative v_ϕ^2/c^2 vs. (ω plot showing the regions of propagation and evanescence).

4.50. We wish to do lower-hybrid heating of a hydrogen plasma column with $\omega_p = 0$ at $r = a$ and $\omega_p = \frac{1}{2}\omega_c$ at the center, in a uniform magnetic field. The antenna launches an X wave with $k_\parallel = 0$.

   (a) Draw a qualitative plot of ω_c, Ω_c, ω_L, and ω_l vs. radius. This graph should not be to scale, but it should show correctly the *relative* magnitudes of these frequencies at the edge and center of the plasma.
   (b) Estimate the thickness of the evanescent layer between ω_l and ω_L (*cf.* the previous problem) if the rf frequency ω is set equal to ω_l at the center.
   (c) Repeat (a) and (b) for ω_p (max) $= 2\omega_c$, and draw a conclusion about this antenna design.

4.51. The electromagnetic ion cyclotron wave (Stix wave) is sometimes used for radiofrequency heating of fusion plasmas. Derive the dispersion relation as follows:

(a) Derive a wave equation in the form of Eq. (4.118) but with displacement current neglected.

(b) Write the x and y components of this equation assuming $k_x = 0$, $k^2 = k_y^2 + k_z^2$, and $k_y k_z E_z \approx 0$.

(c) To evaluate $\mathbf{j}_1 = n_0 e(\mathbf{v}_i - \mathbf{v}_e)$, derive the ion equivalent of Eq. (4.98) to obtain $\mathbf{v}_i$, and make a low-frequency approximation so that $\mathbf{v}_e$ is simply the $\mathbf{E} \times \mathbf{B}$ drift.

(d) Insert the result of (c) into (b) to obtain two simultaneous homogeneous equations for E_x and E_y, using the definition for Ω_p in Eq. (4.49).

(e) Set the determinant to zero and solve to lowest order in Ω_p^2 to obtain

$$\omega^2 = \Omega_c^2 \left[1 + \frac{\Omega_p^2}{c^2} \left(\frac{1}{k_z^2} + \frac{1}{k^2} \right) \right]^{-1}$$

4.52. Compute the damping rate of light waves in a plasma due to electron collisions with ions and neutrals at frequency ν by adding a term $-m\nu\mathbf{v}_{e1}$ to the equation of motion in Eq. (4.83). It is sufficient to show that, for small damping, Eq. (4.85) is replaced by

$$\omega^2 - c^2 k^2 = \omega_p^2 \left(1 - i\nu/\omega_p \right).$$

4.53. The *two-ion hybrid* frequency of S.J. Buchsbaum [Phys. Fluids **3**, 418 (1960)] has the cold-plasma dispersion relation

$$\omega^2 = \Omega_{c1} \Omega_{c2} \frac{\alpha_1 \Omega_{c2} + \alpha_2 \Omega_{c1}}{\alpha_1 \Omega_{c1} + \alpha_2 \Omega_{c2}}$$

in the limit of perpendicular propagation ($\tan^2\theta \to \infty$). Here each ion species has cyclotron frequency Ω_{cj} and fractional density α_j.

(a) Derive the equation by setting $S = 0$ in Eq. (B-29) in Appendix B.

(b) In Fig. P4.53, which of (a) or (b) shows the guiding-center orbits in lower hybrid resonance, and which shows the orbits in two-ion hybrid resonance?

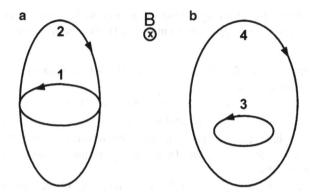

Fig. P4.53

(c) Which orbit in the lower hybrid case corresponds to the ions, and which to the electrons?

(d) Which orbit in the two-ion hybrid case corresponds to the majority species, and which to the minority species? (This is non-trivial.)

Chapter 5
Diffusion and Resistivity

5.1 Diffusion and Mobility in Weakly Ionized Gases

The infinite, homogeneous plasmas assumed in the previous chapter for the equilibrium conditions are, of course, highly idealized. Any realistic plasma will have a density gradient, and the plasma will tend to diffuse toward regions of low density. The central problem in controlled thermonuclear reactions is to impede the rate of diffusion by using a magnetic field. Before tackling the magnetic field problem, however, we shall consider the case of diffusion in the absence of magnetic fields. A further simplification results if we assume that the plasma is weakly ionized, so that charge particles collide primarily with neutral atoms rather than with one another. The case of a fully ionized plasma is deferred to a later section, since it results in a nonlinear equation for which there are few simple illustrative solutions. In any case, partially ionized gases are not rare: High-pressure arcs and ionospheric plasmas fall into this category, and most of the early work on gas discharges involved fractional ionizations between 10^{-3} and 10^{-6}, when collisions with neutral atoms are dominant.

The picture, then, is of a nonuniform distribution of ions and electrons in a dense background of neutrals (Fig. 5.1). As the plasma spreads out as a result of pressure-gradient and electric field forces, the individual particles undergo a random walk, colliding frequently with the neutral atoms. We begin with a brief review of definitions from atomic theory.

5.1.1 Collision Parameters

When an electron, say, collides with a neutral atom, it may lose any fraction of its initial momentum, depending on the angle at which it rebounds. In a head-on

The original version of this chapter was revised. An erratum to this chapter can be found at https://doi.org/10.1007/978-3-319-22309-4_11

© Springer International Publishing Switzerland 2016
F.F. Chen, *Introduction to Plasma Physics and Controlled Fusion*,
DOI 10.1007/978-3-319-22309-4_5

Fig. 5.1 Diffusion of gas
atoms by random collisions

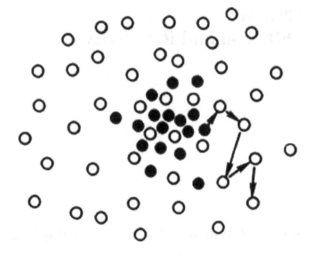

Fig. 5.2 Illustration of the
definition of cross section

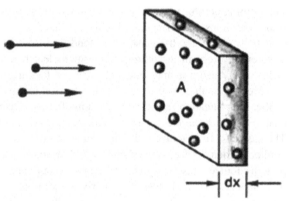

collision with a heavy atom, the electron can lose twice its initial momentum, since
its velocity reverses sign after the collision. The probability of momentum loss can
be expressed in terms of the equivalent cross section σ that the atoms would have if
they were perfect absorbers of momentum.

In Fig. 5.2, electrons are incident upon a slab of area A and thickness dx
containing n_n neutral atoms per m^3. The atoms are imagined to be opaque spheres
of cross-sectional area σ; that is, every time an electron
comes within the area blocked by the atom, the electron loses all of its momentum. The number of atoms in the slab is

$$n_n A\, dx$$

The fraction of the slab blocked by atoms is

$$n_n A \sigma\, dx / A = n_n \sigma\, dx$$

If a flux Γ of electrons is incident on the slab, the flux emerging on the other side is

$$\Gamma' = \Gamma(1 - n_n \sigma \, dx)$$

Thus the change of Γ with distance is

$$d\Gamma/dx = -n_n \sigma \Gamma$$

or

$$\Gamma = \Gamma_0 e^{-n_n \sigma x} \equiv \Gamma_0 e^{-x/\lambda_m} \tag{5.1}$$

In a distance λ_m, the flux would be decreased to $1/e$ of its initial value. The quantity λ_m is the *mean free path* for collisions:

$$\boxed{\lambda_m = 1/n_n \sigma} \tag{5.2}$$

After traveling a distance λ_m, a particle will have had a good probability of making a collision. The mean time between collisions, for particles of velocity v, is given by

$$\tau = \lambda_m/v$$

and the mean frequency of collisions is

$$\tau^{-1} = v/\lambda_m = n_n \sigma v \tag{5.3}$$

If we now average over particles of all velocities v in a Maxwellian distribution, we have what is generally called the collision frequency ν:

$$\nu = n_n \overline{\sigma v} \tag{5.4}$$

5.1.2 Diffusion Parameters

The fluid equation of motion including collisions is, for any species,

$$m n \frac{d\mathbf{v}}{dt} = m n \left[\frac{\partial \mathbf{v}}{\partial t} + (\mathbf{v} \cdot \nabla)\mathbf{v} \right] = \pm en\mathbf{E} - \nabla p - mn\nu\mathbf{v} \tag{5.5}$$

where again the $\pm$ indicates the sign of the charge. The averaging process used to compute ν is such as to make Eq. (5.5) correct; we need not be concerned with the details of this computation. The quantity ν must, however, be assumed to be a constant in order for Eq. (5.5) to be useful. We shall consider a steady state in which $\partial \mathbf{v}/\partial t = 0$. If $\mathbf{v}$ is sufficiently small (or ν sufficiently large), a fluid element will not move into regions of different $\mathbf{E}$ and ∇p in a collision time, and the convective derivative $d\mathbf{v}/dt$ will also vanish. Setting the left-hand side of Eq. (5.5) to zero, we have, for an isothermal plasma,

$$
\begin{aligned}
\mathbf{v} &= \frac{1}{mn\nu}(\pm en\,\mathbf{E} - KT\nabla n) \\
&= \pm\frac{e}{m\nu}\mathbf{E} - \frac{KT}{m\nu}\frac{\nabla n}{n}
\end{aligned}
\tag{5.6}
$$

The coefficients above are called the *mobility* and the *diffusion coefficient*:

$$
\boxed{\mu \equiv |q|/m\nu} \quad \text{Mobility} \tag{5.7}
$$

$$
\boxed{D \equiv KT/m\nu} \quad \text{Diffusion coefficient} \tag{5.8}
$$

These will be different for each species. Note that D is measured in m^2/s. The transport coefficients μ and D are connected by the *Einstein relation*:

$$
\boxed{\mu = |q|D/KT} \tag{5.9}
$$

With the help of these definitions, the flux Γ_j of the jth species can be written

$$
\boxed{\Gamma_j = n\mathbf{v}_j = \pm\mu_j n\,\mathbf{E} - D_j\nabla n} \tag{5.10}
$$

Fick's law of diffusion is a special case of this, occurring when either $\mathbf{E} = 0$ or the particles are uncharged, so that $\mu = 0$:

$$
\boxed{\Gamma = -D\nabla n} \quad \text{Fick's law} \tag{5.11}
$$

This equation merely expresses the fact that diffusion is a random-walk process, in which a net flux from dense regions to less dense regions occurs simply because more particles start in the dense region. This flux is obviously proportional to the gradient of the density. In plasmas, Fick's law is not necessarily obeyed. Because of the possibility of organized motions (plasma waves), a plasma may spread out in a manner which is not truly random.

5.2   Decay of a Plasma by Diffusion

5.2.1   Ambipolar Diffusion

We now consider how a plasma created in a container decays by diffusion to the walls. Once ions and electrons reach the wall, they recombine there. The density near the wall, therefore, is essentially zero. The fluid equations of motion and continuity govern the plasma behavior; but if the decay is slow, we need only keep the time derivative in the continuity equation. The time derivative in the equation of motion, Eq. (5.5), will be negligible if the collision frequency ν is large. We thus have

$$\frac{\partial n}{\partial t} + \nabla \cdot \mathbf{\Gamma}_j = 0 \tag{5.12}$$

with $\mathbf{\Gamma}_j$ given by Eq. (5.10). It is clear that if $\mathbf{\Gamma}_i$ and $\mathbf{\Gamma}_e$ were not equal, a serious charge imbalance would soon arise. If the plasma is much larger than a Debye length, it must be quasineutral; and one would expect that the rates of diffusion of ions and electrons would somehow adjust themselves so that the two species leave at the same rate. How this happens is easy to see. The electrons, being lighter, have higher thermal velocities and tend to leave the plasma first. A positive charge is left behind, and an electric field is set up of such a polarity as to retard the loss of electrons and accelerate the loss of ions. The required $\mathbf{E}$ field is found by setting $\mathbf{\Gamma}_i = \mathbf{\Gamma}_e = \mathbf{\Gamma}$. From Eq. (5.10), we can write

$$\mathbf{\Gamma} = \mu_i n \mathbf{E} - D_i \nabla n = -\mu_e n \mathbf{E} - D_e \nabla n \tag{5.13}$$

$$\mathbf{E} = \frac{D_i - D_e}{\mu_i + \mu_e} \frac{\nabla n}{n} \tag{5.14}$$

The common flux $\mathbf{\Gamma}$ is then given by

$$\begin{aligned}
\mathbf{\Gamma} &= \mu_i \frac{D_i - D_e}{\mu_i + \mu_e} \nabla n - D_i \nabla n \\
&= \frac{\mu_i D_i - \mu_i D_e - \mu_i D_i - \mu_e D_i}{\mu_i + \mu_e} \nabla n \\
&= -\frac{\mu_i D_e + \mu_e D_i}{\mu_i + \mu_e} \nabla n
\end{aligned} \tag{5.15}$$

This is Fick's law with a new diffusion coefficient

$$\boxed{D_a \equiv \frac{\mu_i D_e + \mu_e D_i}{\mu_i + \mu_e}} \tag{5.16}$$

called the *ambipolar diffusion coefficient*. If this is constant, Eq. (5.12) becomes simply

$$\partial n / \partial t = D_a \nabla^2 n \tag{5.17}$$

The magnitude of D_a can be estimated if we take $\mu_e \gg \mu_i$. That this is true can be seen from Eq. (5.7). Since ν is proportional to the thermal velocity, which is proportional to $m^{-1/2}$, μ is proportional to $m^{-1/2}$. Equations (5.16) and (5.9) then give

$$D_a \approx D_i + \frac{\mu_i}{\mu_e} D_e = D_i + \frac{T_e}{T_i} D_i \qquad (5.18)$$

For $T_e = T_i$, we have

$$D_a \approx 2D_i \qquad (5.19)$$

The effect of the ambipolar electric field is to enhance the diffusion of ions by a factor of two, and the diffusion rate of the two species together is primarily controlled by the slower species.

5.2.2   Diffusion in a Slab

The diffusion equation (5.17) can easily be solved by the method of separation of variables. We let

$$n(\mathbf{r}, t) = T(t)S(\mathbf{r}) \qquad (5.20)$$

whereupon Eq. (5.17), with the subscript on D_a understood, becomes

$$S\frac{dT}{dt} = DT\,\nabla^2 S \qquad (5.21)$$

$$\frac{1}{T}\frac{dT}{dt} = \frac{D}{S}\nabla^2 S \qquad (5.22)$$

Since the left side is a function of time alone and the right side a function of space alone, they must both be equal to the same constant, which we shall call $-1/\tau$. The function T then obeys the equation

$$\frac{dT}{dt} = -\frac{T}{\tau} \qquad (5.23)$$

with the solution

$$T = T_0 e^{-t/\tau} \qquad (5.24)$$

The spatial part S obeys the equation

$$\nabla^2 S = -\frac{1}{D\tau}S \qquad (5.25)$$

Fig. 5.3 Density of a
plasma at various times
as it decays by diffusion
to the walls

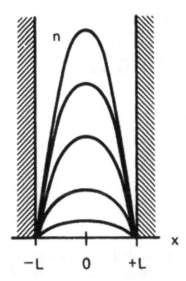

In slab geometry, this becomes

$$\frac{d^2 S}{dx^2} = -\frac{1}{D\tau} S \tag{5.26}$$

with the solution

$$S = A\ \cos\frac{x}{(D\tau)^{1/2}} + B\ \sin\frac{x}{(D\tau)^{1/2}} \tag{5.27}$$

We would expect the density to be nearly zero at the walls (Fig. 5.3) and to have one
or more peaks in between. The simplest solution is that with a single maximum. By
symmetry, we can reject the odd (sine) term in Eq. (5.27). The boundary conditions
$S = 0$ at $x = \pm L$ then requires

$$\frac{L}{(D\tau)^{1/2}} = \frac{\pi}{2}$$

or

$$\tau = \left(\frac{2L}{\pi}\right)^2 \frac{1}{D} \tag{5.28}$$

Combining Eqs. (5.20), (5.24), (5.27), and (5.28), we have

$$n = n_0 e^{-t/\tau} \cos\frac{\pi x}{2L} \tag{5.29}$$

Fig. 5.4 Decay of an
initially nonuniform
plasma, showing the rapid
disappearance of the higher-
order diffusion modes

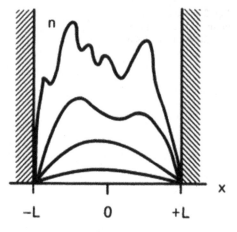

This is called the *lowest diffusion mode*. The density distribution is a cosine, and the
peak density decays exponentially with time. The time constant τ increases with
L and varies inversely with D, as one would expect.

There are, of course, higher diffusion modes with more than one peak. Suppose
the initial density distribution is as shown by the top curve in Fig. 5.4. Such an
arbitrary distribution can be expanded in a Fourier series:

$$n = n_0 \left(\sum_l a_l \cos \frac{\left(l + \frac{1}{2}\right)\pi x}{L} + \sum_m b_m \sin \frac{m\pi x}{L} \right) \tag{5.30}$$

We have chosen the indices so that the boundary condition at $x = \pm L$ is automatically satisfied. To treat the time dependence, we can try a solution of the form

$$n = n_0 \left(\sum_l a_l e^{-t/\tau_l} \cos \frac{\left(l + \frac{1}{2}\right)\pi x}{L} + \sum_m b_m e^{-t/\tau_m} \sin \frac{m\pi x}{L} \right) \tag{5.31}$$

Substituting this into the diffusion equation (5.17), we see that each cosine term
yields a relation of the form

$$-\frac{1}{\tau_l} = -D \left[\left(l + \frac{1}{2}\right)\frac{\pi}{L} \right]^2 \tag{5.32}$$

and similarly for the sine terms. Thus the decay time constant for the lth mode is
given by

$$\tau_l = \left[\frac{L}{\left(l + \frac{1}{2}\right)\pi} \right]^2 \frac{1}{D} \tag{5.33}$$

The fine-grained structure of the density distribution, corresponding to large
l numbers, decays faster, with a smaller time constant τ_l. The plasma decay will

proceed as indicated in Fig. 5.4. First, the fine structure will be washed out by diffusion. Then the lowest diffusion mode, the simple cosine distribution of Fig. 5.3, will be reached. Finally, the peak density continues to decay while the plasma density profile retains the same shape.

5.2.3 Diffusion in a Cylinder

The spatial part of the diffusion equation, Eq. (5.25), reads, in cylindrical geometry,

$$\frac{d^2S}{dr^2} + \frac{1}{r}\frac{dS}{dr} + \frac{1}{D\tau}S = 0 \tag{5.34}$$

This differs from Eq. (5.26) by the addition of the middle term, which merely accounts for the change in coordinates. The need for the extra term is illustrated simply in Fig. 5.5. If a slice of plasma in (a) is moved toward larger x without being allowed to expand, the density would remain constant. On the other hand, if a shell of plasma in (b) is moved toward larger r with the shell thickness kept constant, the density would necessarily decrease as $1/r$. Consequently, one would expect the solution to Eq. (5.34) to be like a damped cosine (Fig. 5.6). This function is called a *Bessel function of order zero,* and Eq. (5.34) is called Bessel's equation (of order zero). Instead of the symbol cos, it is given the symbol J_0. The function $J_0(r/[D\tau]^{1/2})$ is a solution to Eq. (5.34), just as $\cos[x/(D\tau)^{1/2}]$ is a solution to Eq. (5.26). Both cos kx and $J_0(kr)$ are expressible in terms of infinite series and may be found in mathematical tables. Unfortunately, Bessel functions are not yet found in most hand calculators.

To satisfy the boundary condition $n = 0$ at $r = a$, we must set $a/(D\tau)^{1/2}$ equal to the first zero of J_0; namely, 2.4. This yields the decay time constant τ. The plasma again decays exponentially, since the temporal part of the diffusion equation, Eq. (5.23), is unchanged. We have described the lowest diffusion mode in a

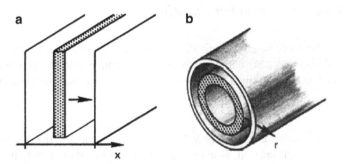

Fig. 5.5 Motion of a plasma slab in rectilinear and cylindrical geometry, illustrating the difference between a cosine and a Bessel function

Fig. 5.6 The Bessel
function of order zero

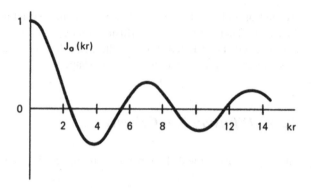

Fig. 5.6 The Bessel
function of order zero

cylinder. Higher diffusion modes, with more than one maximum in the cylinder, will be given in terms of Bessel functions of higher order, in direct analogy to the case of slab geometry.

5.3   Steady State Solutions

In many experiments, a plasma is maintained in a steady state by continuous ionization or injection of plasma to offset the losses. To calculate the density profile in this case, we must add a source term to the equation of continuity:

$$\frac{\partial n}{\partial t} - D\nabla^2 n = Q(\mathbf{r}) \tag{5.35}$$

The sign is chosen so that when Q is positive, it represents a source and contributes to positive $\partial n/\partial t$. In steady state, we set $\partial n/\partial t = 0$ and are left with a Poisson-type equation for $n(\mathbf{r})$.

5.3.1   Constant Ionization Function

In many weakly ionized gases, ionization is produced by energetic electrons in the tail of the Maxwellian distribution. In this case, the source term Q is proportional to the electron density n. Setting $Q = Zn$, where Z is the "ionization function," we have

$$\nabla^2 n = -(Z/D)n \tag{5.36}$$

This is the same equation as that for S, Eq. (5.25). Consequently, the density profile is a cosine or Bessel function, as in the case of a decaying plasma, only in this case the density remains constant. The plasma is maintained against diffusion losses by whatever heat source keeps the electron temperature at its constant value and by a small influx of neutral atoms to replenish those that are ionized.

5.3.2   Plane Source

We next consider what profile would be obtained in slab geometry if there is a localized source on the plane $x = 0$. Such a source might be, for instance, a slit-collimated beam of ultraviolet light strong enough to ionize the neutral gas. The steady state diffusion equation is then

$$\frac{d^2 n}{dx^2} = -\frac{Q}{D}\delta(0)$$ (5.37)

Except at $x = 0$, the density must satisfy $\partial^2 n / \partial x^2 = 0$. This obviously has the solution (Fig. 5.7)

$$n = n_0\left(1 - \frac{|x|}{L}\right)$$ (5.38)

The plasma has a linear profile. The discontinuity in slope at the source is characteristic of δ-function sources.

5.3.3   Line Source

Finally, we consider a cylindrical plasma with a source located on the axis. Such a source might, for instance, be a beam of energetic electrons producing ionization along the axis. Except at $r = 0$, the density must satisfy

$$\frac{1}{r}\frac{\partial}{\partial r}\left(r\frac{\partial n}{\partial r}\right) = 0$$ (5.39)

Fig. 5.7 The triangular density profile resulting from a plane source under diffusion

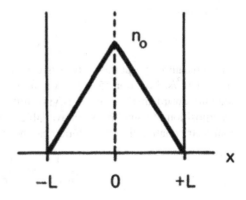

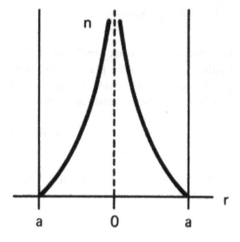

Fig. 5.8 The logarithmic density profile resulting from a line source under diffusion

The solution that vanishes at $r = a$ is

$$n = n_0 \ln(a/r) \tag{5.40}$$

The density becomes infinite at $r = 0$ (Fig. 5.8); it is not possible to determine the density near the axis accurately without considering the finite width of the source.

5.4   Recombination

When an ion and an electron collide, particularly at low relative velocity, they have a finite probability of recombining into a neutral atom. To conserve momentum, a third body must be present. If this third body is an emitted photon, the process is called *radiative recombination*. If it is a particle, the process is called *three-body recombination*. The loss of plasma by recombination can be represented by a negative source term in the equation of continuity. It is clear that this term will be proportional to $n_e n_i = n^2$. In the absence of the diffusion terms, the equation of continuity then becomes

$$\partial n / \partial t = -\alpha n^2 \tag{5.41}$$

The constant of proportionality α is called the *recombination coefficient* and has units of m^3/s. Equation (5.41) is a nonlinear equation for n. This means that the straightforward method for satisfying initial and boundary conditions by linear superposition of solutions is not available. Fortunately, Eq. (5.41) is such a simple nonlinear equation that the solution can be found by inspection. It is

$$\frac{1}{n(\mathbf{r}, t)} = \frac{1}{n_0(\mathbf{r})} + \alpha t \tag{5.42}$$

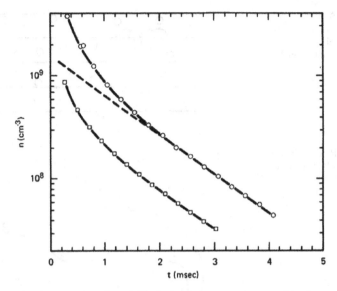

Fig. 5.9 Density decay curves of a weakly ionized plasma under recombination and diffusion (From S. C. Brown, *Basic Data of Plasma Physics,* John Wiley and Sons, New York, 1959)

where $n_0(\mathbf{r})$ is the initial density distribution. It is easily verified that this satisfies Eq. (5.41). After the density has fallen far below its initial value, it decays *reciprocally* with time:

$$n \; \propto 1/\alpha t \qquad\qquad (5.43)$$

This is a fundamentally different behavior from the case of diffusion, in which the time variation is exponential.

Figure 5.9 shows the results of measurements of the density decay in the afterglow of a weakly ionized H plasma. When the density is high, recombination, which is proportional to n^2, is dominant, and the density decays reciprocally. After the density has reached a low value, diffusion becomes dominant, and the decay is thenceforth exponential.

5.5   Diffusion Across a Magnetic Field

The rate of plasma loss by diffusion can be decreased by a magnetic field; this is the problem of confinement in controlled fusion research. Consider a *weakly ionized* plasma in a magnetic field (Fig. 5.10). Charged particles will move along **B** by diffusion and mobility according to Eq. (5.10), since **B** does not affect motion in the parallel direction. Thus we have, for each species,

Fig. 5.10 A charged
particle in a magnetic field
will gyrate about the same
line of force until it makes a
collision

Fig. 5.11 Particle drifts in
a cylindrically symmetric
plasma column do not lead
to losses

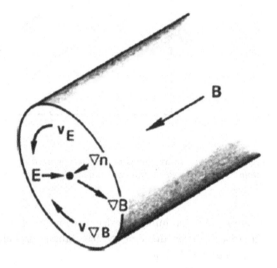

$$\Gamma_z = \pm \mu n E_z - D \frac{\partial n}{\partial z} \qquad (5.44)$$

If there were no collisions, particles would not diffuse at all in the perpendicular
direction—they would continue to gyrate about the same-line of force. There are, of
course, particle drifts across **B** because of electric fields or gradients in **B**, but these
can be arranged to be parallel to the walls. For instance, in a perfectly symmetrie
cylinder (Fig. 5.11), the gradients are all in the radial direction, so that the guiding
center drifts are in the azimuthal direction. The drifts would then be harmless.

When there are collisions, particles migrate across **B** to the walls along the
gradients. They do this by a random-walk process (Fig. 5.12). When an ion, say,
collides with a neutral atom, the ion leaves the collision traveling in a different
direction. It continues to gyrate about the magnetic field in the same direction, but
its phase of gyration is changed discontinuously. (The Larmor radius may also
change, but let us suppose that the ion does not gain or lose energy on the average.)

The guiding center, therefore, shifts position in a collision and undergoes a
random walk. The particles will diffuse in the direction opposite ∇n. The step
length in the random walk is no longer λ_m, as in magnetic-field-free diffusion, but
has instead the magnitude of the Larmor radius r_L. Diffusion across **B** can therefore
be slowed down by decreasing r_L; that is, by increasing **B**.

Fig. 5.12 Diffusion of
gyrating particles by
collisions with neutral
atoms

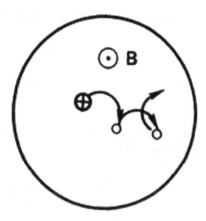

To see how this comes about, we write the perpendicular component of the fluid equation of motion for either species as follows:

$$mn\frac{d\mathbf{v}_\perp}{dt} = \pm en(\mathbf{E} + \mathbf{v}_\perp \times \mathbf{B}) - KT\nabla n - mn\nu\mathbf{v} = 0 \qquad (5.45)$$

We have again assumed that the plasma is isothermal and that ν is large enough for the $d\mathbf{v}_\perp/dt$ term to be negligible. The x and y components are

$$mn\nu v_x = \pm enE_x - KT\frac{\partial n}{\partial x} \pm env_y B$$

$$mn\nu v_y = \pm enE_y - KT\frac{\partial n}{\partial y} \mp env_x B \qquad (5.46)$$

Using the definitions of μ and D, we have

$$v_x = \pm \mu E_x - \frac{D}{n}\frac{\partial n}{\partial x} \pm \frac{\omega_c}{\nu}v_y$$

$$v_y = \pm \mu E_y - \frac{D}{n}\frac{\partial n}{\partial y} \mp \frac{\omega_c}{\nu}v_x \qquad (5.47)$$

Substituting for v_x, we may solve for v_y:

$$v_y(1 + \omega_c^2\tau^2) = \pm \mu E_y - \frac{D}{n}\frac{\partial n}{\partial y} - \omega_c^2\tau^2\frac{E_x}{B} \pm \omega_c^2\tau^2\frac{KT}{eB}\frac{1}{n}\frac{\partial n}{\partial x} \qquad (5.48)$$

where $\tau = \nu^{-1}$. Similarly, v_x is given by

$$v_x(1 + \omega_c^2\tau^2) = \pm \mu E_x - \frac{D}{n}\frac{\partial n}{\partial x} + \omega_c^2\tau^2\frac{E_y}{B} \mp \omega_c^2\tau^2\frac{KT}{eB}\frac{1}{n}\frac{\partial n}{\partial y} \qquad (5.49)$$

The last two terms of these equations contain the $\mathbf{E} \times \mathbf{B}$ and diamagnetic drifts:

$$v_{Ex} = \frac{E_y}{B} \qquad v_{Ey} = -\frac{E_x}{B}$$

$$v_{Dx} = \mp \frac{KT}{eB} \frac{1}{n} \frac{\partial n}{\partial y} \qquad v_{Dy} = \pm \frac{KT}{eB} \frac{1}{n} \frac{\partial n}{\partial x} \tag{5.50}$$

The first two terms can be simplified by defining the perpendicular mobility and diffusion coefficients:

$$\mu_\perp = \frac{\mu}{1 + \omega_c^2 \tau^2} \qquad \boxed{D_\perp = \frac{D}{1 + \omega_c^2 \tau^2}} \tag{5.51}$$

With the help of Eqs. (5.50) and (5.51), we can write Eqs. (5.48) and (5.49) as

$$\mathbf{v}_\perp = \pm \mu_\perp \mathbf{E} - D_\perp \frac{\nabla n}{n} + \frac{\mathbf{v}_E + \mathbf{v}_D}{1 + \left(v^2 / \omega_c^2 \right)} \tag{5.52}$$

From this, it is evident that the perpendicular velocity of either species is composed of two parts. First, there are usual $\mathbf{v}_E$ and $\mathbf{v}_D$ drifts *perpendicular* to the gradients in potential and density. These drifts are slowed down by collisions with neutrals; the drag factor $1 + \left(v^2 / \omega_c^2 \right)$ becomes unity when $v \to 0$. Second, there are the mobility and diffusion drifts *parallel* to the gradients in potential and density. These drifts have the same form as in the $B = 0$ case, but the coefficients μ and D are reduced by the factor $1 + \omega_c^2 \tau^2$.

The product $\omega_c \tau$ is an important quantity in magnetic confinement. When $\omega_c^2 \tau^2 \ll 1$, the magnetic field has little effect on diffusion. When $\omega_c^2 \tau^2 \gg 1$, the magnetic field significantly retards the rate of diffusion across $\mathbf{B}$. The following alternative forms for $\omega_c \tau$ can easily be verified:

$$\omega_c \tau = \omega_c / v = \mu B \cong \lambda_m / r_L \tag{5.53}$$

In the limit $\omega_c^2 \tau^2 \gg 1$, we have

$$D_\perp = \frac{KT}{mv} \frac{1}{\omega_c^2 \tau^2} = \frac{KT v}{m \omega_c^2} \tag{5.54}$$

Comparing with Eq. (5.8), we see that the role of the collision frequency v has been reversed. In diffusion parallel to $\mathbf{B}$, D is proportional to v^{-1}, since collisions retard the motion. In diffusion perpendicular to $\mathbf{B}$, $D_\perp$ is proportional to v, since collisions are needed for cross-field migration. The dependence on m has also been reversed. Keeping in mind that v is proportional to $m^{-1/2}$, we see that $D \propto m^{-1/2}$, while $D_\perp \propto m^{1/2}$. In parallel diffusion, electrons move faster than ions because of their higher thermal velocity; in perpendicular diffusion, electrons escape more slowly because of their smaller Larmor radius.

Disregarding numerical factors of order unity, we may write Eq. (5.8) as

$$D = KT/m\nu \sim v_{\text{th}}^2\tau \sim \lambda_m^2/\tau \tag{5.55}$$

This form, the square of a length over a time, shows that diffusion is a random-walk process with a step length λ_m. Equation (5.54) can be written

$$D_\perp = \frac{KT\nu}{m\omega_c^2} \sim v_{\text{th}}^2\frac{r_L^2}{v_{\text{th}}^2}\nu \sim \frac{r_L^2}{\tau} \tag{5.56}$$

This shows that perpendicular diffusion is a random-walk process with a step length r_L, rather than λ_m.

5.5.1 Ambipolar Diffusion Across B

Because the diffusion and mobility coefficients are anisotropic in the presence of a magnetic field, the problem of ambipolar diffusion is not as straightforward as in the $B = 0$ case. Consider the particle fluxes perpendicular to **B** (Fig. 5.13). Ordinarily, since $\Gamma_{e\perp}$ is smaller than $\Gamma_{i\perp}$, a transverse electric field would be set up so as to aid electron diffusion and retard ion diffusion. However, this electric field can be short-circuited by an imbalance of the fluxes *along* **B**. That is, the negative charge resulting from $\Gamma_{e\perp} < \Gamma_{i\perp}$ can be dissipated by electrons escaping along the field lines. Although the total diffusion must be ambipolar, the perpendicular part of the losses need not be ambipolar. The ions can diffuse out primarily radially, while the electrons diffuse out primarily along **B**. Whether or not this in fact happens depends on the particular experiment. In short plasma columns with the field lines terminating on conducting plates, one would expect the ambipolar electric field to be short-circuited out. Each species then diffuses radially at a different rate. In long, thin plasma columns terminated by insulating plates, one would expect the radial diffusion to be ambipolar because escape along **B** is arduous.

Fig. 5.13 Parallel and perpendicular particle fluxes in a magnetic field

Mathematically, the problem is to solve simultaneously the equations of continuity (5.12) for ions and electrons. It is not the fluxes Γ_j, but the divergences $\nabla \cdot \Gamma_j$ which must be set equal to each other. Separating $\nabla \cdot \Gamma_j$ into perpendicular and parallel components, we have

$$\nabla \cdot \Gamma_i = \nabla_\perp \cdot \left(\mu_{i\perp} n \mathbf{E}_\perp - D_{i\perp} \nabla n \right) + \frac{\partial}{\partial z} \left(\mu_i n E_z - D_i \frac{\partial n}{\partial z} \right)$$

$$\nabla \cdot \Gamma_e = \nabla_\perp \cdot \left(-\mu_{e\perp} n \mathbf{E}_\perp - D_{e\perp} \nabla n \right) + \frac{\partial}{\partial z} \left(-\mu_e n E_z - D_e \frac{\partial n}{\partial z} \right)$$

(5.57)

The equation resulting from setting $\nabla \cdot \Gamma_i = \nabla \cdot \Gamma_e$ cannot easily be separated into one-dimensional equations. Furthermore, the answer depends sensitively on the boundary conditions at the ends of the field lines. Unless the plasma is so long that parallel diffusion can be neglected altogether, there is no simple answer to the problem of ambipolar diffusion across a magnetic field.

5.5.2   Experimental Checks

Whether or not a magnetic field reduces transverse diffusion in accordance with Eq. (5.51) became the subject of numerous investigations. The first experiment performed in a tube long enough that diffusion to the ends could be neglected was that of Lehnert and Hoh in Sweden. They used a helium positive column about 1 cm in diameter and 3.5 m long (Fig. 5.14). In such a plasma, the electrons are continuously lost by radial diffusion to the walls and are replenished by ionization of the neutral gas by the electrons in the tail of the velocity distribution. These fast electrons, in turn, are replenished by acceleration in the longitudinal electric field. Consequently, one would expect E_z to be roughly proportional to the rate of transverse diffusion. Two probes set in the wall of the discharge tube were used to measure E_z as B was varied. The ratio of $E_z(B)$ to $E_z(0)$ is shown as a function of B in Fig. 5.15. At low B fields, the experimental points follow closely the predicted curve, calculated on the basis of Eq. (5.52). At a critical field B_c of about 0.2 T, however, the experimental points departed from theory and, in fact, showed an

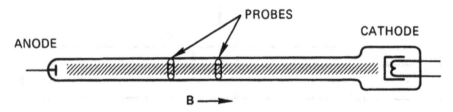

Fig. 5.14 The Lehnert-Hoh experiment to check the effect of a magnetic field on diffusion in a weakly ionized gas

Fig. 5.15 The normalized longitudinal electric field measured as a function of B at two different pressures. Theoretical curves are shown for comparison (From F. C. Hoh and B. Lehnert, *Phys. Fluids* **3**, 600 (1960))

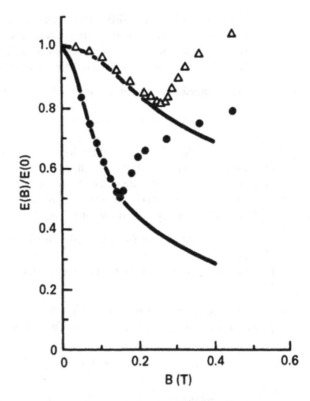

increase of diffusion with B. The critical field B_c increased with pressure, suggesting that a critical value of $\omega_c\tau$ was involved and that something went wrong with the "classical" theory of diffusion when $\omega_c\tau$ was too large.

The trouble was soon found by Kadomtsev and Nedospasov in the U.S.S.R. These theorists discovered that an instability should develop at high magnetic fields; that is, a plasma wave would be excited by the E_z field, and that this wave would cause enhanced radial losses. The theory correctly predicted the value of B_c. The wave, in the form of a helical distortion of the plasma column, was later seen directly in an experiment by Allen, Paulikas, and Pyle at Berkeley. This helical instability of the positive column was the first instance in which "anomalous diffusion" across magnetic fields was definitively explained, but the explanation was applicable only to weakly ionized gases. In the fully ionized plasmas of fusion research, anomalous diffusion proved to be a much tougher problem to solve.

Problems

5.1. The electron–neutral collision cross section for 2-eV electrons in He is about $6\pi a_0^2$, where $a_0 = 0.53 \times 10^{-8}$ cm is the radius of the first Bohr orbit of the hydrogen atom. A positive column with no magnetic field has $p = 1$ Torr of He (at room temperature) and $KT_e = 2$ eV.

(a) Compute the electron diffusion coefficient in m²/s, assuming that $\overline{\sigma v}$ averaged over the velocity distribution is equal to σv for 2-eV electrons.

(b) If the current density along the column is 2 kA/m² and the plasma density is 10^{16} m^{-3}, what is the electric field along the column?

5.2. A weakly ionized plasma slab in plane geometry has a density distribution

$$n(x) = n_0 \cos (\pi x/2L) \qquad -L \le x \le L$$

The plasma decays by both diffusion and recombination. If $L = 0.03$ m, $D = 0.4$ m²/s, and $\alpha = 10^{-15}$ m³/s, at what density will the rate of loss by diffusion be equal to the rate of loss by recombination?

5.3. A weakly ionized plasma is created in a cubical aluminum box of length L on each side. It decays by ambipolar diffusion.

(a) Write an expression for the density distribution in the lowest diffusion mode.

(b) Define what you mean by the decay time constant and compute it if $D_a = 10^{-3}$ m²/s.

5.4. A long, cylindrical positive column has $B = 0.2$ T, $KT_i = 0.1$ eV, and other parameters the same as in Problem 5.1. The density profile is

$$n(r) = n_0 J_0 \left(r/[D\tau]^{1/2} \right)$$

with the boundary condition $n = 0$ at $r = a = 1$ cm. Note: $J_0(z) = 0$ at $z = 2.4$.

(a) Show that the ambipolar diffusion coefficient to be used above can be approximated by $D_{\perp e}$.

(b) Neglecting recombination and losses from the ends of the column, compute the confinement time τ.

5.5. For the density profile of Fig. 5.7, derive an expression for the peak density n_0 in terms of the source strength ϱ in ion-electron pairs per m².

5.6. You do a recombination experiment in a weakly ionized gas in which the main loss mechanism is recombination. You create a plasma of density 10^{20} m^{-3} by a sudden burst of ultraviolet radiation and observe that the density decays to half its initial value in 10 ms. What is the value of the recombination coefficient α? Give units.

5.6   Collisions in Fully Ionized Plasmas

When the plasma is composed of ions and electrons alone, all collisions are Coulomb collisions between charged particles. However, there is a distinct difference between (a) collisions between like particles (ion–ion or electron–electron collisions) and

Fig. 5.16 Shift of guiding
centers of two like particles
making a 90° collision

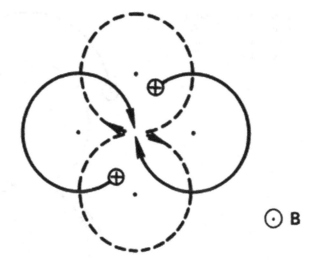

(b) collisions between unlike particles (ion–electron or electron–ion collisions).
Consider two identical particles colliding (Fig. 5.16). If it is a head-on collision, the
particles emerge with their velocities reversed; they simply interchange their orbits,
and the two guiding centers remain in the same places. The result is the same as in a
glancing collision, in which the trajectories are hardly disturbed. The worst that can
happen is a 90° collision, in which the velocities are changed 90° in direction. The
orbits after collision will then be the dashed circles, and the guiding centers will have
shifted. However, it is clear that the "center of mass" of the two guiding centers
remains stationary. For this reason, *collisions between like particles give rise to very
little diffusion.* This situation is to be contrasted with the case of ions colliding with
neutral atoms. In that case, the final velocity of the neutral is of no concern, and the
ion random-walks away from its initial position. In the case of ion–ion collisions,
however, there is a detailed balance in each collision; for each ion that moves
outward, there is another that moves inward as a result of the collision.

   When two particles of opposite charge collide, however, the situation is entirely
different (Fig. 5.17). The worst case is now the 180° collision, in which the particles
emerge with their velocities reversed. Since they must continue to gyrate about the
lines of force in the proper sense, both guiding centers will move in the same
direction. *Unlike-particle collisions give rise to diffusion.* The physical picture is
somewhat different for ions and electrons because of the disparity in mass. The
electrons bounce off the nearly stationary ions and random-walk in the usual
fashion. The ions are slightly jostled in each collision and move about as a result
of frequent bombardment by electrons. Nonetheless, because of the conservation of
momentum in each collision, the rates of diffusion are the same for ions and
electrons, as we shall show.

Fig. 5.17 Shift of guiding
centers of two oppositely
charged particles making a
180° collision

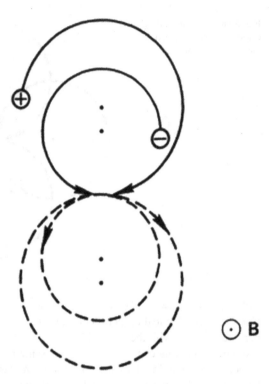

5.6.1 Plasma Resistivity

The fluid equations of motion including the effects of charged-particle collisions
may be written as follows (cf. Eq. (3.47)):

$$Mn\frac{d\mathbf{v}_i}{dt} = en(\mathbf{E} + \mathbf{v}_i \times \mathbf{B}) - \nabla p_i - \nabla \cdot \boldsymbol{\pi}_i + \mathbf{P}_{ie}$$

$$mn\frac{d\mathbf{v}_e}{dt} = -en(\mathbf{E} + \mathbf{v}_e \times \mathbf{B}) - \nabla p_e - \nabla \cdot \boldsymbol{\pi}_e + \mathbf{P}_{ei}$$

(5.58)

The terms $\mathbf{P}_{ie}$ and $\mathbf{P}_{ei}$ represent, respectively, the momentum gain of the ion fluid
caused by collisions with electrons, and vice versa. The stress tensor $\mathbf{P}_j$ has been
split into the isotropic part p_j and the anisotropic viscosity tensor $\boldsymbol{\pi}_j$. Like-particle
collisions, which give rise to stresses within each fluid individually, are contained in
$\boldsymbol{\pi}_j$. Since these collisions do not give rise to much diffusion, we shall ignore the
terms $\nabla \cdot \boldsymbol{\pi}_j$. As for the terms $\mathbf{P}_{ei}$ and $\mathbf{P}_{ie}$, which represent the friction between the
two fluids, the conservation of momentum requires

$$\mathbf{P}_{ie} = -\mathbf{P}_{ei}$$

(5.59)

We can write $\mathbf{P}_{ei}$ in terms of the collision frequency in the usual manner:

$$\mathbf{P}_{ei} = mn(\mathbf{v}_i - \mathbf{v}_e)\nu_{ei} \tag{5.60}$$

and similarly for $\mathbf{P}_{ie}$. Since the collisions are Coulomb collisions, one would expect $\mathbf{P}_{ei}$ to be proportional to the Coulomb force, which is proportional to e^2 (for singly charged ions). Furthermore, $\mathbf{P}_{ei}$ must be proportional to the density of electrons n_e and to the density of scattering centers n_i, which, of course, is equal to n_e. Finally, $\mathbf{P}_{ei}$ should be proportional to the relative velocity of the two fluids. On physical grounds, then, we can write $\mathbf{P}_{ei}$ as

$$\mathbf{P}_{ei} = \eta e^2 n^2 (\mathbf{v}_i - \mathbf{v}_e) \tag{5.61}$$

where η is a constant of proportionality. Comparing this with Eq. (5.60), we see that

$$\boxed{\nu_{ei} = \frac{ne^2}{m}\eta} \tag{5.62}$$

The constant η is the *specific resistivity* of the plasma; that this jibes with the usual meaning of resistivity will become clear shortly.

5.6.2   Mechanics of Coulomb Collisions

When an electron collides with a neutral atom, no force is felt until the electron is close to the atom on the scale of atomic dimensions; the collisions are like billiard-ball collisions. When an electron collides with an ion, the electron is gradually deflected by the long-range Coulomb field of the ion. Nonetheless, one can derive an effective cross section for this kind of collision. It will suffice for our purposes to give an order-of-magnitude estimate of the cross section. In Fig. 5.18, an electron of velocity $\mathbf{v}$ approaches a fixed ion of charge e. In the absence of Coulomb forces, the electron would have a distance of closest approach r_0, called the *impact parameter*. In the presence of a Coulomb attraction, the electron will be deflected by an angle χ, which is related to r_0. The Coulomb force is

$$F = -\frac{e^2}{4\pi\varepsilon_0\, r^2} \tag{5.63}$$

This force is felt during the time the electron is in the vicinity of the ion; this time is roughly

$$T \approx r_0/v \tag{5.64}$$

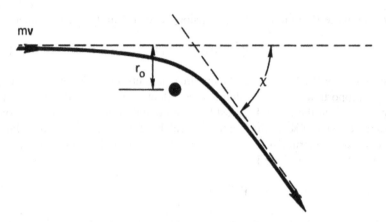

Fig. 5.18 Orbit of an electron making a Coulomb collision with an ion

The change in the electron's momentum is therefore approximately

$$\Delta(mv) = |FT| \approx \frac{e^2}{4\pi\varepsilon_0 r_0 v} \tag{5.65}$$

We wish to estimate the cross section for large-angle collisions, in which $\chi \geq 90°$. For a 90° collision, the change in mv is of the order of mv itself. Thus

$$\Delta(mv) \cong mv \cong e^2/4\pi\varepsilon_0 r_0 v, \quad r_0 = e^2/4\pi\varepsilon_0 \ m v^2 \tag{5.66}$$

The cross section is then

$$\sigma = \pi r_0^2 = e^4/16\pi\varepsilon_0^2 m^2 v^4 \tag{5.67}$$

The collision frequency is, therefore,

$$\nu_{ei} = n\sigma v = ne^4/16\pi\varepsilon_0^2 \ m^2 v^3 \tag{5.68}$$

and the resistivity is

$$\eta = \frac{m}{ne^2}\nu_{ei} = \frac{e^2}{16\pi\varepsilon_0^2 \ mv^3} \tag{5.69}$$

For a Maxwellian distribution of electrons, we may replace v^2 by KT_e/m for our order-of-magnitude estimate:

$$\eta \approx \frac{\pi e^2 m^{1/2}}{(4\pi\varepsilon_0)^2 (KT_e)^{3/2}} \tag{5.70}$$

Equation (5.70) is the resistivity based on large-angle collisions alone. In practice, because of the long range of the Coulomb force, small-angle collisions are much more frequent, and the cumulative effect of many small-angle deflections turns out to be larger than the effect of large-angle collisions. It was shown by Spitzer that Eq. (5.70) should be multiplied by a factor $\ln \Lambda$:

$$\eta \approx \frac{\pi e^2 m^{1/2}}{(4\pi\varepsilon_0)^2 (KT_e)^{3/2}} \ln\Lambda \qquad (5.71)$$

where

$$\Lambda = \overline{\lambda_D/r_0} = 12\pi n\lambda_D^3 \qquad (5.72)$$

This factor represents the maximum impact parameter, in units of r_0 as given by Eq. (5.66), averaged over a Maxwellian distribution. The maximum impact parameter is taken to be λ_D because Debye shielding suppresses the Coulomb field at larger distances. Although Λ depends on n and KT_e, its logarithm is insensitive to the exact values of the plasma parameters. Typical values of $\ln \Lambda$ are given below.

KT_e (eV)	n (m^{-3})	$\ln \Lambda$	
0.2	10^{15}	9.1	(Q-machine)
2	10^{17}	10.2	(lab plasma)
100	10^{19}	13.7	(typical torus)
10^4	10^{21}	16.0	(fusion reactor)
10^3	10^{27}	6.8	(laser plasma)

It is evident that $\ln \Lambda$ varies only a factor of two as the plasma parameters range over many orders of magnitude. For most purposes, it will be sufficiently accurate to let $\ln \Lambda = 10$ regardless of the type of plasma involved.

5.6.3   Physical Meaning of η

Let us suppose that an electric field $\mathbf{E}$ exists in a plasma and that the current that it drives is all carried by the electrons, which are much more mobile than the ions. Let $B = 0$ and $KT_e = 0$, so that $\nabla \cdot \mathbf{P}_e = 0$. Then, in steady state, the electron equation of motion (5.58) reduces to

$$en\,\mathbf{E} = \mathbf{P}_{ei} \qquad (5.73)$$

Since $\mathbf{j} = en(\mathbf{v}_i - \mathbf{v}_e)$, Eq. (5.61) can be written

$$\mathbf{P}_{ei} = \eta en\,\mathbf{j} \qquad (5.74)$$

so that Eq. (5.73) becomes

$$\mathbf{E} = \eta \mathbf{j} \qquad\qquad (5.75)$$

This is simply Ohm's law, and the constant η is just the specific resistivity. The expression for η in a plasma, as given by Eq. (5.71) or Eq. (5.69), has several features which should be pointed out.

(a) In Eq. (5.71), we see that η is *independent of density* (except for the weak dependence in $\ln \Lambda$). This is a rather surprising result, since it means that if a field $\mathbf{E}$ is applied to a plasma, the current $\mathbf{j}$, as given by Eq. (5.75), is independent of the number of charge carriers. The reason is that although j increases with n_e, the frictional drag against the ions increases with n_i. Since $n_e = n_i$ these two effects cancel. This cancellation can be seen in Eqs. (5.68) and (5.69). The collision frequency ν_{ei} is indeed proportional to n, but the factor n cancels out in η. A fully ionized plasma behaves quite differently from a weakly ionized one in this respect. In a weakly ionized plasma, we have $\mathbf{j} = -ne\mathbf{v}_e$, $\mathbf{v}_e = -\mu_e\mathbf{E}$, so that $\mathbf{j} = ne\mu_e\mathbf{E}$. Since μ_e depends only on the density of *neutrals*, the current is proportional to the plasma density n.

(b) Equation (5.71) shows that η is proportional to $(KT_e)^{-3/2}$. As a plasma is heated, the Coulomb cross section decreases, and the resistivity drops rather rapidly with increasing temperature. Plasmas at thermonuclear temperatures (tens of keV) are essentially collisionless; this is the reason so much theoretical research is done on collisionless plasmas. Of course, there must always be some collisions; otherwise, there wouldn't be any fusion reactions either. An easy way to heat a plasma is simply to pass a current through it. The I^2R (or $j^2\eta$) losses then turn up as an increase in electron temperature. This is called *ohmic heating*. The $(KT_e)^{-3/2}$ dependence of η, however, does not allow this method to be used up to thermonuclear temperatures. The plasma becomes such a good conductor at temperatures above 1 keV that ohmic heating is a very slow process in that range.

(c) Equation (5.68) shows that ν_{ei} varies as v^{-3}. The fast electrons in the tail of the velocity distribution make very few collisions. The current is therefore carried mainly by these electrons rather than by the bulk of the electrons in the main body of the distribution. The strong dependence on v has another interesting consequence. If an electric field is suddenly applied to a plasma, a phenomenon known as *electron runaway* can occur. A few electrons which happen to be moving fast in the direction of $-\mathbf{E}$ when the field is applied will have gained so much energy before encountering an ion that they can make only a glancing collision. This allows them to pick up more energy from the electric field and decrease their collision cross section even further. If E is large enough, the cross section falls so fast that these runaway electrons never make a collision. They form an accelerated electron beam detached from the main body of the distribution.

5.6.4 Numerical Values of η

Exact computations of η which take into account the ion recoil in each collision and are properly averaged over the electron distribution were first given by Spitzer. The following result for hydrogen is sometimes called the Spitzer resistivity:

$$\eta_\parallel = 5.2 \times 10^{-5} \frac{Z \ln \Lambda}{T^{3/2}(\text{eV})} \text{ohm-m} \tag{5.76}$$

Here Z is the ion charge number, which we have taken to be 1 elsewhere in this book. Since the dependence on M is weak, these values can also be used for other gases. The subscript $\parallel$ means that this value of η is to be used for motions parallel to **B**. For motions perpendicular to **B**, one should use $\eta_\perp$ given by

$$\eta_\perp = 2.0 \eta_\parallel \tag{5.77}$$

This does not mean that conductivity along **B** is only two times better than conductivity across **B**. A factor like $\omega_c^2 \tau^2$ still has to be taken into account. The factor 2.0 comes from a difference in weighting of the various velocities in the electron distribution. In perpendicular motions, the slow electrons, which have small Larmor radii, contribute more to the resistivity than in parallel motions.

For $KT_e = 100$ eV, Eq. (5.76) yields

$$\eta = 5 \times 10^{-7} \text{ohm-m}$$

This is to be compared with various metallic conductors:

$$\text{copper} \ldots \ldots \ldots \eta = 2 \times 10^{-8} \text{ohm-m}$$
$$\text{stainless steel} \ldots \ldots \eta = 7 \times 10^{-7} \text{ohm-m}$$
$$\text{mercury} \ldots \ldots \ldots \eta = 10^{-6} \text{ohm-m}$$

A 100-eV plasma, therefore, has a conductivity like that of stainless steel.

5.6.5 Pulsed Currents

When a steady-state current is drawn between two electrodes aligned along the magnetic field, electrons are the dominant current carrier, and sheaths are set up at the cathode to limit the current to the set value. When the current is pulsed, however, it takes time to set up the current distribution. It was shown by Stenzel and Urrutia that this time is controlled by whistler waves (R-waves), which must travel the length of the device to communicate the voltage information.

5.7   The Single-Fluid MHD Equations

We now come to the problem of diffusion in a fully ionized plasma. Since the dissipative term $\mathbf{P}_{ei}$ contains the difference in velocities $\mathbf{v}_i - \mathbf{v}_e$, it is simpler to work with a linear combination of the ion and electron equations such that $\mathbf{v}_i - \mathbf{v}_e$ is the unknown rather than $\mathbf{v}_i$ or $\mathbf{v}_e$ separately. Up to now, we have regarded a plasma as composed of two interpenetrating fluids. The linear combination we are going to choose will describe the plasma as a single fluid, like liquid mercury, with a mass density ρ and an electrical conductivity $1/\eta$. These are the equations of magneto-hydrodynamics (MHD).

For a quasineutral plasma with singly charged ions, we can define the mass density ρ, mass velocity $\mathbf{v}$, and current density $\mathbf{j}$ as follows:

$$\rho \equiv n_i M + n_e m \approx n(M + m) \tag{5.78}$$

$$\mathbf{v} \equiv \frac{1}{\rho}(n_i M \mathbf{v}_i + n_e m \mathbf{v}_e) \approx \frac{M \mathbf{v}_i + m \mathbf{v}_e}{M + m} \tag{5.79}$$

$$\mathbf{j} \equiv e(n_i \mathbf{v}_i - n_e \mathbf{v}_e) \approx ne(\mathbf{v}_i - \mathbf{v}_e) \tag{5.80}$$

In the equation of motion, we shall add a term $Mn\mathbf{g}$ for a gravitational force. This term can be used to represent any nonelectromagnetic force applied to the plasma. The ion and electron equations can be written

$$Mn\frac{\partial \mathbf{v}_i}{\partial t} = en(\mathbf{E} + \mathbf{v}_i \times \mathbf{B}) - \nabla p_i + Mn\mathbf{g} + \mathbf{P}_{ie} \tag{5.81}$$

$$mn\frac{\partial \mathbf{v}_e}{\partial t} = -en(\mathbf{E} + \mathbf{v}_e \times \mathbf{B}) - \nabla p_e + mn\mathbf{g} + \mathbf{P}_{ei} \tag{5.82}$$

For simplicity, we have neglected the viscosity tensor $\boldsymbol{\pi}$, as we did earlier. This neglect does not incur much error if the Larmor radius is much smaller than the scale length over which the various quantities change. We have also neglected the $(\mathbf{v} \cdot \nabla)\mathbf{v}$ terms because the derivation would be unnecessarily complicated otherwise. This simplification is more difficult to justify. To avoid a lengthy discussion, we shall simply say that $\mathbf{v}$ is assumed to be so small that this quadratic term is negligible.

We now add Eqs. (5.81) and (5.82), obtaining

$$n\frac{\partial}{\partial t}(M\mathbf{v}_i + m\mathbf{v}_e) = en(\mathbf{v}_i - \mathbf{v}_e) \times \mathbf{B} - \nabla p + n(M + m)\mathbf{g} \tag{5.83}$$

The electric field has cancelled out, as have the collision terms $\mathbf{P}_{ei} = -\mathbf{P}_{ie}$. We have introduced the notation

$$p = p_i + p_e \tag{5.84}$$

for the total pressure. With the help of Eqs. (5.78)–(5.80), Eq. (5.83) can be written simply

$$\rho\frac{\partial \mathbf{v}}{\partial t} = \mathbf{j} \times \mathbf{B} - \nabla p + \rho\mathbf{g} \tag{5.85}$$

This is the single-fluid equation of motion describing the mass flow. The electric field does not appear explicitly because the fluid is neutral. The three body forces on the right-hand side are exactly what one would have expected.

A less obvious equation is obtained by taking a different linear combination of the two-fluid equations. Let us multiply Eq. (5.81) by m and Eq. (5.82) by M and subtract the latter from the former. The result is

$$Mmn\frac{\partial}{\partial t}(\mathbf{v}_i - \mathbf{v}_e) = en(M + m)\mathbf{E} + en(m\mathbf{v}_i + M\mathbf{v}_e) \times \mathbf{B} - m\nabla p_i$$
$$+ M\nabla p_e - (M + m)\mathbf{P}_{ei} \tag{5.86}$$

With the help of Eqs. (5.78), (5.80), and (5.61), this becomes

$$\frac{Mmn}{e}\frac{\partial}{\partial t}\left(\frac{\mathbf{j}}{n}\right) = e\rho\mathbf{E} - (M + m)ne\eta\mathbf{j} - m\nabla p_i + M\nabla p_e + en(m\mathbf{v}_i + M\mathbf{v}_e)$$
$$\times \mathbf{B} \tag{5.87}$$

The last term can be simplified as follows:

$$m\mathbf{v}_i + M\mathbf{v}_e = M\mathbf{v}_i + m\mathbf{v}_e + M(\mathbf{v}_e - \mathbf{v}_i) + m(\mathbf{v}_i - \mathbf{v}_e)$$
$$= \frac{\rho}{n}\mathbf{v} - (M - m)\frac{\mathbf{j}}{ne} \tag{5.88}$$

Dividing Eq. (5.87) by $e\rho$, we now have

$$\mathbf{E} + \mathbf{v} \times \mathbf{B} - \eta\mathbf{j} = \frac{1}{e\rho}\left[\frac{Mmn}{e}\frac{\partial}{\partial t}\left(\frac{\mathbf{j}}{n}\right) + (M - m)\mathbf{j} \times \mathbf{B} + m\nabla p_i - M\nabla p_e\right] \tag{5.89}$$

The $\partial/\partial t$ term can be neglected in slow motions, where inertial (i.e., cyclotron frequency) effects are unimportant. In the limit $m/M \to 0$, Eq. (5.89) then becomes

$$\mathbf{E} + \mathbf{v} \times \mathbf{B} = \eta\mathbf{j} + \frac{1}{en}(\mathbf{j} \times \mathbf{B} - \nabla p_e) \tag{5.90}$$

This is our second equation, called the *generalized Ohm's law*. It describes the electrical properties of the conducting fluid. The $\mathbf{j} \times \mathbf{B}$ term is called the *Hall current* term. It often happens that this and the last term are small enough to be neglected; Ohm's law is then simply

$$\mathbf{E} + \mathbf{v} \times \mathbf{B} = \eta\mathbf{j} \tag{5.91}$$

Equations of continuity for mass ρ and charge σ are easily obtained from the sum and difference of the ion and electron equations of continuity. The set of MHD equations is then as follows:

$$\rho\frac{\partial \mathbf{v}}{\partial t} = \mathbf{j} \times \mathbf{B} - \nabla p + \rho\mathbf{g} \tag{5.85}$$

$$\mathbf{E} + \mathbf{v} \times \mathbf{B} = \eta\mathbf{j} \tag{5.91}$$

$$\frac{\partial \rho}{\partial t} + \nabla \cdot (\rho\mathbf{v}) = 0 \tag{5.92}$$

$$\frac{\partial \sigma}{\partial t} + \nabla \cdot \mathbf{j} = 0 \tag{5.93}$$

Together with Maxwell's equations, this set is often used to describe the equilibrium state of the plasma. It can also be used to derive plasma waves, but it is considerably less accurate than the two-fluid equations we have been using. For problems involving resistivity, the simplicity of the MHD equations outweighs their disadvantages. The MHD equations have been used extensively by astrophysicists working in cosmic electrodynamics, by hydrodynamicists working on MHD energy conversion, and by fusion theorists working with complicated magnetic geometries.

5.8 Diffusion of Fully Ionized Plasmas

In the absence of gravity, Eqs. (5.85) and (5.91) for a steady state plasma become

$$\mathbf{j} \times \mathbf{B} = \nabla p \tag{5.94}$$

$$\mathbf{E} + \mathbf{v} \times \mathbf{B} = \eta\mathbf{j} \tag{5.95}$$

The parallel component of the latter equation is simply

$$E_{\parallel} = \eta_{\parallel}j_{\parallel}$$

which is the ordinary Ohm's law. The perpendicular component is found by taking the cross-product with $\mathbf{B}$:

$$\mathbf{E} \times \mathbf{B} + (\mathbf{v}_{\perp} \times \mathbf{B}) \times \mathbf{B} = \eta_{\perp}\mathbf{j} \times \mathbf{B} = \eta_{\perp}\nabla p$$

$$\mathbf{E} \times \mathbf{B} - \mathbf{v}_{\perp}B^2 = \eta_{\perp}\nabla p$$

$$\mathbf{v}_\perp = \frac{\mathbf{E} \times \mathbf{B}}{B^2} - \frac{\eta_\perp}{B^2}\nabla p \tag{5.96}$$

The first term is just the $\mathbf{E} \times \mathbf{B}$ drift of both species together. The second term is the diffusion velocity in the direction of $-\nabla p$. For instance, in an axisymmetric cylindrical plasma in which $\mathbf{E}$ and ∇p are in the radial direction, we would have

$$v_\theta = -\frac{E_r}{B} \quad v_r = -\frac{\eta_\perp}{B^2}\frac{\partial p}{\partial r} \tag{5.97}$$

The flux associated with diffusion is

$$\mathbf{\Gamma}_\perp = n\mathbf{v}_\perp = -\frac{\eta_\perp n(KT_i + KT_e)}{B^2}\nabla n \tag{5.98}$$

This has the form of Fick's law, Eq. (5.11), with the diffusion coefficient

$$\boxed{D_\perp = \frac{\eta_\perp n\Sigma KT}{B^2}} \tag{5.99}$$

This is the so-called "classical" diffusion coefficient for a fully ionized gas.

Note that $D_\perp$ is proportional to $1/B^2$, just as in the case of weakly ionized gases. This dependence is characteristic of classical diffusion and can ultimately be traced back to the random-walk process with a step length r_L. Equation (5.99), however, differs from Eq. (5.54) for a partially ionized gas in three essential ways. First, $D_\perp$ is *not a constant* in a fully ionized gas; it is proportional to n. This is because the density of scattering centers is not fixed by the neutral atom density but is the plasma density itself. Second, since η is proportional to $(KT)^{-3/2}$, $D_\perp$ decreases with increasing temperature in a fully ionized gas. The opposite is true in a partially ionized gas. The reason for the difference is the velocity dependence of the Coulomb cross section. Third, *diffusion is automatically ambipolar* in a fully ionized gas (as long as like-particle collisions are neglected). $D_\perp$ in Eq. (5.99) is the coefficient for the entire fluid; no ambipolar electric field arises, because both species diffuse at the same rate. This is a consequence of the conservation of momentum in ion-electron collisions. This point is somewhat clearer if one uses the two-fluid equations (see Problem 5.15).

Finally, we wish to point out that *there is no transverse mobility* in a fully ionized gas. Equation (5.96) for $\mathbf{v}_\perp$ contains no component along $\mathbf{E}$ which depends on $\mathbf{E}$. If a transverse $\mathbf{E}$ field is applied to a uniform plasma, both species drift together with the $\mathbf{E} \times \mathbf{B}$ velocity. Since there is no relative drift between the two species, they do not collide, and there is no drift in the direction of $\mathbf{E}$. Of course, there are collisions due to thermal motions, and this simple result is only an approximate one. It comes from our neglect of (a) like-particle collisions, (b) the electron mass, and (c) the last two terms in Ohm's law, Eq. (5.90).

5.9   Solutions of the Diffusion Equation

Since $D_\perp$ is not a constant in a fully ionized gas, let us define a quantity A which *is* constant:

$$A \equiv \eta KT/B^2 \qquad (5.100)$$

We have assumed that KT and B are uniform, and that the dependence of η on n through the $\ln \Lambda$ factor can be ignored. For the case $T_i = T_e$, we then have

$$D_\perp = 2nA \qquad (5.101)$$

With Eq. (5.98), the equation of continuity (5.92) can now be written

$$\partial n/\partial t = \nabla \cdot (D_\perp \nabla n) = A\nabla \cdot (2n\nabla n)$$

$$\partial n/\partial t = A\nabla^2 n^2 \qquad (5.102)$$

This is a nonlinear equation for n, for which there are very few simple solutions.

5.9.1   Time Dependence

If we separate the variables by letting

$$n = T(t)S(\mathbf{r})$$

we can write Eq. (5.102) as

$$\frac{1}{T^2}\frac{dT}{dt} = \frac{A}{S}\nabla^2 S^2 = -\frac{1}{\tau} \qquad (5.103)$$

where $-1/\tau$ is the separation constant. The spatial part of this equation is difficult to solve, but the temporal part is the same equation that we encountered in recombination, Eq. (5.41). The solution, therefore, is

$$\frac{1}{T} = \frac{1}{T_0} + \frac{t}{\tau} \qquad (5.104)$$

At large times t, the density decays as $1/t$, as in the case of recombination. This reciprocal decay is what would be expected of a fully ionized plasma diffusing classically. The exponential decay of a weakly ionized gas is a distinctly different behavior.

5.9.2   Time-Independent Solutions

There is one case in which the diffusion equation can be solved simply. Imagine a long plasma column (Fig. 5.19) with a source on the axis which maintains a steady state as plasma is lost by radial diffusion and recombination. The density profile outside the source region will be determined by the competition between diffusion and recombination. The density falloff distance will be short if diffusion is small and recombination is large, and will be long in the opposite case. In the region outside the source, the equation of continuity is

$$-A\nabla^2 n^2 = -\alpha n^2 \tag{5.105}$$

This equation is linear in n^2 and can easily be solved. In cylindrical geometry, the solution is a Bessel function. In plane geometry, Eq. (5.105) reads

$$\frac{d^2 n^2}{dx^2} = \frac{\alpha}{A} n^2 \tag{5.106}$$

with the solution

$$n^2 = n_0^2 \exp\left[-(\alpha/A)^{1/2} x\right] \tag{5.107}$$

The scale distance is

$$l = (A/\alpha)^{1/2} \tag{5.108}$$

Since A changes with magnetic field while α remains constant, the change of l with B constitutes a check of classical diffusion. This experiment was actually tried on a Q-machine, which provides a fully ionized plasma. Unfortunately, the presence of asymmetric $\mathbf{E} \times \mathbf{B}$ drifts leading to another type of loss—by convection—made the experiment inconclusive.

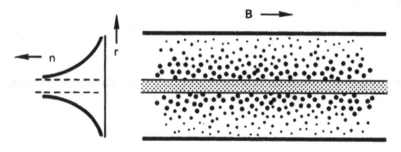

Fig. 5.19 Diffusion of a fully ionized cylindrical plasma across a magnetic field

Finally, we wish to point out a scaling law which is applicable to any fully ionized steady state plasma maintained by a constant source Q in a uniform B field. The equation of continuity then reads

$$-A \nabla^2 n^2 = -\eta KT \nabla^2 \left(n^2/B^2\right) = Q \tag{5.109}$$

Since n and B occur only in the combination n/B, the density profile will remain unchanged as B is changed, and the density itself will increase linearly with B:

$$n \propto B \tag{5.110}$$

One might have expected the equilibrium density n to scale as B^2, since $D_\perp \propto B^{-2}$; but one must remember that $D_\perp$ is itself proportional to n.

5.10 Bohm Diffusion and Neoclassical Diffusion

Although the theory of diffusion via Coulomb collisions had been known for a long time, laboratory verification of the $1/B^2$ dependence of $D_\perp$ in a fully ionized plasma eluded all experimenters until the 1960s. In almost all previous experiments, $D_\perp$ scaled as B^{-1}, rather than B^{-2}, and the decay of plasmas was found to be exponential, rather than reciprocal, with time. Furthermore, the absolute value of $D_\perp$ was far larger than that given by Eq. (5.99). This anomalously poor magnetic confinement was first noted in 1946 by Bohm, Burhop, and Massey, who were developing a magnetic arc for use in uranium isotope separation. Bohm gave the semiempirical formula

$$D_\perp = \frac{1}{16} \frac{KT_e}{eB} \equiv D_B \tag{5.111}$$

This formula was obeyed in a surprising number of different experiments. Diffusion following this law is called *Bohm diffusion*. Since D_B is independent of density, the decay is exponential with time. The time constant in a cylindrical column of radius R and length L can be estimated as follows:

$$\tau \approx \frac{N}{dN/dt} = \frac{n\pi R^2 L}{\Gamma_r 2\pi RL} = \frac{nR}{2\Gamma_r}$$

where N is the total number of ion-electron pairs in the plasma. With the flux Γ_r given by Fick's law and Bohm's formula, we have

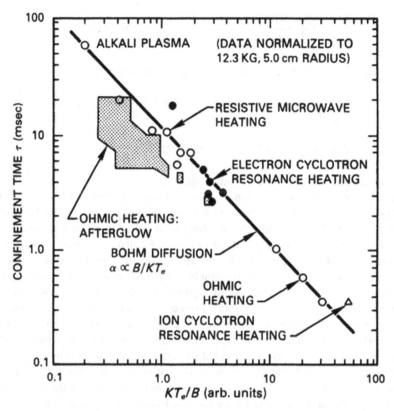

Fig. 5.20 Summary of confinement time measurements taken on various types of discharges in the Model C Stellarator, showing adherence to the Bohm diffusion law (Courtesy of D. J. Grove, Princeton University Plasma Physics Laboratory, sponsored by the U.S. Atomic Energy Commission)

$$\tau \approx \frac{nR}{2D_B \, \partial n/\partial r} \approx \frac{nR}{2D_B n/R} = \frac{R^2}{2D_B} \equiv \tau_B \qquad (5.112)$$

The quantity τ_B is often called the *Bohm time*.

Perhaps the most extensive series of experiments verifying the Bohm formula was done on a half-dozen devices called stellarators at Princeton. A stellarator is a toroidal magnetic container with the lines of force twisted so as to average out the grad-B and curvature drifts described in Sect. 2.2.3. Figure 5.20 shows a compilation of data taken over a decade on many different types of discharges in the Model C stellarator. The measured values of τ lie near a line representing the Bohm time τ_B. Close adherence to Bohm diffusion would have serious consequences for the controlled fusion program. Equation (5.111) shows that D_B increases, rather than decreases, with temperature, and though it decreases with B, it decreases more slowly than expected. In absolute magnitude, D_B is also much larger than $D_\perp$. For instance, for a 100-eV plasma in 1-T field, we have

$$D_B = \frac{1}{16} \frac{(10^2)(1.6 \times 10^{-19})}{(1.6 \times 10^{-19})(1)} = 6.25 \text{m}^2/\text{sec}$$

If the density is 10^{19} m^{-3}, the classical diffusion coefficient is

$$D_\perp = \frac{2nKT\eta_\perp}{B^2} = \frac{(2)(10^{19})(10^2)(1.6 \times 10^{-19})}{(1)^2} \times \frac{(2.0)(5.2 \times 10^{-5})(10)}{(100)^{3/2}}$$

$$= (320)(1.04 \times 10^{-6}) = 3.33 \times 10^{-4} \text{m}^2/\text{sec}$$

The disagreement is four orders of magnitude.

Several explanations have been proposed for Bohm diffusion. First, there is the possibility of magnetic field errors. In the complicated geometries used in fusion research, it is not always clear that the lines of B either close upon themselves or even stay within the chamber. Since the mean free paths are so long, only a slight asymmetry in the magnetic coil structure will enable electrons to wander out to the walls without making collisions. The ambipolar electric field will then pull the ions out. Second, there is the possibility of asymmetric electric fields. These can arise from obstacles inserted into the plasma, from asymmetries in the vacuum chamber, or from asymmetries in the way the plasma is created or heated. The dc $\mathbf{E} \times \mathbf{B}$ drifts then need not be parallel to the walls, and ions and electrons can be carried together to the walls by $\mathbf{E} \times \mathbf{B}$ convection. The drift patterns, called *convective cells*, have been observed. Finally, there is the possibility of oscillating electric fields arising from unstable plasma waves. If these fluctuating fields are random, the $\mathbf{E} \times \mathbf{B}$ drifts constitute a collisionless random-walk process. Even if the oscillating field is a pure sine wave, it can lead to enhanced losses because the phase of the $\mathbf{E} \times \mathbf{B}$ drift can be such that the drift is always outward whenever the fluctuation in density is positive. One may regard this situation as a moving convective cell pattern. Fluctuating electric fields are often observed when there is anomalous diffusion, but in many cases, it can be shown that the fields are not responsible for all of the losses. All three anomalous loss mechanisms may be present at the same time in experiments on fully ionized plasmas.

The scaling of D_B with KT_e and B can easily be shown to be the natural one whenever the losses are caused by $\mathbf{E} \times \mathbf{B}$ drifts, either stationary or oscillating. Let the escape flux be proportional to the $\mathbf{E} \times \mathbf{B}$ drift velocity:

$$\Gamma_\perp = nv_\perp \propto nE/B \tag{5.113}$$

Because of Debye shielding, the maximum potential in the plasma is given by

$$e\phi_{\max} \approx KT_e \tag{5.114}$$

If R is a characteristic scale length of the plasma (of the order of its radius), the maximum electric field is then

$$E_{\text{max}} \approx \frac{\phi_{\text{max}}}{R} \approx \frac{KT_e}{eR} \tag{5.115}$$

This leads to a flux $\Gamma_\perp$ given by

$$\Gamma_\perp \approx \gamma \frac{n}{R} \frac{KT_e}{eB} \approx -\gamma \frac{KT_e}{eB} \nabla n = -D_{\text{B}} \nabla n \tag{5.116}$$

where γ is some fraction less than unity. Thus the fact that D_{B} is proportional to KT_e/eB is no surprise. The value $\gamma = \frac{1}{16}$ has no theoretical justification but is an empirical number agreeing with most experiments to within a factor of two or three.

Recent experiments on toroidal devices have achieved confinement times of order $1000\tau_{\text{B}}$. This was accomplished by carefully eliminating oscillations and asymmetries. However, in toroidal devices, other effects occur which enhance collisional diffusion. Figure 5.21 shows a torus with helical lines of force. The twist is needed to eliminate the unidirectional grad-B and curvature drifts. As a particle follows a B-line, it sees a larger $|B|$ near the inside wall of the torus and a smaller $|B|$ near the outside wall. Some particles are trapped by the magnetic mirror effect and do not circulate all the way around the torus. The guiding centers of these trapped particles trace out banana-shaped orbits as they make successive passes through a given cross section (Fig. 5.21). As a particle makes collisions, it becomes trapped and untrapped successively and goes from one banana orbit to another. The random-walk step length is therefore the width of the banana orbit rather than r_{L}, and the "classical" diffusion coefficient is increased. This is called *neoclassical diffusion*. The dependence of $D_\perp$ on collision frequency v is shown in Fig. 5.22. In the region of small v, banana diffusion is larger than classical diffusion. In the region of large v, there is classical diffusion, but it is modified by currents along **B**. The theoretical rate for neoclassical diffusion has been observed experimentally by Ohkawa at La Jolla, CA, and by Chen at Princeton, NJ.

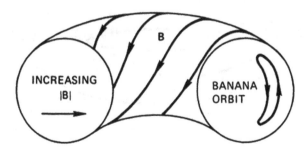

Fig. 5.21 A banana orbit of a particle confined in the twisted magnetic field of a toroidal confinement device. The "orbit" is really the locus of points at which the particle crosses the plane of the paper

Fig. 5.22 Behavior of the
neoclassical diffusion
coefficient with collision
frequency ν

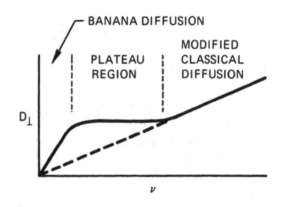

Problems

5.7. Show that the mean free path λ_{ei} for electron–ion collisions is proportional to T_e^2.

5.8. A tokamak is a toroidal plasma container in which a current is driven in the fully ionized plasma by an electric field applied along **B** (Fig. P5.8). How many V/m must be applied to drive a total current of 200 kA in a plasma with $KT_e = 500$ eV and a cross-sectional area of 75 cm^2?

5.9. Suppose the plasma in a fusion reactor is in the shape of a cylinder 1.2 m in diameter and 100 m long. The 5-T magnetic field is uniform except for short mirror regions at the ends, which we may neglect. Other parameters are $KT_i = 20$ keV, $KT_e = 10$ keV, and $n = 10^{21}$ m^{-3} (at $r = 0$). The density profile is found experimentally to be approximately as sketched in Fig. P5.9.

   (a) Assuming classical diffusion, calculate $D_\perp$ at $r = 0.5$ m.
   (b) Calculate dN/dt, the total number of ion-electron pairs leaving the central region radially per second.

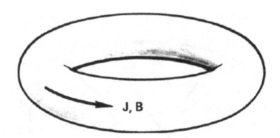

Fig. P5.8

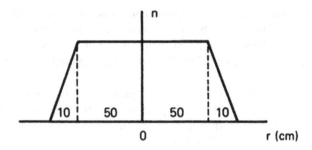

Fig. P5.9

(c) Estimate the confinement time τ by $\tau \approx -N/(dN/dt)$. Note: a rough estimate is all that can be expected in this type of problem. The profile has been simplified and is not realistic.

5.10. Estimate the classical diffusion time of a long plasma cylinder 10 cm in radius, with $n = 10^{21}$ m^{-3}, $KT_e = KT_i = 10$ keV, $B = 5$ T.

5.11. A cylindrical plasma column has a density distribution

$$n = n_0 \left(1 - r^2/a^2\right)$$

where $a = 10$ cm and $n_0 = 10^{19}$ m^{-3}. If $KT_e = 100$ eV, $KT_i = 0$, and the axial magnetic field B_0 is 1 T, what is the ratio between the Bohm and the classical diffusion coefficients perpendicular to B?

5.12. A weakly ionized plasma can still be governed by Spitzer resistivity if $v_{ei} \gg v_{e0}$ where v_{e0} is the electron–neutral collision frequency. Here are some data for the electron–neutral momentum transfer cross section σ_{e0} in square angstroms (Å^2):

	$E = 2$ eV	$E = 10$ eV
Helium	6.3	4.1
Argon	2.5	13.8

For singly ionized He and A plasmas with $KT_e = 2$ and 10 eV (four cases), estimate the fractional ionization $f \equiv n_i/(n_0 + n_i)$ at which $v_{ei} = v_{e0}$, assuming that the value of $\overline{\sigma v}(T_e)$ can be crudely approximated by $\sigma(E)\overline{|v|}(E)$, where $E = KT_e$. (Hint: For v_{e0}, use Eq. (7.11); for v_{ei}, use Eqs. (5.62) and (5.76).)

5.13. The plasma in a toroidal stellarator is ohmically heated by a current along **B** of 10^5 A/m^2. The density is uniform at $n = 10^{19}$ m^{-3} and does not change. The Joule heat ηj^2 goes to the electrons. Calculate the rate of increase of KT_e in eV/μsec at the time when $KT_e = 10$ eV.

5.14. In a θ-pinch, a large current is discharged through a one-turn coil. The rising magnetic field inside the coil induces a surface current in the highly conducting plasma. The surface current is opposite in direction to the coil current and hence keeps the magnetic field out of the plasma. The magnetic field pressure between the coil and the plasma then compresses the plasma.

This can work only if the magnetic field does not penetrate into the plasma during the pulse. Using the Spitzer resistivity, estimate the maximum pulse length for a hydrogen θ-pinch whose initial conditions are $KT_e = 10$ eV, $n = 10^{22}$ m^{-3}, $r = 2$ cm, if the field is to penetrate only 1/10 of the way to the axis.

5.15. Consider an axisymmetric cylindrical plasma with $\mathbf{E} = E_r\hat{\mathbf{r}}$, $\mathbf{B} = B\hat{\mathbf{z}}$, and $\nabla p_i = \nabla p_e = \hat{\mathbf{r}}\,\partial p/\partial r$. If we neglect the $(\mathbf{v} \cdot \nabla)\mathbf{v}$ term, which is tantamount to neglecting the centrifugal force, the steady-state two-fluid equations can be written in the form (Fig. P5.14)

$$en(\mathbf{E} + \mathbf{v}_i \times \mathbf{B}) - \nabla p_i - e^2 n^2 \eta(\mathbf{v}_i - \mathbf{v}_e) = 0$$
$$-en(\mathbf{E} + \mathbf{v}_e \times \mathbf{B}) - \nabla p_e - e^2 n^2 \eta(\mathbf{v}_e - \mathbf{v}_i) = 0$$

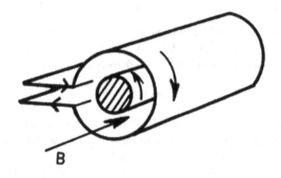

B

Fig. P5.14

(a) From the θ components of these equations, show that $v_{ir} = v_{er}$.

(b) From the r components, show that $v_{j\theta} = v_E + v_{Dj}$ ($j = i, e$).

(c) Find an expression for v_{ir} showing that it does not depend on E_r.

5.16. Use the single-fluid MHD equation of motion and the mass continuity equation to calculate the phase velocity of an ion acoustic wave in an unmagnetized, uniform plasma with $T_e \gg T_i$.

5.17. Calculate the resistive damping of Alfvén waves by deriving the dispersion relation from the single-fluid equations (5.85) and (5.91) and Maxwell's equations (4.72) and (4.77). Linearize and neglect gravity, displacement current, and ∇p.

(a) Show that

$$\frac{\omega^2}{k^2} = c^2 \varepsilon_0 \left(\frac{B_0^2}{\rho_0} - i\omega\eta \right)$$

(b) Find an explicit expression for Im (k) when ω is real and η is small.

5.18. If a cylindrical plasma diffuses at the Bohm rate, calculate the steady state radial density profile $n(r)$, ignoring the fact that it may be unstable. Assume that the density is zero at $r = \infty$ and has a value n_0 at $r = r_0$.

5.19. A cylindrical column of plasma in a uniform magnetic field $\mathbf{B} = B_z\hat{\mathbf{z}}$ carries a uniform current density $\mathbf{j} = j_z\hat{\mathbf{z}}$, where $\hat{\mathbf{z}}$ is a unit vector parallel to the axis of the cylinder.

  (a) Calculate the magnetic field $\mathbf{B}(r)$ produced by this plasma current.
  (b) Write an expression for the grad-B drift of a charged particle with $v_{||} = 0$ in terms of B_z, j_z, r, $v_\perp$, q, and m. You may assume that the field calculated in (a) is small compared to B_z (but not zero).
  (c) If the plasma has electrical resistivity, there is also an electric field $\mathbf{E} = E_z\hat{\mathbf{z}}$. Calculate the azimuthal electron drift due to this field, taking into account the helicity of the $\mathbf{B}$ field.
  (d) Draw a diagram showing the direction of the drifts in (b) and (c) for both ions and electrons in the (r, θ) plane.

Chapter 6
Equilibrium and Stability

6.1 Introduction

If we look only at the motions of individual particles, it would be easy to design a magnetic field which will confine a collisionless plasma. We need only make sure that the lines of force do not hit the vacuum wall and arrange the symmetry of the system in such a way that all the particle drifts $\mathbf{v}_E$, $\mathbf{v}_{\nabla B}$, and so forth are parallel to the walls. From a macroscopic fluid viewpoint, however, it is not easy to see whether a *plasma* will be confined in a magnetic field designed to contain individual particles. No matter how the external fields are arranged, the plasma can generate internal fields which affect its motion. For instance, charge bunching can create $\mathbf{E}$ fields which can cause $\mathbf{E} \times \mathbf{B}$ drifts to the wall. Currents in the plasma can generate $\mathbf{B}$ fields which cause grad-B drifts outward.

We can arbitrarily divide the problem of confinement into two parts: the problem of equilibrium and the problem of stability. A tight-rope walker can easily find an equilibrium, but it is not stable unless he holds a drooping rod. The difference between equilibrium and stability can also be illustrated by a mechanical analogy. Figure 6.1 shows various cases of a marble resting on a hard surface. An equilibrium is a state in which all the forces are balanced, so that a time-independent solution is possible. The equilibrium is stable or unstable according to whether small perturbations are damped or amplified. In case (F), the marble is in a stable equilibrium as long as it is not pushed too far. Once it is moved beyond a threshold, it is in an unstable state. This is called an "explosive instability." In case (G), the marble is in an unstable state, but it cannot make very large excursions. Such an in stability is not very dangerous if the nonlinear limit to the amplitude of the motion is small. The situation with a plasma is, of course, much more complicated than what is seen in Fig. 6.1; to achieve equilibrium requires balancing the forces on each fluid element.

The original version of this chapter was revised. An erratum to this chapter can be found at https://doi.org/10.1007/978-3-319-22309-4_11

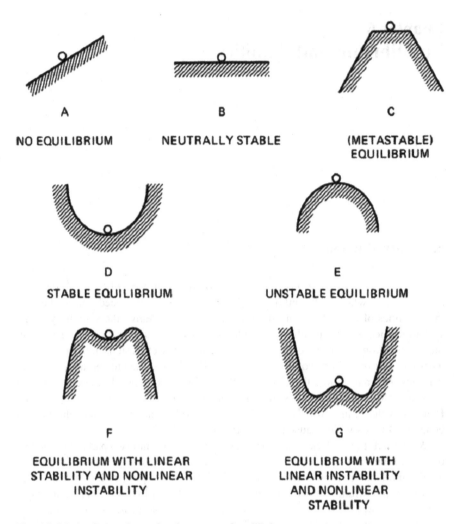

Fig. 6.1 Mechanical analogy of various types of equilibrium

Of the two problems, equilibrium and stability, the latter is easier to treat. One can linearize the equations of motion for small deviations from an equilibrium state. We then have linear equations, just as in the case of plasma waves. The equilibrium problem, on the other hand, is a nonlinear problem like that of diffusion. In complex magnetic geometries, the calculation of equilibria is a tedious process.

6.2   Hydromagnetic Equilibrium

Although the general problem of equilibrium is complicated, several physical concepts are easily gleaned from the MHD equations. For a steady state with $\partial/\partial t = 0$ and $\mathbf{g} = 0$, the plasma must satisfy (cf. Eq. (5.85))

Fig. 6.2 The $\mathbf{j} \times \mathbf{B}$ force of the diamagnetic current balances the pressure-gradient force in steady state

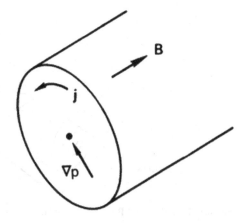

$$\nabla p = \mathbf{j} \times \mathbf{B} \qquad (6.1)$$

and

$$\nabla \times \mathbf{B} = \mu_0 \mathbf{j} \qquad (6.2)$$

From the simple equation (6.1), we can already make several observations.

(a) Equation (6.1) states that there is a balance of forces between the pressure-gradient force and the Lorentz force. How does this come about? Consider a cylindrical plasma with ∇p directed toward the axis (Fig. 6.2). To counteract the outward force of expansion, there must be an azimuthal current in the direction shown. The magnitude of the required current can be found by taking the cross product of Eq. (6.1) with $\mathbf{B}$:

$$\mathbf{j}_\perp = \frac{\mathbf{B} \times \nabla p}{B^2} = (KT_i + KT_e)\frac{\mathbf{B} \times \nabla n}{B^2} \qquad (6.3)$$

This is just the diamagnetic current found previously in Eq. (3.69)! From a single-particle viewpoint, the diamagnetic current arises from the Larmor gyration velocities of the particles, which do not average to zero when there is a density gradient. From an MHD fluid viewpoint, the diamagnetic current is generated by the ∇p force across $\mathbf{B}$; the resulting current is just sufficient to balance the forces on each element of fluid and stop the motion.

(b) Equation (6.1) obviously tells us that $\mathbf{j}$ and $\mathbf{B}$ are each perpendicular to ∇p. This is not a trivial statement when one considers that the geometry may be very complicated. Imagine a toroidal plasma in which there is a smooth radial density gradient so that the surfaces of constant density (actually, constant p) are nested tori (Fig. 6.3). Since $\mathbf{j}$ and $\mathbf{B}$ are perpendicular to ∇p, they must lie on the surfaces of constant p. In general, the lines of force and of current may be twisted this way and that, but they must not cross the constant-p surfaces.

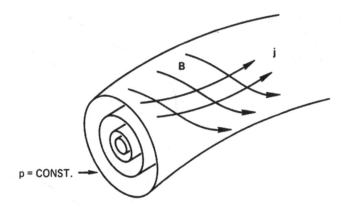

Fig. 6.3 Both the **j** and **B** vectors lie on constant-pressure surfaces

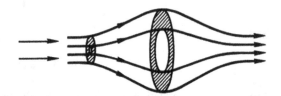

Fig. 6.4 Expansion
of a plasma streaming
into a mirror

(c) Consider the component of Eq. (6.1) along **B**. It says that

$$\partial p/\partial s = 0 \tag{6.4}$$

where s is the coordinate along a line of force. For constant KT, this means that in hydromagnetic equilibrium the density is constant along a line of force. At first sight, it seems that this conclusion must be in error. For, consider a plasma injected into a magnetic mirror (Fig. 6.4). As the plasma streams through, following the lines of force, it expands and then contracts; and the density is clearly not constant along a line of force. However, this situation does not satisfy the conditions of a static equilibrium. The $(\mathbf{v} \cdot \nabla)\mathbf{v}$ term, which we neglected along the way, does not vanish here. We must consider a static plasma with $\mathbf{v} = 0$. In that case, particles are trapped in the mirror, and there are more particles trapped near the midplane than near the ends because the mirror ratio is larger there. This effect just compensates for the larger cross section at the midplane, and the net result is that the density is constant along a line of force.

6.3 The Concept of β

We now substitute Eq. (6.2) into Eq. (6.1) to obtain

$$\nabla p = \mu_0^{-1}(\nabla \times \mathbf{B}) \times \mathbf{B} = \mu_0^{-1}\left[(\mathbf{B} \cdot \nabla)\mathbf{B} - \tfrac{1}{2}\nabla B^2\right] \tag{6.5}$$

Fig. 6.5 In a finite-β
plasma, the diamagnetic
current significantly
decreases the magnetic
field, keeping the sum of the
magnetic and particle
pressures a constant

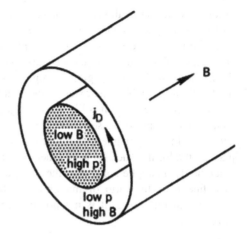

or

$$\nabla\left(p + \frac{B^2}{2\mu_0}\right) = \frac{1}{\mu_0}(\mathbf{B} \cdot \nabla)\mathbf{B} \tag{6.6}$$

In many interesting cases, such as a straight cylinder with axial field, the right-hand side vanishes; **B** does not vary along **B**. In many other cases, the right-hand side is small. Equation (6.6) then says that

$$p + \frac{B^2}{2\mu_0} = \text{constant} \tag{6.7}$$

Since $B^2/2\mu_0$ is the magnetic field pressure, the sum of the particle pressure and the magnetic field pressure is a constant. In a plasma with a density gradient (Fig. 6.5), the magnetic field must be low where the density is high, and vice versa. The decrease of the magnetic field inside the plasma is caused, of course, by the diamagnetic current. The size of the diamagnetic effect is indicated by the ratio of the two terms in Eq. (6.7). This ratio is usually denoted by β:

$$\beta \equiv \frac{\sum nkT}{B^2/2\mu_0} = \frac{\text{Particle pressure}}{\text{Magnetic field pressure}} \tag{6.8}$$

Up to now we have implicitly considered low-β plasmas, in which β is between 10^{-3} and 10^{-6}. The diamagnetic effect, therefore, is very small. This is the reason we could assume a uniform field B_0 in the treatment of plasma waves. If β is low, it does not matter whether the denominator of Eq. (6.8) is evaluated with the vacuum field or the field in the presence of plasma. If β is high, the local value of B can be greatly reduced by the plasma. In that case, it is customary to use the vacuum value of B in the definition of β. High-β plasmas are common in space and MHD energy conversion research. Fusion reactors will have to have β well in excess of 1 % in order to be economical, since the energy produced is proportional to n^2, while the cost of the magnetic container increases with some power of B.

In principle, one can have a $\beta = 1$ plasma in which the diamagnetic current generates a field exactly equal and opposite to an externally generated uniform field. There are then two regions: a region of plasma without field, and a region of field without plasma. If the external field lines are straight, this equilibrium would likely be unstable, since it is like a blob of jelly held together with stretched rubber bands. It remains to be seen whether a $\beta = 1$ plasma of this type can ever be achieved. In some magnetic configurations, the vacuum field has a null inside the plasma; the local value of β would then be infinite there. This happens, for instance, when fields are applied only near the surface of a large plasma. It is then customary to define β as the ratio of maximum particle pressure to maximum magnetic pressure; in this sense, it is not possible for a magnetically confined plasma to have $\beta > 1$.

6.4   Diffusion of Magnetic Field into a Plasma

A problem which often arises in astrophysics is the diffusion of a magnetic field into a plasma. If there is a boundary between a region with plasma but no field and a region with field but no plasma (Fig. 6.6), the regions will stay separated if the plasma has no resistivity, for the same reason that flux cannot penetrate a super-conductor. Any emf that the moving lines of force generate will create an infinite current, and this is not possible. As the plasma moves around, therefore, it pushes the lines of force and can bend and twist them. This may be the reason for the filamentary structure of the gas in the Crab nebula. If the resistivity is finite, however, the plasma can move through the field and vice versa. This diffusion takes a certain amount of time, and if the motions are slow enough, the lines of force

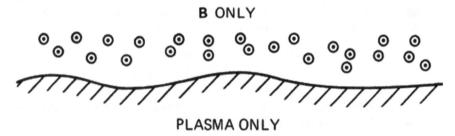

Fig. 6.6 In a perfectly conducting plasma, regions of plasma and magnetic field can be separated by a sharp boundary. Currents on the surface exclude the field from the plasma

need not be distorted by the gas motions. The diffusion time is easily calculated from these equations (cf. Eq. (5.91)):

$$\nabla \times \mathbf{E} = -\dot{\mathbf{B}} \tag{6.9}$$

$$\mathbf{E} + \mathbf{v} \times \mathbf{B} = \eta \mathbf{j} \tag{6.10}$$

For simplicity, let us assume that the plasma is at rest and the field lines are moving into it. Then $\mathbf{v} = 0$, and we have

$$\partial \mathbf{B}/\partial t = -\nabla \times \eta \mathbf{j} \tag{6.11}$$

Since $\mathbf{j}$ is given by Eq. (6.2), this becomes

$$\frac{\partial \mathbf{B}}{\partial t} = -\frac{\eta}{\mu_0} \nabla \times (\nabla \times \mathbf{B}) = -\frac{\eta}{\mu_0} \left[\nabla(\nabla \cdot \mathbf{B}) - \nabla^2 \mathbf{B} \right] \tag{6.12}$$

Since $\nabla \cdot \mathbf{B} = 0$, we obtain a diffusion equation of the type encountered in Chap. 5:

$$\frac{\partial \mathbf{B}}{\partial t} = \frac{\eta}{\mu_0} \nabla^2 \mathbf{B} \tag{6.13}$$

This can be solved by the separation of variables, as usual. To get a rough estimate, let us take L to be the scale length of the spatial variation of $\mathbf{B}$. Then we have

$$\frac{\partial \mathbf{B}}{\partial t} = \frac{\eta}{\mu_0 L^2} \mathbf{B} \tag{6.14}$$

$$\mathbf{B} = \mathbf{B}_0 e^{\pm t/\tau} \tag{6.15}$$

where

$$\tau = \mu_0 L^2 / \eta \tag{6.16}$$

This is the characteristic time for magnetic field penetration into a plasma.

The time τ can also be interpreted as the time for annihilation of the magnetic field. As the field lines move through the plasma, the induced currents cause ohmic heating of the plasma. This energy comes from the energy of the field. The energy lost per m^3 in a time τ is $\eta j^2 \tau$. Since

$$\mu_0 \mathbf{j} = \nabla \times \mathbf{B} \approx \frac{B}{L} \tag{6.17}$$

from Maxwell's equation with displacement current neglected, the energy dissipation is

$$\eta j^2 \tau = \eta \left(\frac{B}{\mu_0 L}\right)^2 \frac{\mu_0 L^2}{\eta} = \frac{B^2}{\mu_0} = 2\left(\frac{B^2}{2\mu_0}\right) \tag{6.18}$$

Thus τ is essentially the time it takes for the field energy to be dissipated into Joule heat.

Problems

6.1. Suppose that an electromagnetic instability limits β to $(m/M)^{1/2}$ in a D–D reactor. Let the magnetic field be limited to 20 T by the strength of materials. If $KT_e = KT_i = 20$ keV, find the maximum plasma density that can be contained.

6.2. In laser-fusion experiments, absorption of laser light on the surface of a pellet creates a plasma of density $n = 10^{27}$ m^{-3} and temperature $T_e \simeq T_i \simeq 10^4$ eV. Thermoelectric currents can cause spontaneous magnetic fields as high as 10^3 T.

    (a) Show that $\omega_c \tau_{ei} \gg 1$ in this plasma, and hence electron motion is severely affected by the magnetic field.

    (b) Show that $\beta \gg 1$, so that magnetic fields cannot effectively confine the plasma.

    (c) How do the plasma and field move so that the seemingly contradictory conditions (a) and (b) can both be satisfied?

6.3. A cylindrical plasma column of radius a contains a coaxial magnetic field $\mathbf{B} = B_0 \hat{\mathbf{z}}$ and has a pressure profile

$$p = p_0 \cos^2(\pi r / 2a)$$

    (a) Calculate the maximum value of p_0.

    (b) Using this value of p_0, calculate the diamagnetic current $\mathbf{j}(r)$ and the total field $\mathbf{B}(r)$.

    (c) Show $j(r)$, $B(r)$, and $p(r)$ on a graph.

    (d) If the cylinder is bent into a torus with the lines of force closing upon themselves after a single turn, this equilibrium, in which the macroscopic forces are everywhere balanced, is obviously disturbed. Is it possible to redistribute the pressure $p(r, \theta)$ in such a way that the equilibrium is restored?

6.4 Consider an infinite, straight cylinder of plasma with a square density profile created in a uniform field B_0 (Fig. P6.4). Show that B vanishes on the axis if $\beta = 1$, by proceeding as follows.

  (a) Using the MHD equations, find $\mathbf{j}_\perp$ in steady state for $KT = \text{constant}$.

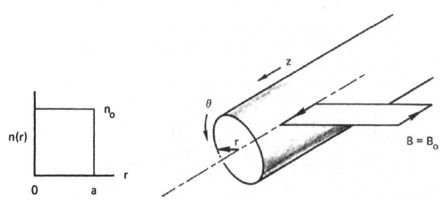

Fig. P6.4

  (b) Using $\nabla \times \mathbf{B} = \mu_0 \mathbf{j}$ and Stokes' theorem, integrate over the area of the loop shown to obtain

$$B_{ax} - B_0 = \mu_0 \sum KT \int_0^\infty \frac{\partial n / \partial r}{B(r)} dr, \quad B_{ax} \equiv B_{r=0}$$

  (c) Do the integral by noting that $\partial n / \partial r$ is a δ-function, so that $B(r)$ at $r = a$ is the average between B_{ax} and B_0.

6.5 A diamagnetic loop is a device used to measure plasma pressure by detecting the diamagnetic effect (Fig. P6.5). As the plasma is created, the diamagnetic current increases, B decreases inside the plasma, and the flux Φ enclosed by the loop decreases, inducing a voltage, which is then time-integrated by an RC circuit (Fig. P6.5).

  (a) Show that

$$\int_{\text{loop}} V \, dt = -N \, \Delta\Phi = -N \int \mathbf{B}_d \cdot d\mathbf{S}, \quad B_d \equiv B - B_0$$

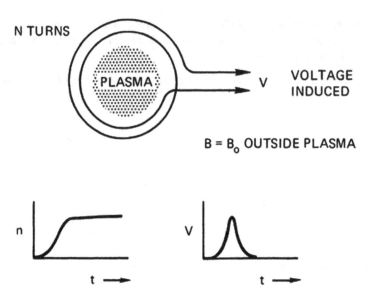

Fig. P6.5

(b)   Use the technique of the previous problem to find $B_d(r)$, but now assume n $(r) = n_0 \exp [-(r/r_0)^2]$. To do the integral, assume $\beta \ll 1$, so that B can be approximated by B_0 in the integral.

(c)   Show that $\int V dt = \frac{1}{2} N \pi r_0^2 \beta B_0$, with β defined as in Eq. (6.8).

6.5 Classification of Instabilities

In the treatment of plasma waves, we assumed an unperturbed state which was one of perfect thermodynamic equilibrium: The particles had Maxwellian velocity distributions, and the density and magnetic field were uniform. In such a state of highest entropy, there is no free energy available to excite waves, and we had to consider waves that were excited by external means. We now consider states that are not in perfect thermodynamic equilibrium, although they are in equilibrium in the sense that all forces are in balance and a time-independent solution is possible. The free energy which is available can cause waves to be self-excited; the equilibrium is then an unstable one. An instability is always a motion which decreases the free energy and brings the plasma closer to true thermodynamic equilibrium.

Instabilities may be classified according to the type of free energy available to drive them. There are four main categories.

Fig. 6.7 Hydrodynamic
Rayleigh–Taylor instability
of a heavy fluid supported
by a light one

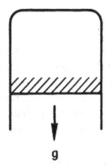 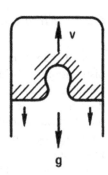

6.5.1   Streaming instabilities

In this case, either a beam of energetic particles travels through the plasma, or a current is driven through the plasma so that the different species have drifts relative to one another. The drift energy is used to excite waves, and oscillation energy is gained at the expense of the drift energy in the unperturbed state.

6.5.2   Rayleigh–Taylor instabilities

In this case, the plasma has a density gradient or a sharp boundary, so that it is not uniform. In addition, an external, non-electromagnetic force is applied to the plasma. It is this force which drives the instability. An analogy is available in the example of an inverted glass of water (Fig. 6.7). Although the plane interface between the water and air is in a state of equilibrium in that the weight of the water is supported by the air pressure, it is an unstable equilibrium. Any ripple in the surface will tend to grow at the expense of potential energy in the gravitation al field. This happens whenever a heavy fluid is supported by a light fluid, as is well known in hydrodynamics.

6.5.3   Universal instabilities

Even when there are no obvious driving forces such as an electric or a gravitational field, a plasma is not in perfect thermodynamic equilibrium as long as it is confined. The plasma pressure tends to make the plasma expand, and the expansion energy can drive an instability. This type of free energy is always present in any finite plasma, and the resulting waves are called *universal instabilities*.

6.5.4 Kinetic instabilities

In fluid theory the velocity distributions are assumed to be Maxwellian. If the distributions are in fact not Maxwellian, there is a deviation from thermodynamic equilibrium; and instabilities can be driven by the anisotropy of the velocity distribution. For instance, if $T_\parallel$ and $T_\perp$ are different, an instability called the modified Harris instability can arise. In mirror devices, there is a deficit of particles with large $v_\parallel / v_\perp$ because of the loss cone; this anisotropy gives rise to a "loss cone instability."

In the succeeding sections, we shall give a simple example of each of these types of instabilities. The instabilities driven by anisotropy cannot be described by fluid theory and a detailed treatment of them is beyond the scope of this book.

Not all instabilities are equally dangerous for plasma confinement. A high-frequency instability near ω_p, for instance, cannot affect the motion of heavy ions. Low-frequency instabilities with $\omega \ll \Omega_c$, however, can cause anomalous ambipolar losses via $\mathbf{E} \times \mathbf{B}$ drifts. Instabilities with $\omega \approx \Omega_c$ not efficiently transport particles across $\mathbf{B}$ but are dangerous in mirror machines, where particles are lost by diffusion in *velocity* space into the loss cone.

6.6 Two-Stream Instability

As a simple example of a streaming instability, consider a uniform plasma in which the ions are stationary and the electrons have a velocity $\mathbf{v}_0$ relative to the ions. That is, the observer is in a frame moving with the "stream" of ions. Let the plasma be cold ($KT_e = KT_i = 0$), and let there be no magnetic field ($B_0 = 0$). The linearized equations of motion are then

$$Mn_0 \frac{\partial \mathbf{v}_{i1}}{\partial t} = en_0 \mathbf{E}_1 \tag{6.19}$$

$$mn_0 \left[\frac{\partial \mathbf{v}_{e1}}{\partial t} + (\mathbf{v}_0 \cdot \nabla)\mathbf{v}_{e1} \right] = -en_0 \mathbf{E}_1 \tag{6.20}$$

The term $(\mathbf{v}_{e1} \cdot \nabla)\mathbf{v}_0$ in Eq. (6.20) has been dropped because we assume $\mathbf{v}_0$ to be uniform. The $(\mathbf{v}_0 \cdot \nabla)\mathbf{v}_1$ term does not appear in Eq. (6.19) because we have taken $\mathbf{v}_{i0} = 0$. We look for electrostatic waves of the form

$$\mathbf{E}_1 = E e^{i(kx - \omega t)} \hat{\mathbf{x}} \tag{6.21}$$

where $\hat{\mathbf{x}}$ is the direction of $\mathbf{v}_0$ and $\mathbf{k}$. Equations (6.19) and (6.20) become

$$-i\omega M n_0 \mathbf{v}_{i1} = en_0 \mathbf{E}_1, \qquad \mathbf{v}_{i1} = \frac{ie}{M\omega} E\hat{\mathbf{x}} \tag{6.22}$$

$$mn_0(-i\omega + ikv_0)\mathbf{v}_{e1} = -en_0\mathbf{E}_1 \qquad \mathbf{v}_{e1} = -\frac{ie}{m}\frac{E\hat{\mathbf{x}}}{\omega - kv_0} \qquad (6.23)$$

The velocities $\mathbf{v}_{j1}$ are in the x direction, and we may omit the subscript x. The ion equation of continuity yields

$$\frac{\partial n_{i1}}{\partial t} + n_0\mathbf{\nabla}\cdot\mathbf{v}_{i1} = 0 \qquad n_{i1} = \frac{k}{\omega}n_0v_{i1} = \frac{ien_0k}{M\omega^2}E \qquad (6.24)$$

Note that the other terms in $\mathbf{\nabla}\cdot(n\mathbf{v}_i)$ vanish because $\mathbf{\nabla}n_0 = \mathbf{v}_{0i} = 0$. The electron equation of continuity is

$$\frac{\partial n_{e1}}{\partial t} + n_0\mathbf{\nabla}\cdot\mathbf{v}_{e1} + (\mathbf{v}_0\cdot\mathbf{\nabla})n_{e1} = 0 \qquad (6.25)$$

$$(-i\omega + ikv_0)n_{e1} + ikn_0v_{e1} = 0 \qquad (6.26)$$

$$n_{e1} = \frac{kn_0}{\omega - kv_0}v_{e1} = -\frac{iekn_0}{m(\omega - kv_0)^2}E \qquad (6.27)$$

Since the unstable waves are high-frequency plasma oscillations, we may not use the plasma approximation but must use Poisson's equation:

$$\varepsilon_0\mathbf{\nabla}\cdot\mathbf{E}_1 = e(n_{i1} - n_{e1}) \qquad (6.28)$$

$$ik\varepsilon_0 E = e(ien_0kE)\left[\frac{1}{M\omega^2} + \frac{1}{m(\omega - kv_0)^2}\right] \qquad (6.29)$$

The dispersion relation is found upon dividing by $ik\varepsilon_0 E$:

$$1 = \omega_p^2\left[\frac{m/M}{\omega^2} + \frac{1}{(\omega - kv_0)^2}\right] \qquad (6.30)$$

Let us see if oscillations with real k are stable or unstable. Upon multiplying through by the common denominator, one would obtain a fourth-order equation for ω. If all the roots ω_j are real, each root would indicate a possible oscillation

$$\mathbf{E}_1 = Ee^{i(kx-\omega_j t)}\hat{\mathbf{x}}$$

If some of the roots are complex, they will occur in complex conjugate pairs. Let these complex roots be written

$$\omega_j = \alpha_j + i\gamma_j \qquad (6.31)$$

Fig. 6.8 The function $F(x,$ $y)$ in the two-stream instability, when the plasma is stable

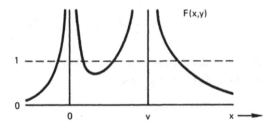

where α and γ are $\text{Re}(\omega)$ and $\text{Im}(\omega)$, respectively. The time dependence is now given by

$$\mathbf{E}_1 = Ee^{i(kx - \alpha_j t)} e^{\gamma_j t} \hat{\mathbf{x}} \qquad (6.32)$$

Positive $\text{Im}(\omega)$ indicates an exponentially growing wave; negative $\text{Im}(\omega)$ indicates a damped wave. Since the roots ω_j occur in conjugate pairs, one of these will always be unstable unless all the roots are real. The damped roots are not self-excited and are not of interest.

The dispersion relation (Eq. (6.30)) can be analyzed without actually solving the fourth-order equation. Let us define

$$x \equiv \omega/\omega_p \quad y \equiv kv_0/\omega_p \qquad (6.33)$$

Then Eq. (6.30) becomes

$$1 = \frac{m/M}{x^2} + \frac{1}{(x - y)^2} \equiv F(x, y). \qquad (6.34)$$

For any given value of y, we can plot $F(x, y)$ as a function of x. This function will have singularities at $x = 0$ and $x = y$ (Fig. 6.8). The intersections of this curve with the line $F(x, y) = 1$ give the values of x satisfying the dispersion relation. In the example of Fig. 6.8, there are four intersections, so there are four real roots ω_j. However, if we choose a smaller value of y, the graph would look as shown in Fig. 6.9. Now there are only two intersections and, therefore, only two real roots. The other two roots must be complex, and one of them must correspond to an unstable wave. Thus, for sufficiently small kv_0, the plasma is unstable. For any given v_0, the plasma is always unstable to long-wavelength oscillations (small k and y). The maximum growth rate predicted by Eq. (6.30) is, for $m/M \ll 1$ (cf. Problem 6.6),

$$\text{Im}\left(\frac{\omega}{\omega_p}\right) \approx \left(\frac{m}{M}\right)^{1/3} \qquad (6.35)$$

Since a small value of kv_0 is required for instability, one can say that for a given k, v_0 has to be sufficiently small for instability. This does not make much physical sense, since v_0 is the source of energy driving the instability. The difficulty comes from our use of the fluid equations. Any real plasma has a finite temperature, and

Fig. 6.9 The function
$F(x, y)$ in the two-stream
instability, when the plasma
is unstable

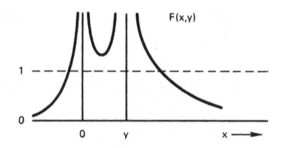

thermal effects should be taken into account by a kinetic-theory treatment. A phenomenon known as Landau damping (Chap. 7) will then occur for $v_0 \lesssim v_{\text{th}}$, and no instability is predicted if v_0 is too small.

This "Buneman" instability, as it is sometimes called, has the following physical explanation. The natural frequency of oscillations in the electron fluid is ω_p, and the natural frequency of oscillations in the ion fluid is $\Omega_p = (m/M)^{1/2}\omega_p$. Because of the Doppler shift of the ω_p oscillations in the moving electron fluid, these two frequencies can coincide in the laboratory frame if kv_0 has the proper value. The density fluctuations of ions and electrons can then satisfy Poisson's equation. Moreover, the electron oscillations can be shown to have *negative energy*. That is to say, the total kinetic energy of the electrons is less when the oscillation is present than when it is absent. In the undisturbed beam, the kinetic energy per m^3 is $\frac{1}{2}mn_0v_0^2$. When there is an oscillation, the kinetic energy is $\frac{1}{2}m(n_0 + n_1)(v_0 + v_1)^2$. When this is averaged over space, it turns out to be less than $\frac{1}{2}mn_0v_0^2$ because of the phase relation between n_1 and v_1 required by the equation of continuity. Consequently, the electron oscillations have negative energy, and the ion oscillations have positive energy. Both waves can grow together while keeping the total energy of the system constant. An instability of this type is used in klystrons to generate microwaves. Velocity modulation due to $\mathbf{E}_1$ causes the electrons to form bunches. As these bunches pass through a microwave resonator, they can be made to excite the natural modes of the resonator and produce microwave power.

Problems

6.6. (a) Derive the dispersion relation for a two-stream instability occurring when there are two cold electron streams with equal and opposite v_0 in a background of fixed ions. Each stream has a density $\frac{1}{2}n_0$.

    (b) Calculate the maximum growth rate.

6.7. A plasma consists of two uniform streams of protons with velocities $+v_0\hat{x}$ and $-v_0\hat{x}$, and respective densities $\frac{2}{3}n_0$ and $\frac{1}{3}n_0$. There is a neutralizing electron fluid with density n_0 and with $v_{0e} = 0$. All species are cold, and there is no magnetic field. Derive a dispersion relation for streaming instabilities in this system.

6.8. A cold electron beam of density δn_0 and velocity u is shot into a cold plasma of density n_0 at rest.

   (a) Derive a dispersion relation for the high-frequency beam-plasma instability that ensues.

   (b) The maximum growth rate γ_m is difficult to calculate, but one can make a reasonable guess if $\delta \ll 1$ by analogy with the electron–ion Buneman instability. Using the result given without proof in Eq. (6.35), give an expression for γ_m in terms of δ.

6.9. Let two cold, counter-streaming ion fluids have densities $\frac{1}{2}n_0$ and velocities $\pm v_0 \hat{y}$ in a magnetic field $B_0 \hat{z}$ and a cold neutralizing electron fluid. The field B_0 is strong enough to confine electrons but not strong enough to affect ion orbits.

   (a) Obtain the following dispersion relation for electrostatic waves propagating in the $\pm \hat{y}$ direction in the frequency range $\Omega_c^2 \ll \omega^2 \ll \omega_c^2$:

$$\frac{\Omega_p^2}{2(\omega - kv_0)^2} + \frac{\Omega_p^2}{2(\omega + kv_0)^2} = \frac{\omega_p^2}{\omega_c^2} + 1$$

   (b) Calculate the dispersion $\omega(k)$, growth rate $\gamma(k)$, and the range of wave numbers of the unstable waves.

6.7   The "Gravitational" Instability

In a plasma, a Rayleigh–Taylor instability can occur because the magnetic field acts as a light fluid supporting a heavy fluid (the plasma). In curved magnetic fields, the centrifugal force on the plasma due to particle motion along the curved lines of force acts as an equivalent "gravitational" force. To treat the simplest case, consider a plasma boundary lying in the y–z plane (Fig. 6.10). Let there be a density gradient

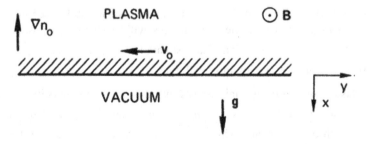

Fig. 6.10 A plasma surface subject to a gravitational instability

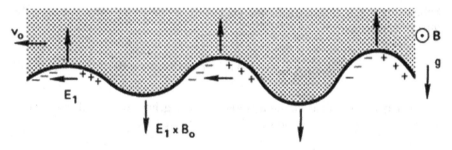

Fig. 6.11 Physical mechanism of the gravitational instability

∇n_0 in the $-x$ direction and a gravitational field $\mathbf{g}$ in the x direction. We may let $KT_i = KT_e. = 0$ for simplicity and treat the low-β case, in which $\mathbf{B}_0$ is uniform. In the equilibrium state, the ions obey the equation

$$Mn_0(\mathbf{v}_0 \cdot \nabla)\mathbf{v}_0 = en\mathbf{v}_0 \times \mathbf{B}_0 + Mn_0\mathbf{g} \qquad (6.36)$$

If g is a constant, v_0 will be also; and $(\mathbf{v}_0 \cdot \nabla)\mathbf{v}_0$ vanishes. Taking the cross product of Eq. (6.36) with $\mathbf{B}_0$, we find, as in Sect. 2.2,

$$\mathbf{v}_0 = \frac{M}{e}\frac{\mathbf{g} \times \mathbf{B}_0}{B_0^2} = -\frac{g}{\Omega_c}\hat{\mathbf{y}} \qquad (6.37)$$

The electrons have an opposite drift which can be neglected in the limit $m/M \to 0$. There is no diamagnetic drift because $KT = 0$, and no $\mathbf{E}_0 \times \mathbf{B}_0$ drift because $\mathbf{E}_0 = 0$.

If a ripple should develop in the interface as the result of random thermal fluctuations, the drift $\mathbf{v}_0$ will cause the ripple to grow (Fig. 6.11). The drift of ions causes a charge to build up on the sides of the ripple, and an electric field develops which changes sign as one goes from crest to trough in the perturbation. As can be seen from Fig. 6.11, the $\mathbf{E}_1 \times \mathbf{B}_0$ drift is always upward in those regions where the surface has moved upward, and downward where it has moved downward. The ripple grows as a result of these properly phased $\mathbf{E}_1 \times \mathbf{B}_0$ drifts.

To find the growth rate, we can perform the usual linearized wave analysis for waves propagating in the y direction: $\mathbf{k} = k\hat{\mathbf{y}}$. The perturbed ion equation of motion is

$$M(n_0 + n_1)\left[\frac{\partial}{\partial t}(\mathbf{v}_0 + \mathbf{v}_1) + (\mathbf{v}_0 + \mathbf{v}_1) \cdot \nabla(\mathbf{v}_0 + \mathbf{v}_1)\right]$$
$$= e(n_0 + n_1)[\mathbf{E}_1 + (\mathbf{v}_0 + \mathbf{v}_1) \times \mathbf{B}_0] + M(n_0 + n_1)\mathbf{g} \qquad (6.38)$$

We now multiply Eq. (6.36) by $1 + (n_1/n_0)$ to obtain

$$M(n_0 + n_1)(\mathbf{v}_0 \cdot \nabla)\mathbf{v}_0 = e(n_0 + n_1)\mathbf{v}_0 \times \mathbf{B}_0 + M(n_0 + n_1)\mathbf{g} \qquad (6.39)$$

Subtracting this from Eq. (6.38) and neglecting second-order terms, we have the *linearized* equation

$$Mn_0 \left[\frac{\partial \mathbf{v}_1}{\partial t} + (\mathbf{v}_0 \cdot \nabla)\mathbf{v}_1 \right] = en_0(\mathbf{E}_1 + \mathbf{v}_1 \times \mathbf{B}_0) \tag{6.40}$$

Note that **g** has cancelled out. Information regarding **g**, however, is still contained in $\mathbf{v}_0$. For perturbations of the form $\exp [i(ky - \omega t)]$, we have

$$M(\omega - kv_0)\mathbf{v}_1 = ie(\mathbf{E}_1 + \mathbf{v}_1 \times \mathbf{B}_0) \tag{6.41}$$

This is the same as Eq. (4.96) except that ω is replaced by $\omega - kv_0$, and electron quantities are replaced by ion quantities. The solution, therefore, is given by Eq. (4.98) with the appropriate changes. For $E_x = 0$ and

$$\Omega_c^2 \gg (\omega - kv_0)^2 \tag{6.42}$$

the solution is

$$v_{ix} = \frac{E_y}{B_0} \quad v_{iy} = -i \frac{\omega - kv_0}{\Omega_c} \frac{E_y}{B_0} \tag{6.43}$$

The latter quantity is the polarization drift in the ion frame. The corresponding quantity for electrons vanishes in the limit $m/M \to 0$. For the electrons, we therefore have

$$v_{ex} = E_y/B_0 \quad v_{ey} = 0 \tag{6.44}$$

The perturbed equation of continuity for ions is

$$\frac{\partial n_1}{\partial t} + \nabla \cdot (n_0 \mathbf{v}_0) + (\mathbf{v}_0 \cdot \nabla)n_1 + n_1 \nabla \cdot \mathbf{v}_0$$
$$+ (\mathbf{v}_1 \cdot \nabla)n_0 + n_0 \nabla \cdot \mathbf{v}_1 + \nabla \cdot (n_1 \mathbf{v}_1) = 0 \tag{6.45}$$

The zeroth-order term vanishes since $\mathbf{v}_0$ is perpendicular to ∇n_0, and the $n_1 \nabla \cdot \mathbf{v}_0$ term vanishes if $\mathbf{v}_0$ is constant. The first-order equation is, therefore,

$$-i\omega n_1 + ikv_0 n_1 + v_{ix}n_0' + ikn_0 v_{iy} = 0 \tag{6.46}$$

where $n_0' = \partial n_0/\partial x$. The electrons follow a simpler equation, since $\mathbf{v}_{e0} = 0$ and $v_{ey} = 0$:

$$-i\omega n_1 + v_{ex}n_0' = 0 \tag{6.47}$$

Note that we have used the plasma approximation and have assumed $n_{i1} = n_{e1}$. This is possible because the unstable waves are of low frequencies (this can be justified *a posteriori*). Equations (6.43) and (6.46) yield

$$(\omega - kv_0)n_1 + i\frac{E_y}{B_0}n_0' + ikn_0\frac{\omega - kv_0}{\Omega_c}\frac{E_y}{B_0} = 0 \qquad (6.48)$$

Equations (6.44) and (6.47) yield

$$\omega n_1 + i\frac{E_y}{B_0}n_0' = 0 \qquad \frac{E_y}{B_0} = \frac{i\omega n_1}{n_0'} \qquad (6.49)$$

Substituting this into Eq. (6.48), we have

$$(\omega - kv_0)n_1 - \left(n_0' + kn_0\frac{\omega - kv_0}{\Omega_c}\right)\frac{\omega n_1}{n_0'} = 0$$

$$\omega - kv_0 - \left(1 + \frac{kn_0}{\Omega_c}\frac{\omega - kv_0}{n_0'}\right)\omega = 0 \qquad (6.50)$$

$$\omega(\omega - kv_0) = -v_0\Omega_c n_0'/n_0 \qquad (6.51)$$

Substituting for v_0 from Eq. (6.37), we obtain a quadratic equation for ω:

$$\omega^2 - kv_0\omega - g\left(n_0'/n_0\right) = 0 \qquad (6.52)$$

The solutions are

$$\omega = \frac{1}{2}kv_0 \pm \left[\frac{1}{4}k^2v_0^2 + g\left(n_0'/n_0\right)\right]^{1/2} \qquad (6.53)$$

There is instability if ω is complex; that is, if

$$-gn_0'/n_0 > \frac{1}{4}k^2v_0^2 \qquad (6.54)$$

From this, we see that instability requires g and n_0'/n_0 to have opposite sign. This is just the statement that the light fluid is supporting the heavy fluid; otherwise, ω is real and the plasma is stable. Since g can be used to model the effects of magnetic field curvature, we see from this that stability depends on the sign of the curvature. Configurations with field lines bending in toward the plasma tend to be stabilizing, and vice versa. For sufficiently small k (long wavelength), the growth rate is given by

$$\boxed{\gamma = \text{Im}\,(\omega) \approx \left[-g\left(n_0'/n_0\right)\right]^{1/2}} \qquad (6.55)$$

Fig. 6.12 A "flute"
instability

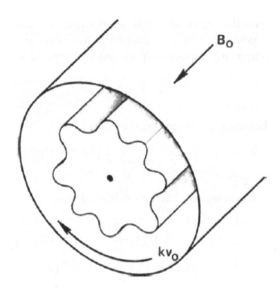

Note that the real part of ω is $\frac{1}{2}kv_0$. Since v_0 is an ion velocity, this is a low-frequency oscillation, as previously assumed. The factor of $\frac{1}{2}$ is merely a consequence of neglecting v_{0e}. The wave is stationary in the frame in which the density-weighted average of all the v_0's is zero, which in this case is the frame moving at $\frac{1}{2}v_0$. The laboratory frame has no particular significance in this case.

This instability, which has $\mathbf{k} \perp \mathbf{B}_0$, is sometimes called a "flute" instability for the following reason. In a cylinder, the waves travel in the θ direction if the forces are in the r direction. The surfaces of constant density then resemble fluted Greek columns (Fig. 6.12).

6.8   Resistive Drift Waves

A simple example of a universal instability is the resistive drift wave. In contrast to gravitational flute modes, drift waves have a small but finite component of $\mathbf{k}$ along $\mathbf{B}_0$. The constant density surfaces. Therefore, resemble flutes with a slight helical twist (Fig. 6.13). If we enlarge the cross section enclosed by the box in Fig. 6.13 and straighten it out into Cartesian geometry it would appear as in Fig. 6.14. The only driving force for the instability is the pressure gradient $KT\,\nabla n_0$ (we assume $KT = $ constant for simplicity). In this case, the zeroth-order drifts (for $\mathbf{E}_0 = 0$) are

$$\mathbf{v}_{i0} = \mathbf{v}_{Di} = \frac{KT_i}{eB_0}\frac{n_0'}{n_0}\hat{\mathbf{y}} \qquad (6.56)$$

$$\mathbf{v}_{e0} = \mathbf{v}_{De} = -\frac{KT_e}{eB_0}\frac{n_0'}{n_0}\hat{\mathbf{y}} \qquad (6.57)$$

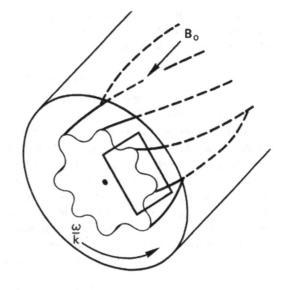

Fig. 6.13 Geometry of a drift instability in a cylinder. The region in the rectangle is shown in detail in Fig. 6.15

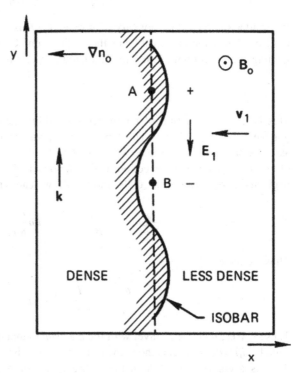

Fig. 6.14 Physical mechanism of a drift wave

From our experience with the flute instability, we might expect drift waves to have a phase velocity of the order of v_{Di} or v_{De}. We shall show that ω/k_y is approximately equal to v_{De}.

Since drift waves have finite k_z, electrons can flow along $\mathbf{B}_0$ to establish a thermodynamic equilibrium among themselves (cf. discussion of Sect. 4.10). They will then obey the Boltzmann relation (Sect. 3.5):

$$n_1/n_0 = e\phi_1/KT_e \tag{6.58}$$

At point A in Fig. 6.14 the density is larger than in equilibrium, n_1 is positive, and therefore ϕ_1 is positive. Similarly, at point B, n_1 and ϕ_1 are negative. The difference in potential means there is an electric field $\mathbf{E}_1$ between A and B. Just as in the case of the flute instability, $\mathbf{E}_1$ causes a drift $\mathbf{v}_1 = \mathbf{E}_1 \times \mathbf{B}_0/B_0^2$ in the x direction. As the wave passes by, traveling in the y direction, an observer at point A will see n_1 and ϕ_1 oscillating in time. The drift $\mathbf{v}_1$ will also oscillate in time, and in fact it is $\mathbf{v}_1$ which causes the density to oscillate. Since there is a gradient ∇n_0 in the $-x$ direction, the drift $\mathbf{v}_1$ will bring plasma of different density to a fixed observer A. A drift wave, therefore, has a motion such that the fluid moves back and forth in the x direction although the wave travels in the y direction.

To be more quantitative, the magnitude of v_{1x} is given by

$$v_{1x} = E_y/B_0 = -ik_y\phi_1/B_0 \tag{6.59}$$

We shall assume v_{1x} does not vary with x and that k_z is much less than k_y; that is, the fluid oscillates incompressibly in the x direction. Consider now the number of guiding centers brought into 1 m^3 at a fixed point A; it is obviously

$$\partial n_1/\partial t = -v_{1x}\partial n_0/\partial x \tag{6.60}$$

This is just the equation of continuity for guiding centers, which, of course, do not have a fluid drift $\mathbf{v}_D$. The term $n_0\,\nabla\cdot\mathbf{v}_1$ vanishes because of our previous assumption. The difference between the density of guiding centers and the density of particles n_1 gives a correction to Eq. (6.60) which is higher order and may be neglected here. Using Eqs. (6.59) and (6.58), we can write Eq. (6.60) as

$$-i\omega n_1 = \frac{ik_y\phi_1}{B_0}n_0' = -i\omega\frac{e\phi_1}{KT_e}n_0 \tag{6.61}$$

Thus we have

$$\frac{\omega}{k_y} = -\frac{KT_e}{eB_0}\frac{n_0'}{n_0} = v_{De} \tag{6.62}$$

These waves, therefore, travel with the *electron* diamagnetic drift velocity and are called *drift waves*. This is the velocity in the y, or azimuthal, direction. In addition, there is a component of $\mathbf{k}$ in the z direction. For reasons not given here, this component must satisfy the conditions

$$k_z \ll k_y \quad v_{thi} \ll \omega/k_z \ll v_{the} \tag{6.63}$$

To see why drift waves are unstable, one must realize that v_{1x} is not quite E_y/B_0 for the ions. There are corrections due to the polarization drift, Eq. (2.66), and the nonuniform $\mathbf{E}$ drift, Eq. (2.59). The result of these drifts is always to make the potential distribution ϕ_1 lag behind the density distribution n_1 (Problem 4.1). This phase shift causes $\mathbf{v}_1$ to be outward where the plasma has already been shifted outward, and vice versa; hence the perturbation grows. In the absence of the phase shift, n_1 and ϕ_1 would be 90° out of phase, as shown in Fig. 6.14, and drift waves would be purely oscillatory.

The role of resistivity comes in because the field $\mathbf{E}_1$ must not be short-circuited by electron flow along $\mathbf{B}_0$. Electron–ion collisions, together with a long distance $\frac{1}{2}\lambda_z$ between crest and trough of the wave, make it possible to have a resistive potential drop and a finite value of $\mathbf{E}_1$. The dispersion relation for resistive drift waves is approximately

$$\boxed{\omega^2 + i\sigma_\parallel(\omega - \omega_*) = 0} \tag{6.64}$$

where

$$\boxed{\omega_* \equiv k_y v_{De}} \tag{6.65}$$

and

$$\sigma_\parallel \equiv \frac{k_z^2}{k_y^2}\Omega_c(\omega_c\tau_{ei}) \tag{6.66}$$

If $\sigma_\parallel$ is large compared with ω, Eq. (6.64) can be satisfied only if $\omega \approx \omega_*$. In that case, we may replace ω by ω_* in the first term. Solving for ω, we then obtain

$$\boxed{\omega \approx \omega_* + (i\omega_*^2/\sigma_\parallel)} \tag{6.67}$$

This shows that $\mathrm{Im}(\omega)$ is always positive and is proportional to the resistivity η. Drift waves are, therefore, unstable and will eventually occur in any plasma with a density gradient. Fortunately, the growth rate is rather small, and there are ways to stop it altogether by making $\mathbf{B}_0$ nonuniform.

Note that Eq. (6.52) for the flute instability and Eq. (6.64) for the drift instability have different structures. In the former, the coefficients are real, and ω is complex when the discriminant of the quadratic is negative; this is typical of a *reactive* instability. In the latter, the coefficients are complex, so ω is always complex; this is typical of a *dissipative* instability.

Problem

6.10 A toroidal hydrogen plasma with circular cross section has major radius $R = 50$ cm, minor radius $a = 2$ cm, $B = 1$ T, $KT_e = 10$ eV, $KT_i = 1$ eV, and $n_0 = 10^{19}\,\mathrm{m}^{-3}$. Taking $n_0/n_0' \simeq a/2$ and $g \simeq (KT_e + KT_i)/MR$, estimate the growth rates of the $m = 1$ resistive drift wave and the $m = 1$ gravitational flute mode. (One can usually apply the slab-geometry formulas to cylindrical geometry by replacing k_y by m/r, where m is the azimuthal mode number.)

6.9   The Weibel Instability[1]

As an example of an instability driven by anisotropy of the distribution function, we give a physical picture (due to B. D. Fried) of the Weibel instability, in which a magnetic perturbation is made to grow. This will also serve as an example of an electromagnetic instability. Let the ions be fixed, and let the electrons be hotter in the y direction than in the x or z directions. There is then a preponderance of fast electrons in the $\pm y$ directions (Fig. 6.15), but equal numbers flow up and down, so that there is no net current. Suppose a field $\mathbf{B} = B_z \hat{\mathbf{z}} \cos kx$ spontaneously arises from noise. The Lorentz force $-e\mathbf{v} \times \mathbf{B}$ then bends the electron trajectories as shown by the dashed curves, with the result that downward-moving electrons congregate at A and upward-moving ones at B. The resulting current sheets $\mathbf{j} = -en_0\mathbf{v}_e$ are phased exactly right to generate a $\mathbf{B}$ field of the shape assumed, and the perturbation grows. Though the general case requires a kinetic treatment, the simple case $v_y = v_0$, $v_x = v_z = 0$ has been solved by Fried from this physical picture, yielding a growth rate $\gamma \approx \omega_p v_0/c$.

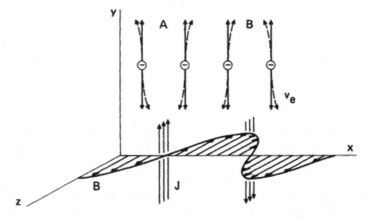

Fig. 6.15 Physical mechanism of the Weibel instability

---

[1] A salute to a good friend. Erich Weibel (1925–1983).

Chapter 7
Kinetic Theory

7.1 The Meaning of $f(\mathbf{v})$

The fluid theory we have been using so far is the simplest description of a plasma; it is indeed fortunate that this approximation is sufficiently accurate to describe the majority of observed phenomena. There are some phenomena, however, for which a fluid treatment is inadequate. For these, we need to consider the velocity distribution function $f(\mathbf{v})$ for each species; this treatment is called kinetic theory. In fluid theory, the dependent variables are functions of only four independent variables: x, y, z, and t. This is possible because the velocity distribution of each species is assumed to be Maxwellian everywhere and can therefore be uniquely specified by only one number, the temperature T. Since collisions can be rare in high-temperature plasmas, deviations from thermal equilibrium can be maintained for relatively long times. As an example, consider two velocity distributions $f_1(v_x)$ and $f_2(v_x)$ in a one-dimensional system (Fig. 7.1). These two distributions will have entirely different behaviors, but as long as the areas under the curves are the same, fluid theory does not distinguish between them.

The density is a function of four scalar variables: $n = n(\mathbf{r}, t)$. When we consider velocity distributions, we have seven independent variables: $f = f(\mathbf{r}, \mathbf{v}, t)$. By $f(\mathbf{r}, \mathbf{v}, t)$, we mean that the number of particles per m^3 at position $\mathbf{r}$ and time t with velocity components between v_x and $v_x + dv_x$, v_y and $v_y + dv_y$, and v_z and $v_z + dv_z$ is

$$f(x, y, z, v_x, v_y, v_z, t) \, dv_x dv_y dv_z$$

The integral of this is written in several equivalent ways:

$$n(\mathbf{r}, t) = \int_{-\infty}^{\infty} dv_x \int_{-\infty}^{\infty} dv_y \int_{-\infty}^{\infty} dv_z f(\mathbf{r}, \mathbf{v}, t) = \int_{-\infty}^{\infty} f(\mathbf{r}, \mathbf{v}, t) d^3 v$$
$$= \int_{-\infty}^{\infty} f(\mathbf{r}, \mathbf{v}, t) d\mathbf{v} \tag{7.1}$$

© Springer International Publishing Switzerland 2016
F.F. Chen, *Introduction to Plasma Physics and Controlled Fusion*,
DOI 10.1007/978-3-319-22309-4_7

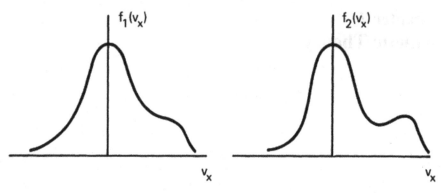

Fig. 7.1 Examples of non-Maxwellian distribution functions

Note that $d\mathbf{v}$ is not a vector; it stands for a three-dimensional volume element in velocity space. If f is normalized so that

$$\int_{-\infty}^{\infty} \hat{f}(\mathbf{r}, \mathbf{v}, t)d\mathbf{v} = 1 \tag{7.2}$$

it is a probability, which we denote by $\hat{f}$. Thus

$$f(\mathbf{r}, \mathbf{v}, t) = n(\mathbf{r}, t)\hat{f}(\mathbf{r}, \mathbf{v}, t) \tag{7.3}$$

Note that $\hat{f}$ is still a function of seven variables, since the shape of the distribution, as well as the density, can change with space and time. From Eq. (7.2), it is clear that $\hat{f}$ has the dimensions $(m/s)^{-3}$; and consequently, from Eq. (7.3), f has the dimensions s^3-m^{-6}.

A particularly important distribution function is the Maxwellian:

$$\hat{f}_m = (m/2\pi KT)^{3/2}\exp(-v^2/v_{th}^2) \tag{7.4}$$

where

$$v \equiv \left(v_x^2 + v_y^2 + v_z^2\right)^{1/2} \quad \text{and} \quad v_{th} \equiv (2KT/m)^{1/2} \tag{7.5}$$

By using the definite integral

$$\int_{-\infty}^{\infty} \exp(-x^2)dx = \sqrt{\pi} \tag{7.6}$$

one easily verifies that the integral of $\hat{f}_m$ over $dv_x \, dv_y \, dv_z$ is unity.

Fig. 7.2 Three-
dimensional velocity space

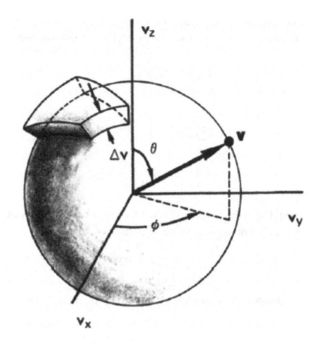

There are several average velocities of a Maxwellian distribution that are
commonly used. In Sect. 1.3, we saw that the root-mean-square velocity is given by

$$\left(\overline{v^2}\right)^{1/2} = (3KT/m)^{1/2} \tag{7.7}$$

The average magnitude of the velocity $|\mathbf{v}|$, or simply $\bar{v}$, is found as follows:

$$\bar{v} = \int_{-\infty}^{\infty} v\hat{f}(\mathbf{v})d^3v \tag{7.8}$$

Since $\hat{f}_m$ is isotropic, the integral is most easily done in spherical coordinates in v
space (Fig. 7.2). Since the volume element of each spherical shell is $4\pi v^2\,dv$,
we have

$$\bar{v} = (m/2\pi KT)^{3/2}\int_0^{\infty} v\left[\exp\left(-v^2/v_{\mathrm{th}}^2\right)\right]4\pi v^2 dv \tag{7.9}$$

$$= \left(\pi v_{\mathrm{th}}^2\right)^{-3/2}4\pi v_{\mathrm{th}}^4\int_0^{\infty}\left[\exp\left(-y^2\right)\right]y^3 dy \tag{7.10}$$

The definite integral has a value $\frac{1}{2}$, found by integration by parts. Thus

$$\bar{v} = 2\pi^{-1/2}v_{\mathrm{th}} = 2(2KT/\pi m)^{1/2} \tag{7.11}$$

The velocity component in a *single direction*, say v_x, has a different average. Of course, $\bar{v}_x$ vanishes for an isotropic distribution; but $|\bar{v}_x|$ does not:

$$|\bar{v}_x| = \int |v_x| \hat{f}_m(\mathbf{v}) d^3v \qquad (7.12)$$

$$= \left(\frac{m}{2\pi KT}\right)^{3/2} \int_{-\infty}^{\infty} dv_y \exp\left(\frac{-v_y^2}{v_{th}^2}\right) \int_{-\infty}^{\infty} dv_z \exp\left(\frac{-v_z^2}{v_{th}^2}\right)$$

$$\times \int_0^{\infty} 2v_x \exp\left(\frac{-v_x^2}{v_{th}^2}\right) dv_x \qquad (7.13)$$

From Eq. (7.6), each of the first two integrals has the value $\pi^{1/2} v_{th}$. The last integral is simple and has the value v_{th}^2. Thus we have

$$|\bar{v}_x| = \left(\pi v_{th}^2\right)^{-3/2} \pi v_{th}^4 = \pi^{-1/2} v_{th} = (2KT/\pi m)^{1/2} \qquad (7.14)$$

The random flux crossing an imaginary plane from one side to the other is given by

$$\Gamma_{random} = \frac{1}{2} n |\bar{v}_x| = \frac{1}{4} n \bar{v} \qquad (7.15)$$

Here we have used Eq. (7.11) and the fact that only half the particles cross the plane in either direction. To summarize: For a Maxwellian,

$$v_{rms} = (3KT/m)^{1/2} \qquad (7.7)$$

$$|\bar{v}| = 2(2KT/\pi m)^{1/2} \qquad (7.11)$$

$$|\bar{v}_x| = (2KT/\pi m)^{1/2} \qquad (7.14)$$

$$\bar{v}_x = 0 \qquad (7.16)$$

For an isotropic distribution like a Maxwellian, we can define another function $g(v)$ which is a function of the scalar magnitude of $\mathbf{v}$ such that

$$\int_0^{\infty} g(v) dv = \int_{-\infty}^{\infty} f(\mathbf{v}) d^3v \qquad (7.17)$$

For a Maxwellian, we see from Eq. (7.9) that

$$g(v) = 4\pi n (m/2\pi KT)^{3/2} v^2 \exp\left(-v^2/v_{th}^2\right) \qquad (7.18)$$

Figure 7.3 shows the difference between $g(v)$ and a one-dimensional Maxwellian distribution $f(v_x)$. Although $f(v_x)$ is maximum for $v_x = 0$, $g(v)$ is zero for $v = 0$.

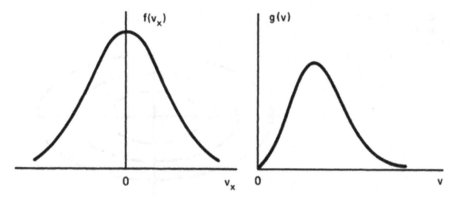

Fig. 7.3 One- and three-dimensional Maxwellian velocity distributions

Fig. 7.4 A spatially
varying one-dimensional
distribution $f(x, v_x)$

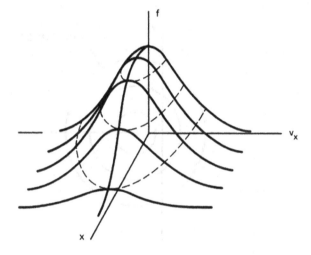

This is just a consequence of the vanishing of the volume in phase space (Fig. 7.2) for $v = 0$. Sometimes $g(v)$ is carelessly denoted by $f(v)$, as distinct from $f(\mathbf{v})$; but $g(v)$ is a different function of its argument than $f(\mathbf{v})$ is of *its* argument. From Eq. (7.18), it is clear that $g(v)$ has dimensions s/m^4.

It is impossible to draw a picture of $f(\mathbf{r}, \mathbf{v})$ at a given time t unless we reduce the number of dimensions. In a one-dimensional system, $f(x, v_x)$ can be depicted as a surface (Fig. 7.4). Intersections of that surface with planes $x =$ constant are the velocity distributions $f(v_x)$. Intersections with planes $v_x =$ constant give density profiles for particles with a given v_x. If all the curves $f(v_x)$ happen to have the same shape, a curve through the peaks would represent the density profile. The dashed curves in Fig. 7.4 are intersections with planes $f =$ constant; these are level curves, or curves of constant f. A projection of these curves onto the $x - v_x$

Fig. 7.5 Contours of
constant f for a
two-dimensional,
anisotropic distribution

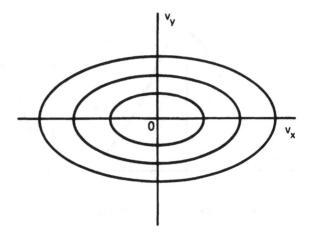

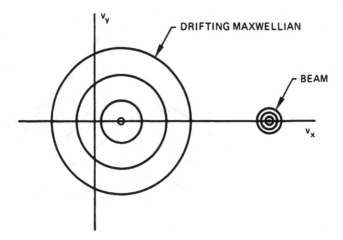

Fig. 7.6 Contours of constant f for a drifting Maxwellian distribution and a "beam" in two dimensions

plane will give a topographical map of f. Such maps are very useful for getting a preliminary idea of how the plasma behaves; an example will be given in the next section.

Another type of contour map can be made for f if we consider $f(\mathbf{v})$ at a given point in space. For instance, if the motion is two dimensional, the contours of $f(v_x, v_y)$ will be circles if f is isotropic in v_x, v_y. An anisotropic distribution would have elliptical contours (Fig. 7.5). A drifting Maxwellian would have circular contours displaced from the origin, and a beam of particles traveling in the x direction would show up as a separate spike (Fig. 7.6).

A loss cone distribution of a mirror-confined plasma can be represented by contours of f in $v_\perp$, $v_\parallel$ space. Figure 7.7 shows how these would look.

Fig. 7.7 Contours of
constant f for a loss-cone
distribution. Here $v_\parallel$ and $v_\perp$
stand for the components
of $\mathbf{v}$ along and
perpendicular to the
magnetic field, respectively

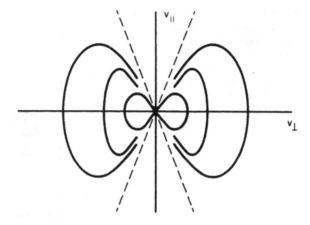

7.2   Equations of Kinetic Theory

The fundamental equation which $f(\mathbf{r}, \mathbf{v}, t)$ has to satisfy is the Boltzmann equation:

$$\frac{\partial f}{\partial t} + \mathbf{v} \cdot \nabla f + \frac{\mathbf{F}}{m} \cdot \frac{\partial f}{\partial \mathbf{v}} = \left(\frac{\partial f}{\partial t} \right)_c \qquad (7.19)$$

Here $\mathbf{F}$ is the force acting on the particles, and $(\partial f/\partial t)_c$ is the time rate of change of f due to collisions. The symbol ∇ stands, as usual, for the gradient in (x, y, z) space. The symbol $\partial/\partial \mathbf{v}$ or ∇_v stands for the gradient in velocity space:

$$\frac{\partial}{\partial \mathbf{v}} = \hat{\mathbf{x}} \frac{\partial}{\partial v_x} + \hat{\mathbf{y}} \frac{\partial}{\partial v_y} + \hat{\mathbf{z}} \frac{\partial}{\partial v_z} \qquad (7.20)$$

The meaning of the Boltzmann equation becomes clear if one remembers that f is a function of seven independent variables. The total derivative of f with time is, therefore

$$\frac{df}{dt} = \frac{\partial f}{\partial t} + \frac{\partial f}{\partial x} \frac{dx}{dt} + \frac{\partial f}{\partial y} \frac{dy}{dt} + \frac{\partial f}{\partial z} \frac{dz}{dt} + \frac{\partial f}{\partial v_x} \frac{dv_x}{dt} + \frac{\partial f}{\partial v_y} \frac{dv_y}{dt} + \frac{\partial f}{\partial v_z} \frac{dv_z}{dt} \qquad (7.21)$$

Here, $\partial f/\partial t$ is the *explicit* dependence on time. The next three terms are just $\mathbf{v} \cdot \nabla f$. With the help of Newton's third law,

$$m \frac{d\mathbf{v}}{dt} = \mathbf{F} \qquad (7.22)$$

the last three terms are recognized as $(\mathbf{F}/m) \cdot (\partial f/\partial \mathbf{v})$. As discussed previously in Sect. 3.3, the total derivative df/dt can be interpreted as the rate of change as seen in

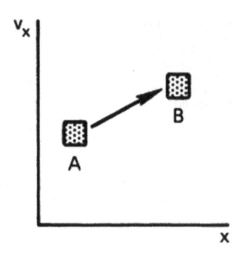

a frame moving with the particles. The difference is that now we must consider the
particles to be moving in six-dimensional $(\mathbf{r}, \mathbf{v})$ space; df/dt is the convective
derivative in phase space. The Boltzmann equation (7.19) simply says that df/dt
is zero unless there are collisions. That this should be true can be seen from the
one-dimensional example shown in Fig. 7.8.

The group of particles in an infinitesimal element $dx \, dv_x$ at A all have velocity v_x
and position x. The density of particles in this phase space is just $f(x, v_x)$. As time
passes, these particles will move to a different x as a result of their velocity v_x and
will change their velocity as a result of the forces acting on them. Since the forces
depend on x and v_x only, all the particles at A will be accelerated the same amount.
After a time t, all the particles will arrive at B in phase space. Since all the particles
moved together, the density at B will be the same as at A. If there are collisions,
however, the particles can be scattered; and f can be changed by the term $(\partial f/\partial t)_c$.

In a sufficiently hot plasma, collisions can be neglected. If, furthermore, the
force $\mathbf{F}$ is entirely electromagnetic, Eq. (7.19) takes the special form

$$\boxed{\frac{\partial f}{\partial t} + \mathbf{v} \cdot \nabla f + \frac{q}{m}(\mathbf{E} + \mathbf{v} \times \mathbf{B}) \cdot \frac{\partial f}{\partial \mathbf{v}} = 0} \qquad (7.23)$$

This is called the *Vlasov equation*. Because of its comparative simplicity, this is the
equation most commonly studied in kinetic theory. When there are collisions with
neutral atoms, the collision term in Eq. (7.19) can be approximated by

$$\left(\frac{\partial f}{\partial t}\right)_c = \frac{f_n - f}{\tau} \qquad (7.24)$$

where f_n is the distribution function of the neutral atoms, and τ is a constant
collision time. This is called a *Krook collision term*. It is the kinetic generalization

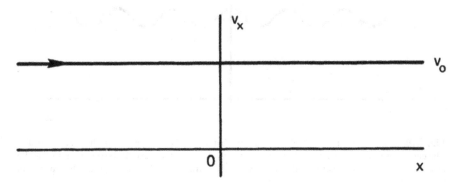

Fig. 7.9 Representation in one-dimensional phase space of a beam of electrons all with the same velocity v_0. The distribution function $f(x, v_x)$ is infinite along the line and zero elsewhere. The line is also the trajectory of individual electrons, which move in the direction of the arrow.

of the collision term in Eq. (5.5). When there are Coulomb collisions, Eq. (7.19) can be approximated by

$$\frac{df}{dt} = -\frac{\partial}{\partial \mathbf{v}} \cdot (f\langle \mathbf{\Delta v}\rangle) \frac{1}{2}\frac{\partial^2}{\partial \mathbf{v}\partial \mathbf{v}} : (f\langle \mathbf{\Delta v}\,\mathbf{\Delta v}\rangle) \tag{7.25}$$

This is called the *Fokker–Planck equation*; it takes into account binary Coulomb collisions only. Here, $\mathbf{\Delta v}$ is the change of velocity in a collision, and Eq. (7.25) is a shorthand way of writing a rather complicated expression.

The fact that df/dt is constant in the absence of collisions means that particles follow the contours of constant f as they move around in phase space. As an example of how these contours can be used, consider the beam-plasma instability of Sect. 6.6. In the unperturbed plasma, the electrons all have velocity v_0, and the contour of constant f is a straight line (Fig. 7.9). The function $f(x, v_x)$ is a wall rising out of the plane of the paper at $v_x = v_0$. The electrons move along the trajectory shown. When a wave develops, the electric field $\mathbf{E}_1$ causes electrons to suffer changes in v_x as they stream along. The trajectory then develops a sinusoidal ripple (Fig. 7.10). This ripple travels at the phase velocity, not the particle velocity. Particles stay on the curve as they move relative to the wave. If E_1 becomes very large as the wave grows, and if there are a few collisions, some electrons will be trapped in the electrostatic potential of the wave. In coordinate space, the wave potential appears as in Fig. 7.11. In phase space, $f(x, v_x)$ will have peaks wherever there is a potential trough (Fig. 7.12). Since the contours of f are also electron trajectories, one sees that some electrons move in closed orbits in phase space; these are just the trapped electrons.

Electron trapping is a nonlinear phenomenon which cannot be treated by straightforward solution of the Vlasov equation. However, electron trajectories can be followed on a computer, and the results are often presented in the form of a plot like Fig. 7.12. An example of a numerical result is shown in Fig. 7.13.

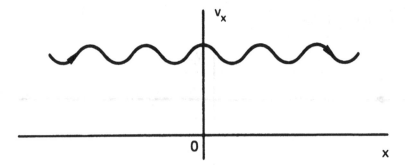

Fig. 7.10 Appearance of the graph of Fig. 7.9 when a plasma wave exists in the electron beam. The entire pattern moves to the right with the phase velocity of the wave. If the observer goes to the frame of the wave, the pattern would stand still, and electrons would be seen to trace the curve with the velocity $v_0 - v_\phi$.

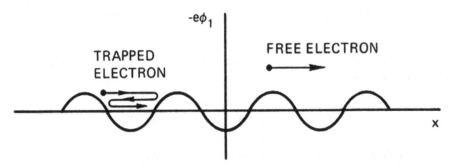

Fig. 7.11 The potential of a plasma wave, as seen by an electron. The pattern moves with the velocity v_ϕ. An electron with small velocity relative to the wave would be trapped in a potential trough and be carried along with the wave.

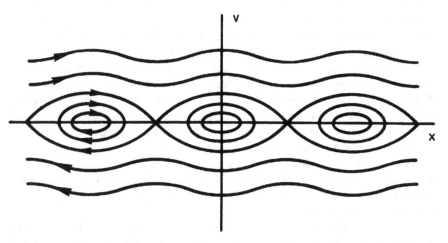

Fig. 7.12 Electron trajectories, or contours of constant f, as seen in the wave frame, in which the pattern is stationary. This type of diagram, appropriate for finite distributions $f(v)$, is easier to understand than the δ-function distribution of Fig. 7.10.

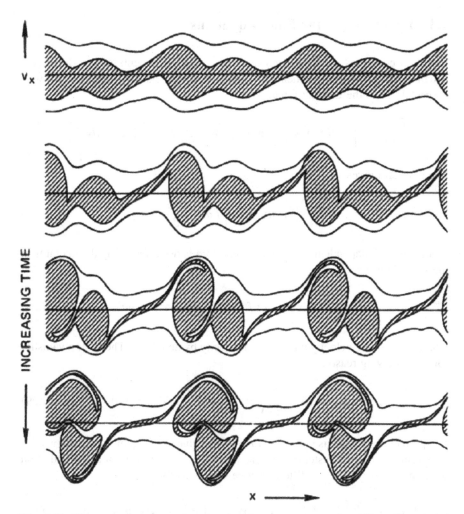

Fig. 7.13 Phase-space contours for electrons in a two-stream instability. The shaded region, initially representing low velocities in the lab frame, is devoid of electrons. As the instability develops past the linear stage, these empty regions in phase space twist into shapes resembling "water bags" [From H. L. Berk, C. E. Nielson, and K. V. Roberts, Phys. Fluids **13**, 986 (1970)].

This is for a two-stream instability in which initially the contours of f have a gap near $v_x = 0$ which separates electrons moving in opposite directions. The development of this uninhabited gap with time is shown by the shaded regions in Fig. 7.13. This figure shows that the instability progressively distorts $f(\mathbf{v})$ in a way which would be hard to describe analytically.

7.3   Derivation of the Fluid Equations

The fluid equations we have been using are simply moments of the Boltzmann equation. The lowest moment is obtained by integrating Eq. (7.19) with **F** specialized to the Lorentz force:

$$\int \frac{\partial f}{\partial t} d\mathbf{v} + \int \mathbf{v} \cdot \nabla f \, d\mathbf{v} + \frac{q}{m} \int (\mathbf{E} + \mathbf{v} \times \mathbf{B}) \cdot \frac{\partial f}{\partial \mathbf{v}} d\mathbf{v} = \int \left(\frac{\partial f}{\partial t} \right)_c d\mathbf{v} \qquad (7.26)$$

The first term gives

$$\int \frac{\partial f}{\partial t} d\mathbf{v} = \frac{\partial}{\partial t} \int f \, d\mathbf{v} = \frac{\partial n}{\partial t} \qquad (7.27)$$

Since **v** is an independent variable and therefore is not affected by the operator ∇, the second term gives

$$\int \mathbf{v} \cdot \nabla f \, d\mathbf{v} = \nabla \cdot \int \mathbf{v} f \, d\mathbf{v} = \nabla \cdot (n\bar{\mathbf{v}}) \equiv \nabla \cdot (n\mathbf{u}) \qquad (7.28)$$

where the average velocity **u** is the fluid velocity by definition. The **E** term vanishes for the following reason:

$$\int \mathbf{E} \cdot \frac{\partial f}{\partial \mathbf{v}} d\mathbf{v} = \int \frac{\partial}{\partial \mathbf{v}} \cdot (f\mathbf{E}) d\mathbf{v} = \int_{S_\infty} f\mathbf{E} \cdot d\mathbf{S} = 0 \qquad (7.29)$$

The perfect divergence is integrated to give the value of $f\mathbf{E}$ on the surface at $v = \infty$. This vanishes if $f \to 0$ faster than v^{-2} as $v \to \infty$, as is necessary for any distribution with finite energy. The $\mathbf{v} \times \mathbf{B}$ term can be written as follows:

$$\int (\mathbf{v} \times \mathbf{B}) \cdot \frac{\partial f}{\partial \mathbf{v}} d\mathbf{v} = \int \frac{\partial}{\partial \mathbf{v}} \cdot (f\mathbf{v} \times \mathbf{B}) d\mathbf{v} - \int f \frac{\partial}{\partial \mathbf{v}} \times (\mathbf{v} \times \mathbf{B}) d\mathbf{v} = 0 \qquad (7.30)$$

The first integral can again be converted to a surface integral. For a Maxwellian, f falls faster than any power of v as $v \to \infty$, and the integral therefore vanishes. The second integral vanishes because $\mathbf{v} \times \mathbf{B}$ is perpendicular to $\partial/\partial \mathbf{v}$. Finally, the fourth term in Eq. (7.26) vanishes because collisions cannot change the total number of particles (recombination is not considered here). Equations (7.27)–(7.30) then yield the *equation of continuity*:

$$\frac{\partial n}{\partial t} + \nabla \cdot (n\mathbf{u}) = 0 \qquad (7.31)$$

The next moment of the Boltzmann equation is obtained by multiplying Eq. (7.19) by $m\mathbf{v}$ and integrating over $d\mathbf{v}$. We have

$$m\int \mathbf{v}\frac{\partial f}{\partial t}d\mathbf{v} + m\int \mathbf{v}(\mathbf{v}\cdot\nabla)f\,d\mathbf{v} + q\int \mathbf{v}(\mathbf{E}+\mathbf{v}\times\mathbf{B})\cdot\frac{\partial f}{\partial\mathbf{v}}d\mathbf{v}$$
$$= \int m\mathbf{v}\left(\frac{\partial f}{\partial t}\right)_c d\mathbf{v} \tag{7.32}$$

The right-hand side is the change of momentum due to collisions and will give the term $\mathbf{P}_{ij}$ in Eq. (5.58). The first term in Eq. (7.32) gives

$$m\int \mathbf{v}\frac{\partial f}{\partial t}d\mathbf{v} = m\frac{\partial}{\partial t}\int \mathbf{v}f\,d\mathbf{v} \equiv m\frac{\partial}{\partial t}(n\mathbf{u}) \tag{7.33}$$

The third integral in Eq. (7.32) can be written

$$\int \mathbf{v}(\mathbf{E}+\mathbf{v}\times\mathbf{B})\cdot\frac{\partial f}{\partial\mathbf{v}}d\mathbf{v} = \int \frac{\partial}{\partial\mathbf{v}}\cdot[f\mathbf{v}(\mathbf{E}+\mathbf{v}\times\mathbf{B})]d\mathbf{v}$$
$$- \int f\mathbf{v}\frac{\partial}{\partial\mathbf{v}}\cdot(\mathbf{E}+\mathbf{v}\times\mathbf{B})d\mathbf{v} - \int f(\mathbf{E}+\mathbf{v}\times\mathbf{B})\cdot\frac{\partial}{\partial\mathbf{v}}\mathbf{v}\,d\mathbf{v} \tag{7.34}$$

The first two integrals on the right-hand side vanish for the same reasons as before, and $\partial\mathbf{v}/\partial\mathbf{v}$ is just the identity tensor $\mathbf{I}$. We therefore have

$$q\int \mathbf{v}(\mathbf{E}+\mathbf{v}\times\mathbf{B})\cdot\frac{\partial f}{\partial\mathbf{v}}d\mathbf{v} = -q\int(\mathbf{E}+\mathbf{v}\times\mathbf{B})f\,d\mathbf{v} = -qn(\mathbf{E}+\mathbf{u}\times\mathbf{B}) \tag{7.35}$$

Finally, to evaluate the second integral in Eq. (7.32), we first make use of the fact that $\mathbf{v}$ is an independent variable not related to ∇ and write

$$\int \mathbf{v}(\mathbf{v}\cdot\nabla)f\,d\mathbf{v} = \int \nabla\cdot(f\mathbf{v}\mathbf{v})d\mathbf{v} = \nabla\cdot\int f\mathbf{v}\mathbf{v}\,d\mathbf{v} \tag{7.36}$$

Since the average of a quantity is $1/n$ times its weighted integral over $\mathbf{v}$, we have

$$\nabla\cdot\int f\mathbf{v}\mathbf{v}\,d\mathbf{v} = \nabla\cdot n\overline{\mathbf{v}\mathbf{v}} \tag{7.37}$$

Now we may separate $\mathbf{v}$ into the average (fluid) velocity $\mathbf{u}$ and a thermal velocity $\mathbf{w}$:

$$\mathbf{v} = \mathbf{u} + \mathbf{w} \tag{7.38}$$

Since $\mathbf{u}$ is already an average, we have

$$\nabla\cdot(n\overline{\mathbf{v}\mathbf{v}}) = \nabla\cdot(n\mathbf{u}\mathbf{u}) + \nabla\cdot(n\overline{\mathbf{w}\mathbf{w}}) + 2\nabla\cdot(n\mathbf{u}\overline{\mathbf{w}}) \tag{7.39}$$

The average $\overline{\mathbf{w}}$ is obviously zero. The quantity $mn\overline{\mathbf{ww}}$ is precisely what is meant by the stress tensor $\mathbf{P}$:

$$\mathbf{P} \equiv mn\overline{\mathbf{ww}} \tag{7.40}$$

The remaining term in Eq. (7.39) can be written

$$\nabla \cdot (n\mathbf{uu}) = \mathbf{u}\nabla \cdot (n\mathbf{u}) + n(\mathbf{u} \cdot \nabla)\mathbf{u} \tag{7.41}$$

Collecting our results from Eqs. (7.33), (7.35), (7.40), and (7.41), we can write Eq. (7.32) as

$$m\frac{\partial}{\partial t}(n\mathbf{u}) + m\mathbf{u}\nabla \cdot (n\mathbf{u}) + mn(\mathbf{u} \cdot \nabla)\mathbf{u} + \nabla \cdot \mathbf{P} - qn(\mathbf{E} + \mathbf{u} \times \mathbf{B}) = \mathbf{P}_{ij} \tag{7.42}$$

The first term in Eq. (7.42) represents *ionization drag*, since $\mathbf{u}$ is small when an ion is first created. Combining the first two terms in Eq. (7.42) with the help of Eq. (7.31), we finally obtain the *fluid equation of motion*:

$$mn\left[\frac{\partial \mathbf{u}}{\partial t} + (\mathbf{u} \cdot \nabla)\mathbf{u}\right] = qn(\mathbf{E} + \mathbf{u} \times \mathbf{B}) - \nabla \cdot \mathbf{P} + \mathbf{P}_{ij} \tag{7.43}$$

This equation, an extension of Eq. (3.44), describes the flow of momentum. To treat the flow of energy, we may take the next moment of Boltzmann equation by multiplying by $\frac{1}{2}m\mathbf{vv}$ and integrating. We would then obtain the *heat flow equation*, in which the coefficient of thermal conductivity κ would arise in the same manner as did the stress tensor $\mathbf{P}$. The equation of state $p \propto \rho^\gamma$ is a simple form of the heat flow equation for $\kappa = 0$.

7.4   Plasma Oscillations and Landau Damping

As an elementary illustration of the use of the Vlasov equation, we shall derive the dispersion relation for electron plasma oscillations, which we treated from the fluid point of view in Sect. 4.3. This derivation will require a knowledge of contour integration. Those not familiar with this may skip to Sect. 7.5. A simpler but longer derivation not using the theory of complex variables appears in Sect. 7.6.

In zeroth order, we assume a uniform plasma with a distribution $f_0(\mathbf{v})$, and we let $\mathbf{B}_0 = \mathbf{E}_0 = 0$. In first order, we denote the perturbation in $f(\mathbf{r}, \mathbf{v}, t)$ by $f_1(\mathbf{r}, \mathbf{v}, t)$:

$$f(\mathbf{r}, \mathbf{v}, t) = f_0(\mathbf{v}) + f_1(\mathbf{r}, \mathbf{v}, t) \tag{7.44}$$

Since $\mathbf{v}$ is now an independent variable and is not to be linearized, the first-order Vlasov equation for electrons is

$$\frac{\partial f_1}{\partial t} + \mathbf{v} \cdot \nabla f_1 - \frac{e}{m}\mathbf{E}_1 \cdot \frac{\partial f_0}{\partial \mathbf{v}} = 0 \tag{7.45}$$

As before, we assume the ions are massive and fixed and that the waves are plane waves in the x direction

$$f_1 \propto e^{i(kx - \omega t)} \tag{7.46}$$

Then Eq. (7.45) becomes

$$-i\omega f_1 + ikv_x f_1 = \frac{e}{m}E_x \frac{\partial f_0}{\partial v_x} \tag{7.47}$$

$$f_1 = \frac{ieE_x}{m}\frac{\partial f_0/\partial v_x}{\omega - kv_x} \tag{7.48}$$

Poisson's equation gives

$$\varepsilon_0 \nabla \cdot \mathbf{E}_1 = ik\varepsilon_0 E_x = -en_1 = -e\iiint f_1 d^3 v \tag{7.49}$$

Substituting for f_1 and dividing by $ik\varepsilon_0 E_x$, we have

$$1 = -\frac{e^2}{km\varepsilon_0}\iiint \frac{\partial f_0/\partial v_x}{\omega - kv_x}d^3 v \tag{7.50}$$

A factor n_0 can be factored out if we replace f_0 by a normalized function $\hat{f}_0$:

$$1 = -\frac{\omega_p^2}{k}\int_{-\infty}^{\infty} dv_z \int_{-\infty}^{\infty} dv_y \int_{-\infty}^{\infty} \frac{\partial \hat{f}_0(v_x, v_y, v_z)/\partial v_x}{\omega - kv_x}\partial v_x \tag{7.51}$$

If f_0 is a Maxwellian or some other factorable distribution, the integrations over v_y and v_z can be carried out easily. What remains is the one-dimensional distribution $\hat{f}_0(v_x)$. For instance, a one-dimensional Maxwellian distribution is

$$\hat{f}_m(v_x) = (m/2\pi KT)^{1/2}\exp(-mv_x^2/2KT) \tag{7.52}$$

The dispersion relation is, therefore,

$$1 = -\frac{\omega_p^2}{k^2}\int_{-\infty}^{\infty} \frac{\partial \hat{f}_0(v_x)/\partial v_x}{v_x - (\omega/k)} \tag{7.53}$$

Since we are dealing with a one-dimensional problem we may drop the subscript x, being careful not to confuse v (which is really v_x) with the total velocity v used earlier:

$$1 = \frac{\omega_p^2}{k^2} \int_{-\infty}^{\infty} \frac{\partial \hat{f}_0 / \partial v}{v - (\omega/k)} \, dv \tag{7.54}$$

Here, $\hat{f}_0$ is understood to be a one-dimensional distribution function, the integrations over v_y and v_z having been made. Equation (7.54) holds for any equilibrium distribution $\hat{f}_0(v)$; in particular, if $\hat{f}_0$ is Maxwellian, Eq. (7.52) is to be used for it.

The integral in Eq. (7.54) is not straightforward to evaluate because of the singularity at $v = \omega/k$. One might think that the singularity would be of no concern, because in practice ω is almost never real; waves are usually slightly damped by collisions or are amplified by some instability mechanism. Since the velocity v is a real quantity, the denominator in Eq. (7.54) never vanishes. Landau was the first to treat this equation properly. He found that even though the singularity lies off the path of integration, its presence introduces an important modification to the plasma wave dispersion relation—an effect not predicted by the fluid theory.

Consider an initial value problem in which the plasma is given a sinusoidal perturbation, and therefore k is real. If the perturbation grows or decays, ω will be complex. The integral in Eq. (7.54) must be treated as a contour integral in the complex v plane. Possible contours are shown in Fig. 7.14 for (a) an unstable wave, with $\text{Im}(\omega) > 0$, and (b) a damped wave, with $\text{Im}(\omega) < 0$. Normally, one would evaluate the line integral along the real v axis by the residue theorem:

$$\int_{C_1} G \, dv + \int_{C_2} G \, dv = 2\pi i R(\omega/k) \tag{7.55}$$

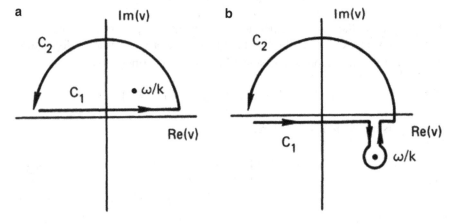

Fig. 7.14 Integration contours for the Landau problem for (**a**) $\text{Im}(\omega) > 0$ and (**b**) $\text{Im}(\omega) < 0$

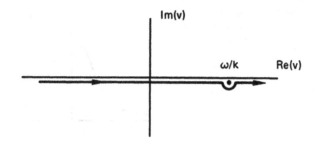

Fig. 7.15 Integration
contour in the complex v
plane for the case of small
$\mathrm{Im}(\omega)$

where G is the integrand, C_1 is the path along the real axis, C_2 is the semicircle at infinity, and $R(\omega/k)$ is the residue at ω/k. This works if the integral over C_2 vanishes. Unfortunately, this does not happen for a Maxwellian distribution, which contains the factor

$$\exp\left(-v^2/v_{\mathrm{th}}^2\right)$$

This factor becomes large for $v \to \pm i\infty$, and the contribution from C_2 cannot be neglected. Landau showed that when the problem is properly treated as an initial value problem the correct contour to use is the curve C_1 passing below the singularity. This integral must in general be evaluated numerically, and Fried and Conte have provided tables for the case when $\hat{f}_0$ is a Maxwellian.

Although an exact analysis of this problem is complicated, we can obtain an approximate dispersion relation for the case of large phase velocity and weak damping. In this case, the pole at ω/k lies near the real v axis (Fig. 7.15). The contour prescribed by Landau is then a straight line along the $\mathrm{Re}(v)$ axis with a small semicircle around the pole. In going around the pole, one obtains $2\pi i$ times half the residue there. Then Eq. (7.54) becomes

$$1 = \frac{\omega_p^2}{k^2}\left[P\int_{-\infty}^{\infty}\frac{\partial \hat{f}_0/\partial v}{v - (\omega/k)}dv + i\pi\frac{\partial \hat{f}_0}{\partial v}\bigg|_{v=\omega/k}\right] \tag{7.56}$$

where P stands for the Cauchy principal value. To evaluate this, we integrate along the real v axis but stop just before encountering the pole. If the phase velocity $v_\phi = \omega/k$ is sufficiently large, as we assume, there will not be much contribution from the neglected part of the contour, since both $\hat{f}_0$ and $\partial \hat{f}_0/\partial v$ are very small there (Fig. 7.16). The integral in Eq. (7.56) can be evaluated by integration by parts:

$$\int_{-\infty}^{\infty}\frac{\partial \hat{f}_0}{\partial v}\frac{dv}{v - v_\phi} = \left[\frac{\hat{f}_0}{v - v_\phi}\right]_{-\infty}^{\infty} - \int_{-\infty}^{\infty}\frac{-\hat{f}_0 dv}{\left(v - v_\phi\right)^2} = \int_{-\infty}^{\infty}\frac{\hat{f}_0 dv}{\left(v - v_\phi\right)^2} \tag{7.57}$$

Since this is just an average of $(v - v_\phi)^{-2}$ over the distribution, the real part of the dispersion relation can be written

Fig. 7.16 Normalized
Maxwellian distribution for
the case $v_\phi \gg v_{th}$

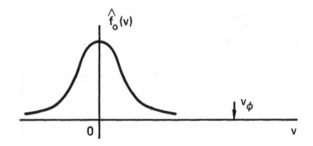

$$1 = \frac{\omega_p^2}{k^2} \overline{(v - v_\phi)^{-2}} \tag{7.58}$$

Since $v_\phi \gg v$ has been assumed, we can expand $(v - v_\phi)^{-2}$:

$$(v - v_\phi)^{-2} = v_\phi^{-2}\left(1 - \frac{v}{v_\phi}\right)^{-2} = v_\phi^{-2}\left(1 + \frac{2v}{v_\phi} + \frac{3v^2}{v_\phi^2} + \frac{4v^3}{v_\phi^3} + \cdots\right) \tag{7.59}$$

The odd terms vanish upon taking the average, and we have

$$\overline{(v - v_\phi)^{-2}} \approx v_\phi^{-2}\left(1 + \frac{3\overline{v^2}}{v_\phi^2}\right) \tag{7.60}$$

We now let $\hat{f}_0$ be Maxwellian and evaluate $\overline{v^2}$. Remembering that v here is an abbreviation for v_x, we can write

$$\frac{1}{2}m\overline{v_x^2} = \frac{1}{2}KT_e \tag{7.61}$$

there being only one degree of freedom. The dispersion relation (7.58) then becomes

$$1 = \frac{\omega_p^2}{k^2}\frac{k^2}{\omega^2}\left(1 + 3\frac{k^2}{\omega^2}\frac{KT_e}{m}\right) \tag{7.62}$$

$$\omega^2 = \omega_p^2 + \frac{\omega_p^2}{\omega^2}\frac{3KT_e}{m}k^2 \tag{7.63}$$

If the thermal correction is small, we may replace ω^2 by ω_p^2 in the second term. We then have

$$\omega^2 = \omega_p^2 + \frac{3KT_e}{m}k^2 \tag{7.64}$$

which is the same as Eq. (4.30), obtained from the fluid equations with $\gamma = 3$.

We now return to the imaginary term in Eq. (7.56). In evaluating this small term, it will be sufficiently accurate to neglect the thermal correction to the real part of ω and let $\omega^2 \approx \omega_p^2$. From Eqs. (7.57) and (7.60), we see that the principal value of the integral in Eq. (7.56) is approximately k^2/ω^2. Equation (7.56) now becomes

$$1 = \frac{\omega_p^2}{\omega^2} + i\pi\frac{\omega_p^2}{k^2}\frac{\partial \hat{f}_0}{\partial v}\bigg|_{v=v_\phi} \tag{7.65}$$

$$\omega^2\left(1 - i\pi\frac{\omega_p^2}{k^2}\left[\frac{\partial \hat{f}_0}{\partial v}\right]_{v=v_\phi}\right) = \omega_p^2 \tag{7.66}$$

Treating the imaginary term as small, we can bring it to the right-hand side and take the square root by Taylor series expansion. We then obtain

$$\boxed{\omega = \omega_p\left(1 + i\frac{\pi}{2}\frac{\omega_p^2}{k^2}\left[\frac{\partial \hat{f}_0}{\partial v}\right]_{v=v_\phi}\right)} \tag{7.67}$$

If $\hat{f}_0$ is a one-dimensional Maxwellian, we have

$$\frac{\partial \hat{f}_0}{\partial v} = (\pi v_{th}^2)^{-1/2}\left(\frac{-2v}{v_{th}^2}\right)\exp\left(\frac{-v^2}{v_{th}^2}\right) = -\frac{2v}{\sqrt{\pi}v_{th}^3}\exp\left(\frac{-v^2}{v_{th}^2}\right) \tag{7.68}$$

We may approximate v_ϕ by ω_p/k in the coefficient, but in the exponent we must keep the thermal correction in Eq. (7.64). The damping is then given by

$$\text{Im}(\omega) = -\frac{\pi}{2}\frac{\omega_p^3}{k^2}\frac{2\omega_p}{k\sqrt{\pi}}\frac{1}{v_{th}^3}\exp\left(\frac{-\omega^2}{k^2 v_{th}^2}\right)$$

$$= -\sqrt{\pi}\omega_p\left(\frac{\omega_p}{kv_{th}}\right)^3\exp\left(\frac{-\omega_p^2}{k^2 v_{th}^2}\right)\exp\left(\frac{-3}{2}\right) \tag{7.69}$$

$$\text{Im}\left(\frac{\omega}{\omega_p}\right) = -0.22\sqrt{\pi}\left(\frac{\omega_p}{kv_{th}}\right)^3\exp\left(\frac{-1}{2k^2\lambda_D^2}\right) \tag{7.70}$$

Since $\text{Im}(\omega)$ is negative, there is a collisionless damping of plasma waves; this is called *Landau damping*. As is evident from Eq. (7.70), this damping is extremely small for small $k\lambda_D$, but becomes important for $k\lambda_D = O(1)$. This effect is connected with f_1, the distortion of the distribution function caused by the wave.

7.5   The Meaning of Landau Damping

The theoretical discovery of wave damping without energy dissipation by collisions is perhaps the most astounding result of plasma physics research. That this is a real effect has been demonstrated in the laboratory. Although a simple physical explanation for this damping is now available, it is a triumph of applied mathematics that this unexpected effect was first discovered purely mathematically in the course of a careful analysis of a contour integral. Landau damping is a characteristic of collisionless plasmas, but it may also have application in other fields. For instance, in the kinetic treatment of galaxy formation, stars can be considered as atoms of a plasma interacting via gravitational rather than electromagnetic forces. Instabilities of the gas of stars can cause spiral arms to form, but this process is limited by Landau damping.

To see what is responsible for Landau damping, we first notice that $\mathrm{Im}(\omega)$ arises from the pole at $v = v_\phi$. Consequently, the effect is connected with those particles in the distribution that have a velocity nearly equal to the phase velocity—the "resonant particles." These particles travel along with the wave and do not see a rapidly fluctuating electric field: They can, therefore, exchange energy with the wave effectively. The easiest way to understand this exchange of energy is to picture a surfer trying to catch an ocean wave (Fig. 7.17). (Warning: this picture is only for directing our thinking along the right lines; it does not correctly explain Eq. (7.70).) If the surfboard is not moving, it merely bobs up and down as the wave goes by and does not gain any energy on the average. Similarly, a boat propelled much faster than the wave cannot exchange much energy with the wave. However, if the surfboard has almost the same velocity as the wave, it can be caught and pushed along by the wave; this is, after all, the main purpose of the exercise. In that case, the surfboard gains energy, and therefore the wave must lose energy and is damped. On the other hand, if the surfboard should be moving slightly faster than the wave, it would push on the wave as it moves uphill; then the wave could gain energy. In a

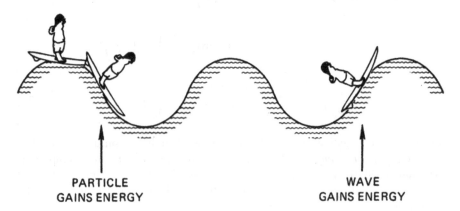

PARTICLE   WAVE
GAINS ENERGY   GAINS ENERGY

Fig. 7.17 Customary physical picture of Landau damping

Fig. 7.18 Distortion of a
Maxwellian distribution
in the region $v \simeq v_\phi$ caused
by Landau damping

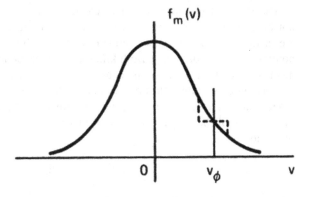

Fig. 7.19 A double-
humped distribution and
the region where
instabilities will develop

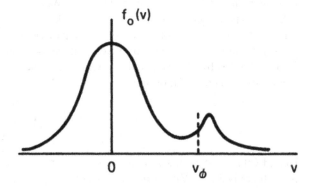

plasma, there are electrons both faster and slower than the wave. A Maxwellian distribution, however, has more slow electrons than fast ones (Fig. 7.18). Consequently, there are more particles taking energy from the wave than vice versa, and the wave is damped. As particles with $v \approx v_\phi$ are trapped in the wave, $f(v)$ is flattened near the phase velocity. This distortion is $f_1(v)$ which we calculated. As seen in Fig. 7.18, the perturbed distribution function contains the same number of particles but has gained total energy (at the expense of the wave).

From this discussion, one can surmise that if $f_0(v)$ contained more fast particles than slow particles, a wave can be excited. Indeed, from Eq. (7.67), it is apparent that $\mathrm{Im}(\omega)$ is positive if $\partial \hat{f}_0 / \partial v$ is positive at $v = v_\phi$. Such a distribution is shown in Fig. 7.19. Waves with v_ϕ in the region of positive slope will be unstable, gaining energy at the expense of the particles. This is just the finite-temperature analogy of the two-stream instability. When there are two cold ($KT = 0$) electron streams in motion, $f_0(v)$ consists of two δ-functions. This is clearly unstable because $\partial f_0 / \partial v$ is infinite; and, indeed, we found the instability from fluid theory. When the streams have finite temperature, kinetic theory tells us that the relative densities and temperatures of the two streams must be such as to have a region of positive $\partial f_0 / \partial v$ between them; more precisely, the total distribution function must have a minimum for instability.

The physical picture of a surfer catching waves is very appealing, but it is not precise enough to give us a real understanding of Landau damping. There are actually two kinds of Landau damping: linear Landau damping, and nonlinear Landau damping. Both kinds are independent of dissipative collisional mechanisms. If a particle is caught in the potential well of a wave, the phenomenon is called "trapping." As in the case of the surfer, particles can indeed gain or lose energy in trapping. However, trapping does not lie within the purview of the linear theory. That this is true can be seen from the equation of motion

$$m\,d^2x/dt^2 = qE(x) \tag{7.71}$$

If one evaluates $E(x)$ by inserting the exact value of x, the equation would be nonlinear, since $E(x)$ is something like $\sin kx$. What is done in linear theory is to use for x the unperturbed orbit; i.e., $x = x_0 + v_0 t$. Then Eq. (7.71) is linear. This approximation, however, is no longer valid when a particle is trapped. When it encounters a potential hill large enough to reflect it, its velocity and position are, of course, greatly affected by the wave and are not close to their unperturbed values. In fluid theory, the equation of motion is

$$m\left[\frac{\partial \mathbf{v}}{\partial t} + (\mathbf{v} \cdot \nabla)\mathbf{v}\right] = q\mathbf{E}(x) \tag{7.72}$$

Here, $\mathbf{E}(x)$ is to be evaluated in the laboratory frame, which is easy; but to make up for it, there is the $(\mathbf{v} \cdot \nabla)\mathbf{v}$ term. The neglect of $(\mathbf{v}_1 \cdot \nabla)\mathbf{v}_1$ in linear theory amounts to the same thing as using unperturbed orbits. In kinetic theory, the nonlinear term that is neglected is, from Eq. (7.45),

$$\frac{q}{m} E_1 \frac{\partial f_1}{\partial v} \tag{7.73}$$

When particles are trapped, they reverse their direction of travel relative to the wave, so the distribution function $f(v)$ is greatly disturbed near $v = \omega/k$. This means that $\partial f_1/\partial v$ is comparable to $\partial f_0/\partial v$, and the term (7.73) is not negligible. Hence, trapping is not in the linear theory.

When a wave grows to a large amplitude, collisionless damping with trapping does occur. One then finds that the wave does not decay monotonically; rather, the amplitude fluctuates during the decay as the trapped particles bounce back and forth in the potential wells. This is *nonlinear* Landau damping. Since the result of Eq. (7.67) was derived from a *linear* theory, it must arise from a different physical effect. The question is: Can untrapped electrons moving close to the phase velocity of the wave exchange energy with the wave? Before giving the answer, let us examine the energy of such electrons.

7.5.1   The Kinetic Energy of a Beam of Electrons

We may divide the electron distribution $f_0(v)$ into a large number of monoenergetic beams (Fig. 7.20). Consider one of these beams: It has unperturbed velocity u and density n_u. The velocity u may lie near v_ϕ, so that this beam may consist of resonant electrons. We now turn on a plasma oscillation $E_1(x, t)$ and consider the kinetic energy of the beam as it moves through the crests and troughs of the wave. The wave is caused by a self-consistent motion of *all* the beams together. If n_u is small enough (the number of beams large enough), the beam being examined has a negligible effect on the wave and may be considered as moving in a given field $E(x, t)$. Let

$$E = E_1 \sin (kx - \omega t) = -d\phi/dx \tag{7.74}$$

$$\phi = (E_1/k) \cos (kx - \omega t) \tag{7.75}$$

The linearized fluid equation for the beam is

$$m\left(\frac{\partial v_1}{\partial t} + u\frac{\partial v_1}{\partial x}\right) = -eE_1 \sin (kx - \omega t) \tag{7.76}$$

A possible solution is

$$v_1 = -\frac{eE_1}{m}\frac{\cos (kx - \omega t)}{\omega - ku} \tag{7.77}$$

This is the velocity modulation caused by the wave as the beam electrons move past. To conserve particle flux, there is a corresponding oscillation in density, given by the linearized continuity equation:

$$\frac{\partial n_1}{\partial t} + u\frac{\partial n_1}{\partial x} = -n_u\frac{\partial v_1}{\partial x} \tag{7.78}$$

Fig. 7.20 Dissection of a distribution $f_0(v)$ into a large number of monoenergetic beams with velocity u and density n_u

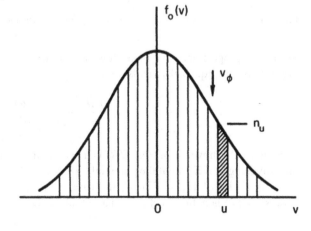

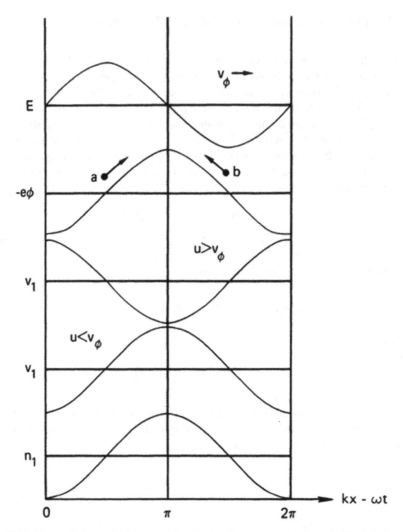

Fig. 7.21 Phase relations of velocity and density for electrons moving in an electrostatic wave

Since v_1 is proportional to $\cos(kx - \omega t)$, we can try $n_1 = \bar{n}_1 \cos(kx - \omega t)$. Substitution of this into Eq. (7.78) yields

$$n_1 = -n_u \frac{eE_1 k}{m} \frac{\cos\left(kx - \omega t\right)}{\left(\omega - ku\right)^2} \tag{7.79}$$

Figure 7.21 shows what Eqs. (7.77) and (7.79) mean. The first two curves show one wavelength of E and of the potential $-e\phi$ seen by the beam electrons. The third curve is a plot of Eq. (7.77) for the case $\omega - ku < 0$, or $u > v_\phi$. This is easily understood: When the electron a has climbed the potential hill, its velocity is small, and vice versa. The fourth curve is v_1 for the case $u < v_\phi$ and it is seen that the sign is reversed. This is because the electron b, moving to the left in the frame of the wave, is decelerated going up to the top of the potential barrier; but since it is moving the opposite way, its velocity v_1 in the *positive* x direction is maximum there. The moving potential hill accelerates electron b to the right, so by the time it reaches the top, it has the maximum v_1. The final curve on Fig. 7.21 shows the density n_1, as given by Eq. (7.79). This does *not* change sign with $u - v_\phi$, because in the frame of the wave, both electron a and electron b are slowest at the top of the potential hill, and therefore the density is highest there. The point is that the relative phase between n_1 and v_1 changes sign with $u - v_\phi$.

We may now compute the kinetic energy W_k of the beam:

$$
\begin{aligned}
W_k &= \frac{1}{2}m(n_u + n_1)(u + v_1)^2 \\
&= \frac{1}{2}m\left(n_u u^2 + n_u v_1^2 + 2u n_1 v_1 + n_1 u^2 + 2n_u u v_1 + n_1 v_1^2\right)
\end{aligned}
\tag{7.80}
$$

The last three terms contain odd powers of oscillating quantities, so they will vanish when we average over a wavelength. The *change* in W_k due to the wave is found by subtracting the first term, which is the original energy. The average energy change is then

$$
\langle \Delta W_k \rangle = \frac{1}{2}m\langle n_u v_1^2 + 2u n_1 v_1 \rangle
\tag{7.81}
$$

From Eq. (7.77), we have

$$
n_u \langle v_1^2 \rangle = \frac{1}{2}n_u \frac{e^2 E_1^2}{m^2 (\omega - ku)^2},
\tag{7.82}
$$

the factor $\frac{1}{2}$ representing $\langle \cos^2(kx - \omega t) \rangle$. Similarly, from Eq. (7.79), we have

$$
2u \langle n_1 v_1 \rangle = n_u \frac{e^2 E_1^2 ku}{m^2 (\omega - ku)^3}
\tag{7.83}
$$

Consequently,

$$
\begin{aligned}
\langle \Delta W_k \rangle &= \frac{1}{4}mn_u \frac{e^2 E_1^2}{m^2 (\omega - ku)^2}\left[1 + \frac{2ku}{(\omega - ku)}\right] \\
&= \frac{n_u}{4}\frac{e^2 E_1^2}{m}\frac{\omega + ku}{(\omega - ku)^3}
\end{aligned}
\tag{7.84}
$$

This result shows that $\langle \Delta W_k \rangle$ depends on the frame of the observer and that it does not change secularly with time. Consider the picture of a frictionless block

Fig. 7.22 Mechanical analogy for an electron moving in a moving potential

Fig. 7.23 The quadratic relation between kinetic energy and velocity causes a symmetric velocity perturbation to give rise to an increased average energy.

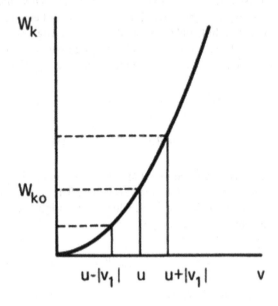

sliding over a washboard-like surface (Fig. 7.22). In the frame of the washboard, ΔW_k is proportional to $-(ku)^{-2}$, as seen by taking $\omega = 0$ in Eq. (7.84). It is intuitively clear that (1) $\langle \Delta W_k \rangle$ is negative, since the block spends more time at the peaks than at the valleys, and (2) the block does not gain or lose energy on the average, once the oscillation is started. Now if one goes into a frame in which the washboard is moving with a steady velocity ω/k (a velocity unaffected by the motion of the block, since we have assumed that n_u is negligibly small compared with the density of the whole plasma), it is still true that the block does not gain or lose energy on the average, once the oscillation is started. But Eq. (7.84) tells us that $\langle \Delta W_k \rangle$ depends on the velocity ω/k, and hence on the frame of the observer. In particular, it shows that a beam has less energy in the presence of the wave than in its absence if $\omega - ku < 0$ or $u > v_\phi$; and it has more energy if $\omega - ku > 0$ or $u < v_\phi$. The reason for this can be traced back to the phase relation between n_1 and v_1. As Fig. 7.23 shows, W_k is a parabolic function of v. As v oscillates between $u - |v_1|$ and $u + |v_1|$, W_k will attain an average value larger than the equilibrium value W_{k0}, *provided* that the particle spends an equal amount of time in each half of the oscillation. This effect is the meaning of the first term in Eq. (7.81), which is positive definite. The second term in that equation is a correction due to the fact that the particle does *not* distribute its time equally. In Fig. 7.21, one sees that both electron a and electron b spend more time at the top of the potential hill than at the bottom, but electron a reaches that point after a period of deceleration, so that v_1 is negative there, while electron b reaches that point after a period of acceleration (to the right), so that v_1 is positive there. This effect causes $\langle \Delta W_k \rangle$ to change sign at $u = v_\phi$.

7.5.2   The Effect of Initial Conditions

The result we have just derived, however, still has nothing to do with linear Landau damping. Damping requires a continuous increase of W_k at the expense of wave energy, but we have found that $\langle \Delta W_k \rangle$ for untrapped particles is constant in time. If neither the untrapped particles nor particle trapping is responsible for linear Landau damping, what is? The answer can be gleaned from the following observation: If $\langle \Delta W_k \rangle$ is positive, say, there must have been a time when it was increasing. Indeed, there are particles in the original distribution which have velocities so close to v_ϕ that at time t they have not yet gone a half-wavelength relative to the wave. For these particles, one cannot take the average $\langle \Delta W_k \rangle$. These particles can absorb energy from the wave and are properly called the "resonant" particles. As time goes on, the number of resonant electrons decreases, since an increasing number will have shifted more than $\tfrac{1}{2}\lambda$ from their original positions. The damping rate, however, can stay constant, since the amplitude is now smaller, and it takes fewer electrons to maintain a constant damping rate.

The effect of the initial conditions is most easily seen from a phase-space diagram (Fig. 7.24). Here, we have drawn the phase-space trajectories of electrons, and also the electrostatic potential $-e\phi_1$ which they see. We have assumed that this electrostatic wave exists at $t = 0$, and that the distribution $f_0(v)$, shown plotted in a plane perpendicular to the paper, is uniform in space and monotonically decreasing with |v| at that time. For clarity, the size of the wave has been greatly exaggerated. Of course, the existence of a wave implies the existence of an $f_1(v)$ at $t = 0$. However, the damping caused by this is a higher-order effect neglected in the linear theory. Now let us go to the wave frame, so that the pattern of Fig. 7.24 does not move, and consider the motion of the electrons. Electrons initially at A start out at the top of the potential hill and move to the right, since they have $v > v_\phi$. Electrons initially at B move to the left, since they have $v < v_\phi$. Those at C and D start at the potential trough and move to the right and left, respectively. Electrons starting on the closed contours E have insufficient energy to go over the potential hill and are trapped. In the limit of small initial wave amplitude, the population of the trapped electrons can be made arbitrarily small. After some time t, short enough that none of the electrons at A, B, C or D has gone more than half a wavelength, the electrons will have moved to the positions marked by open circles. It is seen that the electrons at A and D have gained energy, while those at B and C have lost energy. Now, if $f_0(v)$ was initially uniform in space, there were originally more electrons at A than at C, and more at D than at B. Therefore, there is a net gain of energy by the electrons, and hence a net loss of wave energy. This is linear Landau damping, and it is

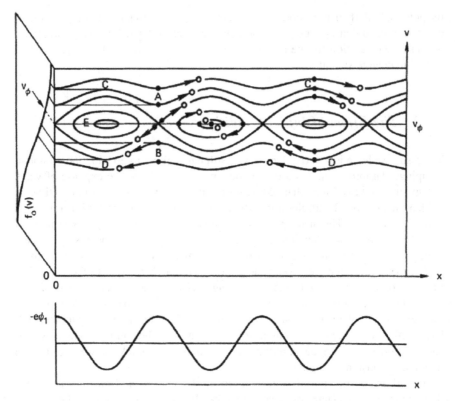

Fig. 7.24 Phase-space trajectories (*top*) for electrons moving in a potential wave (*bottom*). The entire pattern moves to the right. The arrows refer to the direction of electron motion relative to the wave pattern. The equilibrium distribution $f_0(v)$ is plotted in a plane perpendicular to the paper.

critically dependent on the assumed initial conditions. After a long time, the electrons are so smeared out in phase that the initial distribution is forgotten, and there is no further average energy gain, as we found in the previous section. In this picture, both the electrons with $v > v_\phi$ and those with $v < v_\phi$ when averaged over a wavelength, gain energy at the expense of the wave. This apparent contradiction with the idea developed in the picture of the surfer will be resolved shortly.

7.6 A Physical Derivation of Landau Damping

We are now in a position to derive the Landau damping rate without recourse to contour integration. As before, we divide the plasma up into beams of velocity u and density n_u, and examine their motion in a wave

$$E = E_1 \sin(kx - \omega t) \tag{7.85}$$

From Eq. (7.77), the velocity of each beam is

$$v_1 = -\frac{eE_1}{m}\frac{\cos(kx - \omega t)}{\omega - ku} \tag{7.86}$$

This solution satisfies the equation of motion (7.76), but it does not satisfy the initial condition $v_1 = 0$ at $t = 0$. It is clear that this initial condition must be imposed; otherwise, v_1 would be very large in the vicinity of $u = \omega/k$, and the plasma would be in a specially prepared state initially. We can fix up Eq. (7.86) to satisfy the initial condition by adding an arbitrary function of $kx - kut$. The composite solution would still satisfy Eq. (7.76) because the operator on the left-hand side of Eq. (7.76), when applied to $f(kx - kut)$, gives zero. Obviously, to get $v_1 = 0$ at $t = 0$, the function $f(kx - kut)$ must be taken to be $-\cos(kx - kut)$. Thus we have, instead of Eq. (7.86),

$$v_1 = \frac{-eE_1}{m}\frac{\cos(kx - \omega t) - \cos(kx - kut)}{\omega - ku} \tag{7.87}$$

Next, we must solve the equation of continuity (7.78) for n_1, again subject to the initial condition $n_1 = 0$ at $t = 0$. Since we are now much cleverer than before, we may try a Solution of the form

$$n_1 = \bar{n}_1[\cos(kx - \omega t) - \cos(kx - kut)] \tag{7.88}$$

Inserting this into Eq. (7.78) and using Eq. (7.87) for v_1, we find

$$\bar{n}_1 \sin(kx - \omega t) = -n_u \frac{eE_1 k}{m}\frac{\sin(kx - \omega t) - \sin(kx - kut)}{(\omega - ku)^2} \tag{7.89}$$

Apparently, we were not clever enough, since the $\sin(kx - \omega t)$ factor does not cancel. To get a term of the form $\sin(kx - kut)$, which came from the added term in v_1, we can add a term of the form $At \sin(kx - kut)$ to n_1. This term obviously vanishes at $t = 0$, and it will give the $\sin(kx - kut)$ term when the operator on the left-hand side of Eq. (7.78) operates on the t factor. When the operator operates on the $\sin(kx - kut)$ factor, it yields zero. The coefficient A must be proportional to $(\omega - ku)^{-1}$ in order to match the same factor in $\partial v_1/\partial x$. Thus we take

$$n_1 = -n_u \frac{eE_1 k}{m}\frac{1}{(\omega - ku)^2}$$
$$\times [\cos(kx - \omega t) - \cos(kx - kut) - (\omega - ku)t \sin(kx - kut)] \tag{7.90}$$

This clearly vanishes at $t = 0$, and one can easily verify that it satisfies Eq. (7.78).

These expressions for v_1 and n_1 allow us now to calculate the work done by the wave on each beam. The force acting on a unit volume of each beam is

$$F_u = -eE_1 \sin(kx - \omega t)(n_u + n_1) \tag{7.91}$$

and therefore its energy changes at the rate

$$\frac{dW}{dt} = F_u(u + v_1) = -eE_1 \sin(kx - \omega t) \left(\underset{①}{n_u u} + \underset{②}{n_u v_1} + \underset{③}{n_1 u} + \underset{④}{n_1 v_1} \right) \qquad (7.92)$$

We now take the spatial average over a wavelength. The first term vanishes because $n_u u$ is constant. The fourth term can be neglected because it is second order, but in any case it can be shown to have zero average. The terms ② and ③ can be evaluated using Eqs. (7.87) and (7.90) and the identities

$$\langle \sin(kx - \omega t) \cos(kx - kut) \rangle = -\frac{1}{2} \sin(\omega t - kut)$$

$$\langle \sin(kx - \omega t) \sin(kx - kut) \rangle = \frac{1}{2} \cos(\omega t - kut) \qquad (7.93)$$

The result is easily seen to be

$$\left\langle \frac{dW}{dt} \right\rangle_u = \frac{e^2 E_1^2}{2m} n_u \left[\frac{\sin(\omega t - kut)}{\omega - ku} + ku \frac{\sin(\omega t - kut) - (\omega - ku)t \cos(\omega t - kut)}{(\omega - ku)^2} \right] \qquad (7.94)$$

Note that the only terms that survive the averaging process come from the initial conditions.

The total work done on the particles is found by summing over all the beams:

$$\sum_u \left\langle \frac{dW}{dt} \right\rangle_u = \int \frac{f_0(u)}{n_u} \left\langle \frac{dW}{dt} \right\rangle_u du = n_0 \int \frac{\hat{f}_0(u)}{n_u} \left\langle \frac{dW}{dt} \right\rangle_u du \qquad (7.95)$$

Inserting Eq. (7.94) and using the definition of ω_p, we then find for the rate of change of kinetic energy

$$\left\langle \frac{dW_k}{dt} \right\rangle = \frac{\varepsilon_0 E_1^2}{2} \omega_p^2 \left[\int \hat{f}_0(u) \frac{\sin(\omega t - kut)}{\omega - ku} du \right.$$

$$\left. + \int \hat{f}_0(u) \frac{\sin(\omega t - kut) - (\omega - ku)t \cos(\omega t - kut)}{(\omega - ku)^2} ku \, du \right] \qquad (7.96)$$

$$= \frac{1}{2} \varepsilon_0 E_1^2 \omega_p^2 \int_{-\infty}^{\infty} \hat{f}_0(u) du \left\{ \frac{\sin(\omega t - kut)}{\omega - ku} + u \frac{d}{du} \left[\frac{\sin(\omega t - kut)}{\omega - ku} \right] \right\} \qquad (7.97)$$

$$= \frac{1}{2} \varepsilon_0 E_1^2 \omega_p^2 \int_{-\infty}^{\infty} \hat{f}_0(u) du \frac{d}{du} \left[u \frac{\sin(\omega t - kut)}{\omega - ku} \right] \qquad (7.98)$$

This is to be set equal to the rate of loss of wave energy density W_w. The wave energy consists of two parts. The first part is the energy density of the electrostatic field:

$$\langle W_E \rangle = \varepsilon_0 \langle E^2 \rangle / 2 = \varepsilon_0 E_1^2 / 4 \qquad (7.99)$$

The second part is the kinetic energy of oscillation of the particles. If we again divide the plasma up into beams, Eq. (7.84) gives the energy per beam:

$$\langle \Delta W_k \rangle_u = \frac{1}{4} \frac{n_u}{m} \frac{e^2 E_1^2}{(\omega - ku)^2} \left[1 + \frac{2ku}{(\omega - ku)} \right] \qquad (7.100)$$

In deriving this result, we did not use the correct initial conditions, which are important for the resonant particles; however, the latter contribute very little to the total energy of the wave. Summing over the beams, we have

$$\langle \Delta W_k \rangle = \frac{1}{4} \frac{e^2 E_1^2}{m} \int_{-\infty}^{\infty} \frac{f_0(u)}{(\omega - ku)^2} \left[1 + \frac{2ku}{\omega - ku} \right] du \qquad (7.101)$$

The second term in the brackets can be neglected in the limit $\omega/k \gg v_{\text{th}}$, which we shall take in order to compare with our previous results. The dispersion relation is found by Poisson's equation:

$$k \varepsilon_0 E_1 \cos (kx - \omega t) = -e \sum_u n_1 \qquad (7.102)$$

Using Eq. (7.79) for n_1, we have

$$1 = \frac{e^2}{\varepsilon_0 m} \sum_u \frac{n_u}{(\omega - ku)^2} = \frac{e^2}{\varepsilon_0 m} \int_{-\infty}^{\infty} \frac{f_0(u) du}{(\omega - ku)^2} \qquad (7.103)$$

Comparing this with Eq. (7.101), we find

$$\langle \Delta W_k \rangle = \frac{1}{4} \frac{e^2 E_1^2}{m} \frac{\varepsilon_0 m}{e^2} = \frac{\varepsilon_0 E_1^2}{4} = \langle W_E \rangle \qquad (7.104)$$

Thus

$$W_w = \varepsilon_0 E_1^2 / 2 \qquad (7.105)$$

The rate of change of this is given by the negative of Eq. (7.98):

$$\frac{dW_w}{dt} = -W_w \omega_p^2 \int_{-\infty}^{\infty} \hat{f}_0(u) \frac{d}{du} \left[u \frac{\sin (\omega - ku) t}{\omega - ku} \right] du \qquad (7.106)$$

Integration by parts gives

$$\frac{dW_w}{dt} = -W_w\omega_p^2 \left\{ \left[u\hat{f}_0(u) \frac{\sin{(\omega - ku)t}}{\omega - ku} \right]_{-\infty}^{\infty} - \int_{-\infty}^{\infty} u \frac{d\hat{f}_0}{du} \frac{\sin{(\omega - ku)t}}{\omega - ku} du \right\}$$

The integrated part vanishes for well-behaved functions $\hat{f}_0(u)$, and we have

$$\frac{dW_w}{dt} = W_w \frac{\omega}{k}\omega_p^2 \int_{-\infty}^{\infty} \hat{f}'_0(u) \left[\frac{\sin{(\omega - ku)t}}{\omega - ku} \right] du \qquad (7.107)$$

where u has been set equal to ω/k (a constant), since only velocities very close to this will contribute to the integral. In fact, for sufficiently large t, the square bracket can be approximated by a delta function:

$$\delta\left(u - \frac{\omega}{k}\right) = \frac{k}{\pi} \lim_{t \to \infty} \left[\frac{\sin{(\omega - ku)t}}{\omega - ku} \right] \qquad (7.108)$$

Thus

$$\frac{dW_w}{dt} = W_w \omega_p^2 \frac{\pi}{k} \frac{\omega}{k} \hat{f}'_0\left(\frac{\omega}{k}\right) = W_w \pi\omega \frac{\omega_p^2}{k^2} \hat{f}'_0\left(\frac{\omega}{k}\right) \qquad (7.109)$$

Since $\text{Im}(\omega)$ is the growth rate of E_1, and W_w is proportional to E_1^2 we must have

$$dW_w/dt = 2[\text{Im}(\omega)]W_w \qquad (7.110)$$

Hence

$$\text{Im}(\omega) = \frac{\pi}{2}\omega \frac{\omega_p^2}{k^2} \hat{f}'_0\left(\frac{\omega}{k}\right) \qquad (7.111)$$

in agreement with the previous result, Eq. (7.67), for $\omega = \omega_p$. Contour integration is a shortcut to this result but gives no indication of the particle physics involved.

7.6.1   The Resonant Particles

We are now in a position to see precisely which are the resonant particles that contribute to linear Landau damping. Figure 7.25 gives a plot of the factor multiplying $\hat{f}'_0(u)$ in the integrand of Eq. (7.107). We see that the largest contribution comes from particles with $|\omega - ku| < \pi/t$, or $|v - v_\phi|t < \pi/k = \lambda/2$; i.e., those particles in the initial distribution that have not yet traveled a half-wavelength relative to the wave. The width of the central peak narrows with time, as expected. The subsidiary

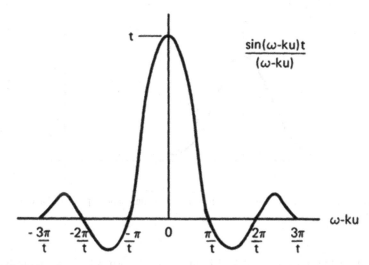

Fig. 7.25 A function which describes the relative contribution of various velocity groups to Landau damping

peaks in the "diffraction pattern" of Fig. 7.25 come from particles that have traveled into neighboring half-wavelengths of the wave potential. These particles rapidly become spread out in phase, so that they contribute little on the average; the initial distribution is forgotten. Note that the width of the central peak is independent of the initial amplitude of the wave; hence, the resonant particles may include both trapped and untrapped particles. This phenomenon is unrelated to particle trapping.

7.6.2   Two Paradoxes Resolved

Figure 7.25 shows that the integrand in Eq. (7.107) is an even function of $\omega - ku$, so that particles going both faster than the wave and slower than the wave add to Landau damping. This is the physical picture we found in Fig. 7.24. On the other hand, the slope of the curve of Fig. 7.25, which represents the factor in the integrand of Eq. (7.106), is an odd function of $\omega - ku$; and one would infer from this that particles traveling faster than the wave give energy to it, while those traveling slower than the wave take energy from it. The two descriptions differ by an integration by parts. Both descriptions are correct; which one is to be chosen depends on whether one wishes to have $\hat{f}_0(u)$ or $\hat{f}_0'(u)$ in the integrand.

A second paradox concerns the question of Galilean invariance. If we take the view that damping requires there be fewer particles traveling faster than the wave than slower, there is no problem as long as one is in the frame in which the plasma is at rest. However, if one goes into another frame moving with a velocity V (Fig. 7.26), there would appear to be more particles faster than the wave than slower, and one would expect the wave to grow instead of decay. This paradox is

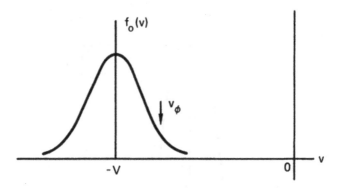

Fig. 7.26 A Maxwellian distribution seen from a moving frame appears to have a region of unstable slope

removed by reinserting the second term in Eq. (7.100), which we neglected. As shown in Sect. 7.5.1, this term can make $\langle \Delta W_k \rangle$ negative. Indeed, in the frame shown in Fig. 7.26, the second term in Eq. (7.100) is not negligible, $\langle \Delta W_k \rangle$ is negative, and the wave appears to have negative energy (that is, there is more energy in the quiescent, drifting Maxwellian distribution than in the presence of an oscillation). The wave "grows," but adding energy to a negative energy wave makes its amplitude decrease.

7.7 BGK and Van Kampen Modes

We have seen that Landau damping is directly connected to the requirement that $f_0(v)$ be initially uniform in space. On the other hand, one can generate undamped electron waves if $f(v, t = 0)$ is made to be constant along the particle trajectories initially. It is easy to see from Fig. 7.24 that the particles will neither gain nor lose energy, on the average, if the plasma is initially prepared so that the density is constant along each trajectory. Such a wave is called a BGK mode, since it was I. B. Bernstein, J. M. Greene, and M. D. Kruskal who first showed that undamped waves of arbitrary ω, k, amplitude, and waveform were possible. The crucial parameter to adjust in tailoring $f(v, t = 0)$ to form a BGK mode is the relative number of trapped and untrapped particles. If we take the small-amplitude limit of a BGK mode, we obtain what is called a Van Kampen mode. In this limit, only the particles with $v = v_\phi$ are trapped. We can change the number of trapped particles by adding to $f(v, t = 0)$ a term proportional to $\delta(v - v_\phi)$. Examination of Fig. 7.24 will show that adding particles along the line $v = v_\phi$ will not cause damping—at a later time, there are just as many particles gaining energy as losing energy. In fact, by choosing distributions with δ-functions at other values of v_ϕ one can generate undamped Van Kampen modes of arbitrary v_ϕ. Such singular initial conditions are, however, not physical. To get a smoothly varying $f(v, t = 0)$, one must sum over Van Kampen modes with a distribution of v_ϕ's. Although each mode is undamped, the total perturbation will show Landau damping because the various modes get out of phase with one another.

7.8 Experimental Verification

Although Landau's derivation of collisionless damping was short and neat, it was not clear that it concerned a physically observable phenomenon until J. M. Dawson gave the longer, intuitive derivation which was paraphrased in Sect. 7.6. Even then, there were doubts that the proper conditions could be established in the laboratory. These doubts were removed in 1965 by an experiment by Malmberg and Wharton. They used probes to excite and detect plasma waves along a collisionless plasma column. The phase and amplitude of the waves as a function of distance were obtained by interferometry. A tracing of the spatial variation of the damped wave is shown in Fig. 7.27. Since in the experiment ω was real but k was complex, the result we obtained in Eq. (7.70) cannot be compared with the data. Instead, a calculation of $\mathrm{Im}(k)/\mathrm{Re}(k)$ for real ω has to be made. This ratio also contains the factor $\exp\left(-v_\phi^2/v_{\mathrm{th}}^2\right)$ which is proportional to the number of resonant electrons in a Maxwellian distribution. Consequently, the logarithm of $\mathrm{Im}(k)/\mathrm{Re}(k)$ should be proportional to $(v_\phi/v_{\mathrm{th}})^2$. Figure 7.28 shows the agreement obtained between the measurements and the theoretical curve.

A similar experiment by Derfler and Simonen was done in plane geometry, so that the results for $\mathrm{Re}(\omega)$ can be compared with Eq. (7.64). Figure 7.29 shows their measurements of $\mathrm{Re}(k)$ and $\mathrm{Im}(k)$ at different frequencies. The dashed curve

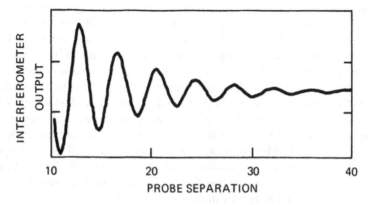

Fig. 7.27 Interferometer trace showing the perturbed density pattern in a damped plasma wave [From J. H. Malmberg and C. B. Wharton, Phys. Rev. Lett. **17**, 175 (1966)]

Fig. 7.28 Verification of Landau damping in the Malmberg–Wharton experiment (*loc. cit.*)

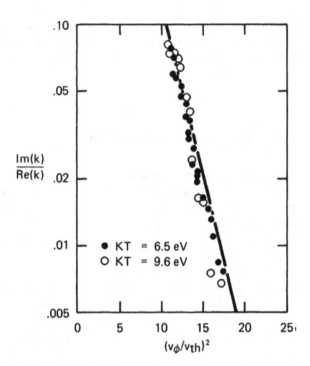

represents Eq. (7.64) and is the same as the one drawn in Fig. 4.5. The experimental points deviate from the dashed curve because of the higher-order terms in the expansion of Eq. (7.59). The theoretical curve calculated from Eq. (7.54), however, fits the data well.

Problems

7.1 Plasma waves are generated in a plasma with $n = 10^{17}$ m^{-3} and $KT_e = 10$ eV. If $k = 10^4$ m^{-1}, calculate the approximate Landau damping rate $|\text{Im}(\omega/\omega_p)|$.

7.2 An electron plasma wave with 1-cm wavelength is excited in a 10-eV plasma with $n = 10^{15}$ m^{-3}. The excitation is then removed, and the wave Landau damps away. How long does it take for the amplitude to fall by a factor of e?

7.3 An infinite, uniform plasma with fixed ions has an electron distribution function composed of (1) a Maxwellian distribution of "plasma" electrons with density n_p and temperature T_p at rest in the laboratory, and (2) a Maxwellian distribution of "beam" electrons with density n_b and temperature T_b centered at $\mathbf{v} = V\hat{\mathbf{x}}$ (Fig. P7.3). If n_b is infinitesimally small, plasma oscillations traveling in the x direction are Landau-damped. If n_b is large, there will be a two-stream instability. The critical n_b at which instability sets in can be found by setting the slope of the total distribution function equal to zero. To keep the algebra simple, we can find an approximate answer as follows.

Fig. 7.29 Experimental measurement of the dispersion relation for plasma waves in plane geometry [From H. Derfler and T. Simonen, *J. Appl. Phys.* **38**, 5018 (1967)]

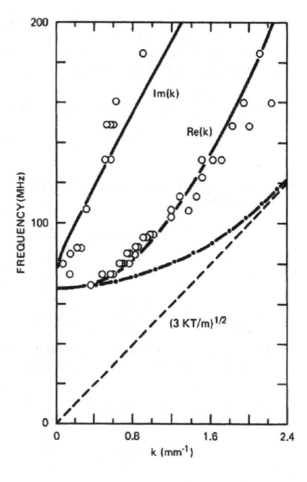

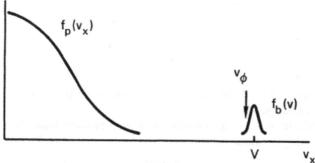

Fig. P7.3 Unperturbed distribution functions $f_p(v_x)$ and $f_b(v_x)$ for the plasma and beam electrons, respectively, in a beam–plasma interaction

(a) Write expressions for $f_p(v)$ and $f_b(v)$, using the abbreviations $v = v_x$, $a^2 = 2KT_p/m$, $b^2 = 2KT_b/m$.

(b) Assume that the phase velocity v_ϕ will be the value of v at which $f_b(v)$ has the largest positive slope. Find v_ϕ and $f_b'(v_\phi)$.

(c) Find $f_p'(v_\phi)$ and set $f_p'(v_\phi) + f_b'(v_\phi) = 0$.

(d) For $V \gg b$, show that the critical beam density is given approximately by

$$\frac{n_b}{n_p} = (2e)^{1/2}\frac{T_b V}{T_p a}\exp\left(-V^2/a^2\right)$$

7.4 To model a warm plasma, assume that the ion and electron distribution functions are given by

$$\hat{f}_{e0}(v) = \frac{a_e}{\pi}\frac{1}{v^2 + a_e^2}$$

$$\hat{f}_{i0}(v) = \frac{a_i}{\pi}\frac{1}{v^2 + a_i^2}$$

(a) Derive the exact dispersion relation in the Vlasov formalism assuming an electrostatic perturbation.

(b) Obtain an approximate expression for the dispersion relation if $\omega \leq \Omega_p$. Under what conditions are the waves weakly damped? Explain physically why $\omega \simeq \Omega_p$ for very large k.

7.5 Consider an unmagnetized plasma with a fixed, neutralizing ion background. The one-dimensional electron velocity distribution is given by

$$f_{0e}(v) = g_0(v) + h_0(v)$$

where

$$g_0(v) = n_p\frac{a_e}{\pi}\frac{1}{v^2 + a_e^2}, \quad h_0(v) = n_b\delta(v - v_0)$$
$$n_0 = n_p + n_b \quad \text{and} \quad n_b \ll n_p$$

(a) Derive the dispersion relation for high-frequency electrostatic perturbations.

(b) In the limit $\omega/k \ll a_e$ show that a solution exists in which $\text{Im}(\omega) > 0$ (i.e., growing oscillations).

7.6 Consider the one-dimensional distribution function

$$f(v) = A \quad |v| < v_m$$
$$f(v) = 0 \quad |v| \geq v_m$$

(a) Calculate the value of the constant A in terms of the plasma density n_0.
(b) Use the Vlasov and Poisson equations to derive an integral expression for electrostatic electron plasma waves.
(c) Evaluate the integral and obtain a dispersion relation $\omega(k)$, keeping terms to third order in the small quantity $k v_m/\omega$.

7.9 Ion Landau Damping

Electrons are not the only possible resonant particles. If a wave has a slow enough phase velocity to match the thermal velocity of ions, ion Landau damping can occur. The ion acoustic wave, for instance, is greatly affected by Landau damping. Recall from Eq. (4.41) that the dispersion relation for ion waves is

$$\frac{\omega}{k} = v_s = \left(\frac{KT_e + \gamma_i KT_i}{M}\right)^{1/2} \tag{7.112}$$

If $T_e \leq T_i$, the phase velocity lies in the region where $f_{0i}(v)$ has a negative slope, as shown in Fig. 7.30a. Consequently, ion waves are heavily Landau-damped if $T_e \leq T_i$. Ion waves are observable only if $T_e \gg T_i$ [Fig. 7.30b], so that the phase velocity lies far in the tail of the ion velocity distribution. A clever way to introduce Landau damping in a controlled manner was employed by Alexeff, Jones, and Montgomery. A weakly damped ion wave was created in a heavy-ion plasma (such as xenon) with $T_e \gg T_i$. A small amount of a light atom (helium) was then added. Since the helium had about the same temperature as the xenon but had much smaller mass, its distribution function was much broader, as shown by the dashed curve in Fig. 7.30b. The resonant helium ions then caused the wave to damp. Ion acoustic waves generated in instabilities are often not seen because of the severe damping by light-ion impurities.

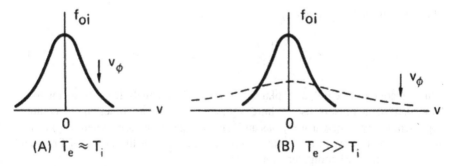

Fig. 7.30 Explanation of Landau damping of ion acoustic waves. For $T_e \approx T_i$, the phase velocity lies well within the ion distribution; for $T_e \gg T_i$, there are very few ions at the phase velocity. Addition of a light ion species (*dashed curve*) increases Landau damping

7.9.1   The Plasma Dispersion Function

To introduce some of the standard terminology of kinetic theory, we now calculate the ion Landau damping of ion acoustic waves in the absence of magnetic fields. Ions and electrons follow the Vlasov equation (7.23) and have perturbations of the form of Eq. (7.46) indicating plane waves propagating in the x direction. The solution for f_1 is given by Eq. (7.48) with appropriate modifications:

$$f_{1j} = -\frac{iq_jE}{m_j}\frac{\partial f_{0j}/\partial v_j}{\omega - kv_j} \tag{7.113}$$

where E and v_j stand for E_{1x}, v_{xj}; and the jth species has charge q_j, mass m_j, and particle velocity v_j. The density perturbation of the jth species is given by

$$n_{1j} = \int_{-\infty}^{\infty} f_{1j}(v_j)dv_j = -\frac{iq_jE}{m_j}\int_{-\infty}^{\infty}\frac{\partial f_{0j}/\partial v_j}{\omega - kv_j}dv_j \tag{7.114}$$

Let the equilibrium distributions f_{0j} be one-dimensional Maxwellians:

$$f_{0j} = \frac{n_{0j}}{v_{\text{th}j}\pi^{1/2}}e^{-v_j^2/v_{\text{th}j}^2} \qquad v_{\text{th}j} \equiv \left(2KT_j/m_j\right)^{1/2} \tag{7.115}$$

Introducing the dummy integration variable $s = v_j/v_{\text{th}j}$, we can write n_{1j} as

$$n_{1j} = \frac{iq_jEn_{0j}}{km_jv_{\text{th}j}^2}\frac{1}{\pi^{1/2}}\int_{-\infty}^{\infty}\frac{(d/ds)\left(e^{-s^2}\right)}{s - \zeta_j}ds \tag{7.116}$$

where

$$\zeta_j \equiv \omega/kv_{\text{th}j} \tag{7.117}$$

We now define the *plasma dispersion function* $Z(\zeta)$:

$$Z(\zeta) = \frac{1}{\pi^{1/2}}\int_{-\infty}^{\infty}\frac{e^{-s^2}}{s - \zeta}ds \qquad I_m(\zeta) > 0 \tag{7.118}$$

This is a contour integral, as explained in Sect. 7.4, and analytic continuation to the lower half plane must be used if $\text{Im}(\zeta) < 0$. $Z(\zeta)$ is a complex function of a complex argument (since ω or k usually has an imaginary part). In cases where $Z(\zeta)$ cannot be approximated by an asymptotic formula, one can use the tables of Fried and Conte or a standard computer subroutine.

To express n_{1j} in terms of $Z(\zeta)$, we take the derivative with respect to ζ:

$$Z'(\zeta) = \frac{1}{\pi^{1/2}} \int_{-\infty}^{\infty} \frac{e^{-s^2}}{(s-\zeta)^2} \, ds$$

Integration by parts yields

$$Z'(\zeta) = \frac{1}{\pi^{1/2}} \left[\frac{-e^{-s^2}}{s-\zeta} \right]_{-\infty}^{\infty} + \frac{1}{\pi^{1/2}} \int_{-\infty}^{\infty} \frac{(d/ds)\left(e^{-s^2}\right)}{s-\zeta} \, ds$$

The first term vanishes, as it must for any well-behaved distribution function. Equation (7.116) can now be written

$$n_{1j} = \frac{iq_j E n_{0j}}{km_j v_{\mathrm{th}j}^2} Z'(\zeta_j) \tag{7.119}$$

Poisson's equation is

$$\varepsilon_0 \nabla \cdot \mathbf{E} = ik\varepsilon_0 E = \sum_j q_j n_{1j} \tag{7.120}$$

Combining the last two equations, separating out the electron term explicitly, and defining

$$\Omega_{pj} \equiv \left(n_{0j} Z_j^2 e^2 / \varepsilon_0 M_j\right)^{1/2} \tag{7.121}$$

We obtain the dispersion relation

$$k^2 = \frac{\omega_p^2}{v_{\mathrm{the}}^2} Z'(\zeta_e) + \sum_j \frac{\Omega_{pj}^2}{v_{\mathrm{th}j}^2} Z'(\zeta_j) \tag{7.122}$$

Electron plasma waves can be obtained by setting $\Omega_{pj} = 0$ (infinitely massive ions). Defining

$$k_{\mathrm{D}}^2 = 2\omega_p^2 / v_{\mathrm{the}}^2 = \lambda_{\mathrm{D}}^{-2}, \tag{7.123}$$

we then obtain

$$k^2 / k_{\mathrm{D}}^2 = \frac{1}{2} Z'(\zeta_e) \tag{7.124}$$

which is the same as Eq. (7.54) when f_{0e} is Maxwellian.

7.9.2 Ion Waves and Their Damping

To obtain ion waves, go back to Eq. (7.122) and use the fact that their phase velocity ω/k is much smaller than v_{the}; hence ζ_e is small, and we can expand $Z(\zeta_e)$ in a power series:

$$Z(\zeta_e) = i\sqrt{\pi}e^{-\zeta_e^2} - 2\zeta_e\left(1 - \frac{2}{3}\zeta_e^2 + \cdots\right) \tag{7.125}$$

The imaginary term comes from the residue at a pole lying near the real s axis [of Eq. (7.56)] and represents electron Landau damping. For $\zeta_e \ll 1$, the derivative of Eq. (7.125) gives

$$Z'(\zeta_e) = -2i\sqrt{\pi}\zeta_e e^{-\zeta_e^2} - 2 + \cdots \simeq -2 \tag{7.126}$$

Electron Landau damping can usually be neglected in ion waves because the slope of $f_e(v)$ is small near its peak. Replacing $Z'(\zeta_e)$ by -2 in Eq. (7.122) gives the ion wave dispersion relation

$$\lambda_D^2\sum_j \frac{\Omega_{pj}^2}{v_{thj}^2}Z'(\zeta_j) = 1 + k^2\lambda_D^2 \simeq 1 \tag{7.127}$$

The term $k^2\lambda_D^2$ represents the deviation from quasineutrality.

We now specialize to the case of a single ion species. Since $n_{0e} = Z_i n_{0i}$, the coefficient in Eq. (7.127) is

$$\lambda_D^2\frac{\Omega_p^2}{v_{thi}^2} = \frac{\varepsilon_0 KT_e}{n_{0e}e^2}\frac{n_{0i}Z^2e^2}{\varepsilon_0 M}\frac{M}{2kT_i} = \frac{1}{2}\frac{ZT_e}{T_i}$$

For $k^2\lambda_D^2 \ll 1$, the dispersion relation becomes

$$\boxed{Z'\left(\frac{\omega}{kv_{thi}}\right) = \frac{2T_i}{ZT_e} = \frac{2}{\theta}} \quad \theta \equiv \frac{ZT_e}{T_i} \tag{7.128}$$

Solving this equation is a nontrivial problem. Suppose we take real k and complex ω to study damping in time. Then the real and imaginary parts of ω must be adjusted so that $Im(Z') = 0$ and $Re(Z') = 2T_i/ZT_e$. There are in general many possible roots ω that satisfy this, all of them having $Im\,\omega < 0$. The least damped, dominant root is the one having the smallest $|Im\,\omega|$. Damping in space is usually treated by taking ω real and k complex. Again we get a series of roots k with $Im\,k > 0$, representing spatial damping. However, the dominant root does not correspond to the same value of ζ_i as in the complex ω case. It turns out that the spatial problem has to be treated with special attention to the excitation mechanism at the boundaries and with more careful treatment of the electron term $Z'(\zeta_e)$.

To obtain an analytic result, we consider the limit $\zeta_i \gg 1$, corresponding to large temperature ratio $\theta \equiv ZT_e/T_i$. The asymptotic expression for $Z'(\zeta_i)$ is

$$Z'(\zeta_i) = -2i\sqrt{\pi}\zeta_i e^{-\zeta_i^2} + \zeta_i^{-2} + \frac{3}{2}\zeta_i^{-4} + \cdots \qquad (7.129)$$

If the damping is small, we can neglect the Landau term in the first approximation. Equation (7.128) becomes

$$\frac{1}{\zeta_i^2}\left(1 + \frac{3}{2}\frac{1}{\zeta_i^2}\right) = \frac{2}{\theta}$$

Since θ is assumed large, ζ_i^2 is large; and we can approximate ζ_i^2 by $\theta/2$ in the second term. Thus

$$\frac{1}{\zeta_i^2}\left(1 + \frac{3}{\theta}\right) = \frac{2}{\theta}, \quad \zeta_i^2 = \frac{3}{2} + \frac{\theta}{2} \qquad (7.130)$$

or

$$\frac{\omega^2}{k^2} = \frac{2KT_i}{M}\left(\frac{3}{2} + \frac{ZT_e}{2T_i}\right) = \frac{ZKT_e + 3KT_i}{M} \qquad (7.131)$$

This is the ion wave dispersion relation (4.41) with $\gamma_i = 3$, generalized to arbitrary Z.

We now substitute Eqs. (7.129) and (7.130) into Eq. (7.128) retaining the Landau term:

$$\frac{1}{\zeta_i^2}\left(1 + \frac{3}{\theta}\right) - 2i\sqrt{\pi}\zeta_i e^{-\zeta_i^2} = \frac{2}{\theta}$$

$$\frac{1}{\zeta_i^2}\left(1 + \frac{3}{\theta}\right) = \frac{2}{\theta}\left(1 + i\sqrt{\pi}\theta\zeta_i e^{-\zeta_i^2}\right)$$

$$\zeta_i^2 = \left(\frac{3 + \theta}{2}\right)\left(1 + i\sqrt{\pi}\theta\zeta_i e^{-\zeta_i^2}\right)^{-1}$$

Expanding the square root, we have

$$\zeta_i \simeq \left(\frac{3 + \theta}{2}\right)^{1/2}\left(1 - \frac{1}{2}i\sqrt{\pi}\theta\zeta_i e^{-\zeta_i^2}\right) \qquad (7.132)$$

The approximate damping rate is found by using Eq. (7.130) in the imaginary term:

$$-\frac{\mathrm{Im}\,\zeta_i}{\mathrm{Re}\,\zeta_i} = -\frac{\mathrm{Im}\,\omega}{\mathrm{Re}\,\omega} = \left(\frac{\pi}{8}\right)^{1/2}\theta(3+\theta)^{1/2}e^{-(3+\theta)/2} \qquad (7.133)$$

where $\theta = ZT_e/T_i$ and $\mathrm{Re}\,\omega$ is given by Eq. (7.131).

This asymptotic expression, accurate for large θ, shows an exponential decrease in damping with increasing θ. When θ falls below 10, Eq. (7.133) becomes inaccurate, and the damping must be computed from Eq. (7.128), which employs the Z-function. For the experimentally interesting region $1 < \theta < 10$, the following simple formula is an analytic fit to the exact solution:

$$-\mathrm{Im}\,\omega/\mathrm{Re}\,\omega = 1.1\theta^{7/4}\exp\left(-\theta^2\right) \qquad (7.134)$$

These approximations are compared with the exact result in Fig. 7.31.

What happens when collisions are added to ion Landau damping? Surprisingly little. Ion-electron collisions are weak because the ion and electron fluids move

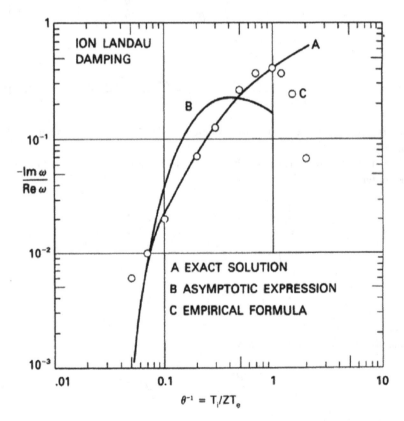

Fig. 7.31 Ion Landau damping of acoustic waves. (*A*) is the exact solution of Eq. (7.128); (*B*) is the asymptotic formula, Eq. (7.133); and (*C*) is the empirical fit, Eq. (7.134), good for $1 < \theta < 10$.

almost in unison, creating little friction between them. Ion–ion collisions (ion viscosity) can damp ion acoustic waves, but we know that sound waves in air can propagate well in spite of the dominance of collisions. Actually, collisions spoil the particle resonances that cause Landau damping, and one finds that the total damping is *less* than the Landau damping unless the collision rate is extremely large. In summary, ion Landau damping is almost always the dominant process with ion waves and it decreases exponentially with the ratio ZT_e/T_i.

Problems

7.7. Ion acoustic waves of 1-cm wavelength are excited in a singly ionized xenon ($A = 131$) plasma with $T_e = 1$ eV and $T_i = 0.1$ eV. If the exciter is turned off, how long does it take for the waves to Landau damp to $1/e$ of their initial amplitude?

7.8. Ion waves with $\lambda = 5$ cm are excited in a singly ionized argon plasma with $n_e = 10^{16}$ m^{-3}, $T_e = 2$ eV, $T_i = 0.2$ eV; and the Landau damping rate is measured. A hydrogen impurity of density $n_H = \alpha n_e$ is then introduced. Calculate the value of α that will double the damping rate.

7.9. In laser fusion experiments one often encounters a hot electron distribution with density n_h and temperature T_h in addition to the usual population with n_e, T_e. The hot electrons can change the damping of ion waves and hence affect such processes as stimulated Brillouin scattering. Assume $Z = 1$ ions with n_i and T_i and define $\theta_e = T_e/T_i$, $\theta_h = T_h/T_i$, $\alpha = n_h/n_i$, $1 - \alpha = n_e/n_i$, $\varepsilon = m/M$ and $k_{Di}^2 = n_i e^2 / \varepsilon_0 K T_i$.

  (a) Write the ion wave dispersion relation for this three-component plasma, expanding the electron Z-functions.

  (b) Show that electron Landau damping is not appreciably increased by n_h if $T_h \gg T_e$.

  (c) Show that ion Landau damping is decreased by n_h, and that the effect can be expressed as an increase in the effective temperature ratio T_e/T_i.

7.10. The dispersion relation for electron plasma waves propagating along $B_0\hat{z}$ can be obtained from the dielectric tensor $\boldsymbol{\varepsilon}$ (Appendix B) and Poisson's equation, $\nabla \cdot (\boldsymbol{\varepsilon} \cdot \mathbf{E}) = 0$, where $\mathbf{E} = -\nabla \phi$. We then have, for a uniform plasma,

$$-\frac{\partial}{\partial z}\left(\varepsilon_{zz}\frac{\partial \phi}{\partial z}\right) = \varepsilon_{zz}k_z^2\phi = 0$$

or $\epsilon_{zz} = 0$. For a cold plasma, Problem 4.4 and Eq. (B.18) in Appendix B give

$$\varepsilon_{zz} = 1 - \frac{\omega_p^2}{\omega^2} \quad \text{or} \quad \omega^2 = \omega_p^2$$

For a hot plasma, Eq. (7.124) gives

$$\varepsilon_{zz} = 1 - \frac{\omega_p^2}{k^2 v_{th}^2} Z'\left(\frac{\omega}{k v_{th}}\right) = 0$$

By expanding the Z-function in the proper limits, show that this equation yields the Bohm–Gross wave frequency [Eq. (4.30)] and the Landau damping rate [Eq. (7.70)].

7.10 Kinetic Effects in a Magnetic Field

When either the dc magnetic field $\mathbf{B}_0$ or the oscillating magnetic field $\mathbf{B}_1$ is finite, the $\mathbf{v} \times \mathbf{B}$ term in the Vlasov equation (7.23) for a collisionless plasma must be included. The linearized equation (7.45) is then replaced by

$$\frac{\partial f_1}{\partial t} + \mathbf{v} \cdot \nabla f_1 + \frac{q}{m}(\mathbf{v} \times \mathbf{B}_0) \cdot \frac{\partial f_1}{\partial \mathbf{v}} = -\frac{q}{m}(\mathbf{E}_1 + \mathbf{v} \times \mathbf{B}_1) \cdot \frac{\partial f_0}{\partial \mathbf{v}} \qquad (7.135)$$

Resonant particles moving along $\mathbf{B}_0$ still cause Landau damping if $\omega/k \simeq v_{th}$, but two new kinetic effects now appear which are connected with the velocity component $\mathbf{v}_\perp$ perpendicular to $\mathbf{B}_0$. One of these is cyclotron damping, which will be discussed later; the other is the generation of cyclotron harmonics, leading to the possibility of the oscillations commonly called Bernstein waves.

Harmonics of the cyclotron frequency are generated when the particles' circular Larmor orbits are distorted by the wave fields $\mathbf{E}_1$ and $\mathbf{B}_1$. These finite-r_L effects are neglected in ordinary fluid theory but can be taken into account to order $k^2 r_L^2$ by including the viscosity π. A kinetic treatment can be accurate even for $k^2 r_L^2 = O(1)$. To understand how harmonics arise, consider the motion of a particle in an electric field:

$$\mathbf{E} = E_x e^{i(kx - \omega t)} \hat{\mathbf{x}} \qquad (7.136)$$

The equation of motion [cf. Eq. (2.10)] is

$$\ddot{x} + \omega_c^2 x = \frac{q}{m} E_x e^{i(kx - \omega t)} \qquad (7.137)$$

If $k r_L$ is not small, the exponent varies from one side of the orbit to the other. We can approximate kx by substituting the undisturbed orbit $x = r_L \sin \omega_c t$ from Eq. (2.7):

$$\ddot{x} + \omega_c^2 x = \frac{q}{m} E_x e^{i(k r_L \sin \omega_c t - \omega t)} \qquad (7.138)$$

The generating function for the Bessel functions $J_n(z)$ is

$$e^{z(t-1/t)/2} = \sum_{n=-\infty}^{\infty} t^n J_n(z) \tag{7.139}$$

Letting $z = kr_L$ and $t = \exp(i\omega_c t)$, we obtain

$$e^{ikr_L \sin \omega_c t} = \sum_{-\infty}^{\infty} J_n(kr_L) e^{in\omega_c t} \tag{7.140}$$

$$\ddot{x} + \omega_c^2 x = \frac{q}{m} E_x \sum_{-\infty}^{\infty} J_n(kr_L) e^{-i(\omega - n\omega_c)t} \tag{7.141}$$

The following solution can be verified by direct substitution:

$$x = \frac{q}{m} E_x \sum_{-\infty}^{\infty} \frac{J_n(kr_L) e^{-i(\omega - n\omega_c)t}}{\omega_c^2 - (\omega - n\omega_c)^2} \tag{7.142}$$

This shows that the motion has frequency components differing from the driving frequency by multiples of ω_c, and that the amplitudes of these components are proportional to $J_n(kr_L)/\left[\omega_c^2 - (\omega - n\omega_c)^2\right]$. When the denominator vanishes, the amplitude becomes large. This happens when $\omega - n\omega_c = \pm\omega_c$, or $\omega = (n \pm 1)\omega_c$, $n = 0, \pm1, \pm2, \ldots$; that is, when the field $E(x, t)$ resonates with any harmonic of ω_c. In the fluid limit $kr_L \to 0$, $J_n(kr_L)$ can be approximated by $(kr_L/2)^n/n!$, which approaches 0 for all n except $n = 0$. For $n = 0$, the coefficient in Eq. (7.142) becomes $(\omega_c^2 - \omega^2)^{-1}$, which is the fluid result [cf. Eq. (4.57)] containing only the fundamental cyclotron frequency.

7.10.1   The Hot Plasma Dielectric Tensor

After Fourier analysis of $f_1(\mathbf{r}, \mathbf{v}, t)$ in space and time, Eq. (7.135) can be solved for a Maxwellian distribution $f_0(\mathbf{v})$, and the resulting expressions $f_1(\mathbf{k}, \mathbf{v}, \omega)$ can be used to calculate the density and current of each species. The result is usually expressed in the form of an equivalent dielectric tensor $\boldsymbol{\epsilon}$, such that the displacement vector $\mathbf{D} = \boldsymbol{\epsilon} \cdot \mathbf{E}$ can be used in the Maxwell's equations $\nabla \cdot \mathbf{D} = 0$ and $\nabla \times \mathbf{B} = \mu_0 \mathbf{D}$ to calculate dispersion relations for various waves (see Appendix B). The algebra is horrendous and therefore omitted. We quote only a restricted result valid for nonrelativistic plasmas with isotropic pressure ($T_\perp = T_\parallel$) and no zero-order drifts $\mathbf{v}_{0j}$; these restrictions are easily removed, but the general formulas are too cluttered for our purposes. We further assume $\mathbf{k} = k_x \hat{\mathbf{x}} + k_z \hat{\mathbf{z}}$, with $\hat{\mathbf{z}}$ being the direction of $\mathbf{B}_0$;

no generality is lost by setting k_y equal to zero, since the plasma is isotropic in the plane perpendicular to $\mathbf{B}_0$. The elements of $\boldsymbol{\epsilon}_R = \boldsymbol{\epsilon}/\epsilon_0$ are then

$$\epsilon_{xx} = 1 + \sum_s \frac{\omega_p^2}{\omega^2} \frac{e^{-b}}{b} \zeta_0 \sum_{-\infty}^{\infty} n^2 I_n(b) Z(\zeta_n)$$

$$\epsilon_{yy} = 1 + \sum_s \frac{\omega_p^2}{\omega^2} \frac{e^{-b}}{b} \zeta_0 \sum_{-\infty}^{\infty} \{n^2 I_n(b) + 2b^2 [I_n(b) - I_n'(b)]\} Z(\zeta_n)$$

$$\epsilon_{xy} = -\epsilon_{yx} = i\sum_s \pm \frac{\omega_p^2}{\omega^2} e^{-b} \zeta_0 \sum_{-\infty}^{\infty} n\left[I_n(b) - I_n'(b)\right] Z(\zeta_n)$$

$$\epsilon_{xz} = \epsilon_{zx} = \sum_s \frac{\omega_p^2}{\omega^2} \frac{e^{-b}}{(2b)^{1/2}} \zeta_0 \sum_{-\infty}^{\infty} n I_n(b) Z'(\zeta_n) \tag{7.143}$$

$$\epsilon_{yz} = -\epsilon_{zy} = -i\sum_s \pm \frac{\omega_p^2}{\omega^2} \left(\frac{b}{2}\right)^{1/2} e^{-b} \zeta_0 \sum_{-\infty}^{\infty} \left[I_n(b) - I_n'(b)\right] Z'(\zeta_n)$$

$$\epsilon_{zz} = 1 - \sum_s \frac{\omega_p^2}{\omega^2} e^{-b} \zeta_0 \sum_{-\infty}^{\infty} I_n(b) \zeta_n Z'(\zeta_n)$$

where $Z(\zeta)$ is the plasma dispersion function of Eq. (7.118), $I_n(b)$ is the nth order Bessel function of imaginary argument, and the other symbols are defined by

$$\omega_{ps}^2 = n_{0s} Z_s^2 e^2 / \epsilon_0 m_s$$

$$\zeta_{ns} = (\omega + n\omega_{cs})/k_z v_{ths} \qquad \zeta_{0s} = \omega/k_z v_{ths}$$

$$\omega_{cs} = |Z_s e B_0 / m_s| \tag{7.144}$$

$$v_{ths}^2 = 2KT_s/m_s$$

$$b_s = \frac{1}{2} k_\perp^2 r_{Ls} = k_x^2 KT_s/m_s \omega_{cs}^2$$

The first sum is over species s, with the understanding that ω_p, b, ζ_0 and ζ_n all depend on s, and that the $\pm$ stands for the sign of the charge. The second sum is over the harmonic number n. The primes indicate differentiation with respect to the argument.

As foreseen, there appear Bessel functions of the finite-r_L parameter b. [The change from $J_n(b)$ to $I_n(b)$ occurs in the integration over velocities.] The elements of $\boldsymbol{\epsilon}$ involving motion along $\hat{\mathbf{Z}}$ contain $Z'(\zeta_n)$, which gives rise to Landau damping when $n = 0$ and $\omega/k_z \simeq v_{th}$. The $n \neq 0$ terms now make possible another collisionless damping mechanism, cyclotron damping, which occurs when $(\omega \pm n\omega_c)/k_z \simeq v_{th}$.

Problem

7.11. In the limit of zero temperature, show that the elements of $\boldsymbol{\epsilon}$ in Eq. (7.143) reduce to the cold-plasma dielectric tensor given in Appendix B.

7.10.2 Cyclotron Damping

When a particle moving along $\mathbf{B}_0$ in a wave with finite k_z has the right velocity, it sees a Doppler-shifted frequency $\omega - k_z v_z$ equal to $\pm n\omega_c$ and is therefore subject to continuous acceleration by the electric field $\mathbf{E}_\perp$ of the wave. Those particles with the "right" phase relative to $\mathbf{E}_\perp$ will gain energy; those with the "wrong" phase will lose energy. Since the energy change is the force times the distance, the faster accelerated particles gain more energy per unit time than what the slower decelerated particles lose. There is, therefore, a net gain of energy by the particles, on the average, at the expense of the wave energy; and the wave is damped. This mechanism differs from Landau damping because the energy gained is in the direction perpendicular to $\mathbf{B}_0$, and hence perpendicular to the velocity component that brings the particle into resonance. The resonance is not easily destroyed by phenomena such as trapping. Furthermore, the mere existence of resonant particles suffices to cause damping; one does not need a negative slope $f_0'(v_z)$, as in Landau damping.

To clarify the physical mechanism of cyclotron damping, consider a wave with $\mathbf{k} = k_x\hat{\mathbf{x}} + k_z\hat{\mathbf{z}}$ with k_z positive. The wave electric field $\mathbf{E}_\perp$ can be decomposed into left- and right-hand circularly polarized components, as shown in Fig. 7.32. For the left-hand component, the vector $\mathbf{E}_\perp$ at positions A, B, and C along the z axis appears as shown in Fig. 7.32a. Since the wave propagates in the $+\hat{\mathbf{z}}$ direction, a stationary electron would sample the vectors at C, B, and A in succession and therefore would see a left-rotating $\mathbf{E}$-field. It would not be accelerated because its Larmor gyration is in the right-hand (clockwise) direction. However, if the electron were moving faster than the wave in the $\hat{\mathbf{z}}$ direction, it would see the vectors at A, B, and C in that order and hence would be resonantly accelerated if its velocity satisfied the condition $\omega - k_z v_z = -\omega_c$. The right-hand component of $\mathbf{E}$ would appear as shown in Fig. 7.32b. Now an electron would see a clockwise rotating $\mathbf{E}$-field if it moved

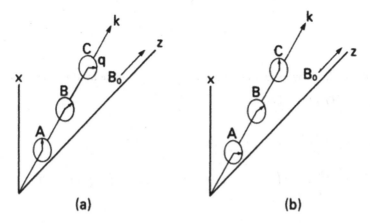

(a) **(b)**

Fig. 7.32 The mechanism of cyclotron damping

more slowly than the wave, so that the vectors at C, B, and A were sampled in succession. This electron would be accelerated if it met the condition $\omega - k_z v_z = +\omega_c$. A plane or elliptically polarized wave would, therefore, be cyclotron damped by electrons moving in either direction in the wave frame.

7.10.3   Bernstein Waves

Electrostatic waves propagating at right angles to $\mathbf{B}_0$ at harmonics of the cyclotron frequency are called Bernstein waves. The dispersion relation can be found by using the dielectric elements given in Eq. (7.143) in Poisson's equation $\nabla \cdot \boldsymbol{\epsilon} \cdot \mathbf{E} = 0$. If we assume electrostatic perturbations such that $\mathbf{E}_1 = -\nabla \phi_1$, and consider waves of the form $\phi_1 = \phi_1 \exp i(\mathbf{k} \cdot \mathbf{r} - \omega t)$, Poisson's equation can be written

$$k_x^2 \epsilon_{xx} + 2k_x k_z \epsilon_{xz} + k_z^2 \epsilon_{zz} = 0 \tag{7.145}$$

Note that we have chosen a coordinate system that has $\mathbf{k}$ lying in the $x - z$ plane, so that $k_y = 0$. We next Substitute for ϵ_{xx}, ϵ_{xz}, and ϵ_{zz} from Eq. (7.143) and express $Z'(\zeta_n)$ in terms of $Z(\zeta_n)$ with the identity

$$Z'(\zeta_n) = -2[1 + \zeta Z(\zeta)] \tag{7.146}$$

Problems

7.12. Prove Eq. (7.146) directly from the integral expressions for $Z(\zeta)$ and $Z'(\zeta)$.

7.13. The principal part of $Z(\zeta)$ for small and large ζ, as used in Eqs. (7.125) and (7.129), is given by

$$Z(\zeta) \simeq -2\zeta \left(1 - \frac{2}{3}\zeta^2 + \cdots \right) \quad |\zeta| \ll 1$$

$$Z(\zeta) \simeq -\zeta^{-1} \left(1 + \frac{1}{2}\zeta^{-2} + \cdots \right) \quad |\zeta| \gg 1$$

Prove these by expanding the denominator in the definition (7.118) of $Z(\zeta)$.

Equation (7.145) becomes

$$k_x^2 + k_z^2 + \sum_s \frac{\omega_p^2}{\omega^2} e^{-b} \zeta_0 \sum_{n=-\infty}^{\infty} I_n(b)$$

$$\times \left[k_x^2 \frac{n^2}{b} Z - 2\left(\frac{2}{b}\right)^{1/2} nk_x k_z (1 + \zeta_n Z) - 2k_z^2 \zeta_n (1 + \zeta_n Z) \right] = 0 \tag{7.147}$$

The expression in the square brackets can be simplified in a few algebraic steps to $2k_z^2 [\zeta_{-n} + \zeta_0^2 Z(\zeta_n)]$ by using the definitions $b = k_x^2 v_{th}^2 / 2\omega_c^2$ and $\zeta_n = (\omega + n\omega_c)/k_z v_{th}$. Further noting that $2k_x^2 \omega_p^2 \zeta_0 / \omega^2 = 2\omega_p^2 / v_{th}^2 \equiv k_D^2$ for each species, we can write Eq. (7.147) as

$$k_x^2 + k_z^2 + \sum_s k_D^2 e^{-b} \left\{ \sum_{n=-\infty}^{\infty} I_n(b)[\zeta_{-n}/\zeta_0 + \zeta_0 Z(\zeta_n)] \right\} = 0 \qquad (7.148)$$

The term ζ_{-n}/ζ_0 is $1 - n\omega_c/\omega$. Since $I_n(b) = I_{-n}(b)$, the term $I_n(b)n\omega_c/\omega$ sums to zero when n goes from $-\infty$ to ∞; hence, ζ_{-n}/ζ_0 can be replaced by 1. Defining $k^2 \equiv k_\perp^2 + k_z^2 = k_x^2 + k_z^2$ we obtain the general dispersion relation for Bernstein waves:

$$1 + \sum_s \frac{k_D^2}{k^2} e^{-b} \left\{ \sum_{n=-\infty}^{\infty} I_n(b)[1 + \zeta_0 Z(\zeta_n)] \right\} = 0. \qquad (7.149)$$

(A)   *Electron Bernstein Waves.* Let us first consider high-frequency waves in which the ions do not move. These waves are not sensitive to small deviations from perpendicular propagation, and we may set $k_z = 0$, so that $\zeta_n \to \infty$. There is, therefore, no cyclotron damping; the gaps in the spectrum that we shall find are not caused by such damping. For large ζ_n, we may replace $Z(\zeta_n)$ by $-1/\zeta_n$, according to the integral of Eq. (7.129). The $n=0$ term in the second sum of Eq. (7.149) then cancels out, and we can divide the sum into two sums, as follows:

$$k^2 + \sum_s k_D^2 e^{-b} \left[\sum_{n=1}^{\infty} I_n(b)(1 - \zeta_0/\zeta_n) + \sum_{n=1}^{\infty} I_{-n}(b)(1 - \zeta_0/\zeta_{-n}) \right] = 0, \quad (7.150)$$

or

$$k^2 + \sum_s k_D^2 e^{-b} \sum_{n=1}^{\infty} I_n(b) \left[2 - \frac{\omega}{\omega + n\omega_c} - \frac{\omega}{\omega - n\omega_c} \right] = 0 \qquad (7.151)$$

The bracket collapses to a single term upon combining over a common denominator:

$$1 = \sum_s \frac{k_D^2}{k^2} e^{-b} \sum_{n=1}^{\infty} I_n(b) \frac{2n^2 \omega_c^2}{\omega^2 - n^2 \omega_c^2} \qquad (7.152)$$

Using the definitions of k_D and b, one obtains the well-known $k_z = 0$ dispersion relation

$$1 = \sum_s \frac{\omega_p^2}{\omega_c^2} \frac{2}{b} e^{-b} \sum_{n=1}^{\infty} \frac{I_n(b)}{(\omega/n\omega_c)^2 - 1} \qquad (7.153)$$

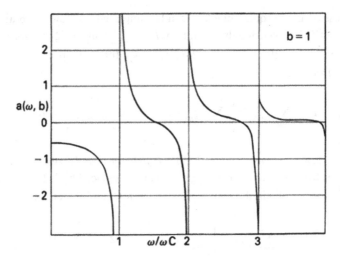

Fig. 7.33 The function $\alpha(\omega, b)$ for electron Bernstein waves [From I. B. Bernstein, *Phys. Rev.* **109**, 10 (1958)]

We now specialize to the case of electron oscillations with $k_z = 0$. Dropping the sum over species, we obtain from Eq. (7.152)

$$\frac{k_\perp^2}{k_D^2} = 2\omega_c^2 \sum_{n=1}^{\infty} \frac{e^{-b}I_n(b)}{\omega^2 - n\omega_c^2} n^2 \equiv \alpha(\omega, b) \tag{7.154}$$

The function $\alpha(\omega, b)$ for one value of b is shown in Fig. 7.33. The possible values of ω are found by drawing a horizontal line at $\alpha(\omega, b) = k_\perp^2/k_D^2 > 0$. It is then clear that possible values of ω lie just above each cyclotron harmonic, and that there is a forbidden gap just below each harmonic.

To obtain the fluid limit, we replace $I_n(b)$ by its small-b value $(b/2)^n/n!$ in Eq. (7.153). Only the $n = 1$ term remains in the limit $b \to 0$, and we obtain

$$1 = \frac{\omega_p^2}{\omega_c^2} \frac{2}{b} \frac{b}{2} \left(\frac{\omega^2}{\omega_c^2} - 1\right)^{-1} = \frac{\omega_p^2}{\omega^2 - \omega_c^2} \tag{7.155}$$

or $\omega^2 = \omega_p^2 + \omega_c^2 = \omega_h^2$, which is the upper hybrid oscillation. As $k_\perp \to 0$, this frequency must be one of the roots. If ω_h falls between two high harmonics of ω_c, the shape of the $\omega - k$ curves changes near $\omega = \omega_h$ to allow this to occur. The $\omega - k$ curves are computed by multiplying Eq. (7.154) by $2\omega_p^2/\omega_c^2$ to obtain $k_\perp^2 r_L^2 = 4\omega_p^2 \alpha(\omega, b)$. The resulting curves for ω/ω_c vs. $k_\perp r_L$ are shown in Fig. 7.34 for various values of ω_p^2/ω_c^2. Note that for each such value, the curves change in character above the corresponding hybrid frequency for that case. At the extreme left of the diagram, where the phase velocity approaches the speed of light

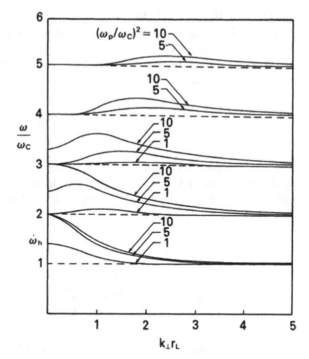

Fig. 7.34 Electron Bernstein wave dispersion relation [Adapted from F. W. Crawford, *J. Appl. Phys.* **36**, 2930 (1965)]

waves in the plasma, these curves must be modified by including electromagnetic corrections.

Electron Bernstein modes have been detected in the laboratory, but inexplicably large spontaneous oscillations at high harmonics of ω_c have also been seen in gas discharges. The story is too long to tell here.

(B) Ion Bernstein Waves. In the case of waves at ion cyclotron harmonics, one has to distinguish between *pure* ion Bernstein waves, for which $k_z = 0$, and *neutralized* ion Bernstein waves, for which k_z has a small but finite value. The difference, as we have seen earlier for lower hybrid oscillations, is that finite k_z allows electrons to flow along $\mathbf{B}_0$ to cancel charge separations. Though the $k_z = 0$ case has already been treated in Eq. (7.153), the distinction between the two cases will be clearer if we go back a step to Eqs. (7.148) and (7.149). Separating out the $n = 0$ term and using Eq. (7.146), we have

$$k_\perp^2 + k_z^2 + \sum_s k_D^2 e^{-b} I_0(b)\left[-\tfrac{1}{2} Z'(\zeta_0)\right] + \sum_s k_D^2 e^{-b} \sum_{n \neq 0} I_n(b)[1 + \zeta_0 Z(\zeta_n)] = 0$$

$$(7.156)$$

The dividing line between pure and neutralized ion Bernstein waves lies in the electron $n = 0$ term. If $\zeta_{0e} \gg 1$ for the electrons, we can use Eq. (7.129) to write $Z'(\zeta_{0e}) \simeq 1/\zeta_{0e}^2$. Since $\omega/k_z \gg v_{\text{the}}$ in this case, electrons cannot flow rapidly enough along $\mathbf{B}_0$ to cancel charge. If $\zeta_{0e} \ll 1$, we can use Eq. (7.126) to write Z' $(\zeta_{0e}) \simeq -2$. In this case we have $\omega/k_z \ll v_{the}$, and the electrons have time to follow the Boltzmann relation (3.73).

Taking first the $\zeta_{0e} \gg 1$ case, we note that $\zeta_{0i} \gg 1$ is necessarily true also, so that the $n=0$ term in Eq. (7.156) becomes

$$-k_z^2 \left[\frac{\omega_p^2}{\omega^2} + \frac{\Omega_p^2}{\omega^2} e^{-b} I_0(b) \right]$$

Here we have taken $b_e \to 0$ and omitted the subscript from b_i. The $n \neq 0$ terms in Eq. (7.156) are treated as before, so that the electron part is given by Eq. (7.155), and the ion part by the ion term in Eq. (7.153). The pure ion Bernstein wave dispersion relation then becomes

$$k_z^2 \left[1 - \frac{\omega_p^2}{\omega^2} - \frac{\Omega_p^2}{\omega^2} e^{-b} I_0(b) \right]$$
$$+ k_\perp^2 \left[1 - \frac{\omega_p^2}{\omega^2 - \omega_c^2} - \frac{\Omega_p^2}{\Omega_c^2} \frac{2}{b} e^{-b} \sum_{n=1}^{\infty} \frac{I_n(b)}{(\omega/n\Omega_c - 1)^2} \right] = 0 \qquad (7.157)$$

Since $\zeta_{0e} \gg 1$ implies small k_z^2, the first term is usually negligible. To examine the fluid limit, we can set the second bracket to zero, separate out the $n=1$ term, and use the small-b expansion of $I_n(b)$, obtaining

$$1 - \frac{\omega_p^2}{\omega^2 - \omega_c^2} - \frac{\Omega_p^2}{\omega^2 - \Omega_c^2} - \sum_{n=2}^{\infty} \frac{n^2 \Omega_p^2 (b/2)^{n-1}}{n! (\omega^2 - n^2\Omega_c^2)} = 0 \qquad (7.158)$$

The sum vanishes for $b=0$, and the remaining terms are equal to the quantity S of Appendix B. The condition $S=0$ yields the upper and lower hybrid frequencies [see the equation following Eq. (4.70)]. Thus, for $k_\perp \to 0$, the low-frequency root approaches ω_l. For finite b, one of the terms in the sum can balance the electron term if $\omega \simeq n\Omega_c$, so there are roots near the ion cyclotron harmonics. The dispersion curves ω/Ω_c vs. $k_\perp r_{Li}$ resemble the electron curves in Fig. 7.34. The lowest two roots for the ion case are shown in Fig. 7.35, together with experimental measurements verifying the dispersion relation.

The lower branches of the Bernstein wave dispersion relation exhibit the *backward-wave* phenomenon, in which the $\omega - k$ curve has a negative slope, indicating that the group velocity is opposite in direction to the phase velocity. That backward waves actually exist in the laboratory has been verified not only by ω vs. k measurements of the type shown in Fig. 7.35, but also by wave interferometer traces which show the motion of phase fronts in the backward direction from receiver to transmitter.

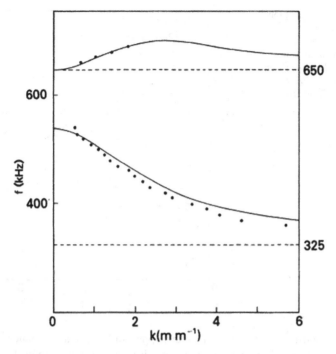

Fig. 7.35 Pure ion Bernstein waves: agreement between theory and experiment in a Q-machine plasma [From J. P. M. Schmitt, *Phys. Rev. Lett.* **31**, 982 (1973)]

Finally, we consider neutralized Bernstein waves, for which ζ_{0e} is small and $Z'(\zeta_{0e}) \simeq -2$. The electron $n = 0$ term in Eq. (7.156) becomes simply k_{De}^2. Assuming that $\zeta_{0i} \gg 1$ still holds, the analysis leading to Eq. (7.157) is unchanged, and Eq. (7.156) becomes

$$
k_z^2\left[1 + \frac{k_D^2}{k_z^2} - \frac{\Omega_p^2}{\omega^2}e^{-b}I_0(b)\right]
$$
$$
+ k_\perp^2\left[1 - \frac{\omega_p^2}{\omega^2 - \omega_c^2} + \frac{\Omega_p^2}{\Omega_c^2}\frac{2}{b}e^{-b}\sum_{n=1}^{\infty}\frac{I_n(b)}{(\omega/n\Omega_c)^2 - 1}\right] = 0 \tag{7.159}
$$

for $k_z^2 \ll k_\perp^2$, an approximate relation for neutralized ion Bernstein waves can be written

$$
1 + k^2\lambda_D^2\left[1 - \frac{\omega_p^2}{\omega^2 - \omega_c^2} + \frac{\Omega_p^2}{\Omega_c^2}\frac{2}{b}e^{-b}\sum_{n=1}^{\infty}\frac{I_n(b)}{(\omega/n\Omega_c - 1)^2}\right] = 0 \tag{7.160}
$$

Note that electron temperature is now contained in λ_D, whereas pure ion Bernstein waves, Eq. (7.157), are independent of KT_e. If $k^2\lambda_D^2$ is small, the bracket in

Fig. 7.36 Neutralized ion
Bernstein modes:
agreement between theory
and experiment in a He
microwave discharge [From
E. Ault and H. Ikezi, *Phys.
Fluids* **13**, 2874 (1970)]

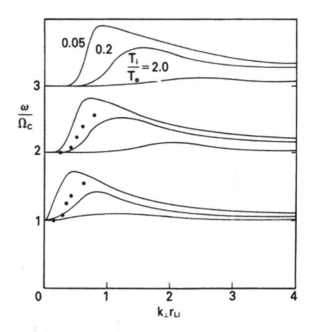

Eq. (7.160) must be large; and this can happen only near a resonance $\omega \simeq n\Omega_c$. Thus the neutralized modes are not sensitive to the lower hybrid resonance $\omega \simeq \omega_l$. Indeed, as $k_\perp r_{Li} \to 0$ the envelope of the dispersion curves approaches the electrostatic ion cyclotron wave relation (4.67), which is the fluid limit for neutralized waves.

Neutralized Bernstein modes are not as well documented in experiment as pure Bernstein modes, but we show in Fig. 7.36 one case in which the former have been seen.

Chapter 8
Nonlinear Effects

8.1 Introduction

Up to this point, we have limited our attention almost exclusively to *linear* phenomena; that is, to phenomena describable by equations in which the dependent variable occurs to no higher than the first power. The entire treatment of waves in Chap. 4, for instance, depended on the process of linearization, in which higher-order terms were regarded as small and were neglected. This procedure enabled us to consider only one Fourier component at a time, with the secure feeling that any nonsinusoidal wave can be handled simply by adding up the appropriate distribution of Fourier components. This works as long as the wave amplitude is small enough that the linear equations are valid.

Unfortunately, in many experiments waves are no longer describable by the linear theory by the time they are observed. Consider, for instance, the case of drift waves. Because they are unstable, drift waves would, according to linear theory, increase their amplitude exponentially. This period of growth is not normally observed—since one usually does not know when to start looking—but instead one observes the waves only after they have grown to a large, steady amplitude. The fact that the waves are no longer growing means that the linear theory is no longer valid, and some *nonlinear* effect is limiting the amplitude. Theoretical explanation of this elementary observation has proved to be a surprisingly difficult problem, since the observed amplitude at saturation is rather small.

A wave can undergo a number of changes when its amplitude gets large. It can change its shape—say, from a sine wave to a lopsided triangular waveform. This is the same as saying that Fourier components at other frequencies (or wave numbers) are generated. Ultimately, the wave can "break," like ocean waves on a beach, converting the wave energy into thermal energy of the particles. A large wave can

The original version of this chapter was revised. An erratum to this chapter can be found at https://doi.org/10.1007/978-3-319-22309-4_11

F.F. Chen, *Introduction to Plasma Physics and Controlled Fusion*,
DOI 10.1007/978-3-319-22309-4_8

267

trap particles in its potential troughs, thus changing the properties of the medium in which it propagates. We have already encountered this effect in discussing nonlinear Landau damping. If a plasma is so strongly excited that a continuous spectrum of frequencies is present, it is in a state of *turbulence*. This state must be described statistically, as in the case of ordinary fluid hydrodynamics. An important consequence of plasma turbulence is *anomalous resistivity*, in which electrons are slowed down by collisions with random electric field fluctuations, rather than with ions. This effect is used for ohmic heating of a plasma (Sect. 5.6.3) to temperatures so high that ordinary resistivity is insufficient.

Nonlinear phenomena can be grouped into three broad categories:

1. *Basically nonlinearizable problems.* Diffusion in a fully ionized gas, for instance, is intrinsically a nonlinear problem (Sect. 5.8) because the diffusion coefficient varies with density. In Sect. 6.1, we have seen that problems of hydromagnetic equilibrium are nonlinear. In Sect. 8.2, we shall give a further example—the important subject of plasma sheaths.

2. *Wave–particle interactions.* Particle trapping (Sect. 7.5) is an example of this and can lead to nonlinear damping. A classic example is the quasilinear effect, in which the equilibrium of the plasma is changed by the waves. Consider the case of a plasma with an electron beam (Fig. 8.1). Since the distribution function has a region where df_0/dv is positive, the system has inverse Landau damping, and plasma oscillations with v_ϕ in the positive-slope region are unstable (Eq. (7.67)). The resonant electrons are the first to be affected by wave–particle interactions, and their distribution function will be changed by the wave electric field. The waves are stabilized when $f_e(v)$ is flattened by the waves, as shown by the dashed line in Fig. 8.1, so that the new equilibrium distribution no longer has a positive slope. This is a typical quasilinear effect. Another example of wave–particle interactions, that of plasma wave echoes, will be given in Sect. 8.6.

3. *Wave–wave interactions.* Waves can interact with each other even in the fluid description, in which individual particle effects are neglected. A single wave can decay by first generating harmonics of its fundamental frequency. These harmonics can then interact with each other and with the primary wave to form other waves at the beat frequencies. The beat waves in turn can grow so large that they can interact and form many more beat frequencies, until the spectrum

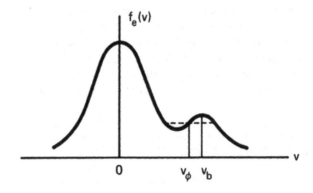

Fig. 8.1 A double-humped, unstable electron distribution

becomes continuous. It is interesting to discuss the direction of energy flow in a turbulent spectrum. In fluid dynamics, long-wavelength modes decay into short-wavelength modes, because the large eddies contain more energy and can decay only by splitting into small eddies, which are each less energetic. The smallest eddies then convert their kinetic motion into heat by viscous damping. In a plasma, usually the opposite occurs. Short-wavelength modes tend to coalesce into long-wavelength modes, which are less energetic. This is because the electric field energy $E^2/2$ is of order $k^2\phi^2/2$, so that if $e\phi$ is fixed (usually by KT_e), the small-k, long-λ modes have less energy. As a consequence, energy will be transferred to small k by instabilities at large k, and some mechanism must be found to dissipate the energy. No such problem exists at large k, where Landau damping can occur. For motions along B_0, nonlinear "modulational" instabilities could cause the energy at small k to be coupled to ions and to heat them. For motions perpendicular to B_0, the largest eddies will have wavelengths of the order of the plasma radius and could cause plasma loss to the walls by convection.

Although problems still remain to be solved in the linear theory of waves and instabilities, the mainstream of plasma research has turned to the much less well understood area of nonlinear phenomena. The examples in the following sections will give an idea of some of the effects that have been studied in theory and in experiment.

8.2 Sheaths

8.2.1 The Necessity for Sheaths

In all practical plasma devices, the plasma is contained in a vacuum chamber of finite size. What happens to the plasma at the wall? For simplicity, let us confine our attention to a one-dimensional model with no magnetic field (Fig. 8.2). Suppose there is no appreciable electric field inside the plasma; we can then let the potential ϕ be zero there. When ions and electrons hit the wall, they recombine and are lost. Since electrons have much higher thermal velocities than ions, they are lost faster

Fig. 8.2 The plasma potential ϕ forms sheaths near the walls so that electrons are reflected. The Coulomb barrier $e\phi_w$ adjusts itself so that equal numbers of ions and electrons reach the walls per second

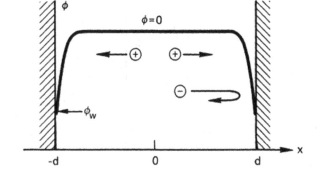

and leave the plasma with a net positive charge. The plasma must then have a potential positive with respect to the wall; i.e., the wall potential ϕ_w is negative. This potential cannot be distributed over the entire plasma, since Debye shielding (Sect. 1.4) will confine the potential variation to a layer of the order of several Debye lengths in thickness. This layer, which must exist on all cold walls with which the plasma is in contact, is called a *sheath*. The function of a sheath is to form a potential barrier so that the more mobile species, usually electrons, is confined electrostatically. The height of the barrier adjusts itself so that the flux of electrons that have enough energy to go over the barrier to the wall is just equal to the flux of ions reaching the wall.

8.2.2   The Planar Sheath Equation

In Sect. 1.4, we linearized Poisson's equation to derive the Debye length. To examine the exact behavior of ϕ (x) in the sheath, we must treat the nonlinear problem; we shall find that there is not always a solution. Figure 8.3 shows the situation near one of the walls. At the plane $x = 0$, ions are imagined to enter the sheath region from the main plasma with a drift velocity u_0. This drift is needed to account for the loss of ions to the wall from the region in which they were created by ionization. For simplicity, we assume $T_i = 0$, so that all ions have the velocity u_0 at $x = 0$. We consider the steady state problem in a collisionless sheath region. The potential ϕ is assumed to decrease monotonically with x. Actually, ϕ could have spatial oscillations, and then there would be trapped particles in the steady state. This does not happen in practice because dissipative processes tend to destroy any such highly organized state.

Fig. 8.3 The potential ϕ in a planar sheath. Cold ions are assumed to enter the sheath with a uniform velocity u_0

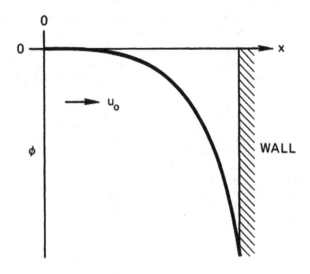

If $u(x)$ is the ion velocity in the sheath, conservation of energy requires

$$\frac{1}{2} M u^2 = \frac{1}{2} M u_0^2 - e\phi(x) \tag{8.1}$$

$$u = \left(u_0^2 - \frac{2e\phi}{M} \right)^{1/2} \tag{8.2}$$

The ion equation of continuity then gives the ion density n_i in terms of the density n_0 in the main plasma:

$$n_0 u_0 = n_i(x)\ u(x) \tag{8.3}$$

$$n_i(x) = n_0 \left(1 - \frac{2e\phi}{M u_0^2} \right)^{-1/2} \tag{8.4}$$

In steady state, the electrons will follow the Boltzmann relation closely:

$$n_e(x) = n_0 \exp(e\phi/KT_e) \tag{8.5}$$

Poisson's equation is then

$$\varepsilon_0 \frac{d^2\phi}{dx^2} = e(n_e - n_i) = e n_0 \left[\exp\left(\frac{e\phi}{KT_e} \right) - \left(1 - \frac{2e\phi}{M u_0^2} \right)^{-1/2} \right] \tag{8.6}$$

The structure of this equation can be seen more clearly if we simplify it with the following changes in notation:

$$\chi \equiv -\frac{e\phi}{KT_e}, \quad \xi \equiv \frac{x}{\lambda_D} = x \left(\frac{n_0 e^2}{\varepsilon_0 KT_e} \right)^{1/2} \quad \mathcal{M} \equiv \frac{u_0}{(KT_e/M)^{1/2}} \tag{8.7}$$

Then Eq. (8.6) becomes

$$\chi'' = \left(1 + \frac{2\chi}{\mathcal{M}^2} \right)^{-1/2} - e^{-\chi} \tag{8.8}$$

where the prime denotes $d/d\xi$. This is the nonlinear equation of a plane sheath, and it has an acceptable solution only if $\mathcal{M}$ is large enough. The reason for the symbol $\mathcal{M}$ will become apparent in the following section on shock waves.

8.2.3   The Bohm Sheath Criterion

Equation (8.8) can be integrated once by multiplying both sides by χ':

$$\int_0^\xi \chi' \chi'' d\xi_1 = \int_0^\xi \left(1 + \frac{2\chi}{\mathcal{M}^2}\right)^{-1/2} \chi' d\xi_1 - \int_0^\xi e^{-\chi} \chi' d\xi_1 \qquad (8.9)$$

where ξ_1 is a dummy variable. Since $\chi = 0$ at $\xi = 0$, the integrations easily yield

$$\frac{1}{2}\left(\chi'^2 - \chi_0'^2\right) = \mathcal{M}^2 \left[\left(1 + \frac{2\chi}{\mathcal{M}^2}\right)^{1/2} - 1\right] + e^{-\chi} - 1. \qquad (8.10)$$

If $\mathbf{E} = 0$ in the plasma, we must set $\chi_0' = 0$ at $\xi = 0$. A second integration to find χ would have to be done numerically; but whatever the answer is, the right-hand side of Eq. (8.10) must be positive for all χ. In particular, for $\chi \ll 1$, we can expand the right-hand terms in Taylor series:

$$\mathcal{M}^2 \left[1 + \frac{\chi}{\mathcal{M}^2} - \frac{1}{2}\frac{\chi^2}{\mathcal{M}^4} + \cdots - 1\right] + 1 - \chi + \frac{1}{2}\chi^2 + \cdots - 1 > 0$$

$$\frac{1}{2}\chi^2\left(-\frac{1}{\mathcal{M}^2} + 1\right) > 0 \qquad (8.11)$$

$$\mathcal{M}^2 > 1 \qquad \text{or} \qquad u_0 > (KT_e/M)^{1/2}$$

This inequality is known as the *Bohm sheath criterion*. It says that ions must enter the sheath region with a velocity greater than the acoustic velocity v_s. To give the ions this directed velocity u_0, there must be a finite electric field in the plasma. Our assumption that $\chi' = 0$ at $\xi = 0$ is therefore only an approximate one, made possible by the fact that the scale of the sheath region is usually much smaller than the scale of the main plasma region in which the ions are accelerated. The value of u_0 is somewhat arbitrary, depending on where we choose to put the boundary $x = 0$ between the plasma and the sheath. Of course, the ion flux $n_0 u_0$ is fixed by the ion production rate, so if u_0 is varied, the value of n_0 at $x = 0$ will vary inversely with u_0. If the ions have finite temperature, the critical drift velocity u_0 will be somewhat lower.

The physical reason for the Bohm criterion is easily seen from a plot of the ion and electron densities vs. χ (Fig. 8.4). The electron density n_e falls exponentially with χ according to the Boltzmann relation. The ion density also falls, since the ions are accelerated by the sheath potential. If the ions start with a large energy, $n_i(\chi)$ falls slowly, since the sheath field causes a relatively minor change in the ions' velocity. If the ions start with a small energy, $n_i(\chi)$ falls fast, and can go below the n_e curve. In that case, $n_e - n_i$ is positive near $\chi = 0$; and Eq. (8.6) tells us that $\phi(x)$ must curve upward, in contradiction to the requirement that the sheath must repel

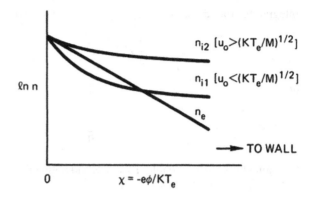

Fig. 8.4 Variation of ion and electron density (logarithmic scale) with normalized potential χ in a sheath. The ion density is drawn for two cases: u_0 greater than and u_0 less than the critical velocity

electrons. In order for this not to happen, the slope of $n_i(\chi)$ at $\chi = 0$ must be smaller (in absolute value) than that of $n_e(\chi)$; this condition is identical with the condition $\mathcal{M}^2 > 1$.

8.2.4   The Child–Langmuir Law

Since $n_e(\chi)$ falls exponentially with χ, the electron density can be neglected in the region of large χ next to the wall (or any negative electrode). Poisson's equation is then approximately

$$\chi'' \approx \left(1 + \frac{2\chi}{\mathcal{M}^2}\right)^{-1/2} \approx \frac{\mathcal{M}}{(2\chi)^{1/2}} \tag{8.12}$$

Multiplying by χ' and integrating from $\xi_1 = \xi_s$ to $\xi_1 = \xi$, we have

$$\tfrac{1}{2}\left(\chi'^2 - \chi'^2_s\right) = \sqrt{2}\mathcal{M}\left(\chi^{1/2} - \chi^{1/2}_s\right) \tag{8.13}$$

where ξ_s is the place where we started neglecting n_e. We can redefine the zero of χ so that $\chi_s = 0$ at $\xi = \xi_s$. We shall also neglect χ'_s, since the slope of the potential curve can be expected to be much steeper in the $n_e = 0$ region than in the finite-n_e region. Then Eq. (8.13) becomes

$$\chi'^2 = 2^{3/2}\mathcal{M}\chi^{1/2}$$
$$\chi' = 2^{3/4}\mathcal{M}^{1/2}\chi^{1/4} \tag{8.14}$$

or

$$d\chi/\chi^{1/4} = 2^{3/4}\mathcal{M}^{1/2}d\xi. \tag{8.15}$$

Integrating from $\xi = \xi_s$ to $\xi = \xi_s + d/\lambda_D = \xi_{\text{wall}}$, we have

$$\frac{4}{3}\chi_w^{3/4} = 2^{3/4}\mathcal{M}^{1/2}d/\lambda_D \tag{8.16}$$

or

$$\mathcal{M} = \frac{4\sqrt{2}}{9}\frac{\chi_w^{3/2}}{d^2}\lambda_D^2 \tag{8.17}$$

Changing back to the variables u_0 and ϕ, and noting that the ion current into the wall is $J = en_0 u_0$, we then find

$$J = \frac{4}{9}\left(\frac{2e}{M}\right)^{1/2}\frac{\varepsilon_0|\phi_w|^{3/2}}{d^2} \tag{8.18}$$

This is just the well-known Child–Langmuir law of space-charge-limited current in a plane diode.

The potential variation in a plasma–wall system can be divided into three parts. Nearest the wall is an electron-free region whose thickness d is given by Eq. (8.18). Here J is determined by the ion production rate, and ϕ_w is determined by the equality of electron and ion fluxes. Next comes a region in which n_e is appreciable; as shown in Sect. 1.4, this region has the scale of the Debye length. Finally, there is a region with much larger scale length, the "presheath," in which the ions are accelerated to the required velocity u_0 by a potential drop $|\phi| \geq \frac{1}{2}KT_e/e$. Depending on the experiment, the scale of the presheath may be set by the plasma radius, the collision mean free path, or the ionization mechanism. The potential distribution, of course, varies smoothly; the division into three regions is made only for convenience and is made possible by the disparity in scale lengths. In the early days of gas discharges, sheaths could be observed as dark layers where no electrons were present to excite atoms to emission. Subsequently, the potential variation has been measured by the electrostatic deflection of a thin electron beam shot parallel to a wall.

Recent theories of discharges in finite cylinders show that the sheaths on the endplates play an essential rôle in moving plasma created at the radial edge towards the axis, so that the final density profile is peaked at the center. It has also been shown that the density profile in such a plasma, including the presheath, tends to fall into a universal shape independent of pressure and discharge diameter.

8.2.5 Electrostatic Probes

The sheath criterion, Eq. (8.11), can be used to estimate the flux of ions to a negatively biased probe in a plasma. If the probe has a surface area A, and if the ions entering the sheath have a drift velocity $u_0 \geq (KT_e/M)^{1/2}$, then the ion current collected is

$$I = n_s eA(KT_e/M)^{1/2} \tag{8.19}$$

The electron current can be neglected if the probe is sufficiently negative (several times KT_e) relative to the plasma to repel all but the tail of the Maxwellian electron distribution. The density n_s is the plasma density at the edge of the sheath. Let us define the sheath edge to be the place where u_0 is exactly $(KT_e/M)^{1/2}$. To accelerate ions to this velocity requires a presheath potential $|\phi| \geq \frac{1}{2}KT_e/e$, so that the sheath edge has a potential

$$\phi_s \simeq -\frac{1}{2}KT_e/e \tag{8.20}$$

relative to the body of the plasma. If the electrons are Maxwellian, this determines n_s:

$$n_s = n_0 e^{e\phi_s/kT_e} = n_0 e^{-1/2} = 0.61 n_0 \tag{8.21}$$

For our purposes it is accurate enough to replace 0.61 with a round number like 1/2; thus, the "saturation ion current" to a negative probe is approximately

$$I_B = \frac{1}{2} n_0 eA(KT_e/M)^{1/2} \tag{8.22}$$

I_B, sometimes called the "Bohm current," gives the plasma density easily, once the temperature is known.

If the Debye length λ_D, and hence the sheath thickness, is very small compared to the probe dimensions, the area of the sheath edge is effectively the same as the area A of the probe surface, regardless of its shape. At low densities, however, λ_D can become large, so that some ions entering the sheath can orbit the probe and miss it. Calculations of orbits for various probe shapes were first made by I. Langmuir and L. Tonks—hence the name "Langmuir probe" ascribed to this method of measurement. Though tedious, these calculations can give accurate determinations of plasma density because an arbitrary definition of sheath edge does not have to be made. By varying the probe voltage, the electron distribution is sampled, and the current–voltage curve of a Langmuir probe can also yield the electron temperature, if the electrons are Maxwellian, or their velocity distribution if they are not. The Langmuir probe is the first plasma diagnostic and is still the simplest and the most localized measurement device. Material electrodes can be inserted only in low-density, cool plasmas. Nonetheless, these include most non-fusion laboratory plasmas, and an extensive literature exists on probe theory. The problem with using a large probe, to which Eq. (8.22) applies, is that it collects from an ill-defined sheath edge that surrounds it, and it may also disturb the plasma. Thin cylindrical probes with radii $\leq \lambda_D$ are more commonly used.

Problems

8.1 A probe whose collecting surface is a square tantalum foil 2×2 mm in area is found to give a saturation ion current of 100 μA in a singly ionized argon plasma (atomic weight $= 40$). If $KT_e = 2$ eV, what is the approximate plasma density? (Hint: Both sides of the probe collect ions!)

8.2 A solar satellite consisting of 10 km^2 of photovoltaic panels is placed in synchronous orbit around the earth. It is immersed in a 1-eV atomic hydrogen plasma at density 10^6 m^{-3}. During solar storms the satellite is bombarded by energetic electrons, which charge it to a potential of -2 kV. Calculate the flux of energetic ions bombarding each m^2 of the panels.

8.3 The sheath criterion of Eq. (8.11) was derived for a cold-ion plasma. Suppose the ion distribution had a thermal spread in velocity around an average drift speed u_0. Without mathematics, indicate whether you would expect the value of u_0 to be above or below the Bohm value, and explain why.

8.4 An ion velocity analyzer consists of a stainless steel cylinder 5 mm in diameter with one end covered with a fine tungsten mesh grid (grid 1). Behind this, inside the cylinder, are a series of insulated, parallel grids. Grid 1 is at "floating" potential—it is not electrically connected. Grid 2 is biased negative to repel all electrons coming through grid 1, but it transmits ions. Grid 3 is the analyzer grid, biased so as to decelerate ions accelerated by grid 2. Those ions able to pass through grid 3 are all collected by a collector plate. Grid 4 is a suppressor grid that turns back secondary electrons emitted by the collector. If the plasma density is too high, a space charge problem occurs near grid 3 because the ion density is so large that a potential hill forms in front of grid 3 and repels ions which would otherwise reach grid 3. Using the Child–Langmuir law, estimate the maximum meaningful He$^+$ current that can be measured on a 4-mm-diam collector if grids 2 and 3 are separated by 1 mm and 100 V.

8.3   Ion Acoustic Shock Waves

When a jet travels faster than sound, it creates a shock wave. This is a basically nonlinear phenomenon, since there is no period when the wave is small and growing. The jet is faster than the speed of waves in air, so the undisturbed medium cannot be "warned" by precursor signals before the large shock wave hits it. In hydrodynamic shock waves, collisions are dominant. Shock waves also exist in plasmas, even when there are no collisions. A magnetic shock, the "bow shock," is generated by the earth as it plows through the interplanetary plasma while dragging along a dipole magnetic field. We shall discuss a simpler example: a collisionless, one-dimensional shock wave which develops from a large-amplitude ion wave.

8.3.1 The Sagdeev Potential

Figure 8.5 shows the idealized potential profile of an ion acoustic shock wave. The reason for this shape will be given presently. The wave is traveling to the left with a velocity u_0. If we go to the frame moving with the wave, the function $\phi(x)$ will be constant in time, and we will see a stream of plasma impinging on the wave from the left with a velocity u_0. For simplicity, let T_i be zero, so that all the ions are incident with the same velocity u_0, and let the electrons be Maxwellian. Since the shock moves much more slowly than the electron thermal speed, the shift in the center velocity of the Maxwellian can be neglected. The velocity of the ions in the shock wave is, from energy conservation,

$$u = \left(u_0^2 - \frac{2e\phi}{M} \right)^{1/2} \tag{8.23}$$

If n_0 is the density of the undisturbed plasma, the ion density in the shock is

$$n_i = \frac{n_0 u_0}{u} = n_0 \left(1 - \frac{2e\phi}{M u_0^2} \right)^{-1/2} \tag{8.24}$$

The electron density is given by the Boltzmann relation. Poisson's equation then gives

$$\varepsilon_0 \frac{d^2\phi}{dx^2} = e(n_e - n_i) = e n_0 \left[\exp\left(\frac{e\phi}{KT_e} \right) - \left(1 - \frac{2e\phi}{M u_0^2} \right)^{-1/2} \right] \tag{8.25}$$

This is, of course, the same equation (Eq. (8.6)) as we had for a sheath. A shock wave is no more than a sheath moving through a plasma. We now introduce the dimensionless variables

Fig. 8.5 Typical potential distribution in an ion acoustic shock wave. The wave moves to the left, so that in the wave frame ions stream into the wave from the left with velocity u_0

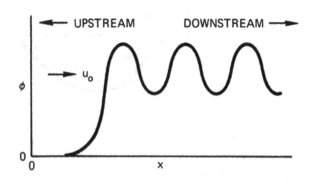

$$\chi \equiv + \frac{e\phi}{KT_e} \qquad \xi \equiv \frac{x}{\lambda_D} \qquad \mathcal{M} \equiv \frac{u_0}{(KT_e/M)^{1/2}} \qquad (8.26)$$

Note that we have changed the sign in the definition of χ so as to keep χ positive in this problem as well as in the sheath problem. The quantity $\mathcal{M}$ is called the *Mach number* of the shock. Equation (8.25) can now be written

$$\frac{d^2\chi}{d\xi^2} = e^\chi - \left(1 - \frac{2\chi}{\mathcal{M}^2}\right)^{-1/2} \equiv -\frac{dV(\chi)}{d\chi} \qquad (8.27)$$

which differs from the sheath equation (8.8) only because of the change in sign of χ.

The behavior of the solution of Eq. (8.27) was made clear by R. Z. Sagdeev, who used an analogy to an oscillator in a potential well. The displacement x of an oscillator subjected to a force $-m \cdot dV(x)/dx$ is given by

$$d^2x/dt^2 = -dV/dx \qquad (8.28)$$

If the right-hand side of Eq. (8.27) is defined as $-dV/d\chi$, the equation is the same as that of an oscillator, with the potential χ playing the role of x, and $d/d\xi$ replacing d/dt. The quasipotential $V(\chi)$ is sometimes called the Sagdeev potential. The function $V(\chi)$ can be found from Eq. (8.27) by integration with the boundary condition $V(\chi) = 0$ at $\chi = 0$:

$$V(\chi) = 1 - e^\chi + \mathcal{M}^2 \left[1 - \left(1 - \frac{2\chi}{\mathcal{M}^2}\right)^{1/2}\right] \qquad (8.29)$$

For $\mathcal{M}$ lying in a certain range, this function has the shape shown in Fig. 8.6. If this were a real well, a particle entering from the left will go to the right-hand side of the

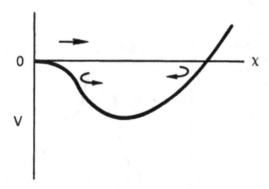

Fig. 8.6 The Sagdeev potential $V(\chi)$. The *upper arrow* is the trajectory of a quasiparticle describing a soliton: it is reflected at the right and returns. The *lower arrows* show the motion of a quasiparticle that has lost energy and is trapped in the potential well. The bouncing back and forth describes the oscillations behind a shock front

Fig. 8.7 The potential in a
soliton moving to the left

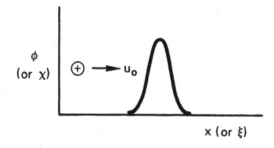

Fig. 8.7 The potential in a soliton moving to the left

well ($x > 0$), reflect, and return to $x = 0$, making a single transit. Similarly, a quasiparticle in our analogy will make a single excursion to positive χ and return to $\chi = 0$, as shown in Fig. 8.7. Such a pulse is called a *soliton*; it is a potential and density disturbance propagating to the left in Fig. 8.7 with velocity u_0.

Now, if a particle suffers a loss of energy while in the well, it will never return to $x = 0$ but will oscillate (in time) about some positive value of x. Similarly, a little dissipation will make the potential of a shock wave oscillate (in space) about some positive value of ϕ. This is exactly the behavior depicted in Fig. 8.5. Actually, dissipation is not needed for this; reflection of ions from the shock front has the same effect. To understand this, imagine that the ions have a small thermal spread in energy and that the height $e\phi$ of the wave front is just large enough to reflect some of the ions back to the left, while the rest go over the potential hill to the right. The reflected ions cause an increase in ion density in the upstream region to the left of the shock front (Fig. 8.5). This means that the quantity

$$\chi' = \frac{1}{n_0}\int_0^\xi (n_e - n_i)d\xi_1 \qquad (8.30)$$

is decreased. Since χ' is the analog of dx/dt in the oscillator problem, our virtual oscillator has lost velocity and is trapped in the potential well of Fig. 8.6.

8.3.2 The Critical Mach Numbers

Solutions of either the soliton type or the wave-train type exist only for a range of $\mathcal{M}$. A lower limit for $\mathcal{M}$ is given by the condition that $V(\chi)$ be a potential well, rather than a hill. Expanding Eq. (8.29) for $\chi \ll 1$ yields

$$\frac{1}{2}\chi^2 - (\chi^2/2\mathcal{M}^2) > 0 \qquad \mathcal{M}^2 > 1 \qquad (8.31)$$

This is exactly the same, both physically and mathematically, as the Bohm criterion for the existence of a sheath (Eq. (8.11)).

An upper limit to $\mathcal{M}$ is imposed by the condition that the function $V(\chi)$ of Fig. 8.6 must cross the χ axis for $\chi > 0$; otherwise, the virtual particle will not be reflected, and the potential will rise indefinitely. From Eq. (8.29), we require

$$e^{\chi} - 1 < \mathcal{M}^2 \left[1 - \left(1 - \frac{2\chi}{\mathcal{M}^2} \right)^{1/2} \right] \qquad (8.32)$$

for some $\chi > 0$. If the lower critical Mach number is surpassed ($\mathcal{M} > 1$), the left-hand side, representing the integral of the electron density from zero to χ, is initially larger than the right-hand side, representing the integral of the ion density. As χ increases, the right-hand side can catch up with the left-hand side if $\mathcal{M}^2$ is not too large. However, because of the square root, the largest value χ can have is $\mathcal{M}^2/2$. This is because $e\phi$ cannot exceed $\frac{1}{2}Mu_0^2$; otherwise, ions would be excluded from the plasma in the downstream region. Inserting the largest value of χ into Eq. (8.32), we have

$$\exp\left(\mathcal{M}^2/2\right) - 1 < \mathcal{M}^2 \qquad \text{or} \qquad \mathcal{M} < 1.6 \qquad (8.33)$$

This is the upper critical Mach number. Shock waves in a cold-ion plasma therefore exist only for $1 < M < 1.6$.

As in the case of sheaths, the physical situation is best explained by a diagram of n_i and n_e vs. χ (Fig. 8.8). This diagram differs from Fig. 8.4 because of the change of sign of ϕ. Since the ions are now decelerated rather than accelerated, n_i will approach infinity at $\chi = \mathcal{M}^2/2$. The lower critical Mach number ensures that the n_i curve lies below the n_e curve at small χ, so that the potential $\phi(x)$ starts off with the right sign for its curvature. When the curve n_{i1} crosses the n_e curve, the soliton $\phi(x)$ (Fig. 8.7) has an inflection point. Finally, when χ is large enough that the areas under the n_i and n_e curves are equal, the soliton reaches a peak, and the n_{i1} and n_e

Fig. 8.8 Variation of ion and electron density (logarithmic scale) with normalized potential χ in a soliton. The ion density is drawn for two cases: Mach number greater than and less than 1.6

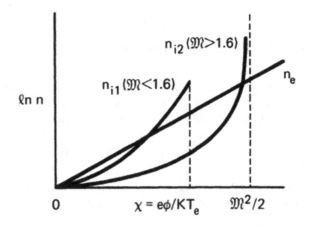

curves are retraced as χ goes back to zero. The equality of the areas ensures that the net charge in the soliton is zero; therefore, there is no electric field outside. If $\mathcal{M}$ is larger than 1.6, we have the curve n_{i2}, in which the area under the curve is too small even when χ has reached its maximum value of $\mathcal{M}^2/2$.

8.3.3 Wave Steepening

If one propagates an ion wave in a cold-ion plasma, it will have the phase velocity given by Eq. (4.42), corresponding to $\mathcal{M} = 1$. How, then, can one create shocks with $\mathcal{M} > 1$? One must remember that Eq. (4.42) was a linear result valid only at small amplitudes. As the amplitude is increased, an ion wave speeds up and also changes from a sine wave to a sawtooth shape with a steep leading edge (Fig. 8.9). The reason is that the wave electric field has accelerated the ions. In Fig. 8.9, ions at the peak of the potential distribution have a larger velocity in the direction of v_ϕ than those at the trough, since they have just experienced a period of acceleration as the wave passed by. In linear theory, this difference in velocity is taken into account, but not the displacement resulting from it. In nonlinear theory, it is easy to see that the ions at the peak are shifted to the right, while those at the trough are shifted to the left, thus steepening the wave shape. Since the density perturbation is in phase with the potential, more ions are accelerated to the right than to the left, and the wave causes a net mass flow in the direction of propagation. This causes the wave velocity to exceed the acoustic speed in the undisturbed plasma, so that $\mathcal{M}$ is larger than unity.

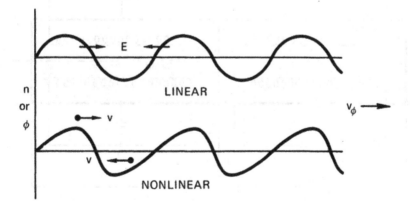

Fig. 8.9 A large-amplitude ion wave steepens so that the leading edge has a larger slope than the trailing edge

8.3.4 Experimental Observations

Ion acoustic shock waves of the form shown in Fig. 8.5 have been generated by R. J. Taylor, D. R. Baker, and H. Ikezi. To do this, a new plasma source, the DP (double-plasma) device, was invented. Figure 8.10 shows schematically how it works. Identical plasmas are created in two electrically isolated chambers by discharges between filaments F and the walls W. The plasmas are separated by the negatively biased grid G, which repels electrons and forms an ion sheath on both sides. A voltage pulse, usually in the form of a ramp, is applied between the two chambers. This causes the ions in one chamber to stream into the other, exciting a large-amplitude plane wave. The wave is detected by a movable probe or particle velocity analyzer P. Figure 8.11 shows measurements of the density fluctuation in the shock wave as a function of time and probe position. It is seen that the wavefront steepens and then turns into a shock wave of the classic shape. The damping of the oscillations is due to collisions.

Problem

8.5 Calculate the maximum possible velocity of an ion acoustic shock wave in an experiment such as that shown in Fig. 8.10, where $T_e = 1.5$ eV, $T_i = 0.2$ eV, and the gas is argon. What is the maximum possible shock wave amplitude in volts?

8.3.5 Double Layers

A phenomenon related to sheaths and ion acoustic shocks is that of the double layer. This is a localized potential jump, believed to occur naturally in the ionosphere,

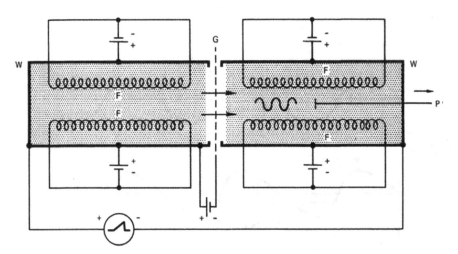

Fig. 8.10 Schematic of a DP machine in which ion acoustic shock waves were produced and detected. [*Cf.* R. J. Taylor, D. R. Baker, and H. Ikezi, *Phys. Rev. Lett.* **24**, 206 (1970).]

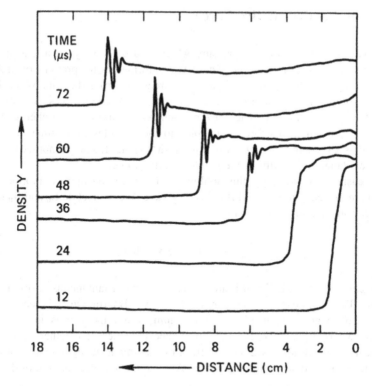

Fig. 8.11 Measurements of the density distribution in a shock wave at various times, showing how the characteristic shape of Fig. 8.5 develops. [From Taylor *et al.*, *loc cit.*]

which neither propagates nor is attached to a boundary. The name comes from the successive layers of net positive and net negative charge that are necessary to create a step in $\phi(x)$. Such a step can remain stationary in space only if there is a plasma flow that Doppler-shifts a shock front down to zero velocity in the lab frame, or if the distribution functions of the transmitted and reflected electrons and ions on each side of the discontinuity are specially tailored so as to make this possible. Double layers have been created in the laboratory in "triple-plasma" devices, which are similar to the DP machine shown in Fig. 8.10, but with a third experimental chamber (without filaments) inserted between the two source chambers. By adjusting the relative potentials of the three chambers, which are isolated by grids, streams of ions or electrons can be spilled into the center chamber to form a double layer there. In natural situations double layers are likely to arise where there are gradients in the magnetic field **B**, not where **B** is zero or uniform as in laboratory simulations. In that case, the $\mu\nabla B$ force (Eq. (2.38)) can play a large role in localizing a double layer away from all boundaries. Indeed, the thermal barrier in tandem mirror reactors is an example of a double layer with strong magnetic trapping. In Sect. 8.11 we shall see that a double layer can arise in "mid-air" when a dense plasma is injected into a diverging magnetic field. Ions accelerated by the potential drop in the double layer can be used to push a spacecraft.

8.4   The Ponderomotive Force

Light waves exert a radiation pressure which is usually very weak and hard to detect. Even the esoteric example of comet tails, formed by the pressure of sunlight, is tainted by the added effect of particles streaming from the sun. When high-powered microwaves or laser beams are used to heat or confine plasmas, however, the radiation pressure can reach several hundred thousand atmospheres! When applied to a plasma, this force is coupled to the particles in a somewhat subtle way and is called the *ponderomotive force*. Many nonlinear phenomena have a simple explanation in terms of the ponderomotive force.

The easiest way to derive this nonlinear force is to consider the motion of an electron in the oscillating **E** and **B** fields of a wave. We neglect dc $\mathbf{E_0}$ and $\mathbf{B_0}$ fields. The electron equation of motion is

$$m\frac{d\mathbf{v}}{dt} = -e[\mathbf{E}(\mathbf{r}) + \mathbf{v} \times \mathbf{B}(\mathbf{r})] \tag{8.34}$$

This equation is exact if **E** and **B** are evaluated at the instantaneous position of the electron. The nonlinearity comes partly from the $\mathbf{v} \times \mathbf{B}$ term, which is second order because both **v** and **B** vanish in the equilibrium, so that the term is no larger than $\mathbf{v_1} \times \mathbf{B_1}$, where $\mathbf{v_1}$ and $\mathbf{B_1}$ are the linear-theory values. The other part of the nonlinearity, as we shall see, comes from evaluating **E** at the actual position of the particle rather than its initial position. Assume a wave electric field of the form

$$\mathbf{E} = \mathbf{E}_s(\mathbf{r}) \cos \omega t \tag{8.35}$$

where $\mathbf{E}_s(\mathbf{r})$ contains the spatial dependence. In first order, we may neglect the $\mathbf{v} \times \mathbf{B}$ term in Eq. (8.34) and evaluate **E** at the initial position $\mathbf{r_0}$. We have

$$m d\mathbf{v_1}/dt = -e\mathbf{E}(\mathbf{r_0}) \tag{8.36}$$

$$\mathbf{v_1} = -(e/m\omega)\mathbf{E}_s \sin \omega t = d\mathbf{r_1}/dt \tag{8.37}$$

$$\delta\mathbf{r_1} = (e/m\omega^2)\mathbf{E}_s \cos \omega t \tag{8.38}$$

It is important to note that in a nonlinear calculation, we cannot write $e^{i\omega t}$ and take its real part later. Instead, we write its real part explicitly as $\cos \omega t$. This is because products of oscillating factors appear in nonlinear theory, and the operations of multiplying and taking the real part do not commute.

Going to second order, we expand $\mathbf{E}(\mathbf{r})$ about $\mathbf{r_0}$:

$$\mathbf{E}(\mathbf{r}) = \mathbf{E}(\mathbf{r_0}) + (\delta\mathbf{r_1} \cdot \nabla)\mathbf{E}|_{r=r_0} + \cdots \tag{8.39}$$

We must now add the term $\mathbf{v}_1 \times \mathbf{B}_1$, where $\mathbf{B}_1$ is given by Maxwell's equation:

$$\nabla \times \mathbf{E} = -\partial \mathbf{B}/\partial t$$
$$\mathbf{B}_1 = -(1/\omega)\nabla \times \mathbf{E}_s|_{r=r_0} \sin \omega t \tag{8.40}$$

The second-order part of Eq. (8.34) is then

$$m d\mathbf{v}_2/dt = -e[(\delta\mathbf{r}_1 \cdot \nabla)\mathbf{E} + \mathbf{v}_1 \times \mathbf{B}_1] \tag{8.41}$$

Inserting Eqs. (8.37), (8.38), and (8.40) into (8.41) and averaging over time, we have

$$m\left\langle \frac{d\mathbf{v}_2}{dt} \right\rangle = -\frac{e^2}{m\omega^2}\frac{1}{2}[(\mathbf{E}_s \cdot \nabla)\mathbf{E}_s + \mathbf{E}_s \times (\nabla \times \mathbf{E}_s)] = \mathbf{f}_{NL} \tag{8.42}$$

Here we used $\langle \sin^2 \omega t \rangle = \langle \cos^2 \omega t \rangle = \frac{1}{2}$. The double cross product can be written as the sum of two terms, one of which cancels the $(\mathbf{E}_s \cdot \nabla)\mathbf{E}_s$ term. What remains is

$$\mathbf{f}_{NL} = -\frac{1}{4}\frac{e^2}{m\omega^2}\nabla E_s^2 \tag{8.43}$$

This is the effective nonlinear force on a single electron. The force per m^3 is $\mathbf{f}_{NL}$ times the electron density n_0, which can be written in terms of ω_p^2. Since $E_s^2 = 2\langle E^2 \rangle$, we finally have for the ponderomotive force the formula

$$\mathbf{F}_{NL} = -\frac{\omega_p^2}{\omega^2}\nabla\frac{\langle \varepsilon_0 E^2 \rangle}{2} \tag{8.44}$$

If the wave is electromagnetic, the second term in Eq. (8.42) is dominant, and the physical mechanism for $\mathbf{F}_{NL}$ is as follows. Electrons oscillate in the direction of $\mathbf{E}$, but the wave magnetic field distorts their orbits. That is, the Lorentz force $-e\mathbf{v} \times \mathbf{B}$ pushes the electrons in the direction of $\mathbf{k}$ (since $\mathbf{v}$ is in the direction of $\mathbf{E}$, and $\mathbf{E} \times \mathbf{B}$ is in the direction of $\mathbf{k}$). The phases of $\mathbf{v}$ and $\mathbf{B}$ are such that the motion does not average to zero over an oscillation, but there is a secular drift along $\mathbf{k}$. If the wave has uniform amplitude, no force is needed to maintain this drift; but if the wave amplitude varies, the electrons will pile up in regions of small amplitude, and a force is needed to overcome the space charge. This is why the effective force $\mathbf{F}_{NL}$ is proportional to the *gradient* of $\langle E^2 \rangle$. Since the drift for each electron is the same, $\mathbf{F}_{NL}$ is proportional to the density—hence the factor ω_p^2/ω^2 in Eq. (8.44).

If the wave is electrostatic, the first term in Eq. (8.42) is dominant. Then the physical mechanism is simply that an electron oscillating along $\mathbf{k} \parallel \mathbf{E}$ moves farther in the half-cycle when it is moving from a strong-field region to a weak-field region than vice versa, so there is a net drift.

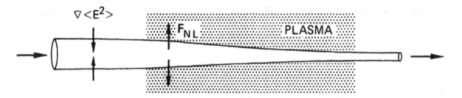

Fig. 8.12 Self-focusing of a laser beam is caused by the ponderomotive force

Although $\mathbf{F}_{NL}$ acts mainly on the electrons, the force is ultimately transmitted to the ions, since it is a low-frequency or dc effect. When electrons are bunched by $\mathbf{F}_{NL}$, a charge-separation field $\mathbf{E}_{cs}$ is created. The total force felt by the electrons is

$$\mathbf{F}_e = -e\mathbf{E}_{cs} + \mathbf{F}_{NL} \qquad (8.45)$$

Since the ponderomotive force on the ions is smaller by $\Omega_p^2/\omega_p^2 = m/M$, the force on the ion fluid is approximately

$$\mathbf{F}_i = e\mathbf{E}_{cs} \qquad (8.46)$$

Summing the last two equations, we find that the force on the plasma is $\mathbf{F}_{NL}$.

A direct effect of $\mathbf{F}_{NL}$ is the self-focusing of laser light in a plasma. In Fig. 8.12 we see that a laser beam of finite diameter causes a radially directed ponderomotive force in a plasma. This force moves plasma out of the beam, so that ω_p is lower and the dielectric constant ε is higher inside the beam than outside. The plasma then acts as a convex lens, focusing the beam to a smaller diameter.

Problems

8.6 A 1-TW laser beam is focused to a spot 50 μm in diameter on a solid target. A plasma is created and heated by the beam, and it tries to expand. The ponderomotive force of the beam, which acts mainly on the region of critical density ($n = n_c$, or $\omega = \omega_p$), pushes the plasma back and causes "profile modification," which is an abrupt change in density at the critical layer.

(a) How much pressure (in N/m^2 and in lbf/in.^{2}) is exerted by the ponderomotive force? (Hint: Note that F_{NL} is in units of N/m^3 and that the gradient length cancels out. To calculate $\langle E^2 \rangle$, assume conservatively that it has the same value as in vacuum, and set the 1-TW Poynting flux equal to the beam's energy density times its group velocity in vacuum.)

(b) What is the total force, in tonnes, exerted by the beam on the plasma?

(c) If $T_i = T_e = 1$ keV, how large a density jump can the light pressure support?

8.7 Self-focusing occurs when a cylindrically symmetric laser beam of frequency ω is propagated through an underdense plasma; that is, one which has

$$n < n_c \equiv \varepsilon_0 m\omega^2/e^2$$

In steady state, the beam's intensity profile and the density depression caused by the beam (Fig. 8.12) are related by force balance. Neglecting plasma heating ($KT \equiv KT_e + KT_i = $ constant), prove the relation

$$n = n_0 e^{-\varepsilon_0 \langle E^2 \rangle / 2n_c KT} \equiv n_0 e^{-\alpha(r)}$$

The quantity $\alpha(0)$ is a measure of the relative importance of ponderomotive pressure to plasma pressure.

8.5   Parametric Instabilities

The most thoroughly investigated of the nonlinear wave–wave interactions are the "parametric instabilities," so called because of an analogy with parametric amplifiers, well-known devices in electrical engineering. A reason for the relatively advanced state of understanding of this subject is that the theory is basically a linear one, but linear about an oscillating equilibrium.

8.5.1   Coupled Oscillators

Consider the mechanical model of Fig. 8.13, in which two oscillators M_1 and M_2 are coupled to a bar resting on a pivot. The pivot P is made to slide back and forth at a frequency ω_0, while the natural frequencies of the oscillators are ω_1 and ω_2. It is clear that, in the absence of friction, the pivot encounters no resistance as long as M_1 and M_2 are not moving. Furthermore, if P is not moving and M_2 is put into motion, M_1 will move; but as long as ω_2 is not the natural frequency of M_1, the amplitude will be small. Suppose now that both P and M_2 are set into motion. The displacement of M_1 is proportional to the product of the displacement of M_2 and the length of the lever arm and, hence, will vary in time as

Fig. 8.13 A mechanical analog of a parametric instability

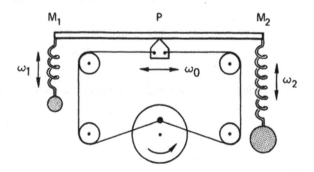

$$\cos \omega_2 t \cos \omega_0 t = \frac{1}{2} \cos\left[(\omega_2 + \omega_0)t\right] + \frac{1}{2}\cos\left[(\omega_2 - \omega_0)t\right] \qquad (8.47)$$

If ω_1 is equal to either $\omega_2 + \omega_0$ or $\omega_2 - \omega_0$, M_1 will be resonantly excited and will grow to large amplitude. Once M_1 starts oscillating, M_2 will also gain energy, because one of the beat frequencies of ω_1 with ω_0 is just ω_2. Thus, once either oscillator is started, each will be excited by the other, and the system is unstable. The energy, of course, comes from the "pump" P, which encounters resistance once the rod is slanted. If the pump is strong enough, its oscillation amplitude is unaffected by M_1 and M_2; the instability can then be treated by a linear theory. In a plasma, the oscillators P, M_1, and M_2 may be different types of waves.

8.5.2 Frequency Matching

The equation of motion for a simple harmonic oscillator x_1 is

$$\frac{d^2 x_1}{dt^2} + \omega_1^2 x_1 = 0 \qquad (8.48)$$

where ω_1 is its resonant frequency. If it is driven by a time-dependent force which is proportional to the product of the amplitude E_0 of the driver, or pump, and the amplitude x_2 of a second oscillator, the equation of motion becomes

$$\frac{d^2 x_1}{dt^2} + \omega_1^2 x_1 = c_1 x_2 E_0 \qquad (8.49)$$

where c_1 is a constant indicating the strength of the coupling. A similar equation holds for x_2:

$$\frac{d^2 x_2}{dt^2} + \omega_2^2 x_2 = c_2 x_1 E_0 \qquad (8.50)$$

Let $x_1 = \bar{x}_1 \cos \omega t$, $x_2 = \bar{x}_2 \cos \omega' t$, and $E_0 = \bar{E}_0 \cos \omega_0 t$. Equation (8.50) becomes

$$\begin{aligned}
\left(\omega_2^2 - \omega'^2\right)\bar{x}_2 \cos \omega' t &= c_2 \bar{E}_0 \bar{x}_1 \cos \omega_0 t \cos \omega t \\
&= c_2 \bar{E}_0 \bar{x}_1 \frac{1}{2}\{\cos\left[(\omega_0 + \omega)t\right] + \cos\left[(\omega_0 - \omega)t\right]\}
\end{aligned} \qquad (8.51)$$

The driving terms on the right can excite oscillators x_2 with frequencies

$$\omega' = \omega_0 \pm \omega \qquad (8.52)$$

In the absence of nonlinear interactions, x_2 can only have the frequency ω_2, so we must have $\omega' = \omega_2$. However, the driving terms can cause a frequency shift so that ω' is only approximately equal to ω_2. Furthermore, ω' can be complex, since there is damping (which has been neglected so far for simplicity), or there can be growth (if there is an instability). In either case, x_2 is an oscillator with finite Q and can respond to a range of frequencies about ω_2. If ω is small, one can see from Eq. (8.52) that both choices for ω' may lie within the bandwidth of x_2, and one must allow for the existence of two oscillators, $x_2(\omega_0 + \omega)$ and $x_2(\omega_0 - \omega)$.

Now let $x_1 = \bar{x}_1 \cos \omega'' t$ and $x_2 = \bar{x}_2 \cos[(\omega_0 \pm \omega)t]$ and insert into Eq. (8.49):

$$\left(\omega_1^2 - \omega''^2\right)\bar{x}_1 \cos \omega'' t$$

$$= c_1 \bar{E}_0 \bar{x}_2 \frac{1}{2}\left(\cos\{[\omega_0 + (\omega_0 \pm \omega)]t\} + \cos\{[\omega_0 - (\omega_0 \pm \omega)]t\}\right) \qquad (8.53)$$

$$= c_1 \bar{E}_0 \bar{x}_2 \frac{1}{2}\{\cos[(2\omega_0 \pm \omega)t] + \cos \omega t\}$$

The driving terms can excite not only the original oscillation $x_1(\omega)$, but also new frequencies $\omega'' = 2\omega_0 \pm \omega$. We shall consider the case $|\omega_0| \gg |\omega_1|$, so that $2\omega_0 \pm \omega$ lies outside the range of frequencies to which x_1 can respond, and $x_1(2\omega_0 \pm \omega)$ can be neglected. We therefore have three oscillators, $x_1(\omega)$, $x_2(\omega_0 - \omega)$, and $x_2(\omega_0 + \omega)$, which are coupled by Eqs. (8.49) and (8.50):

$$\left(\omega_1^2 - \omega^2\right)x_1(\omega) - c_1 E_0(\omega_0)[x_2(\omega_0 - \omega) + x_2(\omega_0 + \omega)] = 0$$

$$\left[\omega_2^2 - (\omega_0 - \omega)^2\right]x_2(\omega_0 - \omega) - c_2 E_0(\omega_0)x_1(\omega) = 0 \qquad (8.54)$$

$$\left[\omega_2^2 - (\omega_0 + \omega)^2\right]x_2(\omega_0 + \omega) - c_2 E_0(\omega_0)x_1(\omega) = 0$$

The dispersion relation is given by setting the determinant of the coefficients equal to zero:

$$\begin{vmatrix} \omega^2 - \omega_1^2 & c_1 E_0 & c_1 E_0 \\ c_2 E_0 & (\omega_0 - \omega)^2 - \omega_2^2 & 0 \\ c_2 E_0 & 0 & (\omega_0 + \omega)^2 - \omega_2^2 \end{vmatrix} = 0 \qquad (8.55)$$

A solution with $\mathrm{Im}(\omega) > 0$ would indicate an instability.

For small frequency shifts and small damping or growth rates, we can set ω and ω' approximately equal to the undisturbed frequencies ω_1 and ω_2. Equation (8.52) then gives a frequency matching condition:

$$\omega_0 \approx \omega_2 \pm \omega_1 \qquad (8.56)$$

When the oscillators are waves in a plasma, ωt must be replaced by $\omega t - \mathbf{k} \cdot \mathbf{r}$. There is then also a wavelength matching condition

$$\mathbf{k}_0 \approx \mathbf{k}_2 \pm \mathbf{k}_1 \qquad (8.57)$$

describing spatial beats; that is, the periodicity of points of constructive and destructive interference in space. The two conditions Eqs. (8.56) and (8.57) are easily understood by analogy with quantum mechanics. Multiplying the former by Planck's constant $\hbar$, we have

$$\hbar\omega_0 = \hbar\omega_2 \pm \hbar\omega_1 \qquad (8.58)$$

E_0 and x_2 may, for instance, be electromagnetic waves, so that $\hbar\omega_0$ and $\hbar\omega_2$ are the photon energies. The oscillator x_1 may be a Langmuir wave, or plasmon, with energy $\hbar\omega_1$. Equation (8.54) simply states the conservation of energy. Similarly, Eq. (8.53) states the conservation of momentum $\hbar\mathbf{k}$.

For plasma waves, the simultaneous satisfaction of Eqs. (8.52) and (8.53) in one-dimensional problems is possible only for certain combinations of waves. The required relationships are best seen on an ω–k diagram (Fig. 8.14). Figure 8.14a shows the dispersion curves of an electron plasma wave ω_2 (Bohm–Gross wave)

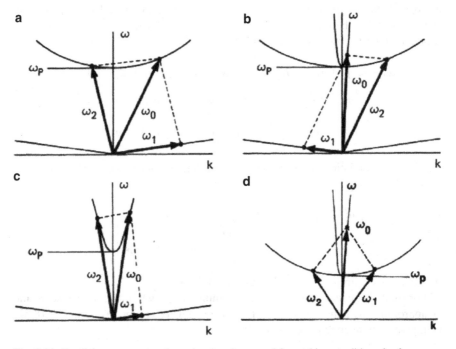

Fig. 8.14 Parallelogram constructions showing the ω- and k-matching conditions for four parametric instabilities: (**a**) electron decay instability, (**b**) parametric decay instability, (**c**) stimulated Brillouin backscattering instability, and (**d**) two-plasmon decay instability. In each case, ω_0 is the incident wave, and ω_1 and ω_2 the decay waves. The *straight lines* are the dispersion relation for ion waves; the *narrow parabolas*, that for light waves; and the *wide parabolas*, that for electron waves

and an ion acoustic wave ω_1 (cf. Fig. 4.13). A large-amplitude electron wave (ω_0, $\mathbf{k}_0$) can decay into a backward moving electron wave (ω_2, $\mathbf{k}_2$) and an ion wave (ω_1, $\mathbf{k}_1$). The parallelogram construction ensures that $\omega_0 = \omega_1 + \omega_2$ and $\mathbf{k}_0 = \mathbf{k}_1 + \mathbf{k}_2$. The positions of ($\omega_0$, $\mathbf{k}_0$) and (ω_2, $\mathbf{k}_2$) on the electron curve must be adjusted so that the difference vector lies on the ion curve. Note that an electron wave cannot decay into two other electron waves, because there is no way to make the difference vector lie on the electron curve.

Figure 8.14b shows the parallelogram construction for the "parametric decay" instability. Here, (ω_0, $\mathbf{k}_0$) is an incident electromagnetic wave of large phase velocity ($\omega_0/k_0 \approx c$). It excites an electron wave and an ion wave moving in opposite directions. Since $|\mathbf{k}_0|$ is small, we have $|\mathbf{k}_1| \approx -|\mathbf{k}_2|$ and $\omega_0 = \omega_1 + \omega_2$ for this instability.

Figure 8.14c shows the ω–k diagram for the "parametric backscattering" instability, in which a light wave excites an ion wave and another light wave moving in the opposite direction. This can also happen when the ion wave is replaced by a plasma wave. By analogy with similar phenomena in solid state physics, these processes are called, respectively, "stimulated Brillouin scattering" and "stimulated Raman scattering."

Figure 8.14d represents the two-plasmon decay instability of an electromagnetic wave. Note that the two decay waves are both electron plasma waves, so that frequency matching can occur only if $\omega_0 \simeq 2\omega_p$. Expressed in terms of density, this condition is equivalent to $n \simeq n_c/4$, when n_c is the critical density (Eq. (4.88)) associated with ω_0. This instability can therefore be expected to occur only near the "quarter-critical" layer of an inhomogeneous plasma.

8.5.3 Instability Threshold

Parametric instabilities will occur at any amplitude if there is no damping, but in practice even a small amount of either collisional or Landau damping will prevent the instability unless the pump wave is rather strong. To calculate the threshold, one must introduce the damping rates Γ_1 and Γ_2 of the oscillators x_1 and x_2. Equation (8.48) then becomes

$$\frac{d^2 x_1}{dt^2} + \omega_1^2 x_1 + 2\Gamma_1 \frac{dx_1}{dt} = 0 \tag{8.59}$$

For instance, if x_1 is the displacement of a spring damped by friction, the last term represents a force proportional to the velocity. If x_1 is the electron density in a plasma wave damped by electron–neutral collisions, Γ_1 is $v_c/2$ (cf. Problem 4.5). Examination of Eqs. (8.49), (8.50), and (8.54) will show that it is all right to use exponential notation and let $d/dt \to -i\omega$ for x_1 and x_2, as long as we keep E_0 real and allow $\bar{x}_1$ and $\bar{x}_2$ to be complex. Equations (8.49) and (8.50) become

$$(\omega_1^2 - \omega^2 - 2i\omega\Gamma_1)x_1(\omega) = c_1 x_2 E_0$$
$$\left[\omega_2^2 - (\omega - \omega_0)^2 - 2i(\omega - \omega_0)\Gamma_2\right]x_2(\omega - \omega_0) = c_2 x_1 E_0$$

(8.60)

We further restrict ourselves to the simple case of two waves—that is, when $\omega \simeq \omega_1$ and $\omega_0 - \omega \simeq \omega_2$ but $\omega_0 + \omega$ is far enough from ω_2 to be nonresonant—in which case the third row and column of Eq. (8.55) can be ignored. If we now express x_1, x_2, and E_0 in terms of their peak values, as in Eq. (8.53), a factor of 1/2 appears on the right-hand sides of Eq. (8.60). Discarding the nonresonant terms and eliminating $\bar{x}_1$ and $\bar{x}_2$ from Eq. (8.60), we obtain

$$\left(\omega^2 - \omega_1^2 + 2i\omega\Gamma_1\right)\left[(\omega_0 - \omega)^2 - \omega_2^2 - 2i(\omega_0 - \omega)\Gamma_2\right] = \tfrac{1}{4}c_1 c_2 \bar{E}_0^2 \qquad (8.61)$$

At threshold, we may set $\mathrm{Im}(\omega) = 0$. The lowest threshold will occur at exact frequency matching; i.e., $\omega = \omega_1$, $\omega_0 - \omega = \omega_2$. Then Eq. (8.61) gives

$$c_1 c_2 \left(\bar{E}_0^2\right)_{\text{thresh}} = 16\omega_1\omega_2\Gamma_1\Gamma_2 \qquad (8.62)$$

The threshold goes to zero with the damping of *either* wave.

Problems

8.8 Prove that stimulated Raman scattering cannot occur at densities above $n_c/4$.

8.9 Stimulated Brillouin scattering is observed when a Nd-glass laser beam ($\lambda = 1.06\ \mu\text{m}$) irradiates a solid D_2 target ($Z = 1$, $M = 2M_H$). The backscattered light is red-shifted by 21.9 Å. From x-ray spectra, it is determined that $KT_e = 1$ keV. Assuming that the scattering occurs in the region where $\omega_p^2 \ll \omega^2$, and using Eq. (4.41) with $\gamma_i = 3$, make an estimate of the ion temperature.

8.10 For stimulated Brillouin scattering (SBS), we may let x_1 in Eq. (8.60) stand for the ion wave density fluctuation n_1, and x_2 for the reflected wave electric field E_2. The coupling coefficients are then given by

$$c_1 = \varepsilon_0 k_1^2 \omega_p^2 / \omega_0 \omega_2 M$$
$$c_2 = \omega_p^2 \omega_2 / n_0 \omega_0$$

and threshold pump intensity in a homogeneous plasma is given by Eq. (8.62). This is commonly expressed in terms of $\langle v_{\text{osc}}^2 \rangle$, the rms electron oscillation velocity caused by the pump wave (cf. Eq. (8.37)):

$$v_{\text{osc}} \equiv eE_0/m\omega_0$$

The damping rate Γ_2 can be found from Problem (4.37b) for $v/\omega \ll 1$.

(a) Show that, for $T_i \ll T_e$ and $v_e^2 \equiv KT_e/m$, the SBS threshold is given by

$$\frac{\langle v_{osc}^2 \rangle}{v_e^2} = \frac{4\Gamma_1 \nu}{\omega_1 \omega_2}$$

where $\omega_1 = k_1 v_s$ and Γ_1 is the ion Landau damping rate given by Eq. (7.133).

(b) Calculate the threshold laser intensity I_0 in W/cm^2 for SBS of CO_2 (10.6 μm) light in a uniform hydrogen plasma with $T_e = 100$ eV, $T_i = 10$ eV, and $n_0 = 10^{23}$ m^{-3} (Hint: Use the Spitzer resistivity to evaluate ν_{ei}.)

8.11 The growth rate of stimulated Brillouin scattering in a homogeneous plasma far above threshold can be computed from Eq. (8.61) by neglecting the damping terms. Let $\omega = \omega_s + i\gamma$ and assume $\gamma^2 \ll \omega_s^2$ and $n \ll n_c$. Show that

$$\gamma = \frac{\overline{v}_{osc}}{2c} \left(\frac{\omega_0}{\omega_s}\right)^{1/2} \Omega_p$$

where $\overline{v}_{osc}$ is the peak oscillating velocity of the electrons.

8.5.4 Physical Mechanism

The parametric excitation of waves can be understood very simply in terms of the ponderomotive force (Sect. 8.4). As an illustration, consider the case of an electromagnetic wave (ω_0, k_0) driving an electron plasma wave (ω_2, k_2) and a low-frequency ion wave (ω_1, k_1) (Fig. 8.14b). Since ω_1 is small, ω_0 must be close to ω_p. However, the behavior is quite different for $\omega_0 < \omega_p$ than for $\omega_0 > \omega_p$. The former case gives rise to the "oscillating two-stream" instability (which will be treated in detail), and the latter to the "parametric decay" instability.

Suppose there is a density perturbation in the plasma of the form $n_1 \cos k_1 x$; this perturbation can occur spontaneously as one component of the thermal noise. Let the pump wave have an electric field $E_0 \cos \omega_0 t$ in the x direction, as shown in Fig. 8.15. In the absence of a dc field B_0, the pump wave follows the relation $\omega_0^2 = \omega_p^2 + c^2 k_0^2$, so that $k_0 \approx 0$ for $\omega_0 \approx \omega_p$. We may therefore regard E_0 as spatially uniform. If ω_0 is less than ω_p, which is the resonant frequency of the cold electron fluid, the electrons will move in the direction opposite to E_0, while the ions do not move on the time scale of ω_0. The density ripple then causes a charge separation, as shown in Fig. 8.15. The electrostatic charges create a field E_1, which oscillates at the frequency ω_0. The ponderomotive force due to the total field is given by Eq. (8.44):

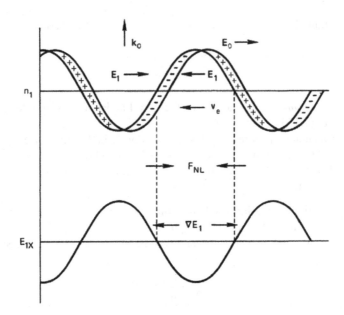

Fig. 8.15 Physical mechanism of the oscillating two-stream instability

$$\mathbf{F}_{NL} = -\frac{\omega_p^2}{\omega_0^2}\nabla\frac{\left\langle (E_0 + E_1)^2 \right\rangle}{2}\varepsilon_0 \tag{8.63}$$

Since E_0 is uniform and much larger than E_1, only the cross term is important:

$$\mathbf{F}_{NL} = -\frac{\omega_p^2}{\omega_0^2}\frac{\partial}{\partial x}\frac{\langle 2E_0 E_1 \rangle}{2}\varepsilon_0 \tag{8.64}$$

This force does not average to zero, since E_1 changes sign with E_0. As seen in Fig. 8.15, F_{NL} is zero at the peaks and troughs of n_1 but is large where ∇n_1 is large. This spatial distribution causes $\mathbf{F}_{NL}$ to push electrons from regions of low density to regions of high density. The resulting dc electric field drags the ions along also, and the density perturbation grows. The threshold value of F_{NL} is the value just sufficient to overcome the pressure $\nabla n_{i1}(KT_i + KT_e)$, which tends to smooth the density. The density ripple does not propagate, so that $\text{Re}(\omega_1) = 0$. This is called the *oscillating two-stream instability* because the sloshing electrons have a time-averaged distribution function which is double-peaked, as in the two-stream instability (Sect. 6.6).

If ω_0 is larger than ω_p, this physical mechanism does not work, because an oscillator driven faster than its resonant frequency moves opposite to the direction of the applied force (this will be explained more clearly in the next section). The directions of $\mathbf{v}_e$, $\mathbf{E}_1$, and $\mathbf{F}_{NL}$ are then reversed on Fig. 8.15, and the ponderomotive force moves ions from dense regions to less dense regions. If the density perturbation did not move, it would decay. However, if it were a traveling ion acoustic wave, the inertial delay between the application of the force F_{NL} and the change of ion positions causes the density maxima to move into the regions into which F_{NL} is pushing the ions. This can happen, of course, only if the phase velocity of the ion wave has just the right value. That this value is v_s can be seen from the fact that the phase of the force F_{NL} in Fig. 8.15 (with the arrows reversed now) is exactly the same as the phase of the electrostatic restoring force in an ion wave, where the potential is maximum at the density maximum and vice versa. Consequently, F_{NL} adds to the restoring force and augments the ion wave. The electrons, meanwhile, oscillate with large amplitude if the pump field is near the natural frequency of the electron fluid; namely, $\omega_2^2 = \omega_p^2 + \frac{3}{2}k^2 v_{th}^2$. The pump cannot have exactly the frequency ω_2 because the beat between ω_0 and ω_2 must be at the ion wave frequency $\omega_1 = kv_s$, so that the expression for F_{NL} in Eq. (8.64) can have the right frequency to excite ion waves. If this frequency matching is satisfied, viz., $\omega_1 = \omega_0 - \omega_2$, both an ion wave and an electron wave are excited at the expense of the pump wave. This is the mechanism of the *parametric decay instability*.

8.5.5 The Oscillating Two-Stream Instability

We shall now actually derive this simplest example of a parametric instability with the help of the physical picture given in the last section. For simplicity, let the temperatures T_i and T_e and the collision rates v_i and v_e all vanish. The ion fluid then obeys the low-frequency equations

$$Mn_0 \frac{\partial v_{i1}}{\partial t} = en_0 E = F_{NL} \tag{8.65}$$

$$\frac{\partial n_{i1}}{\partial t} = n_0 \frac{\partial v_{i1}}{\partial x} = 0 \tag{8.66}$$

Since the equilibrium is assumed to be spatially homogeneous, we may Fourier-analyze in space and replace $\partial/\partial x$ by ik. The last two equations then give

$$\frac{\partial^2 n_{i1}}{\partial t^2} + \frac{ik}{M} F_{NL} = 0 \tag{8.67}$$

with F_{NL} given by Eq. (8.64). To find E_1, we must consider the motion of the electrons, given by

$$m\left(\frac{\partial v_e}{\partial t} + v_e \frac{\partial}{\partial x} v_e\right) = -e(E_0 + E_1) \tag{8.68}$$

where E_1 is related to the density n_{e1} by Poisson's equation

$$ik\varepsilon_0 E_1 = -en_{e1} \tag{8.69}$$

We must realize at this point that the quantities E_1, v_e, and n_{e1} each has two parts: a high-frequency part, in which the electrons move independently of the ions, and a low-frequency part, in which they move along with the ions in a quasineutral manner. To lowest order, the motion is a high-frequency one in response to the spatially uniform field E_0:

$$\frac{\partial v_{e0}}{\partial t} = -\frac{e}{m}E_0 = -\frac{e}{m}\hat{E}_0 \cos \omega_0 t \tag{8.70}$$

Linearizing about this oscillating equilibrium, we have

$$\frac{\partial v_{e1}}{\partial t} + ikv_{e0}v_{e1} = -\frac{e}{m}E_1 = -\frac{e}{m}(E_{1h} + E_{1l}) \tag{8.71}$$

where the subscripts h and l denote the high- and low-frequency parts. The first term consists mostly of the high-frequency velocity v_{eh}, given by

$$\frac{\partial v_{eh}}{\partial t} = -\frac{e}{m}E_{1h} = \frac{n_{eh}e^2}{ik\varepsilon_0 m} \tag{8.72}$$

where we have used Eq. (8.69). The low-frequency part of Eq. (8.71) is

$$ikv_{e0}v_{eh} = -\frac{e}{m}E_{1l}$$

The right-hand side is just the ponderomotive term used in Eq. (8.65) to drive the ion waves. It results from the low-frequency beat between v_{e0} and v_{eh}. The left-hand side can be recognized as related to the electrostatic part of the ponderomotive force expression in Eq. (8.42).

The electron continuity equation is

$$\frac{\partial n_{e1}}{\partial t} + ikv_{e0}n_{e1} + n_0 ikv_{e1} = 0 \tag{8.73}$$

We are interested in the high-frequency part of this equation. In the middle term, only the low-frequency density n_{el} can beat with v_{e0} to give a high-frequency term, if we reject phenomena near $2\omega_0$ and higher harmonics. But $n_{el} = n_{i1}$ by quasineutrality so we have

$$\frac{\partial n_{eh}}{\partial t} + ikn_0 v_{eh} + ikv_{e0}n_{i1} = 0 \tag{8.74}$$

Taking the time derivative, neglecting $\partial n_{i1}/\partial t$, and using Eqs. (8.70) and (8.72), we obtain

$$\frac{\partial^2 n_{eh}}{\partial t^2} + \omega_p^2 n_{eh} = \frac{ike}{m} n_{i1} E_0 \tag{8.75}$$

Let n_{eh} vary as $\exp(-i\omega t)$:

$$\left(\omega_p^2 - \omega^2\right) n_{eh} = \frac{ike}{m} n_{i1} E_0 \tag{8.76}$$

Equations (8.69) and (8.76) then give the high-frequency field:

$$E_{1h} = -\frac{e^2}{\varepsilon_0 m} \frac{n_{i1} E_0}{\omega_p^2 - \omega^2} \approx -\frac{e^2}{\varepsilon_0 m} \frac{n_{i1} E_0}{\omega_p^2 - \omega_0^2} \tag{8.77}$$

In setting $\omega \approx \omega_0$ we have assumed that the growth rate of n_{i1} is very small compared with the frequency of E_0. The ponderomotive force follows from Eq. (8.64):

$$F_{\mathrm{NL}} \approx \frac{\omega_p^2}{\omega_0^2} \frac{e^2}{m} \frac{ikn_{i1}}{\omega_p^2 - \omega_0^2} \langle E_0^2 \rangle \tag{8.78}$$

Note that both E_{1h} and F_{NL} change sign with $\omega_p^2 - \omega_0^2$. This is the reason the oscillating two-stream instability mechanism does not work for $\omega_0^2 > \omega_p^2$. The maximum response will occur for $\omega_0^2 \approx \omega_p^2$, and we may neglect the factor (ω_p^2/ω_0^2). Equation (8.67) can then be written

$$\frac{\partial^2 n_{i1}}{\partial t^2} \approx \frac{e^2 k^2}{2Mm} \frac{\hat{E}_0^2 n_{i1}}{\omega_p^2 - \omega_0^2} \tag{8.79}$$

Since the low-frequency perturbation does not propagate in this instability, we can let $n_{i1} = \bar{n}_{i1} \exp \gamma t$, where γ is the growth rate. Thus

$$\gamma^2 \approx \frac{e^2 k^2}{2Mm} \frac{\hat{E}_0^2}{\omega_p^2 - \omega_0^2} \tag{8.80}$$

and γ is real if $\omega_0^2 < \omega_p^2$. The actual value of γ will depend on how small the denominator in Eq. (8.77) can be made without the approximation $\omega^2 \approx \omega_0^2$. If damping is finite, $\omega_p^2 - \omega^2$ will have an imaginary part proportional to $2\Gamma_2 \omega_p$, where Γ_2 is the damping rate of the electron oscillations. Then we have

$$\gamma \propto \hat{E}_0 / \Gamma_2^{1/2} \tag{8.81}$$

Far above threshold, the imaginary part of ω will be dominated by the growth rate γ rather than by Γ_2. One then has

$$\gamma^2 \propto \frac{\hat{E}_0^2}{\gamma} \qquad \gamma \propto \left(\hat{E}_0\right)^{2/3} \tag{8.82}$$

This behavior of γ with E_0 is typical of all parametric instabilities. An exact calculation of γ and of the threshold value of E_0 requires a more careful treatment of the frequency shift $\omega_p - \omega_0$ than we can present here.

To solve the problem exactly, one solves for n_{i1} in Eq. (8.76) and substitutes into Eq. (8.79):

$$\frac{\partial^2 n_{i1}}{\partial t^2} = -\frac{ike}{M} n_{eh} E_0 \tag{8.83}$$

Equations (8.75) and (8.83) then constitute a pair of equations of the form of Eqs. (8.49) and (8.50), and the solution of Eq. (8.55) can be used. The frequency ω_1 vanishes in that case because the ion wave has $\omega_1 = 0$ in the zero-temperature limit.

8.5.6 The Parametric Decay Instability

The derivation for $\omega_0 > \omega_p$ follows the same lines as above and leads to the excitation of a plasma wave and an ion wave. We shall omit the algebra, which is somewhat lengthier than for the oscillating two-stream instability, but shall instead describe some experimental observations. The parametric decay instability is well documented, having been observed both in the ionosphere and in the laboratory. The oscillating two-stream instability is not often seen, partly because $\mathrm{Re}(\omega) = 0$ and partly because $\omega_0 < \omega_p$ means that the incident wave is evanescent. Figure 8.16 shows the apparatus of Stenzel and Wong, consisting of a plasma source similar to that of Fig. 8.10, a pair of grids between which the field E_0 is generated by an oscillator, and a probe connected to two frequency spectrum analyzers. Figure 8.17 shows spectra of the signals detected in the plasma. Below threshold, the high-frequency spectrum shows only the pump wave at 400 MHz, while the low-frequency spectrum shows only a small amount of noise. When the pump wave amplitude is increased slightly, an ion wave at 300 kHz appears in the low-frequency spectrum; and at the same time, a sideband at 399.7 MHz appears in the high-frequency spectrum. The latter is an electron plasma wave at the difference frequency. The ion wave then can be observed to beat with the pump wave to give a small signal at the sum frequency, 400.3 MHz.

This instability has also been observed in ionospheric experiments. Figure 8.18 shows the geometry of an ionospheric modification experiment performed with the

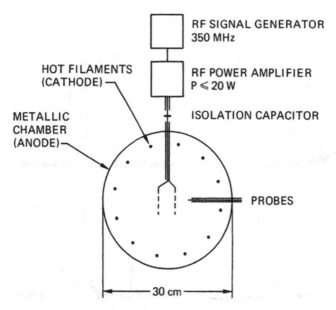

Fig. 8.16 Schematic of an experiment in which the parametric decay instability was verified. [From A. Y. Wong *et al.*, *Plasma Physics and Controlled Nuclear Fusion Research, 1971*, I, 335 (International Atomic Energy Agency, Vienna, 1971).]

large radio telescope at Platteville, Colorado. A 2-MW radiofrequency beam at 7 MHz is launched from the antenna into the ionosphere. At the layer where $\omega_0 \gtrsim \omega_p$, electron and ion waves are generated, and the ionospheric electrons are heated. In another experiment with the large dish antenna at Arecibo, Puerto Rico, the ω and $\mathbf{k}$ of the electron waves were measured by probing with a 430-MHz radar beam and observing the scattering from the grating formed by the electron density perturbations.

Problems

8.12 In laser fusion, a pellet containing thermonuclear fuel is heated by intense laser beams. The parametric decay instability can enhance the heating efficiency by converting laser energy into plasma wave energy, which is then transferred to electrons by Landau damping. If an iodine laser with 1.3-μm wavelength is used, at what plasma density does parametric decay take place?

8.13 (a) Derive the following dispersion relation for an ion acoustic wave in the presence of an externally applied ponderomotive force F_{NL}:

$$\left(\omega^2 + 2i\Gamma\omega - k^2 v_s^2\right)n_1 = ikF_{NL}/M,$$

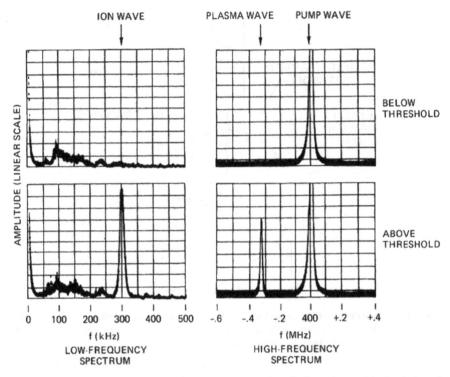

ION WAVE   PLASMA WAVE   PUMP WAVE

BELOW
THRESHOLD

ABOVE
THRESHOLD

AMPLITUDE (LINEAR SCALE)

0   100   200   300   400   500   -.6   -.4   -.2   400   +.2   +.4

f (kHz)   f (MHz)

LOW-FREQUENCY   HIGH-FREQUENCY
SPECTRUM   SPECTRUM

Fig. 8.17 Oscillograms showing the frequency spectra of oscillations observed in the device of Fig. 8.16. When the driving power is just below threshold, only noise is seen in the low-frequency spectrum and only the driver (pump) signal in the high-frequency spectrum. A slight increase in power brings the system above threshold, and the frequencies of a plasma wave and an ion wave simultaneously appear. [Courtesy of R. Stenzel, UCLA.]

where Γ is the damping rate of the undriven wave (when $F_{NL} = 0$). (Hint: introduce a "collision frequency" v in the ion equation of motion, evaluate Γ in terms of v, and eventually replace v by its Γ-equivalent.)

(b) Evaluate F_{NL} for the case of stimulated Brillouin scattering in terms of the amplitudes E_0 and E_2 of the pump and the backscattered wave, respectively, thus recovering the constant c_1 of Problem (8.10). (Hint: cf. Eq. (8.64).)

8.14 In Fig. 8.17 it is seen that the upper sideband at $\omega_0 + \omega_1$ is missing. Indeed, in most parametric processes the upper sideband is observed to be smaller than the lower sideband. Using simple energy arguments, perhaps with a quantum mechanical analogy, explain why this should be so.

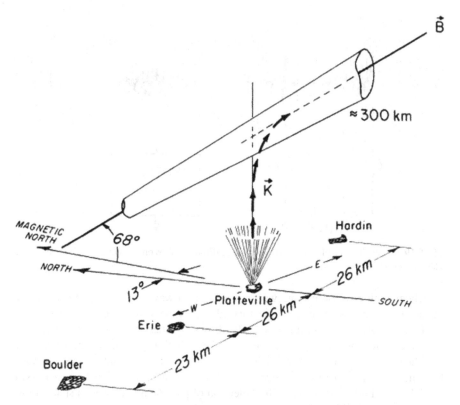

Fig. 8.18 Geometry of an ionospheric modification experiment in which radiofrequency waves were absorbed by parametric decay. [From W. F. Utlaut and R. Cohen, *Science* **174**, 245 (1971).]

8.6   Plasma Echoes

Since Landau damping does not involve collisions or dissipation, it is a reversible process. That this is true is vividly demonstrated by the remarkable phenomenon of plasma echoes. Figure 8.19 shows a schematic of the experimental arrangement. A plasma wave with frequency ω_1 and wavelength λ_1 is generated at the first grid and propagated to the right. The wave is Landau-damped to below the threshold of detectability. A second wave of ω_2 and λ_2 is generated by a second grid a distance l from the first one. The second wave also damps away. If a third grid connected to a receiver tuned to $\omega = \omega_2 - \omega_1$ is moved along the plasma column, it will find an echo at a distance $l' = l\omega_2/(\omega_2 - \omega_1)$. What happens is that the resonant particles causing the first wave to damp out retains information about the wave in their distribution function. If the second grid is made to reverse the change in the resonant particle distribution, a wave can be made to reappear. Clearly, this process can occur only in a very nearly collisionless plasma. In fact, the echo amplitude has been used as a sensitive measure of the collision rate. Figure 8.20 gives a physical picture of why echoes occur. The same basic mechanism lies behind observations

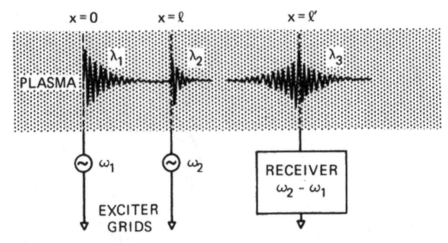

Fig. 8.19 Schematic of a plasma echo experiment. [From A. Y. Wong and D. R. Baker, *Phys. Rev.*
188, 326 (1969).]

of echoes with electron plasma waves or cyclotron waves. Figure 8.20 is a plot of
distance vs. time, so that the trajectory of a particle with a given velocity is a
straight line. At $x = 0$, a grid periodically allows bunches of particles with a spread
in velocity to pass through. Because of the velocity spread, the bunches mix
together, and after a distance l, the density, shown at the right of the diagram,
becomes constant in time. A second grid at $x = l$ alternately blocks and passes
particles at a higher frequency. This selection of particle trajectories in space–time
then causes a bunching of particles to reoccur at $x = l'$.

The relation between l' and l can be obtained from this simplified picture, which
neglects the influence of the wave electric field on the particle trajectories. If $f_1(v)$ is
the distribution function at the first grid and it is modulated by $\cos \omega_1 t$, the
distribution at $x > 0$ will be given by

$$f(x, v, t) = f_1(v) \cos \left(\omega_1 t - \frac{\omega_1}{v} x \right) \tag{8.84}$$

The second grid at $x = l$ will further modulate this distribution by a factor
containing ω_2 and the distance $x - l$:

$$f(x, v, t) = f_{12}(v) \cos \left(\omega_1 t - \frac{\omega_1}{v} x \right) \cos \left[\omega_2 t - \frac{\omega_2}{v}(x - l) \right] \tag{8.85}$$

$$
\begin{aligned}
= f_{12}(v) \frac{1}{2} \Big\{ &\cos \left[(\omega_2 + \omega_1)t - \frac{\omega_2(x - l) + \omega_1 x}{v} \right] \\
&+ \cos \left[(\omega_2 - \omega_1)t - \frac{\omega_2(x - l) - \omega_1 x}{v} \right] \Big\}
\end{aligned}
\tag{8.86}
$$

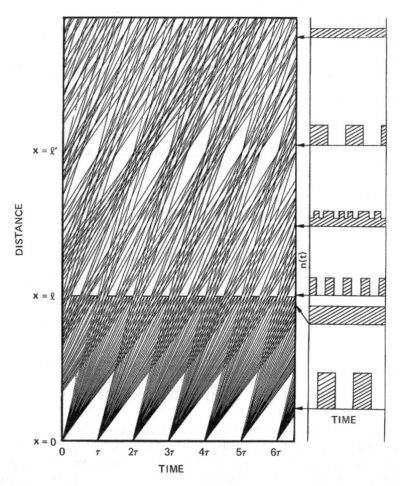

Fig. 8.20 Space–time trajectories of gated particles showing the bunching that causes echoes. The density at various distances is shown at the *right*. [From D. R. Baker, N. R. Ahern, and A. Y. Wong, *Phys. Rev. Lett.* **20**, 318 (1968).]

The echo comes from the second term, which oscillates at $\omega = \omega_2 - \omega_1$ and has an argument independent of v if

$$\omega_2(x - l) = \omega_1 x$$

or

$$x = \omega_2 l / (\omega_2 - \omega_1) \equiv l' \qquad (8.87)$$

Fig. 8.21 Measurements of
echo amplitude profiles for
various separations
l between the driver grids.
The *solid circles* correspond
to the case $\omega_2 < \omega_1$, for
which no echo is expected.
[From Baker, Ahern, and
Wong, *loc. cit.*]

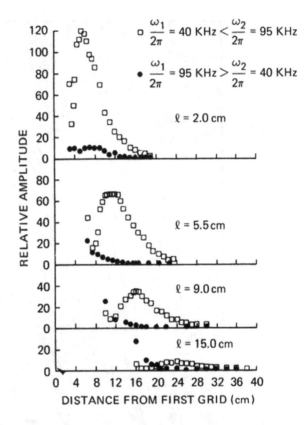

The spread in velocities, therefore, does not affect the second term at $x = l'$, and the phase mixing has been undone. When integrated over velocity, this term gives a density fluctuation at $\omega = \omega_2 - \omega_1$. The first term is undetectable because phase mixing has smoothed the density perturbations. It is clear that l' is positive only if $\omega_2 > \omega_1$. The physical reason is that the second grid has less distance in which to unravel the perturbations imparted by the first grid, and hence must operate at a higher frequency.

Figure 8.21 shows the measurements of Baker, Ahern, and Wong on ion wave echoes. The distance l' varies with l in accord with Eq. (8.87). The solid dots, corresponding to the case $\omega_2 < \omega_1$, show the absence of an echo, as expected. The echo amplitude decreases with distance because collisions destroy the coherence of the velocity modulations.

8.7 Nonlinear Landau Damping

When the amplitude of an electron or ion wave excited, say, by a grid is followed in space, it is often found that the decay is not exponential, as predicted by linear theory, if the amplitude is large. Instead, one typically finds that the amplitude

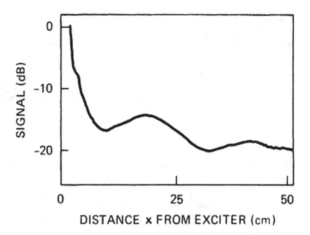

Fig. 8.22 Measurement of the amplitude profile of a nonlinear electron wave showing nonmonotonic decay. [From R. N. Franklin, S. M. Hamberger, H. Ikezi, G. Lampis, and G. J. Smith, *Phys. Rev. Lett.* **28**, 1114 (1972).]

decays, grows again, and then oscillates before settling down to a steady value. Such behavior for an electron wave at 38 MHz is shown in Fig. 8.22. Although other effects may also be operative, these oscillations in *amplitude* are exactly what would be expected from the nonlinear effect of particle trapping discussed in Sect. 7.5. Trapping of a particle of velocity v occurs when its energy in the wave frame is smaller than the wave potential; that is, when

$$|e\phi| > \frac{1}{2}m(v - v_\phi)^2$$

Small waves will trap only these particles moving at high speeds near v_ϕ. To trap a large number of particles in the main part of the distribution (near $v = 0$) would require

$$|q\phi| = \frac{1}{2}mv_\phi^2 = \frac{1}{2}m(\omega/k)^2 \tag{8.88}$$

When the wave is this large, its linear behavior can be expected to be greatly modified. Since $|\phi| = |E/k|$, the condition Eq. (8.88) is equivalent to

$$\omega \cong \omega_B, \quad \text{where} \quad \omega_B^2 \equiv |qkE/m| \tag{8.89}$$

The quantity ω_B is called the *bounce frequency* because it is the frequency of oscillation of a particle trapped at the bottom of a sinusoidal potential well (Fig. 8.27). The potential is given by

$$\phi = \phi_0(1 - \cos kx) = \phi_0\left(\frac{1}{2}k^2x^2 + \cdots\right) \tag{8.90}$$

Fig. 8.23 A trapped
particle bouncing in the
potential well of a wave

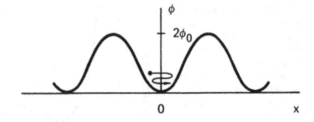

The equation of motion is

$$m\frac{d^2x}{dt^2} = -m\omega^2x = qE = -q\frac{d\phi}{dx} = -qk\phi_0 \sin kx \tag{8.91}$$

The frequency ω is not constant unless x is small, $\sin kx \approx kx$, and ϕ is approximately parabolic. Then ω takes the value ω_B defined in Eq. (8.89). When the resonant particles are reflected by the potential, they give kinetic energy back to the wave, and the amplitude increases. When the particles bounce again from the other side, the energy goes back into the particles, and the wave is damped. Thus, one would expect oscillations in amplitude at the frequency ω_B in the wave frame. In the laboratory frame, the frequency would be $\omega' = \omega_B + kv_\phi$; and the amplitude oscillations would have wave number $k' = \omega'/v_\phi = k[1 + (\omega_B/\omega)]$. The condition $\omega_B \gtrsim \omega$ turns out to define the breakdown of linear theory even when other processes besides particle trapping are responsible.

Another type of nonlinear Landau damping involves the beating of two waves. Suppose there are two high-frequency electron waves (ω_1, k_1) and (ω_2, k_2). These would beat to form an amplitude envelope traveling at a velocity $(\omega_2 - \omega_1)/(k_2 - k_1) \approx d\omega/dk = v_g$. This velocity may be low enough to lie within the ion distribution function. There can then be an energy exchange with the resonant ions. The potential the ions see is the effective potential due to the ponderomotive force (Fig. 8.24), and Landau damping or growth can occur. This damping provides an effective way to heat ions with high-frequency waves, which do not ordinarily interact with ions. If the ion distribution is double-humped, it can excite the electron waves. Such an instability is called a *modulational instability*.

Problems

8.15 Make a graph to show clearly the degree of agreement between the echo data of Fig. 8.21 and Eq. (8.87).

8.16 Calculate the bounce frequency of a deeply trapped electron in a plasma wave with 10-V rms amplitude and 1-cm wavelength (Fig. 8.23).

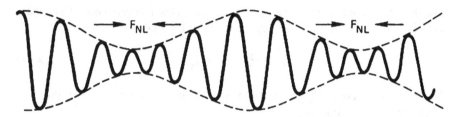

Fig. 8.24 The ponderomotive force caused by the envelope of a modulated wave can trap particles and cause wave-particle resonances at-the group velocity

8.8   Equations of Nonlinear Plasma Physics

There are two nonlinear equations that have been treated extensively in connection with nonlinear plasma waves: The Korteweg–de Vries equation and the nonlinear Schrödinger equation. Each concerns a different type of nonlinearity. When an ion acoustic wave gains large amplitude, the main nonlinear effect is wave steepening, whose physical explanation was given in Sect. 8.3.3. This effect arises from the $\mathbf{v} \cdot \nabla \mathbf{v}$ term in the ion equation of motion and is handled mathematically by the Korteweg–de Vries equation. The wave-train and soliton solutions of Figs. 8.5 and 8.7 are also predicted by this equation.

When an electron plasma wave goes nonlinear, the dominant new effect is that the ponderomotive force of the plasma waves causes the background plasma to move away, causing a local depression in density called a *caviton*. Plasma waves trapped in this cavity then form an isolated structure called an *envelope soliton* or *envelope solitary wave*. Such solutions are described by the nonlinear Schrödinger equation. Considering the difference in both the physical model and the mathematical form of the governing equations, it is surprising that solitons and envelope solitons have almost the same shape.

8.8.1   The Korteweg–de Vries Equation

This equation occurs in many physical situations including that of a weakly nonlinear ion wave:

$$\frac{\partial U}{\partial \tau} + U \frac{\partial U}{\partial \xi} + \frac{1}{2} \frac{\partial^3 U}{\partial \xi^3} = 0 \qquad (8.92)$$

where U is amplitude, and τ and ξ are timelike and spacelike variables, respectively. Although several transformations of variables will be necessary before this form is obtained, two physical features can already be seen. The second term in Eq. (8.92) is easily recognized as the convective term $\mathbf{v} \cdot \nabla \mathbf{v}$ leading to wave steepening.

The third term arises from wave dispersion; that is, the k dependence of the phase velocity. For $T_i = 0$, ion waves obey the relation (Eq. (4.48))

$$\omega^2 = k^2 c_s^2 \left(1 + k^2 \lambda_D^2\right)^{-1} \tag{8.93}$$

The dispersive term $k^2 \lambda_D^2$ arises from the deviation from exact neutrality. By Taylor-series expansion, one finds

$$\omega = k c_s - \frac{1}{2} k^3 c_s \lambda_D^2 \tag{8.94}$$

showing that the dispersive term is proportional to k^3. This is the reason for the third derivative term in Eq. (8.92). Dispersion must be kept in the theory to prevent very steep wavefronts (corresponding to very large k) from spuriously dominating the nonlinear behavior.

The Korteweg–de Vries equation admits of a solution in the form of a soliton; that is, a single pulse which retains its shape as it propagates with some velocity c (not the velocity of light!). This means that U depends only on the variable $\xi - c\tau$ rather than ξ or τ separately. Defining $\zeta \equiv \xi - c\tau$, so that $\partial/\partial \tau = -c\, d/d\zeta$ and $\partial/\partial \xi = d/d\zeta$, we can write Eq. (8.92) as

$$-c\frac{dU}{d\zeta} + U\frac{dU}{d\zeta} + \frac{1}{2}\frac{d^3 U}{d\zeta^3} = 0 \tag{8.95}$$

This can be integrated with the dummy variable ζ':

$$-c\int_\zeta^\infty \frac{dU}{d\zeta'}\, d\zeta' + \frac{1}{2}\int_\zeta^\infty \frac{dU^2}{d\zeta'}\, d\zeta' + \frac{1}{2}\int_\zeta^\infty \frac{d}{d\zeta'}\left(\frac{d^2 U}{d\zeta'^2}\right) d\zeta' = 0. \tag{8.96}$$

If $U(\zeta)$ and its derivatives vanish at large distances from the soliton ($|\zeta| \to \infty$), the result is

$$cU - \frac{1}{2}U^2 - \frac{1}{2}\frac{d^2 U}{d\zeta^2} = 0 \tag{8.97}$$

Multiplying each term by $dU/d\zeta$, we can integrate once more, obtaining

$$\frac{1}{2}cU^2 - \frac{1}{6}U^3 - \frac{1}{4}\left(\frac{dU}{d\zeta}\right)^2 = 0 \tag{8.98}$$

or

$$\left(\frac{dU}{d\zeta}\right)^2 = \frac{2}{3}U^2(3c - U) \tag{8.99}$$

This equation is satisfied by the soliton solution

$$U(\zeta) = 3c\,\mathrm{sech}^2\left[(c/2)^{1/2}\zeta\right] \tag{8.100}$$

as one can verify by direct substitution, making use of the identities

$$\frac{d}{dx}(\mathrm{sech}\,x) = -\mathrm{sech}\,x\,\tanh x \tag{8.101}$$

and

$$\mathrm{sech}^2x + \tanh^2x = 1 \tag{8.102}$$

Equation (8.100) describes a structure that looks like Fig. 8.7, reaching a peak at $\zeta = 0$ and vanishing at $\zeta \to \pm\infty$. The soliton has speed c, amplitude $3c$, and half-width $(2/c)^{1/2}$. All are related, so that c specifies the energy of the soliton. The larger the energy, the larger the speed and amplitude, and the narrower the width. The occurrence of solitons depends on the initial conditions. If the initial disturbance has enough energy and the phases are right, a soliton can be generated; otherwise, a large-amplitude wave will appear. If the initial disturbance has the energy of several solitons and the phases are right, an N-soliton solution can be generated. Since the speed of the solitons increases with their size, after a time the solitons will disperse themselves into an ordered array, as shown in Fig. 8.25.

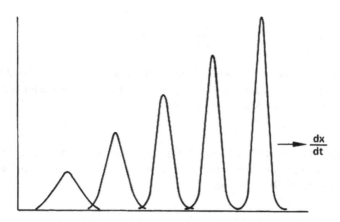

Fig. 8.25   A train of solitons, generated at the *left*, arrayed according to the relation among speed, height, and width

We next wish to show that the Korteweg–de Vries equation describes large-amplitude ion waves. Consider the simple case of one-dimensional waves with cold ions. The fluid equations of motion and continuity are

$$\frac{\partial v_i}{\partial t} + v_i \frac{\partial v_i}{\partial x} = -\frac{e}{m}\frac{\partial \phi}{\partial x} \tag{8.103}$$

$$\frac{\partial n_i}{\partial t} + \frac{\partial}{\partial x}(n_i v_i) = 0 \tag{8.104}$$

Assume Boltzmann electrons (Eq. (3.73)); Poisson's equation is then

$$\varepsilon_0 \frac{\partial^2 \phi}{\partial x^2} = e\left(n_0 e^{e\phi/KT_e} - n_i\right) \tag{8.105}$$

The following dimensionless variables will make all the coefficients unity:

$$\begin{aligned}
x' &= x/\lambda_D = x(n_0 e^2/\varepsilon_0 KT_e)^{1/2} \\
t' &= \Omega_p t = t(n_0 e^2/\varepsilon_0 M)^{1/2} \\
\chi &= e\phi/KT_e \qquad n' = n_i/n_0 \\
v' &= v/v_s = v(M/KT_e)^{1/2}
\end{aligned} \tag{8.106}$$

Our set of equations becomes

$$\frac{\partial v'}{\partial t'} + v' \frac{\partial v'}{\partial x'} = \frac{\partial \chi}{\partial x'} \tag{8.107}$$

$$\frac{\partial n'}{\partial t'} + \frac{\partial}{\partial x'}\left(n' v'\right) = 0 \tag{8.108}$$

$$\frac{\partial^2 \chi}{\partial x'^2} = e^\chi - n' \tag{8.109}$$

If we were to transform to a frame moving with velocity $v' = \mathcal{M}$, we would recover Eq. (8.27). As shown following Eq. (8.27), this set of equations admits of soliton solutions for a range of Mach numbers $\mathcal{M}$.

Problem

8.17 Reduce Eqs. (8.107)–(8.109) to Eq. (8.27) by assuming that χ, n', and v' depend only on the variable $\xi' \equiv x' - \mathcal{M}t'$. Integrate twice as in Eqs. (8.96)–(8.98) to obtain

$$\frac{1}{2}\left(d\chi/d\xi'\right)^2 = e^\chi - 1 + \mathcal{M}\left[\left(\mathcal{M}^2 - 2\chi\right)^{1/2} - \mathcal{M}\right]$$

Show that soliton solutions can exist only for $1 < \mathcal{M} < 1.6$ and $0 < \chi_{max} < 1.3$.

To recover the $K\!-\!dV$ equation, we must expand in the wave amplitude and keep one order higher than in the linear theory. Since for solitons the amplitude and speed are related, we can choose the expansion parameter to be the Mach number excess δ, defined to be

$$\delta \equiv \mathcal{M} - 1 \tag{8.110}$$

We thus write

$$
\begin{aligned}
n' &= 1 + \delta n_1 + \delta^2 n_2 + \cdots \\
\chi' &= \delta \chi_1 + \delta^2 \chi_2 + \cdots \\
v' &= \delta v_1 + \delta^2 v_2 + \cdots
\end{aligned}
\tag{8.111}
$$

We must also transform to the scaled variables[1]

$$\xi = \delta^{1/2}\left(x' - t'\right) \qquad \tau = \delta^{3/2} t' \tag{8.112}$$

so that

$$\frac{\partial}{\partial t'} = \delta^{3/2}\frac{\partial}{\partial \tau} - \delta^{1/2}\frac{\partial}{\partial \xi}$$

$$\frac{\partial}{\partial x'} = \delta^{1/2}\frac{\partial}{\partial \xi}$$

Substituting Eqs. (8.111) and (8.113) into Eq. (8.109), we find that the lowest-order terms are proportional to δ, and these give

$$\chi_1 = n_1 \tag{8.114}$$

Doing the same in Eqs. (8.107) and (8.108), we find that the lowest-order terms are proportional to $\delta^{3/2}$, and these give

$$\frac{\partial v_1}{\partial \xi} = \frac{\partial \chi_1}{\partial \xi} = \frac{\partial n_1}{\partial \xi} \tag{8.115}$$

Since all vanish as $\xi \to \infty$, integration gives

$$n_1 = \chi_1 = v_1 \equiv U \tag{8.116}$$

Thus our normalization is such that all the linear perturbations are equal and can be called U. We next collect the terms proportional to δ^2 in Eq. (8.109) and to $\delta^{5/2}$ in Eqs. (8.107) and (8.108). This yields the set

---

[1] It is not necessary to explain why; the end will justify the means.

$$\frac{\partial^2 \chi_1}{\partial \xi^2} = \chi_2 - n_2 + \frac{1}{2}\chi_1^2 \tag{8.117}$$

$$\frac{\partial v_1}{\partial \tau} - \frac{\partial v_2}{\partial \xi} + v_1\frac{\partial v_1}{\partial \xi} = -\frac{\partial \chi_2}{\partial \xi} \tag{8.118}$$

$$\frac{\partial n_1}{\partial \tau} - \frac{\partial n_2}{\partial \xi} + \frac{\partial}{\partial \xi}(v_2 + n_1 v_1) \tag{8.119}$$

Solving for n_2 in Eq. (8.117) and for $\partial v_2/\partial \xi$ in Eq. (8.118), we substitute into Eq. (8.119):

$$\frac{\partial n_1}{\partial \tau} + \frac{\partial^3 \chi_1}{\partial \xi^3} - \frac{\partial \chi_2}{\partial \xi} - \frac{1}{2}\frac{\partial \chi_1^2}{\partial \xi} + \frac{\partial v_1}{\partial \tau} + v_1\frac{\partial v_1}{\partial \xi} + \frac{\partial \chi_2}{\partial \xi} + \frac{\partial}{\partial \xi}(n_1 v_1) = 0 \tag{8.120}$$

Fortunately, χ_2 cancels out, and replacing all first-order quantities by U results in

$$\frac{\partial U}{\partial \tau} + U\frac{\partial U}{\partial \xi} + \frac{1}{2}\frac{\partial^3 U}{\partial \xi^3} = 0 \tag{8.121}$$

which is the same as Eq. (8.92). Thus, ion waves of amplitude one order higher than linear are described by the Korteweg–de Vries equation.

Problem

8.18 A soliton with peak amplitude 12 V is excited in a hydrogen plasma with $KT_e = 10$ eV and $n_0 = 10^{16}$ m^{-3}. Assuming that the Korteweg–de Vries equation describes the soliton, calculate its velocity (in m/s) and its full width at half maximum (in mm). (Hint: First show that the soliton velocity c is equal to unity in the normalized units used to derive the K–dV equation.)

8.8.2 The Nonlinear Schrödinger Equation

This equation has the standard dimensionless form

$$i\frac{\partial \psi}{\partial t} + p\frac{\partial^2 \psi}{\partial x^2} + q|\psi|^2\psi = 0 \tag{8.122}$$

where ψ is the wave amplitude, $i = (-1)^{1/2}$, and p and q are coefficients whose physical significance will be explained shortly. Equation (8.122) differs from the usual Schrödinger equation

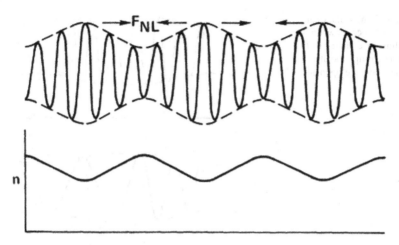

Fig. 8.26 The ponderomotive force of a plasma wave with nonuniform intensity causes ions to flow toward the intensity minima. The resulting density ripple traps waves in its troughs, thus enhancing the modulation of the envelope

$$i\hbar \frac{\partial \psi}{\partial t} + \frac{\hbar^2}{2m} \frac{\partial^2 \psi}{\partial x^2} - V(x,t)\psi = 0 \qquad (8.123)$$

in that the potential $V(x, t)$ depends on ψ itself, making the last term nonlinear. Note, however, that V depends only on the magnitude $|\psi|^2$ and not on the phase of ψ. This is to be expected, as far as electron plasma waves are concerned, because the nonlinearity comes from the ponderomotive force, which depends on the gradient of the wave intensity.

Plane wave solutions of Eq. (8.122) are modulationally unstable if $pq > 0$; that is, a ripple on the envelope of the wave will tend to grow. The picture is the same as that of Fig. 8.24 even though we are considering here fluid, rather than discrete particle, effects. For plasma waves, it is easy to see how the ponderomotive force can cause a modulational instability. Figure 8.26 shows a plasma wave with a rippled envelope. The gradient in wave intensity causes a ponderomotive force which moves both electrons and ions toward the intensity minima, forming a ripple in the plasma density. Plasma waves are trapped in regions of low density because their dispersion relation

$$\omega^2 = \omega_p^2 + \frac{3}{2}k^2 v_{th}^2 \qquad (4.30)$$

permits waves of large k to exist only in regions of small ω_p. The trapping of part of the k spectrum further enhances the wave intensity in the regions where it was already high, thus causing the envelope to develop a growing ripple.

The reason the sign of pq matters is that p and q for plasma waves turn out to be proportional, respectively, to the *group dispersion* dv_g/dk and the *nonlinear frequency shift* $\delta\omega \propto \partial\omega/\partial|\psi|^2$. We shall show later that

Fig. 8.27 Modulational
instability occurs when the
nonlinear frequency shift
and the group velocity
dispersion have opposite
signs

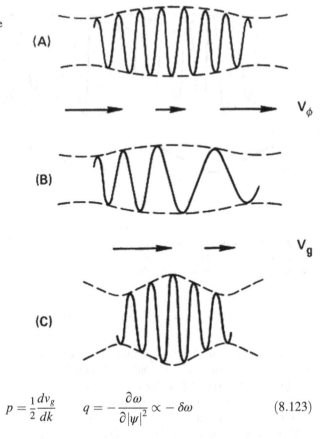

$$p = \frac{1}{2}\frac{dv_g}{dk} \qquad q = -\frac{\partial \omega}{\partial |\psi|^2} \propto -\delta\omega \qquad\qquad (8.123)$$

Modulational instability occurs when $pq > 0$; that is, when $\delta\omega$ and dv_g/dk have
opposite signs. Figure 8.27 illustrates why this is so. In Fig. 8.27a, a ripple in the
wave envelope has developed as a result of random fluctuations. Suppose $\delta\omega$ is
negative. Then the phase velocity ω/k, which is proportional to ω, becomes somewhat smaller in the region of high intensity. This causes the wave crests to pile up on
the left of Fig. 8.27b and to spread out on the right. The local value of k is therefore
large on the left and small on the right. If dv_g/dk is positive, the group velocity will be
larger on the left than the right, so the wave energy will pile up into a smaller space.
Thus, the ripple in the envelope will become narrower and larger, as in Fig. 8.27c. If
$\delta\omega$ and dv_g/dk were of the same sign, this modulational instability would not happen.

Although plane wave solutions to Eq. (8.123) are modulationally unstable when
$pq > 0$, there can be solitary structures called *envelope solitons* which are stable.
These are generated from the basic solution

$$w(x, t) = \left(\frac{2A}{q}\right)^{1/2} \mathrm{sech}\left[\left(\frac{A}{p}\right)^{1/2} x\right] e^{iAt} \qquad\qquad (8.124)$$

Fig. 8.28 An envelope
soliton

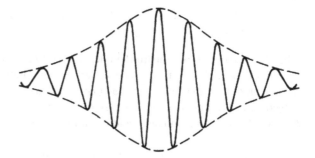

where A is an arbitrary constant which ties together the amplitude, width, and
frequency of the packet. At any given time, the disturbance resembles a simple
soliton (Eq. (8.100)) (though the hyperbolic secant is not squared here), but the
exponential factor makes $w(x, t)$ oscillate between positive and negative values. An
envelope soliton moving with a velocity V has the more general form (Fig. 8.28)

$$\psi(x,t) = \left(\frac{2A}{q}\right)^{1/2} \text{sech}\left[\left(\frac{A}{p}\right)^{1/2}(x - x_0 - Vt)\right] \exp i\left(At + \frac{V}{2p}x - \frac{V^2}{4p}t + \theta_0\right)$$

$$(8.125)$$

where x_0 and θ_0 are the initial position and phase. It is seen that the magnitude of
V also controls the number of wavelengths inside the envelope at any given time.

Problems

8.19 Show by direct substitution that Eq. (8.124) is a solution of Eq. (8.122).

8.20 Verify Eq. (8.125) by showing that, if $w(x, t)$ is a solution of Eq. (8.122), then

$$\psi = w(x - x_0 - Vt, t)\exp\left[i\left(\frac{V}{2p}x - \frac{V^2}{4p}t + \theta_0\right)\right]$$

is also a solution.

We next wish to show that the nonlinear Schrödinger equation describes large-
amplitude electron plasma waves. The procedure is to solve self-consistently for the
density cavity that the waves dig by means of their ponderomotive force and for the
behavior of the waves in such a cavity. The high-frequency motion of the electrons
is governed by Eqs. (4.28), (4.18), and (4.19), which we rewrite respectively in one
dimension as

$$\frac{\partial u}{\partial t} = -\frac{e}{m}E - \frac{3KT_e}{mn_0}\frac{\partial n}{\partial x}$$

$$(8.126)$$

$$\frac{\partial n}{\partial t} + n_0\frac{\partial u}{\partial x} = 0$$

$$(8.127)$$

$$\frac{\partial E}{\partial x} = -\varepsilon_0^{-1} en \tag{8.128}$$

where n_0 is the uniform unperturbed density; and E, n, and u are, respectively, the perturbations in electric field, electron density, and fluid velocity. These equations are linearized, so that nonlinearities due to the $\mathbf{u} \cdot \nabla \mathbf{u}$ and $\nabla \cdot (n\mathbf{u})$ terms are not considered. Taking the time derivative of Eq. (8.127) and the x derivative of Eq. (8.126), we can eliminate u and E with the help of Eq. (8.128) to obtain

$$\frac{\partial^2 n}{\partial t^2} - \frac{3KT_e}{m}\frac{\partial^2 n}{\partial x^2} + \frac{n_0 e^2}{m\varepsilon_0}n = 0 \tag{8.129}$$

We now replace n_0 by $n_0 + \delta n$ to describe the density cavity; this is the only nonlinear effect considered. Equation (8.129) is of course followed by any of the linear variables. It will be convenient to write it in terms of u and use the definition of ω_p; thus

$$\frac{\partial^2 u}{\partial t^2} - \frac{3KT_e}{m}\frac{\partial^2 u}{\partial x^2} + \omega_p^2\left(1 + \frac{\delta n}{n_0}\right)u = 0 \tag{8.130}$$

The velocity u consists of a high-frequency part oscillating at ω_0 (essentially the plasma frequency) and a low-frequency part u_l describing the quasineutral motion of electrons following the ions as they move to form the density cavity. Both fast and slow *spatial* variations are included in u_l.

Let

$$u(x, t) = u_l(x, t)e^{-i\omega_0 t} \tag{8.131}$$

Differentiating twice in time, we obtain

$$\frac{\partial^2 u}{\partial t^2} = \left(\ddot{u}_l - 2i\omega_0\dot{u}_l - \omega_0^2 u_l\right)e^{-i\omega_0 t}$$

where the dot stands for a time derivative on the slow time scale. We may therefore neglect $\ddot{u}_l$, which is much smaller than $\omega_0^2 u_l$:

$$\frac{\partial^2 u}{\partial t^2} = -\left(\omega_0^2 u_l + 2i\omega_0\dot{u}_l\right)e^{-i\omega_0 t}$$

Substituting into Eq. (8.130) gives

$$\left[2i\omega_0\dot{u}_l + \frac{3KT_e}{m}\frac{\partial^2 u_l}{\partial x^2} + \left(\omega_0^2 - \omega_p^2 - \omega_p^2\frac{\delta n}{n_0}\right)u_l\right]e^{-i\omega_0 t} = 0 \tag{8.133}$$

We now transform to the dimensionless variables

$$t' = \omega_p t \qquad \omega' = \omega/\omega_p \qquad x' = x/\lambda_D$$
$$u' = u(KT_e/m)^{-1/2} \qquad \delta n' = \delta n/n_0 \tag{8.134}$$

obtaining

$$\left[i\omega_0' \frac{\partial u_l'}{\partial t'} + \frac{3}{2}\frac{\partial^2 u_l'}{\partial x'^2} + \frac{1}{2}\left(\omega_0'^2 - 1 - \delta n'\right)u_l' \right] e^{-i\omega_0' t'} = 0$$

Defining the frequency shift Δ

$$\Delta \equiv \left(\omega_0 - \omega_p\right)/\omega_p = \omega_0' - 1 \tag{8.135}$$

and assuming $\Delta \ll 1$, we have $\omega_0'^2 - 1 \approx 2\Delta$. We may now drop the primes (these being understood), convert back to $u(x, t)$ via Eq. (8.131), and approximate ω_0' by 1 in the first term to obtain

$$i\frac{\partial u}{\partial t} + \frac{3}{2}\frac{\partial^2 u}{\partial x^2} + \left(\Delta - \frac{1}{2}\delta n\right)u = 0 \tag{8.136}$$

Here it is understood that $\partial/\partial t$ is the time derivative on the slow time scale, although u contains both the $\exp(-i\omega_0 t)$ factor and the slowly varying coefficient u_l. We have essentially derived the nonlinear Schrödinger equation (8.122), but it remains to evaluate δn in terms of $|u_l|^2$.

The low-frequency equation of motion for the electrons is obtained by neglecting the inertia term in Eq. (4.28) and adding a ponderomotive force term from Eq. (8.44)

$$0 = -enE - KT_e\frac{\partial n}{\partial x} - \frac{\omega_p^2}{\omega_0^2}\frac{\partial}{\partial x}\frac{\langle \varepsilon_0 E^2\rangle}{2}. \tag{8.137}$$

Here we have set $\gamma_e = 1$ since the low-frequency motion should be isothermal rather than adiabatic. We may set

$$\langle E^2 \rangle \simeq \frac{m^2\omega_0^2}{e^2}\langle u^2 \rangle \tag{8.138}$$

by solving the high-frequency Eq. (8.126) without the thermal correction. With $E = -\nabla\phi$ and $\chi = e\phi/KT_e$, Eq. (8.137) becomes

$$\frac{\partial}{\partial x}(\chi - \ln n) - \frac{1}{2}\frac{m}{KT_e}\frac{\partial}{\partial x}\langle u^2 \rangle = 0 \tag{8.139}$$

Integrating, setting $n = n_0 + \delta n$, and using the dimensionless units Eq. (8.134), we have

$$\frac{1}{2}\langle u^2 \rangle = \frac{1}{4}|u|^2 = \chi - \ln(1 + \delta n) \cong \chi - \delta n \qquad (8.140)$$

We must now eliminate χ by solving the cold-ion Eqs. (8.103) and (8.104). Since we are now using the electron variables Eq. (8.134), and since $\Omega_p = \epsilon\omega_p$, $v_s = \epsilon(KT_e/m)^{1/2}$, where $\epsilon \equiv (m/M)^{1/2}$, the dimensionless form of the ion equations is

$$\frac{1}{\epsilon}\frac{\partial u_i}{\partial t} + u_i\frac{\partial u_i}{\partial x} + \frac{\partial \chi}{\partial x} = 0 \qquad (8.141)$$

$$\frac{1}{\epsilon}\frac{\partial \delta n_i}{\partial t} + \frac{\partial}{\partial x}[(1 + \delta n_i)u_i] = 0 \qquad (8.142)$$

Here we have set $n_i' = (n_0 + \delta n_i)/n_0 = 1 + \delta n_i'$ and have dropped the prime. If the soliton is stationary in a frame moving with velocity V, the perturbations depend on x and t only through the combination $\xi = x - x_0 - Vt$. Thus

$$\frac{\partial}{\partial x} = \frac{\partial}{\partial \xi} \qquad\qquad \frac{\partial}{\partial t} = -V\frac{\partial}{\partial \xi}$$

and we obtain after linearization

$$-\frac{V}{\epsilon}\frac{\partial u_i}{\partial \xi} + \frac{\partial \chi}{\partial \xi} = 0 \qquad u_i = \frac{\epsilon}{V}\chi \qquad (8.143)$$

$$-\frac{V}{\epsilon}\frac{\partial \delta n_i}{\partial \xi} + \frac{\partial u_i}{\partial \xi} = 0 \qquad \delta n_i = \frac{\epsilon}{V}u_i \qquad (8.144)$$

From this and the condition of quasineutrality for the slow motions, we obtain

$$\delta n_e = \delta n_i = \frac{\epsilon^2}{V^2}\chi. \qquad (8.145)$$

Substituting for χ Eq. (8.140), where δn is really δn_e, we find

$$\delta n_e = \frac{1}{4}|u|^2\left(\frac{V^2}{\epsilon^2} - 1\right)^{-1}. \qquad (8.146)$$

Upon inserting this into Eq. (8.136), we finally have

$$i\frac{\partial u}{\partial t} + \frac{3}{2}\frac{\partial^2 u}{\partial x^2} + \left[\Delta - \frac{1}{8}\left(\frac{V^2}{\epsilon^2} - 1\right)^{-1}|u|^2\right]u = 0. \qquad (8.147)$$

Comparing with Eq. (8.122), we see that this is the nonlinear Schrödinger equation if Δ can be neglected and

$$p = \frac{3}{2} \qquad q = -\frac{1}{8}\left(\frac{m/M}{V^2 - m/M}\right) \tag{8.148}$$

Finally, it remains to show that p and q are related to the group dispersion and nonlinear frequency shift as stated in Eq. (8.123). This is true for $V^2 \ll m/M$. In dimensionless units, the Bohm–Gross dispersion relation (4.30) reads

$$\omega'^2 = 1 + \delta n' + 3k'^2 \tag{8.149}$$

where $k' = k\lambda_{\mathrm{D}}$, and we have normalized ω to ω_{p0} (the value outside the density cavity). The group velocity is

$$v'_g = \frac{d\omega'}{dk'} = \frac{3k'}{\omega'} \tag{8.150}$$

so that

$$\frac{dv'_g}{dk'} = \frac{3}{\omega'} \cong 3$$

and

$$p = \frac{1}{2}\frac{dv'_g}{dk'} = \frac{3}{2} \tag{8.151}$$

For $V^2 \ll c^2$, Eq. (8.146) gives

$$\delta n' = -\frac{1}{4}\left|u'\right|^2$$

so that Eq. (8.144) can be written

$$\omega'^2 = 1 - \frac{1}{4}\left|u'\right|^2 + 3k'^2 \tag{8.152}$$

Then

$$2\omega'\,d\omega' = -\frac{1}{4}d\left|u'\right|^2$$

$$\delta\omega' \propto \frac{d\omega'}{d|u'|^2} \cong -\frac{1}{8} \tag{8.153}$$

From Eq. (8.148), we have, for $V^2 \ll \epsilon^2$,

$$q \cong \frac{1}{8} = -\frac{d\omega'}{d|u'|^2}$$

as previously stated.

If the condition $V^2 \ll \epsilon^2$ is not satisfied, the ion dynamics must be treated more carefully; one has coupled electron and ion solitons which evolve together in time. This is the situation normally encountered in experiment and has been treated theoretically.

In summary, a Langmuir-wave soliton is described by Eq. (8.125), with $p = 3/2$ and $q = 1/8$ and with $\psi(x, t)$ signifying the low-frequency part of $u(x, t)$, where u, x, and t are all in dimensionless units. Inserting the exp $(-i\omega_0 t)$ factor and letting x_0 and θ_0 be zero, we can write Eq. (8.125) as follows:

$$u(x,t) = 4A^{1/2}\mathrm{sech}\left[\left(\frac{2A}{3}\right)^{1/2}(x - Vt)\right] \exp\left\{-i\left[\left(\omega_0 + \frac{V^2}{6} - A\right)t - \frac{V}{3}x\right]\right\}$$

$$(8.154)$$

The envelope of the soliton propagates with a velocity V, which is so far unspecified. To find it accurately involves simultaneously solving a Korteweg–de Vries equation describing the motion of the density cavity, but the underlying physics can be explained much more simply. The electron plasma waves have a group velocity, and V must be near this velocity if the wave energy is to move along with the envelope. In dimensionless units, this velocity is, from Eq. (8.150),

$$V \cong v'_g = \frac{3k'}{\omega'} \cong 3k' \qquad (8.155)$$

The term $i(V/3)x$ in the exponent of Eq. (8.154) is therefore just the ikx factor indicating propagation of the waves inside the envelope. Similarly, the factor $-i(V^2/6)t$ is just $-i(3/2)k'^2 t'$, which can be recognized from Eq. (8.149) as the Bohm–Gross frequency for $\delta n' = 0$, the factor ½ coming from expansion of the square root. Since $\omega_0 \simeq \omega_p$, the terms $\omega_0 + (V^2/6)$ represent the Bohm–Gross frequency, and A is therefore the frequency shift (in units of ω_p) due to the cavity in $\delta n'$. The soliton amplitude and width are given in Eq. (8.154) in terms of the shift A, and the high-frequency electric field can be found from Eq. (8.138).

Cavitons have been observed in devices similar to that of Fig. 8.16. Figures 8.29 and 8.30 show two experiments in which structures like the envelope solitons discussed above have been generated by injecting high-power rf into a quiescent plasma. These experiments initiated the interpretation of laser-fusion data in terms of "profile modification," or the change in density profile caused by the ponderomotive force of laser radiation near the critical layer, where $\omega_0 \simeq \omega_p$, ω_0 being the laser frequency.

Fig. 8.29 A density cavity, or "caviton," dug by the ponderomotive force of an rf field near the critical layer. The high-frequency oscillations (not shown) were probed with an electron beam. [From H. C. Kim, R. L. Stenzel, and A. Y. Wong, *Phys. Rev. Lett.* **33**, 886 (1974).]

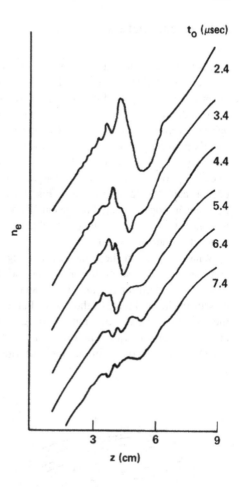

t_0 (μsec)

Problems

8.21   Check that the relation between the frequency shift A and the soliton amplitude in Eq. (8.154) is reasonable by calculating the average density depression in the soliton and the corresponding average change in ω_p. (Hint: Use Eq. (8.146) and assume that the sech^2 factor has an average value of $\simeq \frac{1}{2}$ over the soliton width.)

8.22   A Langmuir-wave soliton with an envelope amplitude of 3.2 V peak-to-peak is excited in a 2-eV plasma with $n_0 = 10^{15}$ m^{-3}. If the electron waves have $k\lambda_D = 0.3$, find (a) the full width at half maximum of the envelope (in mm), (b) the number of wavelengths within this width, and (c) the frequency shift (in MHz) away from the linear-theory Bohm–Gross frequency.

8.23   A density cavity in the shape of a square well is created in a one-dimensional plasma with $KT_e = 3$ eV. The density outside the cavity is $n_0 = 10^{16}$ m^{-3}, and that inside is $n_i = 0.4 \times 10^{16}$ m^{-3}. If the cavity is long enough that boundary resonances can be ignored, what is the wavelength of the shortest electron plasma wave that can be trapped in the cavity?

8.9   Reconnection

In a collisionless plasma, electrons and ions gyrate around magnetic field lines and are tied to them. If these field lines are brought close together, the plasma will become very dense, and hence highly collisional. In that case, plasma can diffuse across field lines and change the magnetic geometry. Due to kinetic effects, this can happen even without collisions. Reconnection of magnetic field lines occurs in the earth's field on the night side, as shown in Fig. 8.31. The plasma of the solar wind pushes the earth's dipole field away from the sun into a magnetotail, where field lines in opposite directions are brought close to one another. The magnetic configuration can change, as shown in Fig. 8.32. In the magnetotail, the left-hand part becomes a loop connected to the earth's poles, while the right-hand part breaks off to connect to the interplanetary field. At the magnetopause, the lines break and connect to the solar wind.

What happens in the thin reconnecting layer shown in Fig. 8.33 is extremely complicated because both collisional and collisionless damping mechanisms, as well as finite-Larmor-radius effects can occur there. There are many models for the conditions in the layer. In the Sweet-Parker model, B and entrained n diffuse into the layer by resistive diffusion, building up a high density there which must escape at a large velocity v_0 of the order of the Alfvén speed. H. Petschek showed that MHD shocks can occur in fast reconnection, but these do not stay long enough to

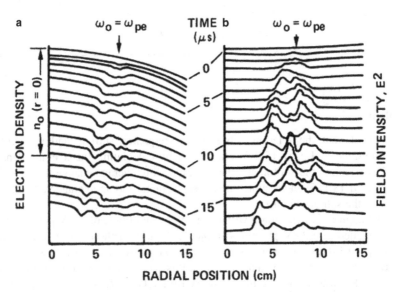

Fig. 8.30 Coupled electron and ion wave solitons. In (**a**) the low-frequency density cavities are seen to propagate to the left. In (**b**) the high-frequency electric field, as measured by wire probes, is found to be large at the local density minima. [From H. Ikezi, K. Nishikawa, H. Hojo, and K. Mima, *Plasma Physics and Controlled Nuclear Fusion Research*, 1974, *II*, 609, International Atomic Energy Agency, Vienna, 1975.]

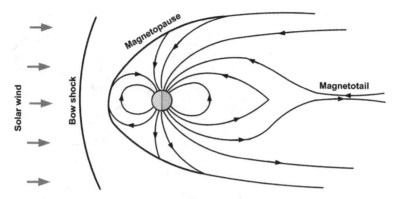

Fig. 8.31 Schematic diagram of the earth's magnetic field, as it is blown by the solar wind into a magnetotail

Fig. 8.32 Reconnection of field lines

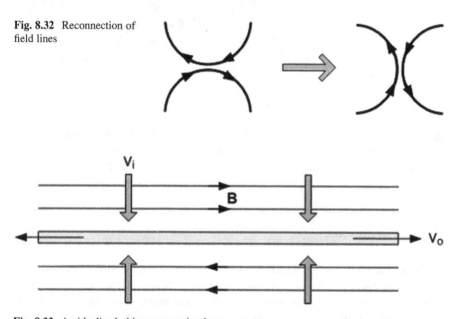

Fig. 8.33 An idealized, thin reconnection layer

have much effect. With the advent of fast computers, W. Daughton has studied three-dimensional reconnection and the generation of turbulence. Reconnection occurs on the surface of the sun itself. These regions manifest themselves as sunspots.

A controlled experiment on reconnection was built at Princeton by M. Yamada and H. Ji, with theoretical support by R. Kulsrud. A diagram of the magnetic field in their machine is shown in Fig. 8.34. The two large rings shown vertically carry current to generate the field lines shown. As the current in the rings is decreased, the

Fig. 8.34 Schematic of the field lines in Yamada's reconnection machine

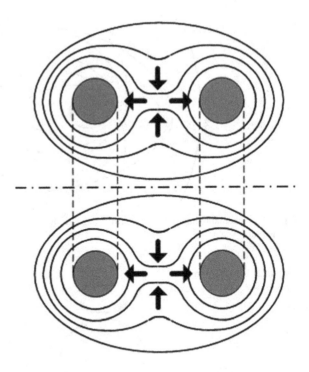

field lines move in the direction of the arrows, and a reconnection sheet of horizontal cross section forms. If the current is increased, the field lines move opposite to the arrows, and the reconnection sheet has a vertical cross section.

8.10 Turbulence

In Chap. 4 we treated waves in a plasma by linearizing the equations of motion, thus limiting the discussion to waves of small amplitude. Sinusoidal waves cannot grow indefinitely, however; nonlinear effects change their shapes, limit their amplitudes, and ultimately turning them into random turbulence. Figure 8.35 shows a jet of air forming vortices and then breaking into turbulence. In pipes carrying a liquid, the flow is smooth at low velocities, when the velocity has a smooth profile from slow at the edge to fast at the center, as seen in Fig. 8.36 (top). At high velocities, the flow breaks up into turbulence, as in Fig. 8.36 (bottom). The flow slows down, and energy goes into noise. Nonlinearity can happen in a plasma in other ways. For instance, in the plasma oscillation of Fig. 4.2, the amplitude can become so large that the excursion of an electron sheet can overlap the next wavelength. In early experiments in which a current was drawn between an anode and a cathode in a magnetized cylinder, probes inserted into the plasma always detected turbulent fluctuations like those shown in Fig. 8.37. The basic waves that grew into this

Fig. 8.35 Turbulence in a jet of air, made visible by smoke. [from M. Van Dyke, *An Album of Fluid Motion*, The Parabolic Press, Stanford, CA (1982)]

LAMINAR FLOW

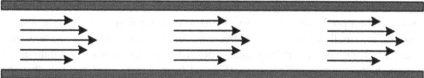

TURBULENT FLOW

Fig. 8.36 Turbulence in a water pipe

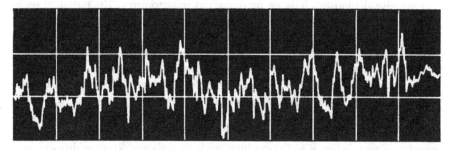

Fig. 8.37 Turbulent potential fluctuations observed in a fusion plasma (Data by the author)

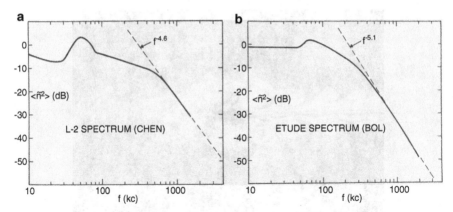

Fig. 8.38 Turbulence spectra in (**a**) the L-2, a reflex arc (2000 G, 10^{12} cm^{-3}); and (**b**) the Etude, a stellarator (6700 G, 10^{13} cm^{-3}). Here, frequency is measured rather than k

nonlinear state were not found for years. We now know of many instabilities that can cause this.

A turbulent field can be decomposed into a spectrum of eddies of different sizes. After excitation, the distribution of sizes—the turbulent spectrum $S(k)$—will settle down to a form predictable by theory. In two-dimensional systems, such as a sheet of graphene, small eddies will coalesce into larger eddies. In general, however, large eddies will break up into small eddies with a given size distribution. In ordinary hydrodynamics, Kolmogoroff found from dimensional analysis that the spectrum should vary as $k^{-5/3}$. In laboratory plasmas, the turbulence is driven by instabilities such as the Rayleigh-Taylor or drift-wave instabilities, and electron motions are dominated by their $\mathbf{E} \times \mathbf{B}$ drifts. Early experiments at Princeton showed consistent spectra varying as k^{-5} even in entirely different machines, as shown in Fig. 8.38. Because plasma moves very differently from normal fluids, it is not surprising that the exponent of k is quite different. The boundary-dependent peaks in Fig. 8.38 are from the low-frequency instabilities driving the turbulence. By contrast, the spectrum in an early tokamak (PLT) at Princeton, shown in Fig. 8.39, varies as $k^{-2.8}$.

In the modern era, electrostatic probes cannot be inserted into a tokamak, and non-invasive diagnostics must be used, such as reflectometry. The turbulent spectrum will vary greatly with position because of sawtooth oscillations in the interior and Edge Localized Modes near the boundary. Generalizations such as those from the early days can no longer be made. With the advent of fast computers, turbulence in the three-dimensional geometry of a tokamak can be simulated. Figure 8.40 shows such a computation. The magnetic field lines are no longer in the well-ordered magnetic island structure but are completely scrambled.

Basic experiments on plasma turbulence have been done on Walter Gekelman's LAPD (Large Area Plasma Device) machine at the University of California, Los Angeles. Shown in Fig. 8.41 the machine produces a plasma 60 cm in diameter and 17 m long, ionized by electrons emitted from a hot cathode and accelerated through

Fig. 8.39 Turbulence spectrum in the PLT tokamak, taken by Mazzucato at three radial positions

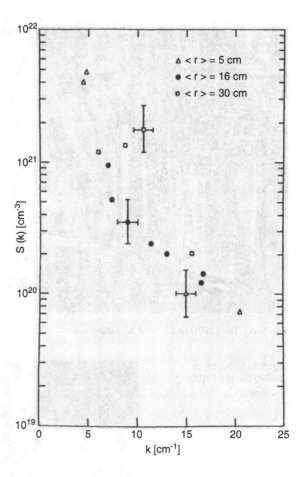

Fig. 8.40 Computer simulation of turbulent field lines in a tokamak. [Courtesy of W.W. Lee, Princeton Plasma Physics Lab]

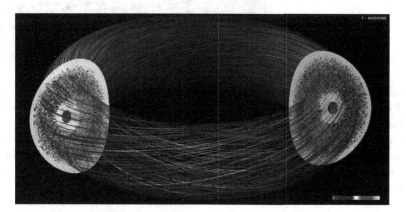

Fig. 8.41 The LAPD machine of W. Gekelman

Fig. 8.42 Flux ropes observed in the LAPD machine

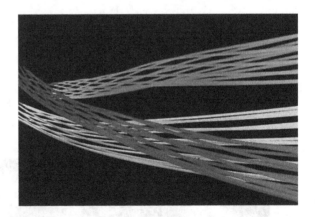

a mesh anode. Pulsed for 10–20 ms at 1 Hz, the plasma reaches $n \leq 4 \times 10^{18}$ m^{-3} and $KT_e \leq 8$ eV in $B \leq 0.3$ T (3 kG). The machine is ideally suited for studies of Alfvén waves, which require long length and high B-field.

Fig. 8.42 gives an example of nonlinear effects observed in LAPD which would be hard to predict theoretically. By measuring the currents' B-fields, a current in the plasma was found to filament into curving flux ropes which themselves are composed of smaller filaments.

8.11   Sheath Boundaries

Laboratory plasmas are not infinitely long cylinders, as many theories assume. Plasmas of finite length in a B-field have sheaths at the endplates which control the density profiles. As an example, consider the discharge in Fig. 8.43, which is short enough that gradients in the z direction can be neglected but long enough that the end sheaths do not overlap. The magnetic field is strong enough to confine the electrons to move rapidly only in the z direction, and to move in the r direction only via collisions. But the B-field is weak enough that we can neglect the curvature of the ion orbits inside the radius a. We now describe the *Simon short-circuit effect*, which causes the electrons to behave as if they could cross the magnetic field rapidly although they actually do not.

In Fig. 8.44a, the top of each figure is the radial boundary; the bottom is the interior of the discharge. We assume that radiofrequency power is applied at $r = a$, so that initially n is higher in tube ① than in tube ②. The sheath at the endplate adjusts itself to allow the electron loss to be equal to the ion loss at the Bohm velocity c_s at each r, thus keeping the plasma neutral. The sheath drop $\phi_p(r)$ is therefore given by

$$n\left(\frac{KT_e}{2\pi m}\right)^{\!\!1/2} e^{-e\phi_p/KT_e} = n\left(\frac{KT_e}{M}\right)^{\!\!1/2} \quad \therefore \quad \frac{e\phi_p}{KT_e} = \ln\left(\frac{M}{2\pi m}\right)^{\!\!1/2}, \qquad (8.156)$$

where the bracket on the l.h.s. is the random electron velocity in one dimension, and the one on the r.h.s. is the acoustic speed. Since ϕ_p is independent of n, the sheath drop is normally independent of radius. However, when ions are injected at $r = a$, they diffuse inward from tube ① to tube ②. Electrons can't follow them, but the sheath drop on tube ② can increase slightly to trap more electrons to keep tube ② neutral. The sheath on tube ② is becomes thicker than the one on tube ①, as shown in Fig. 8.44a. This mechanism allows electrons to fall into thermal equilibrium and follow the Boltzmann relation for over all radii:

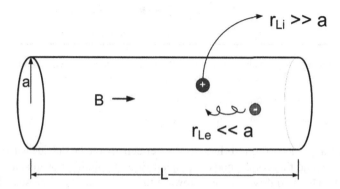

Fig. 8.43 Model of a finite-length laboratory plasma

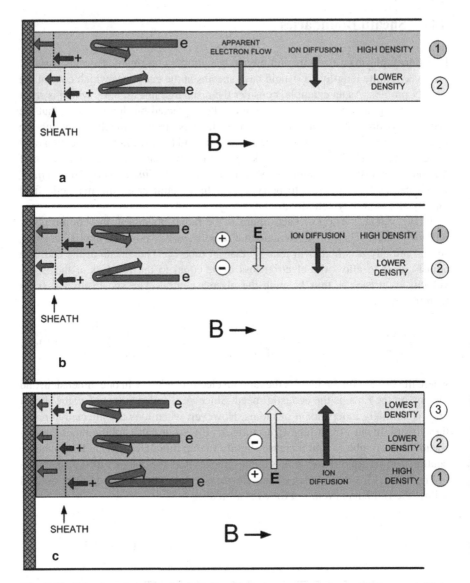

Fig. 8.44 Behavior of sheaths (**a**) initially, (**b**) during approach to equilibrium, and (**c**) in equilibrium

$$n(r) = n_0 e^{e\phi(r)/KT_e(r)}. \qquad (3.73)$$

During the approach to equilibrium, the higher density on the outside causes ϕ to be higher there, and this corresponds to a radial E-field pointing inwards, as shown in Fig. 8.44b. The field drives the ions toward the center at a fast rate scaled to KT_e. There cannot be an E-field along **B** because that would cause a large electron current, and the sheaths have stopped that. Therefore, the ions cannot leave the center except at their low thermal velocities. The ion density builds up there until an outward E-field following Eq. (3.73) drives outward any ions that are created in the interior. If all ions are created at the boundary, $n(r)$ would be flat. The equilibrium sheaths and fields are shown in Fig. 8.44c. A 2011 theory by Curreli and Chen showed that a discharge with the geometry of Fig. 8.43 with sheaths at the endplates will fall into an equilibrium with a "universal" density profile which is independent of pressure and plasma radius a.

Chapter 9
Special Plasmas

9.1 Non-Neutral Plasmas

9.1.1 Pure Electron Plasmas

The concepts of plasma physics have so far been couched in terms of well-behaved, quasineutral plasmas, but there are other plasmas with special properties. Particle accelerators have a single species, but the kinematic effects are so large that the collective effects are not important. It is possible to generate single-species plasmas at low densities such that the electric fields are manageable. Ronald Davidson, Malmberg and O'Neil, Dan Dubin, and others have developed this interesting topic. Consider an infinite cylinder of electrons of uniform density n_0 in a uniform coaxial magnetic field $\mathbf{B}$ (Fig. 9.1). A large electric field $\mathbf{E} = E_r(r)\hat{\mathbf{r}}$ will arise, (where E_r is negative), and a typical fluid element of electrons will drift in a circular orbit, since everything is azimuthally symmetric. Those with small, off-axis orbits will drift around the axis in "diocotron" orbits, as shown at the right. We wish to calculate the rotation frequency ω_r:

$$\omega_r(r) \equiv v_\theta(r)/r. \tag{9.1}$$

In equilibrium, the inward and outward forces will balance:

$$
\begin{aligned}
mv_\theta^2(r)/r &= -eE_r(r) - ev_\theta(r)B \\
mr\omega_r^2 &= -eE_r - ev_\theta B = -eBr\omega_r - eE_r
\end{aligned}
\tag{9.2}
$$

The radial E-field is found from Poisson's equation:

© Springer International Publishing Switzerland 2016
F.F. Chen, *Introduction to Plasma Physics and Controlled Fusion*,
DOI 10.1007/978-3-319-22309-4_9

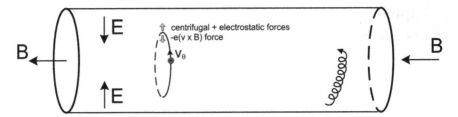

Fig. 9.1 A non-neutral plasma of electrons. (Adapted from R.C. Davidson, *Physics of Nonneutral plasmas* (Addison-Wesley, Redwood City, CA, 1990)).

$$\frac{1}{r}\frac{d}{dr}(rE_r) = -en_e/\varepsilon_0$$

$$rE_r = \frac{-en_e}{\varepsilon_0}\int_0^r r'dr' = \frac{-en_e r^2}{2\varepsilon_0} \tag{9.3}$$

$$E_r = -\frac{r}{2}\frac{en_e}{\varepsilon_0} = -\frac{r}{2}\frac{m}{e}\omega_p^2 \qquad r \le a$$

Equation (9.2) then becomes

$$\omega_r^2 + \omega_c\omega_r + \frac{eE_r}{mr} = \omega_r^2 + \omega_c\omega_r - \tfrac{1}{2}\omega_p^2 = 0$$

$$\omega_r = \tfrac{1}{2}\omega_c\left[-1 \pm \left(1 - 2\omega_p^2/\omega_c^2\right)^{1/2}\right]. \tag{9.4}$$

Since ω_r is independent of radius, we see that such an pure electron plasma rotates as a solid body. When $\omega_p^2/\omega_c^2 > \tfrac{1}{2}$, there is no solution; otherwise, there are two solutions. For $2\omega_p^2/\omega_c^2 \ll 1$, expanding the square root yields the frequencies $\omega_r \approx -\omega_c$, and the lower frequency

$$\omega_r \approx \omega_p^2/2\omega_c \equiv \omega_D. \tag{9.5}$$

Called the *diocotron* frequency, ω_D has an orbit like the one at the right in Fig. 9.1.

With appropriate modifications, these equations can also describe pure ion plasmas. Both types of single-species plasmas have been produced and studied in the laboratory.

9.1.2 Experiments

A device called a Malmberg-Penning trap, shown in Fig. 9.2, is commonly used to study single-species plasmas. This useful device evolved from the work of

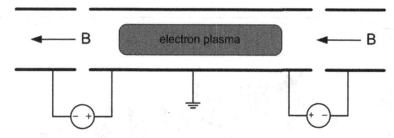

Fig. 9.2 A Malmberg-Penning trap for electrons

Wolfgang Paul, Hans Dehmelt, and Norman Ramsey,[1] who won the Nobel Prize in 1989. The trap consists of three coaxial, conducting cylinders, with the ends biased negatively to trap electrons and positively to trap ions. Conservation of canonical angular momentum can be used to show that the particles are trapped unless they make a collision with a stray neutral atom or with a error field due to imperfect machining of the container. The physical reason for this can be seen from Fig. 5.16, which shows that like-particle collisions do not cause diffusion.

As an example of what can be done with these plasmas, F. Anderegg et al. studied diffusion due to long-range interactions. Recall that Spitzer diffusion (Eq. (5.71)) had an arbitrary cutoff related to the Debye length. Ion-ion diffusion was quantified in an ion trap with LIF (laser-induced fluorescence) diagnostics. In the data shown in Fig. 9.3, one sees that the measured diffusion coefficient is an order of magnitude larger than "classical" but is closer when collisions in large-orbit $\mathbf{E} \times \mathbf{B}$ drifts are included.

9.2   Solid, Ultra-Cold Plasmas

By freezing the plasma in an ion trap to sub-Kelvin temperatures, it is possible to create liquid and solid plasmas. As the thermal motions of the particles decrease, their thermal energies become comparable to the Coulomb energy W_c between particles. This condition can be quantified by defining a coupling parameter Γ, the ratio between the average W_c and KT. Two singly charged particles separated by a distance a have a Coulomb energy e^2/a. Hence

$$\Gamma = e^2/aKT. \tag{9.6}$$

---

[1] The author's thesis adviser.

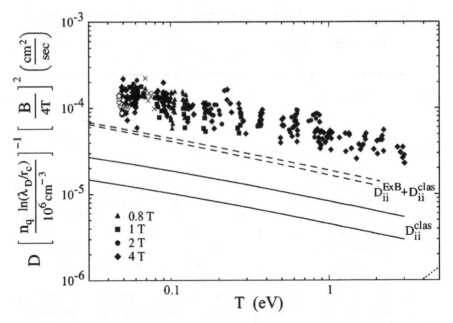

Fig. 9.3 Ion-ion diffusion coefficient compared with theory. The double lines show the expected range. (Adapted from F. Anderegg et al., Phys. Rev. Lett. **78**, 2128 (1997))

The average interparticle distance a should vary as $n^{1/3}$, but the coefficient depends on the shape of the enclosing volume. The value of a can be approximated by the Wigner-Seitz radius:

$$a = (3/4\pi n)^{1/3} \approx 0.62n^{1/3}. \tag{9.7}$$

When Γ, as defined by Eqs. (9.6) and (9.7), becomes large, the plasma first turns into a liquid as fluid elements become highly correlated. At $\Gamma > 174$, the plasma becomes a solid crystal such as the one shown in Fig. 9.4. To achieve such a large value of Γ, the plasma is cooled to a temperature of 10 mK (10 milliKelviin, or 0.01 K) by laser cooling. In this process, a laser is tuned to a frequency just below that of a transition in the atom. A laser photon is absorbed only by atoms moving rapidly toward the laser with a blue Doppler shift. The atom's momentum is lowered by that of the photon. The photon is later re-emitted in a random direction at a lower frequency, so that the fastest atoms are removed, and the plasma is cooled.

The Penning trap is a useful device in which unusual plasmas have been created. For instance, a positronium plasma has been formed, consisting of positrons and electrons, with no ions. Also possible is the creation of anti-hydrogen, whose atoms consist of negative antiprotons and positrons.

Fig. 9.4 A plasma crystal (from C.F. Driscoll et al., Physica C: Superconductivity **369**, Nos. 1–4, 21 (2002).) This is produced in a Penning trap such as the one in Fig. 9.2

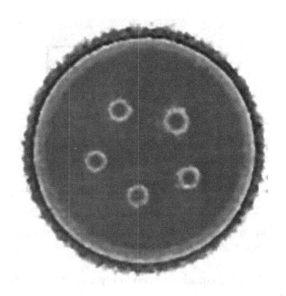

9.3 Pair-ion Plasmas

One would expect that a plasma with equal masses would behave quite differently from a normal plasma with slow ions immersed in a sea of fast electrons. Though it is possible to make a positronium plasma, the recombination rate is so fast that there is no time to do experiments. Fullerenes, stable molecules of 60 carbon atoms arranged in a hollow sphere, move and recombine slowly because of the large mass. It is thus possible to produce a long-lived pair-ion plasma with C_{60}. In Fig. 9.5, neutral C_{60} is injected from an oven. A ring of fast electrons up to 150 eV driven through the C_{60} forms C_{60}^+ ions by ionization and C_{60}^- ions by attachment. A magnetic field confines the electrons but allows the heavy ions to diffuse into the center to form a fullerene pair-ion plasma. The plasma then passes through a hole into a chamber for experimentation.

Consider a singly charged pair-ion plasma with a common mass M and temperature KT. The equations of motion for the ions of charge $+$ or $-$ are

$$Mn_0 \frac{dv_\pm}{dt} = \pm n_0 E - \nabla p. \tag{9.8}$$

Assuming waves of the form $\exp[i(kz - \omega t)]$ and $E = -\nabla \phi = -ik\phi$, we have

$$-i\omega M n_0 v_\pm = \mp ikn_0 e\phi - 3KTikn_\pm, \tag{9.9}$$

$$v_\pm = \frac{k}{\omega}\left(\pm \frac{e\phi}{M} + c_s^2 \frac{n_\pm}{n_0}\right), \quad \text{where} \quad c_s^2 \equiv 3KT/M. \tag{9.10}$$

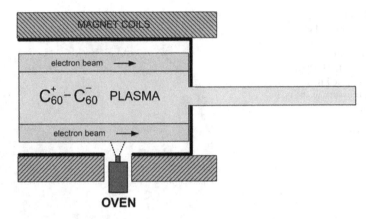

Fig. 9.5 Schematic of a fullerene plasma source. Adapted from W. Oohara and R. Hatakeyama, Phys. Plasmas **14**, 055704 (2007)

From the equation of continuity $\partial n/\partial t + \nabla \cdot (nv) = 0$, we have

$$-i\omega n_\pm + ikn_0 v_\pm = 0, \qquad \frac{n_\pm}{n_0} = \frac{k}{\omega} v_\pm. \qquad (9.11)$$

Using this in Eq. (9.9), we find

$$-i\omega v_\pm = \mp ik(e/M)\phi - ik(k/\omega)c_s^2 v_\pm,$$

$$v_\pm\left(1 - \frac{k^2}{\omega^2}c_s^2\right) = \pm\frac{k}{\omega}\frac{e\phi}{M}. \qquad (9.12)$$

There are two solutions, depending on whether ϕ is zero. If ϕ is zero, the $+$ and $-$ ions move together without creating an E-field, and we have an ordinary sound wave:

$$\omega^2/k^2 = c_s^2. \qquad (9.13)$$

If ϕ is not zero, Eq. (9.12) says that $v_+ = -v_-$. Then

$$\frac{n_\pm}{n_0} = \frac{k}{\omega} v_\pm = \pm k^2 \frac{e\phi/M}{\omega^2 - k^2 c_s^2}. \qquad (9.14)$$

Poisson's equation then yields

$$\varepsilon_0 k^2 \phi = e(n_+ - n_-) = en_0 k^2 \frac{2k^2 e\phi/M}{\omega^2 - k^2 c_s^2}$$

$$1 = \frac{en_0}{\varepsilon_0} \frac{2e/M}{\omega^2 - k^2 c_s^2} = \frac{2\Omega_p^2}{\omega^2 - k^2 c_s^2},$$

$$\omega^2 = k^2 c_s^2 + 2\Omega_p^2 \tag{9.15}$$

where Ω_p is the ion plasma frequency $(ne^2/\varepsilon_0 M)^{1/2}$. Thus, a pair-ion plasma supports an ion acoustic wave and an ion plasma wave. The latter (Eq. (9.15)) is the analog of the Bohm-Gross wave of Eq. (4.30), but with a factor two because of the two ion species. These two waves are connected by an intermediate-frequency wave which can be derived only with kinetic theory.

9.4 Dusty Plasmas

In Chap. 5 we considered three-component plasmas consisting ions, electrons, and neutral atoms in partially ionized plasmas. In general, there can be contaminants of macroscopic size, "dust", made of other atomic species. In outer space, comet tails are dusty plasmas, as are some nebulas, such as the Orion nebula, and planetary rings, such as the one on Saturn. On earth, dusty plasmas can exist in flames; rocket exhausts; thermonuclear explosions; atmospheric-pressure plasmas (Sect. 9.6); and, importantly, in plasma processing (see Chap. 10). We shall find that dust has two main effects. First, it introduces low-frequency waves in the motions of the charged dust. Second, it changes the quasineutrality condition so that n_e is no longer equal to Zn_i, thus modifying the normal waves in the plasma.

Dust grains have sizes from tens of nanometers to hundreds of microns. Since electrons impinge on them much more often than ions do, the grains will have a negative surface potential V_s. Consider a spherical grain of radius a and charge $q < 0$. The capacitance of the sphere (with distant walls) is

$$C = 4\pi\varepsilon_0 a. \tag{9.16}$$

The surface charge q is then

$$q = CV_s = 4\pi\varepsilon_0 a V_s. \tag{9.17}$$

The value of either q or V_s depends on the Debye length λ_D in the background plasma.

If λ_D is $\ll a$, the grain is a small, isolated particle like a spherical Langmuir probe (Sect. 8.2.5). An ion will be attracted to the grain and will either strike it or orbit around it, depending on its orbital angular momentum around the grain. Enough electrons will be collected to ensure that the net current is zero. The required value of V_s is the "floating potential" of probe theory, which we need not discuss here. On the other hand, if λ_D is $\gg a$, the grains are a third charged component of the plasma along with the ions and electrons. For instance, in a Cs

Fig. 9.6 Langmuir probe
traces of a Q-machine
plasma with (*bottom trace*)
and without (*top trace*) dust.
(A. Barkan, N. D'Angelo,
and R. L. Merlino, Phys.
Rev. Lett. **73**, 3093 (1994))

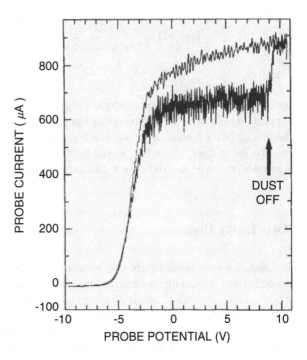

plasma with $KT = 0.21$ eV, λ_D is 34 μm at $n = 10^{10}$ cm^{-3} while $a \approx 1$ μm, so that $\lambda_D \gg a$ is satisfied.

Assume the latter condition, and let the dust grains have a charge $Z_d < 0$ and a density n_d. Charge neutrality requires

$$n_i = n_e - Z_d n_d. \tag{9.18}$$

Thus, the "electron" density is lowered by the presence of dust. This has been observed by Barkan et al. in a potassium Q-machine (Fig. 4.14). The dust, consisting of aluminum silicate particles of 5-μm average radius, was dropped into the plasma through a mesh on a rotating cylinder surrounding the plasma. The dust density was about 5×10^4 cm^{-3}. Figure 9.6 shows Langmuir probe traces of the plasma with and without the dust. In the presence of dust, the electron saturation current is seen to be lowered by the slow velocities of the heavy dust.

Studies of how dust is charged include many minor effects too detailed to be described here; for instance, secondary electron emission upon ion or electron impact, photoemission of electrons (often the dominant mechanism in cosmic plasmas), and field emission. It is not surprising that dust particles can arrange themselves in crystal arrays. Figure 9.7 shows how a picture of such a crystal can be obtained. A plasma is formed by RF applied to the bottom electrode, and dust is introduced by a shaker that is not shown; The dust is illuminated by a laser beam spread into a plane by a cylindrical lens, and a camera records the dust through a hole in the upper, grounded electrode. The dust is suspended above the RF

Fig. 9.7 Schematic of a setup to photograph the motion of dust grains in an RF plasma. The laser beam is spread into a sheet by a lens, shown rotated 90°. The dust dispenser is not shown. (Adapted from G. E. Morfill and H. Thomas, J. Vac. Sci. Technol. A **14**, 490 (1996))

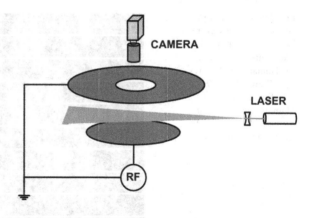

Fig. 9.8 Picture of a dust crystal taken by Morfill et al. (*loc. cit.*) with the apparatus shown in Fig. 9.7. The grains are trapped in the sheath on the lower electrode

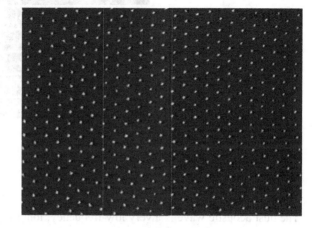

electrode, which is negative relative to the plasma. A still picture of a dust crystal array is shown in Fig. 9.8.

By adding straight barriers to make a channel on the lower electrode, linear arrays of dust particles can be produced, as shown in Fig. 9.9. These can be pushed by the laser from the right. Only the first particle is pushed, and the others maintain the crystal spacing.

When the dust density is sufficiently high, the charged dust in a plasma can be considered as an additional fluid component exhibiting collective effects. The presence of charged dust in a plasma modifies all of the wave modes in Chap. 4, even with a DC magnetic field, and introduces new "dust waves" involving the motions of the charged dust. The dispersion relations for the plasma waves are modified through the quasi-neutrality condition (Eq. (9.18)), an example of which is the "dust ion-acoustic wave", which is analyzed below. We begin by considering the dust acoustic wave, which is a very low frequency, longitudinal, compressional wave involving the dynamics of the dust particles in a plasma.

Fig. 9.9 A linear array of
dust particles pushed by a
laser from the right. The
frames are 100 ms apart.
(A. Homann et al., Phys.
Rev. E **56**, 7138 (1997);
P.K. Shukla and A. A.
Mamun, *Intro. to Dusty
Plasma Physics* (IOP Press,
Bristol, UK, 2002.)

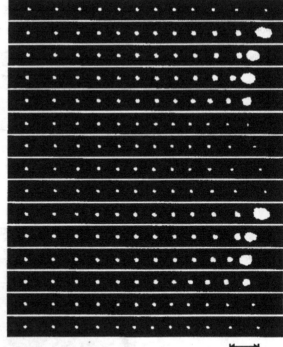

1 mm

9.4.1   Dust Acoustic Waves

The dust acoustic wave is a very low-frequency, longitudinal compressional wave
involving the motions of the dust particles. Because the heavy dust moves more
slowly than the ions and electrons, the latter have time to relax into Maxwellian
distributions:

$$n_e = n_{e0}\exp(e\phi/KT_e) = n_{e0}(1 + e\phi/KT_e + \ldots), \quad n_{e1} = n_{e0}(e\phi/KT_e)$$

$$n_i = n_{i0}\exp(-e\phi/KT_i) = n_{i0}(1 - e\phi/KT_i + \ldots), \quad n_{i1} = n_{i0}(-e\phi/KT_i)$$

$$(9.19)$$

Using the subscript d for dust, we can write the 1-D dust equations of motion and
continuity as:

$$\frac{\partial \mathbf{v}_d}{\partial t} = -\frac{q_d}{M_d}\nabla\phi - v_{thd}^2\frac{\nabla n_{d1}}{n_{d0}} \tag{9.20}$$

$$\frac{\partial n_{d1}}{\partial t} + n_{d0}\nabla \cdot \mathbf{v}_d = 0, \tag{9.21}$$

$$\text{where} \quad v_{thd}^2 \equiv 3KT_d/M_d, \tag{9.22}$$

and the subscript 1 has been dropped from v_{d1} since $v_{d0}=0$. Poisson's equation is

$$\nabla^2\phi = \frac{1}{\varepsilon_0}(en_{e1} - en_{i1} - q_d n_{d1}). \tag{9.23}$$

The dust charge is taken to be constant. To express n_{d1} in terms of ϕ, take the time derivative of Eq. (9.21) and use Eq. (9.20):

$$\frac{\partial^2 n_{d1}}{\partial t^2} = -n_{d0}\nabla \cdot \frac{\partial \mathbf{v}_d}{\partial t} = -n_{d0}\nabla \cdot \left(-\frac{q_d}{M_d}\nabla\phi - v_{thd}^2 \frac{\nabla n_{d1}}{n_{d0}}\right)$$

$$= n_{d0}\frac{q_d}{M_d}\nabla^2\phi + v_{thd}^2\nabla^2 n_{d1} \tag{9.24}$$

$$\text{or} \quad \left(\frac{\partial^2}{\partial t^2} - v_{thd}^2\nabla^2\right)n_{d1} = \frac{q_d n_{d0}}{M_d}\nabla^2\phi.$$

Using Eq. (9.19) in Eq. (9.23) gives

$$\nabla^2\phi = \left(\frac{n_{e0}e^2}{\varepsilon_0 KT_e} + \frac{n_{i0}e^2}{\varepsilon_0 KT_i}\right)\phi - \frac{eZ_d}{\varepsilon_0}n_{d1} = \left(k_{De}^2 + k_{Di}^2\right)\phi - \frac{eZ_d}{\varepsilon_0}n_{d1}, \tag{9.25}$$

where $k_{Dj}^2 \equiv n_{j0}e^2/\varepsilon_0 KT_j$, so that, with $k_D^2 \equiv k_{De}^2 + k_{Di}^2$, Eq. (9.25) simplifies to

$$\nabla^2\phi = k_D^2\phi - (eZ_d/\varepsilon_0)n_{d1} \tag{9.26}$$

Remembering that n_{d1} is the density fluctuation of a dust wave, we can let n_{d1} and ϕ take the usual form $\exp[i(kz - \omega t)]$. Equations (9.24) and (9.26) then become, respectively,

$$\left(\omega^2 - k^2 v_{thd}^2\right)n_{d1} = \frac{eZ_d n_{d0}}{M_d}k^2\phi \tag{9.27}$$

and

$$\left(k^2 + k_D^2\right)\phi = \frac{eZ_d}{\varepsilon_0}n_{d1}. \tag{9.28}$$

Thus,

$$k^2 + k_D^2 = \frac{n_{d0}e^2 Z_d^2}{\varepsilon_0 M_d}\frac{k^2}{\left(\omega^2 - k^2 v_{thd}^2\right)}. \tag{9.29}$$

The coefficient on the r.h.s can be called the square of the "dust plasma frequency" ω_{pd} in analogy with Eq. (4.25). The dispersion relation for these low-frequency *dust acoustic waves* is then

$$1 + \frac{k_D^2}{k^2} - \frac{\omega_{pd}^2}{\omega^2 - k^2 v_{thd}^2} = 0, \quad \omega_{pd} \equiv \left(\frac{n_{d0} e^2 Z_d^2}{\varepsilon_0 M_d} \right)^{1/2}, \tag{9.30}$$

where k_D refers to the electron-ion plasma. In laboratory plasmas, typically $KT_e \gg KT_i$, so that $k_D^2 \approx k_{Di}^2 = \lambda_{Di}^{-2} = n_{i0} e^2 / \varepsilon_0 KT_i$ and

$$\omega^2 - k^2 v_{thd}^2 = \frac{k^2 \omega_{pd}^2}{k^2 + k_{Di}^2} = \frac{k^2 \lambda_{Di}^2 \omega_{pd}^2}{1 + k^2 \lambda_{Di}^2}. \tag{9.31}$$

We can define

$$c_d \equiv \omega_{pd} \lambda_{Di} = \left(\frac{n_{d0}}{n_{i0}} Z_d^2 \frac{KT_i}{M_d} \right)^{1/2} \tag{9.32}$$

as the dust acoustic speed, in analogy with $c_s = \omega_p \lambda_D$. For cold dust, we can neglect v_{thd}, and then have for the phase velocity

$$\frac{\omega}{k} \approx \frac{c_d}{\left(1 + k^2 \lambda_{Di}^2 \right)^{1/2}}. \tag{9.33}$$

For $k^2 \lambda_{Di}^2 \ll 1$, the waves are non-dispersive sound-like waves with $\omega/k \approx c_d$. These dust plasma waves, with $c_d \propto 1/M_d^{1/2}$, propagate very slowly. For example, in a typical laboratory dusty plasma with dust particles of 1 μm size, $KT_e = 2.5$ eV, $KT_i = 0.025$ eV, $Z_d \approx 2000$, $M_d \approx 10^{-15}$ kg, and $n_{d0}/n_{i0} \sim 10^{-5}$, the dust acoustic speed c_d is ~1 cm/s. Dust acoustic waves can be seen by illuminating the dust with a thin sheet of laser light as shown in Fig. 9.7. Figure 9.10 shows a frame from a video of a dust acoustic wave.

9.4.2   Dust Ion-acoustic Waves

Dust has a physically reasonable effect on the normal ion waves of Sect. 4.6. The ion equations of motion and continuity are:

$$\frac{\partial \mathbf{v}_i}{\partial t} = -\frac{e}{M} \nabla \phi - v_{thi}^2 \frac{\nabla n_{i1}}{n_{i0}} \tag{9.34}$$

$$\frac{\partial n_{i1}}{\partial t} + n_{i0} \nabla \cdot \mathbf{v}_i = 0. \tag{9.35}$$

Inserting Eq. (9.34) into the time derivative of Eq. (9.35) gives

Fig. 9.10 A single-frame video image of a dust acoustic wave propagating from *right* to *left*. The bright vertical features are the wave crests imaged from scattered laser light (courtesy of R. Merlino)

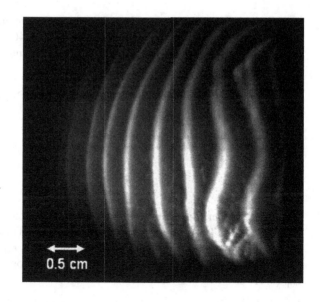

0.5 cm

$$\left(\frac{\partial^2}{\partial t^2} - v_{thi}^2 \nabla^2\right) n_{i1} = n_{i0}\frac{e}{M}\nabla^2\phi. \tag{9.36}$$

Again let the waves have the usual form $\exp[i(kz - \omega t)]$, so that

$$(\omega^2 - k^2 v_{thi}^2) n_{i1} = \frac{e n_{i0}}{M} k^2 \phi. \tag{9.37}$$

Since the phase velocity ω/k will be scaled to KT_e, as in ion sound waves, and $KT_i \ll KT_e$, the v_{thi}^2 term can be neglected, so that

$$\omega^2 n_{i1} = \frac{e n_{i0}}{M} k^2 \phi, \text{ and, similarly, } \omega^2 n_{d1} = \frac{q_d n_{d0}}{M_d} k^2 \phi. \tag{9.38}$$

Using Eq. (9.19) for n_{e1} and Eq. (9.38) for n_{i1} and n_{d1} in Eq. (9.23) yields

$$-k^2\phi = \left(k_{De}^2\phi - \frac{k^2}{\omega^2}\omega_{pi}^2\phi - \frac{k^2}{\omega^2}\omega_{pd}^2\phi\right),$$

$$1 + \frac{k_{De}^2}{k^2} = \frac{\omega_{pi}^2}{\omega^2} + \frac{\omega_{pd}^2}{\omega^2} \tag{9.39}$$

The phase velocity is then given by

$$\frac{\omega^2}{k^2} = \frac{\omega_{pi}^2 + \omega_{pd}^2}{k^2 + k_{De}^2} \approx \frac{\omega_{pi}^2 \lambda_{De}^2}{1 + k^2 \lambda_{De}^2}. \tag{9.40}$$

Since $Z_d n_{d0}$ is at most n_{i0}, the ratio $\omega_{pd}^2/\omega_{pi}^2 = (n_{d0}/n_{i0})(Z_d^2 M/M_d)$ is less than $Z_d M/M_d$. We can then neglect the ω_{pd}^2 term and have done so. Equation (9.40) justifies the definition of c_d as a phase velocity in Eq. (9.32).

If n_{0e} is not so small that λ_D is as large as the wavelength of the wave, we can neglect the $k^2\lambda_{De}^2$ term in the denominator of Eq. (9.40). Also, q_d is usually much larger than e, so n_e can be much smaller than n_i. The numerator is then

$$\omega_{pi}^2\lambda_{De}^2 = \frac{n_{0i}e^2}{\varepsilon_0 M}\frac{\varepsilon_0 KT_e}{n_{0e}e^2} = \frac{n_{0i}}{n_{0e}}c_s^2 \gg c_s^2.$$

Equation (9.40), finally, gives the phase velocity

$$\frac{\omega}{k} = \left(\frac{n_{0i}}{n_{0e}}\right)^{1/2} c_s. \tag{9.41}$$

This is the dispersion relation for dust-modified ion acoustic waves, or dust ion acoustic waves. The dust density does not appear explicitly here, but it determines the n_{i0}/n_{e0} ratio via the quasineutrality condition $n_{i0} = n_{e0} + |Z_d|n_{d0}$ (Eq. (9.18)) for negatively charged dust. In terms of the dust fraction $\delta \equiv n_{d0}/n_{i0}$, we can write Eq. (9.41) as

$$\frac{\omega^2}{k^2} = \left(\frac{n_{i0}}{n_{i0} - |Z_d|n_{d0}}\right)c_s^2 = \frac{c_s^2}{1 - \delta|Z_d|}.$$

Thus, the phase velocity is

$$\frac{\omega}{k} = \frac{c_s}{\left(1 - \delta|Z_d|\right)^{1/2}}, \qquad \delta = n_{d0}/n_{i0}. \tag{9.42}$$

This shows that dust increases the velocity of ion acoustic waves, with the consequence that Landau damping of those waves is decreased.

9.5 Helicon Plasmas

So far, we have mainly treated plasmas consisting of charged species: negative ones, such as electrons and charged dust; and positive ones, such as ions and positrons. Such fully ionized plasmas, however, have to be specially prepared in the laboratory in fusion devices, Q-machines, and such. Most laboratory plasmas are partially ionized and include neutral atoms. Collisions with neutrals were considered in Chap. 5 on diffusion. Plasmas that have practical applications, such as semiconductor etching and magnetic sputtering, are partially ionized. The granddaddy of three-component plasmas is the helicon plasma, which includes not only neutrals but also a magnetic field. Though helicons are complicated, they have been studied exhaustibly worldwide and are well understood.

Helicon plasmas are ionized by helicon waves, which are basically whistler waves (R waves) confined to a cylinder. Their frequencies generally lie between ω_c and the lower hybrid frequency ω_l (Eq. (4.71)). To satisfy the boundary conditions on the cylinder, a second wave has to be generated near the boundary. This second wave is an electron cyclotron wave traveling obliquely to the B-field; it is called the Trivelpiece-Gould (TG) mode (Fig. 4.21). Let β be the total k such that $\beta^2 = k_\perp^2 + k_z^2$. The R wave dispersion relation for propagation at an angle θ to **B** is (Problem 9.2)

$$\frac{c^2\beta^2}{\omega^2} = 1 - \frac{\omega_p^2/\omega^2}{1 - (\omega_c/\omega)\cos\theta} \xrightarrow[\omega_c \gg \omega]{} \frac{\omega_p^2}{\omega\omega_c\cos\theta}. \qquad (9.45)$$

Defining $k \equiv k_z = \beta\cos\theta$, we have

$$\beta = \frac{\omega^2}{c^2}\frac{1}{\beta}\frac{\omega_p^2}{\omega\omega_c\cos\theta} = \frac{\omega}{c^2}\frac{\omega_p^2}{\omega_c}\frac{1}{k} = \frac{\omega}{k}\varepsilon_0\mu_0\frac{n_0 e^2}{\varepsilon_0 m}\frac{m}{eB_0}$$

Thus,

$$\beta = \frac{\omega}{k}\frac{n_0 e\mu_0}{B_0}. \qquad (9.46)$$

This is the basic equation for helicon waves. It shows that the density increases linearly with B_0, and the electron mass m has cancelled out to this order. The frequency is much lower than the electron frequencies ω_c and ω_p. When terms in m are kept, a second wave, the TG mode, is obtained. The relation between the helicon (H) and TG waves has been clarified in papers by D. Arnush, who also wrote a computer program HELIC for the properties of these waves. Damping by electron-neutral collisions is important in the exact theory. The relation between the H and TG waves is shown in Fig. 9.11. There, k (the wave number parallel to B_0) is plotted against β, the cylindrical wave number in the radial direction, for azimuthal wave number $m = 1$. Above a minimum, there are two solutions for k for given n and B_0, the one with large β being the TG mode.

Helicon discharges have been studied experimentally in many machines with long, uniform magnetic fields of order 0.1 T. The first such machine, built by R.W. Boswell in Australia, reached a density of almost 10^{20} m^{-3} on axis. Different types of antennas have been used, the most efficient being helical ones matching the helicity of $m = +1$ waves obeying Eq. (9.46). For reasons not well understood, $m = -1$ waves rotating in the opposite direction do not propagate as well. The coupling of radiofrequency (RF) energy from the antenna to the plasma has been found to involve parametric instabilities. Instabilities such as the drift-wave insta-bility (Fig. 6.14) have been studied in helicon discharges. Magnetic confinement of electrons (but not of argon ions, which have Larmor orbits larger than the discharge radius), combined with efficient antenna coupling, enables helicons to achieve

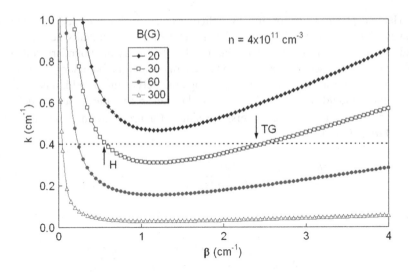

Fig. 9.11 The dispersion relation for $m = 1$ helicon and TG waves at one density and four magnetic fields (F.F. Chen, Plasma Sources Sci. Technol. **24**, 014001 (2015))

Fig. 9.12 A large helicon discharge inside its magnet coils (R.T.S. Chen, R.A. Breun, S. Gross, N. Hershkowitz, M.J. Hsieh, and J. Jacobs, Plasma Sources Sci. Technol. **4**, 337 (1995))

higher densities than in other RF plasmas at the same power. However, the expense of a DC B-field has so far prevented helicons from being accepted by industry. To overcome this, arrays of short helicon discharges with permanent magnets have been proposed for producing large, uniform, high-density plasmas for plasma processing (see Sect. 10.3).

Figure 9.12 shows a large helicon discharge. Note that At high B-field and high power, the plasma can shrink into a dense blue core which is almost fully ionized.

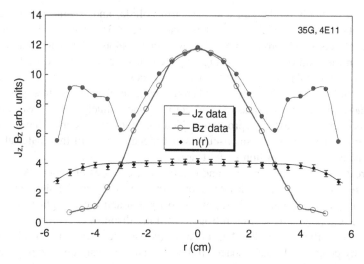

Fig. 9.13 Radial profiles of $n(r)$, $B_z(r)$, and $J_z(r)$ in a helicon discharge at $B = 3.5$ mT and $n = 4 \times 10^{17}$ m^{-3} (D.D. Blackwell, T.G. Madziwa, D. Arnush, and F.F. Chen, Phys. Rev. Lett. **88**, 145002 (2002))

Helicon devices proposed for semiconductor etching and spacecraft propulsion will be shown in Chap. 10.

The phase velocity of helicon waves along **B** is usually comparable to the velocities of "primary" electrons... those that do the ionization. This fact gave rise to a hypothesis that helicons accelerate the ionizing electrons by inverse Landau damping. However, measurements of electron velocity distributions showed no such population of fast electrons. It turns out that most of the RF energy goes into the TG mode near the radial boundary, where the antenna is; and only a small amount goes directly into the helicon mode, which peaks near the axis. The TG wave then couples to helicon waves to form a combined TG-helicon wave. The TG wave parametrically decays into electron cyclotron waves, as mentioned above, and the latter heats the electrons by collisional damping. Though the TG wave is an essential part of the RF coupling, it is not easily detected because it is normally localized to a thin layer near the surface. By lowering the B-field to thicken the layer, and measuring the RF current rather than the RF B-field, the TG wave can be detected, as shown in Fig. 9.13. The current profile shows large peaks due to the TG mode, but these are not seen in the B-field profile.

9.6 Plasmas in Space

As we leave the earth, we first encounter plasmas in the ionosphere, where auroras are born about 100 km above the earth. Further out are the Van Allen radiation belts, filled with plasma from the solar wind. Further yet is the magnetosphere where the B-fields of the sun and the earth collide, and, on the night side, the magnetotail, where reconnection takes place (Fig. 8.31). The sun has important plasma effects in its prominences, flares, and sunspots. The outer planets have plasma in their atmospheres and rings. Space plasma physics is the study of these plasmas in the solar system. Plasma astrophysics, on the other hand, concerns plasmas in the rest of the universe. Galaxies contain plasmas in stars, gas clouds, and black holes. Pulsars, quasars, and active galactic nuclei are all astrophysical plasmas, some with extreme properties. Neutron stars, for instance, have densities of order 4×10^{17} kg/m^3. A sugar-cube sized piece (1 cm^3) of this material would weigh as much as 1.3 million Eiffel towers.

Auroras are produced by ions and electrons of energies of about 100–1000 eV which reach down to altitudes of about 100 km. They excite the oxygen and nitrogen atoms in the atmosphere, giving off green and brown light from oxygen and blue and purple light from nitrogen. Striking displays of Aurora Borealis are observed in northern latitudes. Aurora Australis also occurs in Antarctica but is seen mostly by penguins. In outer space all that we can see must be in the plasma state, but this is only a tiny fraction of a universe containing dark energy, dark matter, and black holes. Nonetheless, astrophysics cannot be studied without the language of plasma physics.

9.7   Atmospheric-Pressure Plasmas

When a voltage is applied between two electrodes in air, a spark will form between the electrodes, forming a plasma, when the voltage is high enough. This breakdown, studied in 1889 by F. Paschen, was perhaps the first plasma experiment. The breakdown voltage depends on the gas, the pressure p, and the distance d between the electrodes. Figure 9.14 shows this relationship in a modern experiment in argon, where contaminants such as moisture are absent. When pd is small, electrons' mean free paths are longer than d, and large voltages are required to accelerate ions to energies that can release secondary electrons from the surface. When pd is large, electrons lose energy in collisions with the neutral gas, and the voltage rises again. Thus there is a minimum in the Paschen curve. After breakdown between parallel plates, the current is not uniform but tends to flow in streamers, each reaching its own equilibrium.

In such an equilibrium, each time an ion strikes the negative plate (the cathode), γ electrons are released, γ being the secondary electron emission coefficient. These electrons are accelerated by the sheath into the plasma and collide with neutrals to form new electron-ion pairs. Each new ion will strike the cathode to release more secondary electrons, thus creating an avalanche. The density will exponentiate. Let α be the probability for an ionizing collision in a unit distance. Clearly α is proportional to $1/\lambda_{\mathrm{mfp}}$, where λ_{mfp} is the mean free path for ionization. One electron will generate $e^{\alpha d}$ ion-electron pairs in a distance d. Upon colliding with the cathode,

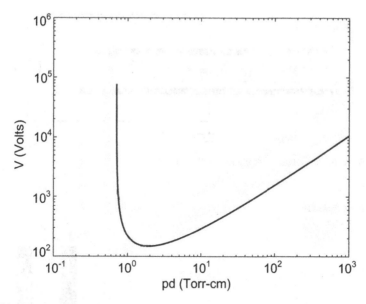

Fig. 9.14 A Paschen curve for argon

these ions will generate $\gamma(e^{\alpha d} - 1)$ new electrons. In equilibrium this has to reproduce exactly the original electron. Thus steady state requires

$$\gamma\left(e^{\alpha d} - 1\right) = 1. \tag{9.47}$$

This is called a *Townsend discharge*. The symbols α and γ are also used to denote two regimes of RF discharges, a low-current and a high-current regime, with a discontinuous jump between the two.

9.7.1 Dielectric Barrier Discharges

To prevent sparking at atmospheric pressures, a dielectric barrier can be inserted between the two electrodes. In Fig. 9.15, the electrodes are covered with insulating dielectric, and high-voltage pulses create the plasma by capacitive coupling. In Fig. 9.16, a dielectric barrier separates the electrodes, but the E-field extends into the space above to ionize the plasma. The substrate to be treated is passed horizontally through the plasma.

Dielectric barrier discharges are used in xenon lamps and for the pixels in plasma display panels (TV screens), for instance.

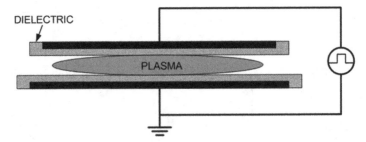

Fig. 9.15 A dielectric barrier discharge of Type 1

Fig. 9.16 A dielectric
barrier discharge of Type 2

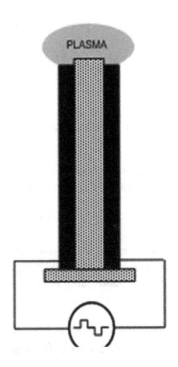

9.7.2 *RF Pencil Discharges*

Another type of discharge that operates at atmospheric pressure is in the shape of a pencil, as shown in Fig. 9.17. The plasma is excited with RF or microwaves to prevent arcing. In one commercial application, a mixture of helium and oxygen is injected, and 200 W of RF at the industrial frequency of 13.56 MHz is applied between the center tube and ground. A plasma beam about two inches long is formed with density up to 10^{13} cm^{-3}, compared to normal air density of 3×10^{19} cm^{-3}. Since no vacuum system is required, these beams can be used medically for cauterizing skin and for some dental procedures. The beam can also

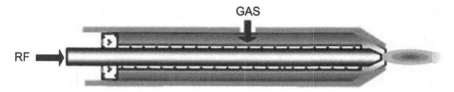

Fig. 9.17 Schematic of a pencil-type atmospheric plasma

be scanned, line by line, over a large surface that requires plasma treatment, such as cleaning or deposition. To treat a wire-shaped object, an atmospheric plasma can be made inside a cylinder through which the wire is passed. Conversely, a catheter can be sterilized with a plasma created with a wire inside it.

A duodenoscope is a small medical instrument inserted into the small intestine to treat such conditions of the bile ducts and main pancreatic duct. These devises have very small openings which are difficult to sterilize. In 2015, numerous deaths were caused by a superbug known by the acronym CRE which survives normal sterilization methods. Since bacteria would be killed by a plasma at 1 or 2 eV, duodenoscopes could be cleaned with an atmospheric plasma, since the small mean free path of electrons in air would allow them to enter very small cavities. In recent years atmospheric pencil plasmas have become widely used in the medical profession.

Problems

9.1 A C_{60} pair-ion plasma is created with a temperature $KT = 0.3$ eV. Describe the sheath at the end walls that intercept the magnetic field lines.

9.2 Modify the R- and L-wave dispersion relations for propagation at an angle θ to **B**.

Chapter 10
Plasma Applications

10.1 Introduction

The effect of plasmas had been noticed as early as 1901, when Marconi found that radio waves could cross the Atlantic in spite of the curvature of the earth. We now know that the waves were reflected by the ionosphere. The study of plasmas probably began with Irving Langmuir's experiments on sheaths in 1928, and it was he who coined the name plasma in a blood-free context. Practical use of plasmas began in the late 1940s with E.O. Lawrence's invention of the calutron (named for the University of California) for the separation of U^{235} from U^{238} for use in atomic bombs. It was the effort to tame the H-bomb into a steady source of electricity—hydrogen fusion—that spawned modern plasma physics. More on that later.

Today, many objects used in everyday life are made or treated with plasmas. These are not the fully ionized plasmas needed for fusion but are partially ionized plasmas with electron temperatures below about 4 eV, the so-called low-temperature plasmas. About 12 % of electricity generated in the U.S. is used for lighting, and over 60 % of lamps involve low-temperature plasmas. Fluorescent lights use an argon or neon plasma containing mercury to generate invisible ultraviolet light, which then excites a phosphor coating that glows visibly. Fluorescents are being replaced by more efficient LEDs (light-emitting diodes), which contain solid-state plasmas in p-n junctions. The latter are also at the heart of solar cells. Semiconductors in electronic devices are made with the use of plasma etching and deposition. Plastic sheets are made hydrophylic or hydrophobic depending on whether they are to be printed on, such as in food packaging. The pixels in our TVs and computer screens are etched with plasmas. Windows are glazed with plasmas to transmit or reflect specific wavelengths. Heat barriers in jet engines are made with plasma deposition. Thrusters in spacecraft are plasma ejectors (Sect. 10.4). In

The original version of this chapter was revised. An erratum to this chapter can be found at https://doi.org/10.1007/978-3-319-22309-4_11

medicine, plasmas are used for sterilization and for hardening or knee implants by ion implantation, for instance. The subject of plasma chemistry has been developed for these applications. All lasers contain plasmas, ranging from the huge megajoule lasers at Livermore in the U.S., in Osaka in Japan, and at Bordeaux in France, to the laser pointers used in lectures.

Many phenomena in basic physics either involve plasmas or may involve plasmas when finally solved. Lightning strikes are gas discharges between charged clouds or between a cloud and ground. Ball lightning is a glowing sphere of plasma that is on the ground and lasts many seconds. Because of its unpredictability, it is unexplained. The geodynamo inside the earth that creates its magnetic field involves motions of liquids, but probably not gaseous plasmas. The BICEP2 experiment (Background Imaging of Cosmic Extragalactic Polarization) seeks to find evidence of primordial gravitational waves from the Big Bang in the cosmic microwave background. Dust is believed to be involved. Dusty plasmas have been added to this edition.

10.2 Fusion Energy

All of the world's energy depends on the sun. Our fossil fuels come from trees that grew in sunlight millions of years ago. Sunlight causes evaporation and makes clouds, and the rain or snow from them gives us hydropower. To minimize the CO_2 released by burning coal, oil, or gas, efforts are made to develop solar and wind power, both of which derive energy from the sun. The temperature in the sun's interior is about 1 keV. At this temperature hydrogen atoms live for a million years before they fuse into helium to release energy. What if we can make our own miniature sun on Earth? We can't wait a million years, so we must increase the temperature to speed up the process. We can also increase the inertia in a head-on collision with heavier isotopes: deuterium (D) with one proton and one neutron, and tritium (T) with one proton and two neutrons. The inertia has to overcome the electric repulsion of ions with like charges. For the sun, Hans Bethe invented the carbon cycle, in which protons can be made to fuse in a series of reactions involving carbon, each of which requires less energy. Better carbon cycles have been found since then, but no one has found a cycle that can work on our time scales. One might think that we could accelerate deuterons in a particle accelerator and put a solid piece of tritium ice in the beam, but that won't work because the deuterons lose more energy in off-angle scatters than they gain in head-on fusion collisions. There are two main ways to recover the energy from D–T fusion collisions: magnetic confinement, and inertial fusion.

10.2.1 Magnetic Fusion

By trapping a plasma in a magnetic field, the ions and electrons are in thermal equilibrium with Maxwellian distributions, so that the energy gained or lost in elastic collisions is returned to the thermal distribution. Only a few collisions result in fusion. The idea is to create a 50–50 % DT plasma at around 30 keV so that there are enough high-energy ions in the tail of the distribution to fuse, generating more than the energy used to create the plasma. Since the plasma is in thermal equilibrium, these are called *thermonuclear reactions*. There are other nuclei besides D and T that can be used; the principal reactions are:

$$D + T \rightarrow \alpha + n + 17.6\,\text{MeV}$$
$$D + D \rightarrow {}^{3}\text{He} + n + 3.27\,\text{MeV}$$
$$D + D \rightarrow T + p + 4.05\,\text{MeV}$$
$$D + {}^{3}\text{He} \rightarrow \alpha + p + 18.34\,\text{MeV}$$
$$D + {}^{6}\text{Li} \rightarrow 2\alpha + 22.4\,\text{MeV}$$
$$p + {}^{7}\text{Li} \rightarrow 2\alpha + 17.2\,\text{MeV}$$
$$p + {}^{6}\text{Li} \rightarrow \alpha + {}^{3}\text{He} + 4.0\,\text{MeV}$$
$$p + {}^{11}\text{B} \rightarrow 3\alpha + 8.7\,\text{MeV}$$

Here p is a proton and n a neutron. The two D–D reactions occur with about equal probability, but one of them yields a neutron. Since thick shielding would be required to protect personnel from neutrons, aneutronic reactions are much preferred. Furthermore, neutrons not only require shielding but they also carry most of the energy, which must be captured as heat to be used in a steam turbine to produce electricity. The D–^{3}He reaction is aneutronic, but ^{3}He does not occur naturally. It is produced in the first D–D reaction, which is neutronic. The ^{6}Li reactions use the 8 % isotope of lithium. The p-^{7}Li reaction is aneutronic, but it occurs only 20 % of the time. This leaves the p-^{11}B reaction, which is the best choice because it is entirely aneutronic and ^{7}Li is a plentiful element. Though its reactivity is relatively low, p-^{11}B has such promise that it is being pursued by a private enterprise. Charged products can, in principle at least, produce electricity without going through a Carnot cycle.

Figure 10.1 shows the fusion probability for the various reactions. It is clear that the D–T reaction is by far the best. Shown in Fig. 10.2, the D–T reaction produces a neutron and a helium nucleus, or α-particle. The mass difference between the D + T and the resulting $\alpha + n$ is converted into 17.6 MeV of energy by $E = Mc^{2}$. Most of the energy is carried off by the 14-MeV neutron, while the 3.5-MeV α-particle is trapped in the magnetic field which confines the plasma. This confinement is far from perfect, and the mean times for leakage of ions and of plasma energy are called respectively τ_{p} and τ_{E}. Neglecting the difference between τ_{p} and τ_{E}, we show in Fig. 10.3 the required $n\tau$ product for breakeven and ignition. Breakeven occurs when the fusion energy produced is equal to that used in creating the plasma. This is called the *Lawson criterion*. Ignition occurs when the α-particles are trapped in the B-field long enough that they can maintain the plasma's temperature without further input of energy. The Joint European Tokamak (JET) in England is, at the time of this writing, on the verge of achieving breakeven. The ITER tokamak in France (originally an acronym of International Thermonuclear Experimental Reactor), is designed to achieve ignition some years after it begins D–T operation in 2027.

Fig. 10.1 Reactivity of
various fusion collisions
vs. ion temperature

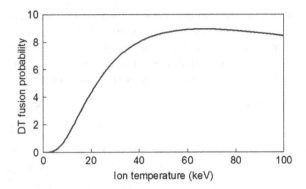

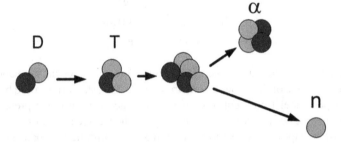

Fig. 10.2 Diagram of the D–T reaction

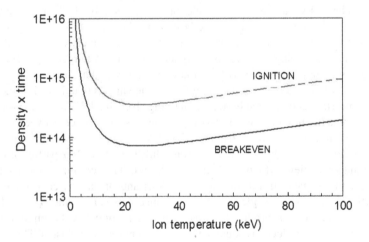

Fig. 10.3 The $n\tau$ product for D–T fusion, in units of cm^{-3} s, vs. KT_i

There are two principal ways to surpass the Lawson criterion: magnetic confinement and inertial confinement. In magnetic confinement, the minimum $n\tau$ in Fig. 10.3 is achieved by holding a dense plasma in a magnetic field for a time τ. In inertial confinement, a much denser plasma is held for a very short time by compressing it with a laser or other pulsed power source. We consider magnetic confinement first. Ions and electrons gyrate in Larmor orbits but are free to move along lines of **B**. To trap them, one can use magnetic mirrors or cusps (see Figs. 2.7, 2.8, and 2.14), or close the lines in a torus. Large mirror machines have been built with special devices to slow the loss of particles which scatter into the loss cone, but these have not been as successful as tori. In a torus, the B-field lines must not close on themselves but must randomly trace out magnetic surfaces. Instabilities then arise, such as the gravitational instability in Sect. 6.7 and the drift-wave instability in Sect. 6.8. The former is driven by centrifugal force, and the latter by the pressure gradient which must exist somewhere in a confined plasma. An ion temperature gradient instability has also been troublesome. These instabilities have been brought under control by shear in the magnetic field and other more subtle means, and it is now possible to satisfy the Lawson criterion.

10.2.1.1   Pinches and Pulsed Power

Inertial confinement research began with pinches. These are plasmas created with a pulsed current, whose B-field compresses the plasma to higher density until the plasma pressure is balanced by the magnetic field pressure. Figure 10.4 shows a simple z-pinch (zed-pinch in the U.K.), like the one described theoretically as early as 1934 by W.H. Bennett. This type of pinch suffers from two instabilities: a sausage instability, shown in Fig. 10.5, and a kink instability, shown in Fig. 10.6. In the sausage instability, if a bulge develops in a linear pinch, the B-field is weakened with the new field lines. The plasma pressure is then able to push the lines further out. The opposite case of a reduced radius neck developing in the plasma is shown in Fig. 10.5. In the kink instability (Fig. 10.6), a kink, or bend, in the plasma will cause the B-field to be stronger on the inside of the curve than on the outside, and the magnetic pressure difference will enhance the kink. By adding a DC magnetic field to the pinch, these instabilities can be slowed down, but such pinches could not achieve confinement times adequate for fusion.

In 1952 James Tuck at Los Alamos in the U.S. made a toroidal z-pinch called the Perhapsatron. To drive the current, the plasma was threaded through an iron core and was the secondary winding of a transformer. This was one of the first toroidal experiments. In the U.K., a large toroidal "zed-pinch" called Zeta was built and exhibited at the 1958 Geneva conference, at which each nation showed what it had been doing while fusion was a classified subject. Neutrons were observed from Zeta which indicated that it had produced fusion reactions. However, the DT reactions were found to come from collisions on the wall, not the interior, much to the chagrin of my good friend Peter Thonemann.

Z-pinch. The z-pinch is so simple that it was not easy to give up on them in spite of their unstable nature. One solution was to drive a current through an array of hundreds of fine metal wires, as was done by T.W.L. Sanford et al. in 1995. Results from these and later wire array experiments are discussed in the context of a series

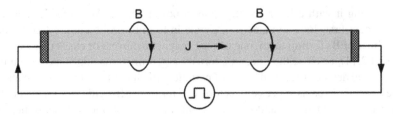

Fig. 10.4 A Z-pinch, or Bennett pinch

Fig. 10.5 Mechanism
of a sausage instability

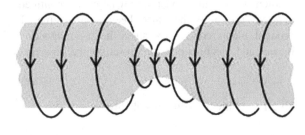

Fig. 10.6 Mechanism
of a kink instability

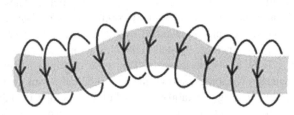

of distinct physical mechanisms by Malcolm Haines *et al.* in 2005. The wires not
only provided a hard core for the pinches, but they also stabilized the kinking in
neighboring wires. Rayleigh–Taylor instabilities then made regular indentations
along the linear plasmas. Related experiments were done with imploding cylindrical liners. Though the main purpose of these experiments was to create X-rays, a
capsule containing deuterium, struck by such implosions, could produce 10^{10} DD
fusion neutrons. A very large experiment, the Z-machine, at Sandia in New Mexico
in the U.S., focused a number of these pinches onto a small wire-array target
enclosing a DD-filled capsule inside a *hohlraum* (defined later) to produce fusion.

The power to the pinches was delivered through thick coaxial "cables" which
used demineralized (high resistivity) water as a dielectric between metal cylinders.
Water has a high dielectric constant of 80. The power was stored in compact
megajoule capacitor banks. The capacitors are charged in parallel, slowly, and
discharged in series, fast, in order to pulse-charge the water-dielectric coaxial
cables in under 1 μs. This process is shown in Fig. 10.7.

A spectacular discharge on the surface of the demineralized water tank in the
Z-machine is shown in Fig. 10.8. Such discharges occur only at late time in normal
operation and do not affect the electrical energy delivered to the load during the
main power pulse.

CHARGING

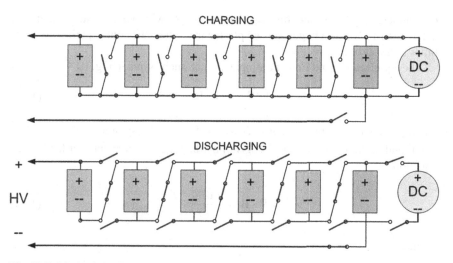

DISCHARGING

Fig. 10.7 Mechanism of charging and discharging a Marx bank

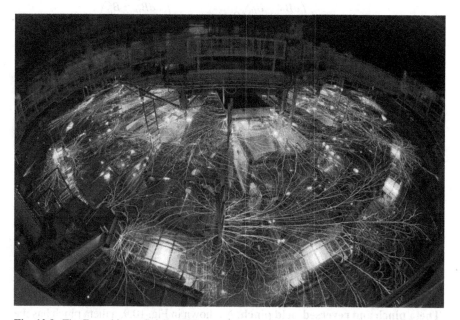

Fig. 10.8 The Z-machine at Sandia, NM

Smaller pulsed-power systems at universities can deliver ~1 TW of power in 100-ns pulses. A megajoule of energy is only the amount required to boil half a cup of water, but to deliver this energy in 100 ns takes a large machine. The Z-machine, along with high-power lasers, represent significant advances in technology, but they do not directly relate to fusion energy. Electric power must be steady and

continuously available, and this is difficult to supply with pulsed systems operating
at less than 10 Hz. The largest pulsed plasma machines carry out experiments a few
times a day at best.

The equilibrium condition of a simple z-pinch such as that in Fig. 10.4 is given by

$$\nabla p = \mathbf{J} \times \mathbf{B}, \tag{10.1}$$

where p is the plasma pressure, $\mathbf{B} = B_\theta \hat{\theta}$, and $\mathbf{J} = J_z \hat{z}$. By symmetry, derivatives in
the $\hat{\theta}$ and $\hat{z}$ directions vanish. $\mathbf{J}$ is given by the time-independent fourth Maxwell
equation

$$\mu_0 \mathbf{J} = \nabla \times \mathbf{B} = \left(\frac{\partial B_\theta}{\partial r} + \frac{B_\theta}{r} \right) \hat{z}, \tag{10.2}$$

So that

$$\mu_0 \mathbf{J} \times \mathbf{B} = (\nabla \times \mathbf{B}) \times \mathbf{B} = (\nabla \times \mathbf{B})_z \hat{z} \times B_\theta \hat{\theta}$$
$$= \left(\frac{dB_\theta}{dr} + \frac{B_\theta}{r} \right) B_\theta (-\hat{r}) = -\left(\frac{1}{2} \frac{dB_\theta^2}{dr} + \frac{B_\theta^2}{r} \right) \hat{r}$$

The z-pinch equilibrium is thus given by

$$\frac{d}{dr} \left(p + \frac{B_\theta^2}{2\mu_0} \right) + \frac{B_\theta^2}{\mu_0 r} = 0. \tag{10.3}$$

Analogously, equilibrium of a θ-pinch is given by

$$\frac{d}{dr} \left(p + \frac{B_z^2}{2\mu_0} \right) = 0, \tag{10.4}$$

without the extra azimuthal term. This shows that the plasma pressure is balanced
by the magnetic-field pressure.

Problem

10.1. A z-pinch of radius a has a uniform current $\mathbf{J} = J_z \hat{z}$ and a plasma pressure p
(r) which is balanced by the $\mathbf{J} \times \mathbf{B}$ force. Derive the parabolic form of $p(r)$.

Theta pinch and reversed-field pinch. As shown in Fig. 10.9, a theta pinch has the
plasma current going in the theta direction. After preionization, a pulse in the theta-
pinch coils drives an azimuthal current $\mathbf{J}$, creating a plasma trapped in the B-field of
the coils. By Newton's third law, this action creates a reaction: the current $\mathbf{J}$ is in the
direction to produce a B-field on axis opposing the B-field of the coils. With
sufficient current, the internal B-field can become larger than that from the coils,
and we have a *reversed-field pinch*. The plasma can extend as far as the separatrix,
which divides the internal field lines from those that extend beyond the coils.

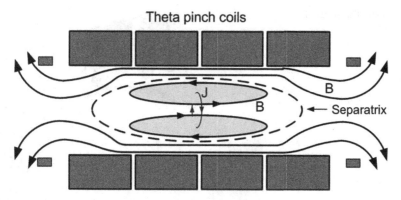

Fig. 10.9 Schematic of a reversed-field pinch (RFP)

The plasma is kept from sliding axially by the $J_\theta \times B_r$ force at its ends. At a critical current, the B-field on axis can be zero, and we have created a $\beta = \infty$ plasma!

The plasma in Fig. 10.9 suffers from instabilities. If the coils are long or far away, the plasma can tilt or slide in the horizontal direction. A hydromagnetic instability can occur at the ends because of the *unfavorable curvature* there. Though the pinch is necessarily pulsed, it lasts much longer than the instability growth time. The effect of curvature is illustrated in Fig. 10.10, where the animals represent plasma pressure from above. When the curvature is concave to the plasma, more plasma pressure can be supported, and *visa versa*. The curvature drift is described in Sect. 2.3.2 for a convex curvature. In Fig. 10.9, the sharp bends at the ends of the plasma are highly unstable, but fortunately the unstable fields (see Fig. 6.11) are short-circuited by the ions, which have large Larmor orbits and can cross the B-field. The finite-Larmor-radius effect was discussed in Sect. 2.4.

The Taylor state. Consider now a toroidal z-pinch such as the Zeta machine mentioned previously. The plasma is held in a toroidal B-field, and a current is driven through the plasma to produce a twist, turning the field lines into helices. It was found that, after a period of violent shaking, the plasma settled into a quiescent state. J.B. Taylor of the U.K. found that, if the helicity of the B-field is conserved, this relaxed state is a force-free, minimum energy equilibrium following the equation

$$\nabla \times \mathbf{B} = \lambda\mathbf{B}, \qquad (10.5)$$

where λ is a constant. The long derivation is omitted here, but Eq. (10.5) is an equation that occurs often in different fields of physics.

10.2.1.2 Magnetic Mirrors

It is possible to trap charged particles between magnetic mirrors, which reflect particles with finite velocity $v_\perp$ perpendicular **B**. In Eq. (2.46) it was shown that the

Fig. 10.10 The effect of curvature

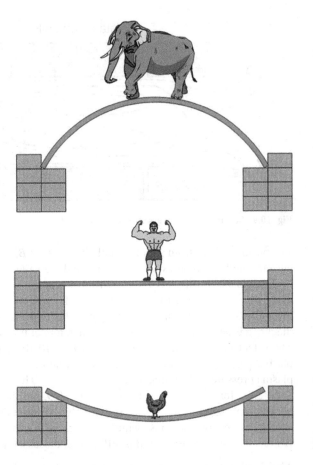

magnetic moment $\mu = \frac{1}{2}mv_\perp^2 /B$ of a particle is conserved. The energy W of a particle can then be written with constant μ:

$$W = \tfrac{1}{2}mv^2 = \tfrac{1}{2}mv_\parallel^2 + \tfrac{1}{2}mv_\perp^2 = \tfrac{1}{2}mv_\parallel^2 + \mu B.$$

Consider a barely trapped particle that is turned around at $B = B_{max}$. Let $v_\parallel = v_\parallel$ at $B = B_0$ and $v_\parallel = 0$ at $B = B_{max}$. Then we have

$$W = \tfrac{1}{2}mv_\parallel^2 + \mu B_0 \quad \text{at} \quad B = 0$$
$$W = 0 + \mu B_{max} \quad \text{at} \quad B = B_{max}$$

(10.6)

Since energy is conserved, this gives

$$1 = \frac{B_0}{B_{max}} + \frac{\frac{1}{2}mv_\parallel^2}{\mu B_{max}} = \frac{B_0}{B_{max}}\left[1 + \frac{v_\parallel^2}{v_\perp^2}\right]_0,$$

(10.7)

$$R_m \equiv \frac{B_{max}}{B_0} = 1 + \frac{v_\parallel^2}{v_\perp^2} = \frac{v^2}{v_\perp^2}.$$

$R_{\rm m}$ is called the *mirror ratio*, and particles starting with $v/v_\perp > \sqrt{R_m}$ are in the *loss cone* (Fig. 2.9) and are not confined by the mirror (Fig. 10.11).

The end losses from a simple mirror are so large that many modifications have been made to minimize these losses. Before we get to these, consider the stability of the plasma. At the center of the mirror, the field lines are convex to the plasma, and Rayleigh–Taylor instabilities can occur. At the throats of the mirror, the field lines are concave to the plasma, providing a stabilizing effect. If the machine is long enough to hold a useful volume of plasma, however, the unstable region dominates. One way to stabilize that part is to add "Ioffe bars", named after the inventor, which are four conductors carrying current in the axial direction, as shown in Fig. 10.12. The azimuthal fields from the bars, added to the mirror field, form twisted magnetic surfaces which have the minimum-B property. That is, the field strength |B| increases in every direction, forming a magnetic well for the plasma. This is a very stable arrangement, but the plasma still leaks out the ends of the mirror. In addition, mirrors suffer from another instability, the cyclotron-ion instability.

If one links the currents in the coils and the bars into a single conductor, a "baseball coil" is obtained, as shown in Fig. 10.13. A very large baseball-type magnetic mirror was built at Livermore Laboratory in the U.S. This heavy device (shown in

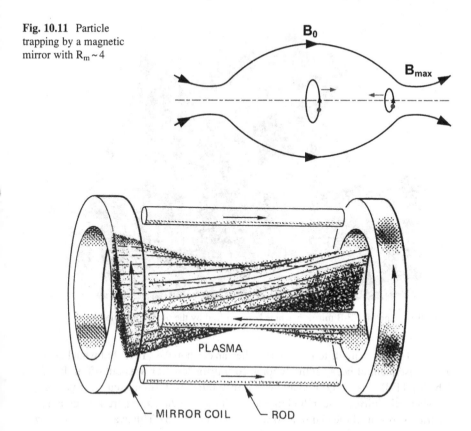

Fig. 10.11 Particle trapping by a magnetic mirror with $R_{\rm m} \sim 4$

Fig. 10.12 A magnetic mirror with Ioffe bars

Fig. 10.13 Diagram of a
baseball coil

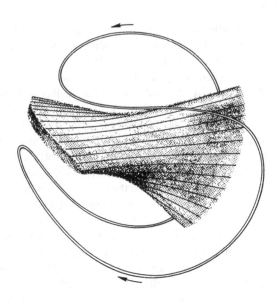

Fig. 10.14 The MFTF-B mirror machine being moved by the old Roman method

Fig. 10.14, was being lifted into place when an earthquake struck, but what happened
is inconsequential because the funding for mirror research was cut off at that time.
The MFTF-B was never used and was turned into a walk-in museum for visitors.

Magnetic mirror research continued in Tsukuba, Japan, where a large axisymmetric mirror machine Gamma 10 was built (Fig. 10.15). This was a *tandem mirror*

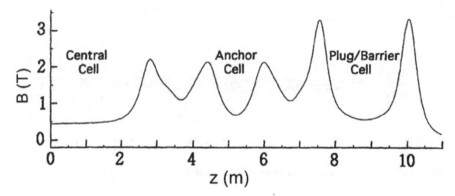

Fig. 10.15 Magnetic field configuration in a tandem mirror

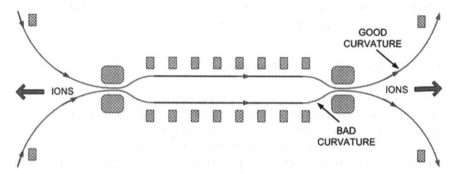

Fig. 10.16 Formation of an ion beam outside a simple mirror

consisting of several mirrors in series. For instance, one mirror could have minimum-B stabilization, while the end mirror could be a short one with a large mirror ratio for confinement. The main interest in mirrors was the possibility of direct conversion of plasma energy into electricity without going through a heat cycle. As ions collide, some will enter the loss cone and escape. These will be accelerated in the z direction because as B decreases, the conservation of μ means that $v_\perp$ will decrease, and hence $v_\parallel$ must increase. An ion beam will be ejected, as shown in Fig. 10.16, and it will be neutralized by electrons, which can escape easily by their frequent collisions. The ion beam will be sorted by energy and collected by a series of bins, as seen in Fig. 10.17, with the more energetic ones going farther before drifting sideways. Thus, the bins provide a DC current. Since the ions enter the bins at low energy, little heat is lost; and this would be an efficient way to convert fusion energy directly to electricity.

Mirrors are fueled by tangential injection of ions at the midplane. Mirrors are of course not immune to instability. In addition to the flute interchange instability in the central cell, there are instabilities driven by the anisotropy of the distribution functions, which have an empty loss cone. For instance, there is an Alfvén ion-cyclotron

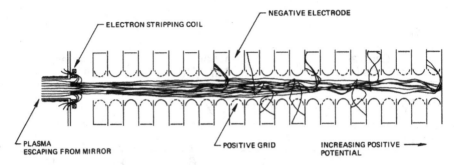

ELECTRON STRIPPING COIL

NEGATIVE ELECTRODE

PLASMA
ESCAPING FROM MIRROR

POSITIVE GRID

INCREASING POSITIVE ⟶
POTENTIAL

Fig. 10.17 Conceptual scheme for direct conversion of ion energy to electricity

instability. Compared with toroidal devices, mirrors do not confine plasmas as well and are not as suitable for fusion. However, they are useful for industrial applications where confinement is not as important as the ejected beams. They can also produce high-β plasmas for experiments not possible in low-β devices.

10.2.1.3 Reversed-Field Configurations

The possibility of creating a high-β plasma spawned a number of large experiments on reversed-field configurations (RFCs). An RFC is a high-β plasma requiring no toroidal field or conducting boundaries. A diagram of an RFC is shown in Fig. 10.18. Though the plasma is pulsed, many instabilities can grow faster than the pulse length. At the ends of the plasma is a region of bad curvature where gravitational instabilities are stabilized by finite ion Larmor radius r_{Li}. This is not simple, since r_{Li} varies rapidly in the nonuniform field. The most dangerous instability is thought to be the $n = 2$ tilt mode, shown in Fig. 10.19, where n is the azimuthal mode number. There is also an $n = 1$ rotational instability. Much of the work on FRCs is theoretical, but these instabilities have been observed. The plasma lasts microseconds, and the total temperature $KT_e + KT_i$ can reach 800 eV.

In the confined region of Fig. 10.18, plasma pressure is needed to balance the magnetic pressure. It can be shown that the average β is given by $\langle \beta \rangle = 1 - \frac{1}{2}(r_s/r_w)^2$, where r_w is the wall radius. Since $r_s/r_w \leq 1$, $\langle \beta \rangle$ must be $> \frac{1}{2}$, and this is an intrinsically high-β device.

FRCs are translatable; that is, they can be pushed magnetically from one chamber to another. For fusion purposes, an FRC can, in principle, be translated into chamber with a DC magnetic field, a conducting wall for stabilization, and even "blankets" for capturing the neutrons and converting their energy into heat.

10.2.1.4 Stellarators

Confinement of plasma in a torus eliminates endlosses but introduces new problems. It is convenient to classify tori is by their aspect ratios. In a circular torus with

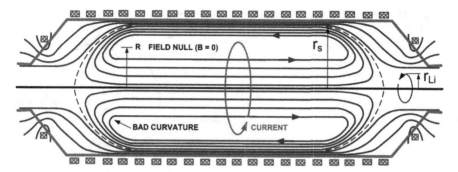

Fig. 10.18 Geometry of an RFC. The *dashed line* is the separatrix, with maximum radius r_s

Fig. 10.19 Drawing of a tilt instability in an FRC

Fig. 10.20 A circular torus
with aspect ratio R/a

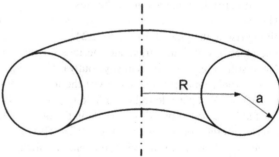

circular cross section, we can define R as the major radius and a as the minor radius, as shown in Fig. 10.20. The *aspect ratio* is defined as R/a. Proceeding from large to small aspect ratio, we start with stellarators.

Toroidal confinement began around 1951 when Lyman Spitzer, Jr., and Martin Schwarzschild built the figure-8 shaped Model A-1 machine at Princeton University. Being an astronomer, Spitzer named it a stellarator. The idea of a figure-8 came to Spitzer during a long ride on a ski lift at Garmisch-Partenkirchen. This was a solution to the problem of vertical drifts in a torus, as shown in Fig. 10.21. The

Fig. 10.21 Vertical drifts
in a torus

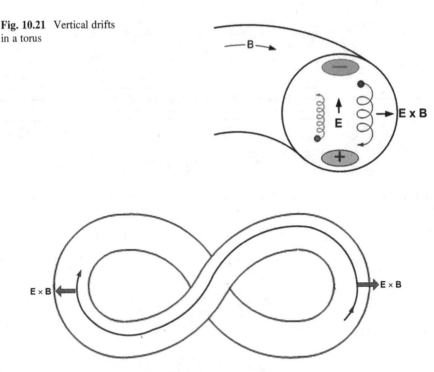

Fig. 10.22 Principle of a figure-8 stellarator

problem is this. In a torus **B** must be stronger on the inside (near the major axis, to the left here) than on the outside because the field lines are crowded together there. This causes the Larmor radii to be larger on one side of each orbit than on the other, giving rise to the grad-B drift [Eq. (2.24)], which depends on the sign of the charge. The electrons drift upwards, and the ions downwards, creating a vertical E-field. The resulting $\mathbf{E} \times \mathbf{B}$ drift is always outward, causing a loss of plasma.

To cancel this drift, one can twist the torus into a figure-8 shape, as shown in Fig. 10.22. As a particle follows a field line, its $\mathbf{E} \times \mathbf{B}$ drift is towards the outside in one half, and back towards the inside in the other half. A demonstration model of a figure-8 stellarator is shown in Fig. 10.23. This was built for the Atoms for Peace conference in Geneva in 1958, an event at which all nations revealed their secret work on fusion. An electron gun could be inserted into the model to accelerate electrons that traced out the field lines. An A-2 stellarator (fondly called the Etude) was operated by Kees Bol. The B-1 stellarator was run by the author in 1954, showing that electrons (not plasma) could be confined for millions of traverses around the 8, in spite of errors in the fabrication of the coils. A B-2 stellarator, shown in Fig. 10.24, was shipped to Geneva also, together with its power supplies and controls so that it could be operated normally. Meanwhile, T. Stix and R. Palladino at Princeton constructed a figure-8 stellarator in the form of a square, which they named B-64, or 8^2.

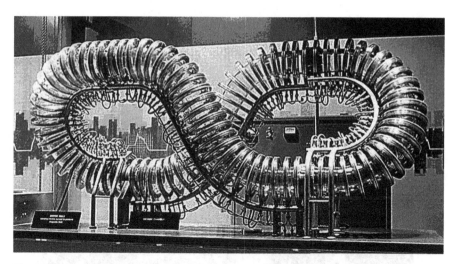

Fig. 10.23 Exhibit model of a figure-8 stellarator

Fig. 10.24 The B-2 stellarator at Princeton

It was soon realized that the twist needed to cancel the vertical drifts could be produced without a figure-8 machine. What was needed was another set of coils wound helically around the torus, as shown in Fig. 10.25. This method had the additional advantage that the helical current could be varied. Th B-series of machines was replaced by the C-stellarator, a much larger machine with two important innovations: ion cyclotron heating, and divertors for capturing the escaping plasma. Previously, the plasma was heated only by ohmic heating, in which a current was induced around the torus by a transformer of which the plasma was the secondary winding. Since the plasma has a resistance caused by collisions of the

Fig. 10.25   A stellarator
with helical windings
[Google Images, 2015]

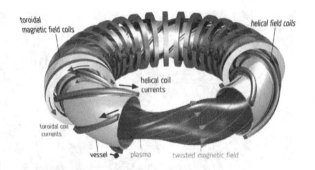

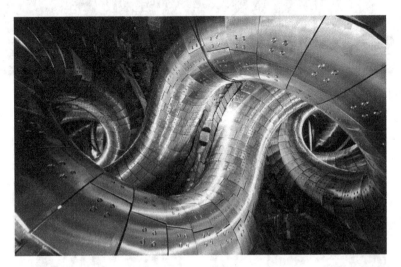

Fig. 10.26   The LHD stellarator in Japan [Google Images, 2015]

current-carrying electrons with ions, it was heated by the I^2R losses. The current
created a poloidal B-field so that the field lines went through a poloidal angle ι (iota)
each time they went around the torus the long way. A kink instability occurs when
iota exceeds 2π. This is called the Kruskal-Shafranov limit. These difficulties have
been overcome in tokamaks, which differ from stellarators in that the poloidal field
is generated by the plasma current itself, and not by external coils. Modern
stellarators use electron-cyclotron, lower-hybrid, and neutral-beam heating, as in
tokamaks, and do not depend on ohmic heating.

Stellarators have been evolving since 1951, and variations with names such as
torsatron, heliotron, heliac, and helias have been built in different countries. The
most spectacular of these, the LHD (Large Helical Device) in Japan, is shown in
Fig. 10.26. The entire vacuum chamber was shaped to follow the magnetic field. In
Germany, D. Pfirsch and H. Schlüter combined the toroidal and helical coils into a
series of 20 planar and 50 non-planar coils, several of each shape, to form the
Wendelstein 7-X, shown in Fig. 10.27. This machine is under construction in

Fig. 10.27 Drawing of the Wendelstein 7-X stellarator [T. Klinger, Max-Planck Institute for Plasma Physics, Greifswald, Germany]

Greifswald, Germany. As with the LHD, the vacuum chamber is not round. To make room for all these coils, stellarators must have large aspect ratio.

With the large advances in the development and understanding of tokamaks, stellarators are no longer preferred for plasma confinement in fusion experiments. Nonetheless, stellarators have advantages which may make them more suitable than tokamaks for reactors. For instance, their poloidal field is created by helical windings and is not dependent on the plasma current, whose shape is not in direct control. Stellarators are also immune from "disruptions", which terminate tokamak plasmas unexpectedly and are not yet well understood.

10.2.1.5   Tokamaks

Invented in Russia, TOKAMAK is a Russian acronym for *toroidal chamber with an axial magnetic field*. Since tokamaks have become the favored form of toroidal fusion devices, so much is known about their behavior that only a few general characteristics can be given here. In tokamak literature, the rotational transform ι (iota) of stellarators is replaced by its reciprocal q. Thus, a field line that comes back to the same poloidal position after going around the torus twice the long way has $\iota = \frac{1}{2}$ or $q = 2$. A typical q profile is shown in Fig. 10.28. The rational-q surfaces have field lines which close upon themselves and are special. The region $q < 1$ corresponds to $\iota > 1$ and is unstable. In tokamaks this instability has the form of sawtooth oscillations, as shown in Fig. 10.29 for T_i and T_e *vs.* time at $q = 1$. As the temperature rises at the center, the resistivity lowers, and the current density increases, thus driving the temperature higher until the configuration can no longer be sustained. Then the sawtooth crashes, ejecting a hot plasma outwards. Unlike stellarators, tokamaks have a self-organized plasma that generates its own behavior to achieve an equilibrium.

It is clear that the profile $q(r)$ should depend on the current profile $J(r)$. An example of this variation is shown in Fig. 10.30. It is seen that more peaked $J(r)$'s give higher $q(a)$'s. The particles, both ions and electrons, travel in interesting orbits on these magnetic surfaces.

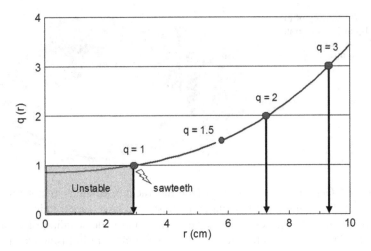

Fig. 10.28 A typical q profile in a tokamak

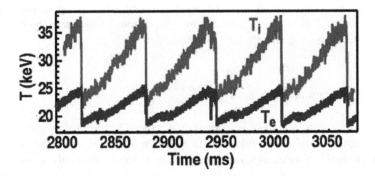

Fig. 10.29 Sawtooth oscillations in a tokamak

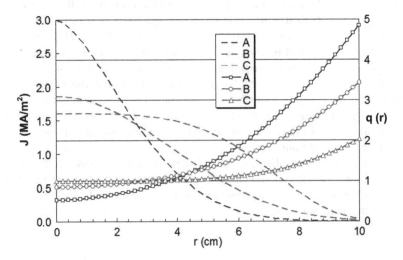

Fig. 10.30 The q profile for three different J profiles

Islands and Bananas

At the radii where q is a rational number, the field lines form magnetic islands. An example of this is shown in Fig. 10.31. In the islands, a particle will hop from one island to the next, eventually returning to the initial island in a different position. Its intersections with this cross section will trace out the islands. However, some particles can't go all the way around the torus. Consider the trajectory shown in Fig. 10.32. Since the B-field is stronger on the inside of the torus (nearer the major axis) than on the outside, a particle with small $v_\parallel$ can be mirror-trapped and reflected back. Projected onto a cross-sectional plane, this orbit resembles that in Fig. 10.33 and is appropriately called a *banana* orbit. These orbits can cause enhanced diffusion

Fig. 10.31 Magnetic islands at the $q = 3/2$ surface

Fig. 10.32 Mirror-trapping of a particle in a torus

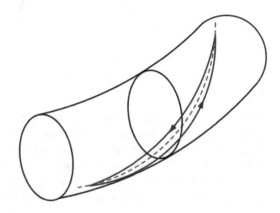

Fig. 10.33 A banana orbit.
A particle with larger $v_\parallel/v_\perp$
would follow the dashed
orbit

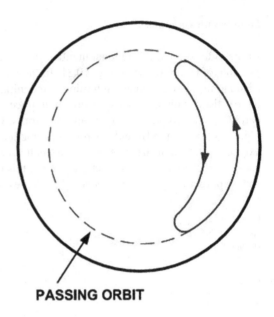

PASSING ORBIT

Fig. 10.34 Mechanism
of the Ware Pinch

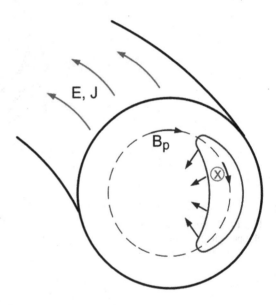

of plasma. Normally, a collision can result in a random-walk step the size of a Larmor
diameter (Fig. 5.17); but now a particle can jump from one banana to another. The
result is that diffusion depends no longer on the toroidal field B_ϕ, but on the weaker
poloidal field B_θ. These effects contribute to "neoclassical diffusion" (see Fig. 5.22).

An effect associated with bananas is the Ware Pinch, illustrated in Fig. 10.34.
Particles in banana orbits have an $\mathbf{E}_\phi \times \mathbf{B_p}$ drift which is always inward, just as in a

linear z-pinch. Thus, bananas tend to move inward, countering collisional diffusion outward. The inward velocity of a Ware pinch is given by

$$v_{Ware} = (2 \pm 0.5)A^{-1/2}E_\phi/B_\theta, \qquad (10.8)$$

where A is the aspect ratio, and the range covers details such as Z_{eff}.

Bootstrap Current

Another interesting effect in tokamaks is the *bootstrap current*, a current that arises, enhancing the toroidal current, as the plasma diffuses outward, as if the tokamak is pulling itself up "by its own bootstraps". This expression, originated around 1781, means to overcome an impediment without outside help. Figure 10.35a is a reminder of Eq. (2.17) showing the velocity of its guiding center when a particle is pushed by a force perpendicular to **B**:

$$\mathbf{v}_f = \frac{1}{q}\frac{\mathbf{F} \times \mathbf{B}}{B^2}. \qquad (10.9)$$

In Fig. 10.35b, the black arrows show the outward pressure force on the plasma in a monotonic density profile. This causes the azimuthal drift of electrons, which is innocuous. The toroidal current generates an azimuthal $\mathbf{B_p}$ (blue arrows). The electrons drift in *this* field $\mathbf{B_p}$, driven by ∇p, is the bootstrap current. It is always in the same direction as the main current and hence adds it. The bootstrap current can be as large as 70–90 % of the total current, and the tokamak is really pulling itself up by its own bootstraps.

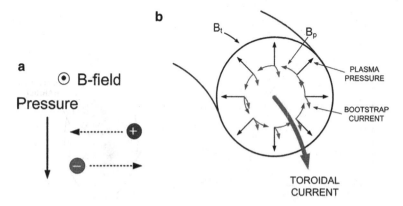

Fig. 10.35 Origin of the bootstrap current [the small *red arrows* pointing out of the paper in (**b**)]

The H-Mode

In 1982, Friedrich (Fritz) Wagner was heating the ASDEX tokamak with neutral beams when, at a certain threshold power, the plasma density suddenly doubled, and instabilities quieted down. This is now called the H (high)-mode. The density and temperature of the plasma did not fall to near zero at the edge but stopped falling at a pedestal level (Fig. 10.36a) as if there were a transport barrier there (Fig. 10.36b). Later experiments showed that there is a highly sheared azimuthal flow at the barrier which prevented normal outward diffusion of plasma. The plasma was instead lost in bursts called ELMs (Edge Localized Modes), a name which suggests that these were not well understood. The H-mode density is a factor 2–3 above previous values, now called the L (low)-mode. Other large tokamaks have also produced the H-mode, and now they all operate with it.

ELMs are thought to be "peeling-ballooning" modes driven by the large boot-strap current caused by the sharp pressure gradient in the thin pedestal layer. Peeling is a form of the kink mode. Ballooning is a bubbling out of the plasma into a weaker field, where extrusion can grow more easily. The pedestal raises the β of the plasma, and thus the fusion power, which increases as β^2. The H-mode was an entirely unanticipated gift of nature. The conditions can be adjusted so that these large outward bursts of plasma are replaced by a stream of small bursts, called grassy ELMs. Even better, it is possible eliminate ELMs altogether. This has been achieved in the ASDEX Upgrade in Germany and the DIII-D in the U.S. with neutral beams counter-injected relative to the plasma current. Figure 10.37 shows the ELM activity as seen by the oscillations in the deuterium α-line D_α. As the power is raised, the tokamak goes from the L-mode into the H-mode. At first, there are ELMs, but a long period ELM-free H-mode ensues. The average density, however, is low in the ELM-free mode, as seen in Fig. 10.38. Apparently, the radial flux at high density cannot be transported to the open field lines leading to the divertor without ELMs.

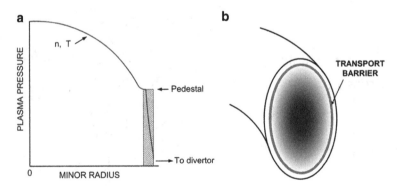

Fig. 10.36 The H-mode barrier

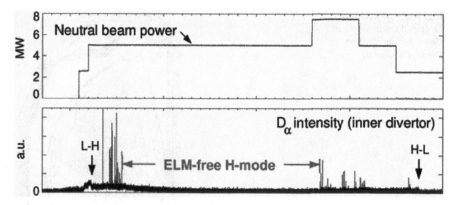

Fig. 10.37 An ELM-free discharge in ASDEX-U [Suttrop et al., Plasma Phys. Control. Fusion **45**, 1399 (2003)]

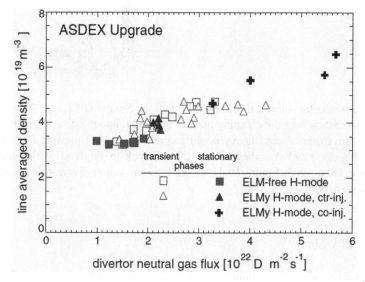

Fig. 10.38 Density increase from ELM-free to ELMy H-mode with counter- and co-injection in ASDEX-U [Suttrop et al., loc. cit.]

D-Shapes and Divertors

A toroidal plasma suffers from instabilities and other particle loss mechanisms that do not occur in cylinders. Since these effects are caused by the fact that one side of the plasma is closer to the major axis than the other, we can increase the volume of the plasma harmlessly by making the tokamak taller rather than wider. This has led to tokamaks with D-shaped cross sections, which are now commonly adopted. An example is shown in Fig. 10.39. At the top and bottom of the plasma there are "divertors", which capture the plasma exhaust. The confined part of the plasma is

Fig. 10.39 Diagram of a
D-shaped tokamak (from
General Atomics, San
Diego, California)

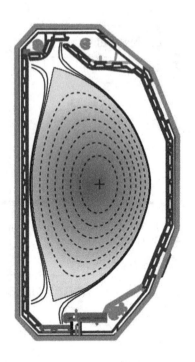

enclosed by the last closed flux surface, called the Scrape Off Layer. The surfaces outside of that capture the escaping plasma and lead it into the divertors, where high temperature materials and heavy cooling condense the hot plasma. Figure 10.40 shows the elongated chamber of the JET tokamak in England, which has been operating since 1983. Figure 10.41 shows a water-cooled divertor designed for a future tokamak.

The Density Limit

In assembling data from numerous tokamaks, M. Greenwald found that their average plasma density always fell below a hard limit which was proportional to the power input. This is shown in Fig. 10.42. It was thought that as the power was raised, the wall was bombarded by escaping ions, thus releasing high-Z impurities near the boundary. These would radiate and cool the plasma, increasing the resistivity there and thus changing the current profile into an unstable form. However, the density limit does not depend on the impurity level or the power. If the density is slowly raised in the H-mode until it reaches the limit, the H-mode disrupts unstably into the L-mode. If the density is allowed fall slowly, it decays at the rate that keeps the discharge at marginal stability. The density limit is not well understood.

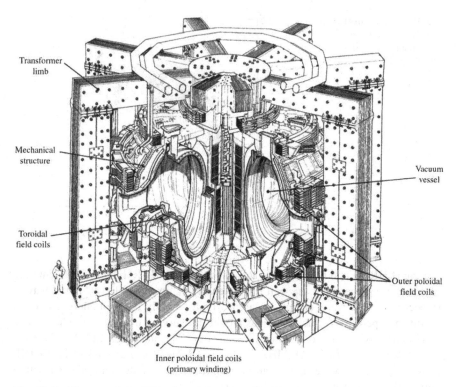

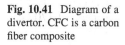

Fig. 10.40 Diagram of the JET tokamak in the U.K. [J. Wesson, *Tokamaks*, Oxford Science Publications (1987), p 658]

Fig. 10.41 Diagram of a divertor. CFC is a carbon fiber composite

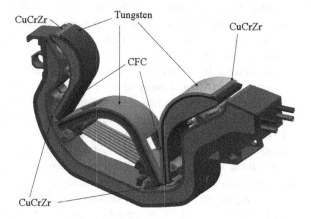

Heating and Current Drive

There are four main ways to heat a fusion plasma: ion-cyclotron resonance heating (ICRH). electron-cyclotron resonance heating (ECRH), lower-hybrid resonance

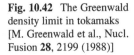

Fig. 10.42 The Greenwald
density limit in tokamaks
[M. Greenwald et al., Nucl.
Fusion **28**, 2199 (1988)]

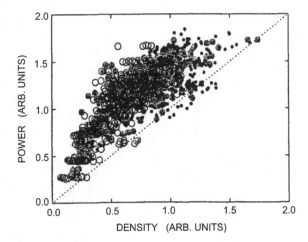

heating (LHRH), and neutral beam injection (NBI). In **ICRH** and **LHRH**, internal antennas are placed near the wall to excite ion cyclotron and lower hybrid waves, respectively. The wave energy is damped preferentially into ion energy. Though useful in experiments, such antennas would not survive in a reactor. These three waves are of quite disparate frequencies. For instance, at $B_0 = 3\text{T}$ and $n = 1 \times 10^{20}$ m^{-3}, f_{ci} is 1.1 MHz, f_{LH} is 230 MHz, and f_{ce} is 84 GHz. ECRH, being in the microwave range, can be carried in waveguides and injected with a horn without requiring an internal antenna.

ECRH power, of order 100 GHz, is produced by gyrotrons, which produce electron cyclotron radiation by injecting an electron beam to gyrate in a strong magnetic field. A picture of a gyrotron is shown in Fig. 10.43. The magnet is at the bottom, and the top part collects the depleted beam. A bank of these, at 170 GHz and 1 MW, will be used in ITER. As shown in Problem 4.47, the cyclotron resonance can be reached only from the inside of the torus, where space in the central column is at a premium. For this reason, ECRH is often applied at $2\omega_c$.

Large tokamaks are mainly heated by the injection of neutral atoms (**NBI**) which can penetrate into the interior of the plasma before being ionized. These atoms must therefore have a large injection energy, typically hundreds of keV. For instance, the JT-60U tokamak in Japan has developed a 500-keV, 22-A ion source for an injector. To make a neutral beam one can start with a positive ion and add an electron, or start with a negative ion and strip an electron, which is easier. Neutral beam injectors typically start with negative deuterium ions (D$^-$), which are then passed through a gas such as lithium or cesium to be turned into neutrals. Neutral beams can be injected radially or, usually, tangentially in the same direction as the tokamak current.

Figure 10.44 shows the DIII-D tokamak at General Atomic in San Diego, California. The two large cylinders on the left are the neutral beam injectors. A beam dump on the far side captures the un-ionized beam. Neutral beam injectors tend to be larger than the tokamak itself and to dominate the laboratory. When

Fig. 10.43 A large
gyrotron [ITER.org]

neutral beams are injected in the direction of the tokamak current, they contribute to
that current when ionized. Varying the angle of injection affords a way to control
the current profile. This is called non-inductive current drive, since it is steady-state
and does not involve pulsing a transformer.

Electron cyclotron current drive is another way to achieve steady-state operation
of a tokamak. To do this, waves at a frequency corresponding to ω_c somewhere in
the interior of the plasma are injected from the inside of the torus. We can use
Fig. 4.36 if we add a line representing the injected ω. If an X-wave is sent in from
the outside of the torus, ω lies to the right, and the whole diagram moves to the right
towards ω as ω_p increases. The wave will meet the cutoff at ω_R and be reflected. But
if the wave is injected from inside the torus, where ω_c is larger than ω_L, it can travel
to the radius where $\omega = \omega_c$ and give electron cyclotron current drive with its
z component. The space in the central column very limited, and the need to inject
from the inside is problematic.

Fig. 10.44 The DIII-D tokamak in the U.S.

Disruptions

Tokamaks are subject to catastrophic events known as disruptions. There are many possible causes for this loss of confinement, but it is a global instability that causes a thermal quench—a sudden cooling of the plasma. The resistivity then increases, and the toroidal voltage can then accelerate electron runaways to MeVs. Large "halo" currents flow poloidally, partly in the plasma and partly in the wall. Large $\mathbf{J} \times \mathbf{B}$ forces are induced in the walls as $\mathbf{J}$ decreases. The D-shaped plasma "falls" vertically, as shown in Fig. 10.45. It is sometimes possible to anticipate a disruption when oscillations appear in Mirnov pickup coils near the edge. In that case, the disruption can be prevented by injection of a neutral gas such as He or Ne through a fast valve. It is clear that disruptions must be avoided in a reactor, and this can be done by operating the tokamak below its maximum power. With superconducting magnets, tokamaks can run almost steady state. The EAST (Experimental Advanced Superconducting Tokamak) in China can run pulses as long as 1000 s without disrupting.

10.2.1.6 Spheromaks and Spherical Tokamaks

When the aspect ratio of a tokamak is reduce to the order of unity, the machine becomes a spherical tokamak. If the central column is removed and the plasma is injected, it forms a self-organized structure called a spheromak. A smoke ring is an example of a self-organized structure. Instabilities, especially the kink instabilities, are suppressed at low aspect ratios, probably because of the short connection length between good and bad curvature regions.

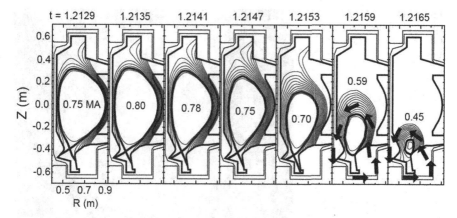

Fig. 10.45 Vertical motion of the plasma during a tokamak disruption [R. S. Granetz et al., Nucl. Fusion **36**, 545 (1996)]

Fig. 10.46 Drawing of a
spherical tokamak showing
the regions of good and bad
curvature of one field line
[from F.F. Chen, *An
Indispensable Truth*
(Springer, 2011), p. 376;
adapted from original by
S. Prager, Univ. of
Wisconsin]

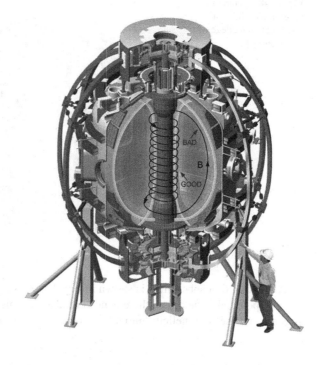

A diagram of the NSTX spherical tokamak experiment at Princeton is shown in Fig. 10.46. A more compact machine, the MegAmpere Spherical Tokamak MAST is shown in Fig. 10.47. START (Small Tight Aspect Ratio Tokamak), a predecessor of MAST, achieved a beta of 40 %, an order of magnitude higher than the "Troyon limit" for ordinary tokamaks.

Fig. 10.47 The MAST
spherical tokamak [Culham
Centre for Fusion Energy,
U.K.]

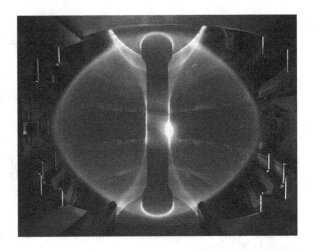

Fig. 10.48 Diagram of a
spheromak [Univ. of
Washington, College of
Engineering]

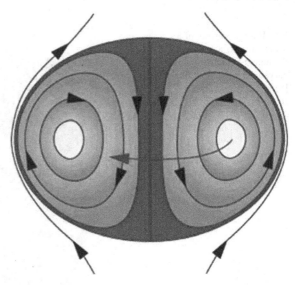

A diagram of a spheromak is shown in Fig. 10.48. A toroidal current creates an entirely poloidal B-field, and there is no central core. This is formed by injecting a plasma with the right helicity into a flux conserver, as shown in Fig. 10.49.

10.3   Plasma Accelerators

In Fig. 7.17 we saw how a particle can be accelerated by a wave, thus causing damping of the wave. In 1979 John Dawson, in a paper with T. Tajima, proposed that this effect could be used to accelerate particles on purpose for experiments on

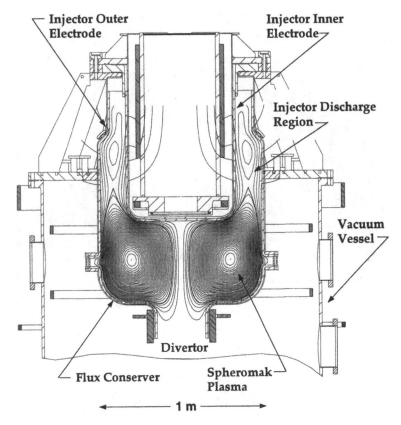

Fig. 10.49 Formation of a spheromak [E.B. Hooper et al., Lawrence Livermore National Laboratory Report UCRL-JC-132034]

nuclear physics and the structure of matter. Originally, Dawson envisioned accelerating particles traveling at an angle to a wave, as shown in Figs. 10.50, 10.51, thus gaining velocity faster than that of the wave. This idea spawned many experiments, principally in the U.S., Japan, and England, to develop a new type of accelerator. So far, only straight acceleration at zero angle has been tried.

There were two early ideas on plasma accelerators: *beatwave* and *wakefield*. In the beatwave case, two lines of a laser at (ω_0, k_0) and (ω_2, k_2) are set to resonate at ω_p, which is assumed to be much smaller than the laser frequencies. The beat frequencies are

$$\omega_1 = \omega_0 - \omega_2 \equiv \Delta\omega \simeq \omega_p$$
$$k_1 = k_0 - k_2 \equiv \Delta k \equiv k_p \qquad (10.10)$$

This excites a plasma wave with a phase velocity $v_\phi = \omega_1/k_1 \approx \omega_p/k_p$. Since plasma waves in a cold plasma do not depend on wavelength [Eq. (4.30)], k_p is allowed to

Fig. 10.50 Surfing at an angle to a wave [Google images]

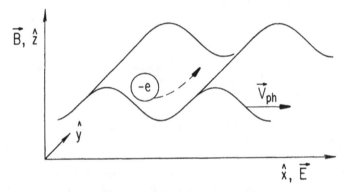

Fig. 10.51 Scheme of a surfatron accelerator [T. Katsouleas and J.M. Dawson, Phys. Rev. Lett. **51**, 392 (1983)]

have the above value. The phase and group velocities of the light waves are given by Eqs. (4.86) and (4.87), in which to subscript 0 pertains to the light waves. Thus

$$v_{\phi 0}^2 = \frac{\omega_0^2}{k_0^2} = c^2 + \frac{\omega_p^2}{k_0^2}, \quad v_{g0} = c^2/v_{\varphi 0} = c\left(1 + \omega_p^2/\omega_0^2\right)^{-1/2}, \qquad (10.11)$$

and similarly for ω_2. In the usual limit $\omega_p \ll \omega_0$ so that $v_{g0} = d\omega_0/dk_0 \approx \Delta\omega/\Delta k$, we see from Eqs. (10.10 and 10.11) that v_{g0} of the light waves is nearly equal to the v_ϕ of the plasma wave. This assures that a laser pulse can continue to push particles trapped in the plasma wave over long distances.

At high plasma densities the assumption $\omega_p \ll \omega_0$ no longer holds true. The *critical density* n_c at which $\omega_p = \omega_0$ is defined by

$$\frac{n}{n_c} \equiv \frac{\omega_p^2}{\omega_0^2}. \tag{10.12}$$

The number n_c is a useful one in laser experiments. In terms of n_c, Eq. (10.11) can be written

$$v_\varphi \simeq v_{g0} = c(1 - n/n_c)^{1/2} \quad (n \ll n_c). \tag{10.13}$$

For example, $n_c \approx 1.0 \times 10^{25}/\mathrm{m}^3$ for the 10.6-μm CO_2 laser line, $n_c \approx 1.21 \times 10^{25}/\mathrm{m}^3$ for the 9.6-μm line, and $n_c \approx 1.29 \times 10^{25}/\mathrm{m}^3$ for the 9.3-μm line. The beat frequency of $3 \times 10^{12}/\mathrm{s}$ between the 10.6- and 9.6-μm lines corresponds to the plasma frequency at $n = 1.1 \times 10^{23}/\mathrm{m}^3$. This is a high density that can be produced by a pulsed plasma or even by the laser beams themselves. Beat waves were produced at UCLA in 1985, and acceleration of electrons by beat waves in 1993. Figure 10.52 shows two oscillations differing in frequency by 5 % and their beat wave, which has 20 times the wavelength, according to Eq. (10.10).

In wakefield acceleration, plasma waves are excited by a short pulse of electrons or photons. The waves have plasma frequency f_p and wavelength $\lambda \approx c/f_p$, since drive pulse travels at $v \approx c$. If the plasma wave has sufficient amplitude, electrons will be swept up by the wave and accelerated to $v \approx c$, gaining mass rather than velocity in the relativistic limit. It is also possible to inject electrons in the right phase by photoemission from a solid with a pulsed laser synchronized with the wave (Fig. 10.53).

The relativity factor γ of a plasma wave can be expressed simply in terms of ω_p:

$$\gamma \equiv \left(1 - v_\phi^2/c^2\right)^{-1/2} \tag{10.14}$$

Using Eqs. (10.12) and (10.13), we have

$$\gamma = [1 - (1 - n/n_c)]^{-1/2} = (n/n_c)^{-1/2} = \omega_0/\omega_p. \tag{10.15}$$

Real experiments involve beams, not the one-dimensional, infinite plane waves described so far. In two dimensions, one finds that in addition to regions of acceleration and deceleration, the wave has regions of focusing and defocusing. As computers advanced, it was possible to do cell-by-cell computer simulation to see the nonlinear development. One result is shown in Fig. 10.54. Here, a very short, intense laser or electron pulse is sent through a plasma. The ponderomotive force of the pulse ejects all the electrons, leaving a bubble of the slow-moving ions. As the pulse passes, the electrons are attracted back by the ion charge, forming a negative layer around the bubble and converging into an electron bunch behind the bubble.

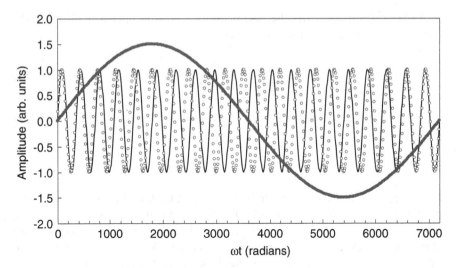

Fig. 10.52   The beat wave of two oscillations differing in frequency by 5 %

Fig. 10.53   Electrons surfing on the plasma wave behind a drive pulse

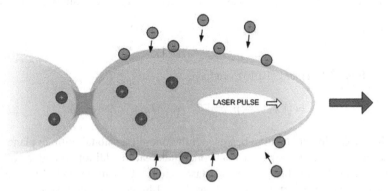

Fig. 10.54   A plasma bubble created by a short laser or electron pulse [adapted from Scientific American, February, 2006]

This has not been seen in experiment, but particle acceleration by the wakefield effect has been achieved in several countries with the world's largest lasers. The most successful of these experiments was done at the Stanford Linear Accelerator

Fig. 10.55 Plot of energy vs. radius of an electron pulse in vacuum (*left*) and in a plasma (*right*) [M.J. Hogan et. al., Phys. Rev. Lett. **95**, 054802 (2005)]

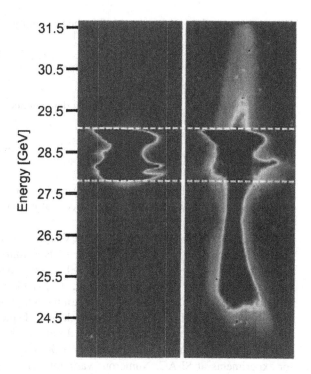

Center (SLAC) by a team led by C. Joshi of UCLA. Figure 10.55 shows a highly relativistic 28.5-GeV electron pulse, 50 femtoseconds long, with 20 kA peak current. Electrons that have lost energy in forming the wake form the tail of the distribution, and the few electrons that have been accelerated to higher energy by the wake field are seen at the top. Note that this is not a spatial distribution, since all is traveling a the speed of light.

How large can the plasma wave (ω_p, k_p) get? Consider the linearized Poisson equation with stationary ions:

$$\varepsilon_0 \nabla^2 \phi = -k_p^2 \varepsilon_0 \phi = -e n_{e1} \equiv -en$$
$$e\phi = ne^2 / \varepsilon_0 k_p^2 \tag{10.16}$$

where $n \equiv n_{e1}$ for this discussion. The maximum ϕ occurs when $n = n_0$, the background density. Hence,

$$e\phi_{max} = \frac{n_0 e^2}{\varepsilon_0 m} \frac{m}{k_p^2} = \frac{\omega_p^2}{k_p^2} m \approx mc^2, \tag{10.17}$$

for highly relativistic waves. Let us define

$$\delta \equiv \phi / \phi_{max} = n_1 / n_0. \tag{10.18}$$

The magnitude of the plasma-wave E is then given by

$$|E| = |k_p \phi| = |k_p \phi_{\text{max}} \delta| = \frac{\delta \omega_p^2}{e \, k_p} m \simeq \frac{\delta}{e} \omega_p mc, \qquad (10.19)$$

where, again, $\omega_p/k_p \approx c$. Extracting the n dependence, we have

$$|E| = \frac{\delta}{e} \omega_p mc = \frac{\delta}{e} \left(\frac{me^2}{\varepsilon_0} \right)^{1/2} n_0^{1/2} c = \delta \left(\frac{m}{\varepsilon_0} \right)^{1/2} n_0^{1/2} c = .096 \, \delta \, n_0^{1/2} \, \text{V/m}. \quad (10.20)$$

This is an extremely high field. For instance, if $n_0 = 10^{24}/\text{m}^3$, E is of order 1 GeV/cm. In principle, this would allow linear accelerators to be shortened by three orders of magnitude.

Wakefield experiments in the 10's of GeV regime require a preformed plasma to prevent head-erosion of the laser pulse if it has to ionize the plasma also. Such a plasma has to have $n > 10^{16}$ cm^{-3} over ≥ 1 m and have a radius greater than the blow-out radius (Fig. 10.54) of order $c/\omega_p \simeq 17$ μm. Gases such as Li, Rb, or Cs can be used, but lithium has the advantage that it has a high second ionization potential of 75.6 eV, so that Li^{++} ions can be neglected. Typically, a plasma of density 5×10^{16} cm^{-3}, 1.5 m long, can be created with a 200 mJ laser pulse (4 TW for 50 fs). Such target plasmas were developed by K.A. Marsh and C. Clayton at UCLA for experiments at SLAC. Numerous variations of wakefield schemes have been developed with the aim of producing high quality electron beams with low emittance (transverse momentum). For instance, a low frequency laser, such as CO_2 can be used to form the wake, and a high frequency 800 nm laser used for injection of electrons. Finally, these methods can also be applied to positrons for electron-positron colliders.

10.3.1   Free-Electron Lasers

A related subject is that of the free-electron laser (FEL), which is the opposite of plasma accelerators in that an electron beam is used to create radiation rather than vice versa. A diagram of an FEL appears in Fig. 10.56. An array of permanent magnets, called a wiggler, is shown linearly, though it is usually helical. A relativistic electron beam is injected from the left and is wiggled by the Lorentz force from the magnets. It also creates a plasma and waves in the plasma. Without going into the mathematics, one can see that motion of the beam and the plasma can emit radiation, and when phased properly, this radiation can grow at the expense of beam energy. The depleted beam is caught in the beam dump at the right.

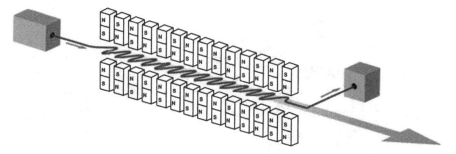

Fig. 10.56 Diagram of a free-electron laser (FEL)

10.4 Inertial Fusion

In the 1970s the advent of powerful lasers motivated physicists to explore the possibility of achieving fusion in short bursts rather than in a steady-state magnetized plasma. Among these were John Nuckolls and Ray Kidder at Livermore (now Lawrence Livermore National Laboratory); Keith Brueckner at University of California, San Diego; Keeve "Kip" Siegel in Michigan, and Chiyoe Yamanaka in Osaka, Japan. Siegel founded KMS Fusion but died in 1975 while testifying in Congress. Inertial fusion can also be achieved without lasers with collapsing magnetized metal "liners" in a z-pinch, or with ion beams, as in Fig. 10.8, but the greatest progress has been with lasers. The largest such experiment is the National Ignition Facility (NIF) at Livermore. This program could not have started without the LASNEX code written by Nuckolls, and laser fusion research still relies substantially on computer simulation.

10.4.1 Glass Lasers

In laser fusion, a fuel pellet is compressed by laser energy to a density $\rho \sim 1000$ times the density of solid DT (0.2 g/cm^3), or about 20 times that of lead, and temperature of order 10 keV. The Lawson criterion for breakeven in laser fusion works out to be

$$\rho r \geq 1 \, \text{g/cm}^2, \tag{10.21}$$

where r is the compressed radius. For $\rho = 200$ g/cm^3, r is ~50 μm.

There are two ways to achieve laser fusion. In *direct-drive* fusion, laser energy is directed as uniformly as possible over the surface of a fuel pellet, as shown in Fig. 10.57. The plastic ablator material is heated by the laser light and blasts off, pushing the layer of frozen DT to a density satisfying the Lawson criterion. To avoid Rayleigh–Taylor instabilities (Fig. 6.11), which would make small dimples

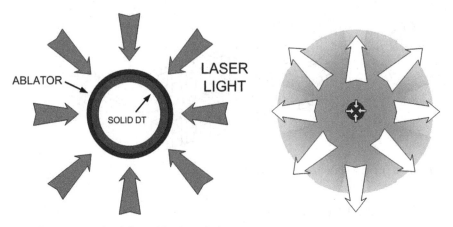

Fig. 10.57 Schematic of direct-drive laser fusion

grow into large gouges, the laser light is carefully smoothed out by random phase plates. The layer of frozen DT is serendipitously smoothed out by the small amount of heat from the slow decay of tritium into ^{3}He, an electron, and a neutrino. In the U.S., this approach is followed at the Laboratory for Laser Energetics in Rochester, NY.

In *indirect-drive* fusion, the capsule is compressed by X-rays inside a gold cylinder called a *hohlraum* ("empty space" in German). Figure 10.58 shows a hohlraum with laser beams entering holes at the ends. The beams must not hit the sides of the hole, or they would create plasma that will refract or reflect the beams. Once inside the hohlraum, the beams strike the walls, creating a bath of X-rays, which will then heat the capsule and compress it. This is a miniature version of an H-bomb. Several views of hohlraums are shown in Fig. 10.59.

In the U.S., indirect-drive fusion is explored at the NIF at Livermore. This facility is a marvel of engineering. Light from a single oscillator is preamplified and then split into 192 beams. The beams are amplified in Nd-glass plates lit by LED flashlamps powered by 422 MJ of capacitors. Faraday rotators and polarizers are inserted to prevent reflected light from being amplified backward and destroying the front end. Since parametric instabilities are less important at shorter wavelengths, the 1.05 μm Nd-YAG (neodymium-doped yttrium aluminum garnet) beams are frequency tripled by KDP (potassium dihydrogen phosphate) crystals to 351 nm in the green. A single dust particle on one of the glass plates will absorb enough laser light to damage the plate, which must be replaced. Refrigerator-sized replacement units are stored below beam lines and can be inserted into place. An early picture of the NIF laser bay is shown in Fig. 10.60.

The laser beams are mirrored in a switchyard to enter into a spherical target chamber, shown in Fig. 10.61. They must all reach the target within a few picoseconds. Light travels only 0.3 mm in a picosecond. Hence the lengths in this system the size of a football field have to be kept constant with temperature control.

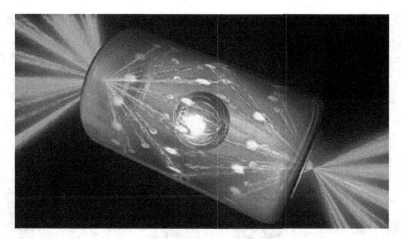

Fig. 10.58 Drawing of a hohlraum [courtesy of Lawrence Livermore Laboratory via Google Images]

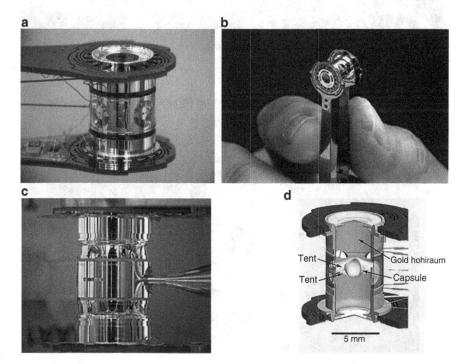

Fig. 10.59 Hohlraums: (**a**) a hohlraum in its holder; (**b**) size of a hohlraum; (**c**) a "rugby" hohlraum with an ellipsoidal cavity; (**d**) suspension of a capsule in a hohlraum with thin plastic membranes or "tents"

Fig. 10.60 The laser bay of the National Ignition Facility [courtesy of Lawrence Livermore Laboratory via Google Images]

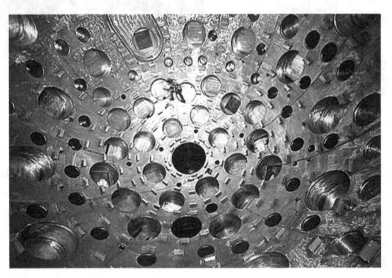

Fig. 10.61 The NIF target chamber, seen from inside [courtesy of Lawrence Livermore Laboratory via Google Images]. Note the figure of a man above center

The beams carry 15 MJ of 351 nm light, giving a peak power of 430 TW. The total electricity generation in the U.S. is about 0.5 TW on average.

Impressive as these numbers are, fusion with glass lasers is not an energy source. The goal of ignition, in which the alpha particles from the DT reaction keeps the plasma hot, has not been achieved. The most that has been achieved by 2015 is fuel gain, in which the fusion energy out is larger than the laser energy in, with a laser efficiency of about 1%. Furthermore, the NIF can pulse only twice a day to allow the glass to cool. A fusion reactor would require pulsing at 10 Hz. Each shot destroys a target such as that shown in Fig. 10.59. The hohlraum cannot simply be dropped from the top of the chamber; the DT ice would melt. The target has to be shot into the lasers' focus in exactly the right orientation. After each shot, the debris has to be cleared so as not to interfere with the next laser shot. Heavy products like gold take a long time to reach the wall. Liquid Li walls or FLiBe (fluorine, lithium, beryllium) waterfalls have been suggested to catch the gamma rays and other products but they cannot clear the whole volume rapidly. The NIF cannot lead to a fusion reactor. It is funded by the Nuclear Weapons Stockpile Stewardship program in addition to the Department of Energy and serves as a safe way to study such topics as the equation of state under extreme conditions without using underground nuclear explosions.

In addition to accurate focusing, the timing of the laser pulses is crucial. A simple intense laser pulse can drive the fuel into a center hotspot until the ρr criterion is exceeded, but one can do much better than that. Here are two ways. In *fast ignition*, the compression and heating phases follow each other, just as in a gasoline engine. First, one laser pulse compresses the fuel to the required density; then a second high-intensity, ultrashort pulse heats the core. The compression pulse can have less energy and uniformity than usual, so that the energy gain is 10–20 times higher than normal (Fig. 10.62).

In *shock ignition*, the general principle is to launch a series of shocks which overtake one another at the central spot. In NIF, this is done in just two steps with a single laser. Figure 10.63a shows the timing compared with that of a conventional single pulse. First, there is a low-velocity compression, which entrains more fuel mass and saves energy compared the normal procedure. This is followed by a strong

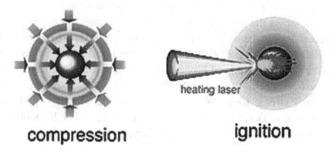

compression   **ignition**

Fig. 10.62 Stages of fast ignition [modified from Google Images; originally from Los Alamos National Laboratory's *Dense Plasma Theory*]

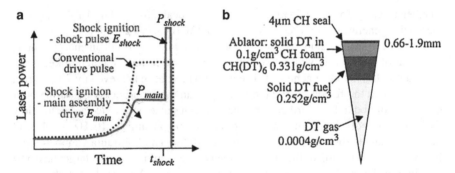

Fig. 10.63 (**a**) Time sequence of a shock ignition pulse, compared to a conventional pulse. The total time has to be short enough to avoid heating the hohlraum [L.J. Perkins et al., Phys. Rev. Lett. **103**, 045004 (2009)]. (**b**) Design of a capsule for shock ignition

Fig. 10.64 Computed fusion energy from shock ignition compared with normal compression [L.J. Perkins et al., *loc. cit.*]

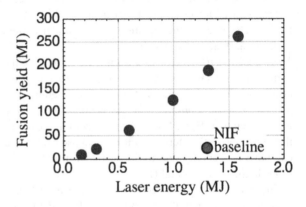

shock pulse timed to arrive when the compression reaches stagnation at a velocity of about 4×10^7 cm/s. The shock contains 20–30 % of the energy of the pulse. The hope is to generate 120–150 MJ of fusion power using only 1–1.6 MJ of laser power. Figure 10.63b is a possible design of a capsule suitable for shock ignition, and Fig. 10.64 shows the expected fusion energy *vs.* laser energy.

To improve the twice-a-day repetition rate, plans are to construct a High Repetition Rate Advanced Petawatt Laser System (HAPLS), designed in collaboration with the Czech Institute of Physics. The laser would produce 1 PW (10^{15} W) in 30 fs pulses at 10 Hz. In collaboration which Czech scientists, this would be installed at the Extreme Light Infrastructure Beamlines facility (ELI-Beamlines), under construction in Dolní Breźaňy near Prague in the Czech Republic. Such high rep-rate lasers are already available in KrF (krypton-fluoride) lasers.

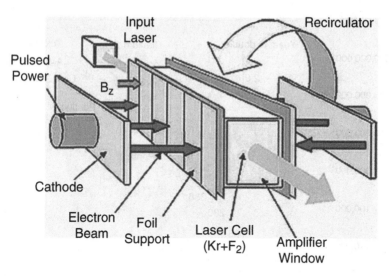

Fig. 10.65 Diagram of a krypton-fluoride laser amplifier [J.D. Sethian et al., Phys. Plasmas **10**, 2142 (2003)]

10.4.2 KrF Lasers

Krypton fluoride lasers, such as the Electra laser at the Naval Research Laboratory in the U.S., can pulse at 5 Hz at 400–700 J to give almost steady-state power. The technology can be extrapolated to 10^6 shots at 50–150 kJ for laser fusion. In the diagram of Fig. 10.65, an electron beam produced by pulsed power (Sect. 10.2.1.1) passes through a 1-mil (0.001 in.) Ti foil to ionize a cell containing Kr and F_2. The recirculator reflects the electrons to use them a second time. The KrF wavelength of 248 nm is even shorter than the 3ω, 351 nm wavelength of Nd-YAG lasers and is suitable for minimizing parametric instabilities. Though KrF lasers have much lower power than glass lasers, they can provide the rep rate needed for fusion.

10.5 Semiconductor Etching

All the computers, iPhones, iPads, and tablets are based on a semiconductor chip about 1 cm^2 in size and containing billions of transistors. The doubling of the transistor count every 2 years is known as Moore's Law, shown in Fig. 10.66. A "chip" can contain several cores, or Central Processing Units (CPUs). The progress is so rapid that in 1989 the largest Intel chip contained 1.18 million transistors with a critical dimension (CD) of 1 μm for its smallest feature. In 2014, the count increased to 5.56 *billion* transistors with a CD of 22 nm (0.022 μm).

Chips cannot be made without plasmas. The transistors lie on several layers, and connections are made between layers with copper or aluminum conductors

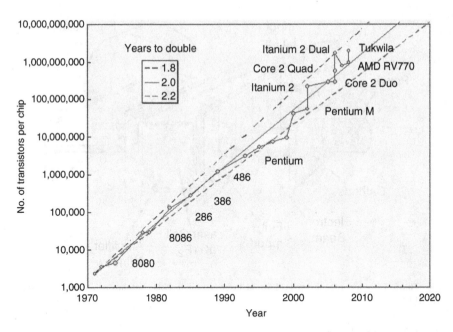

Fig. 10.66 Moore's law of transistors on a chip

Fig. 10.67 Cross section of a transistor on a polycrystalline substrate

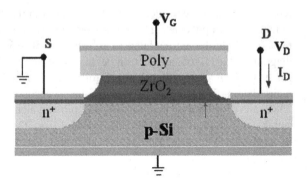

deposited into channels called *vias*. The transistors themselves are made by repeated plasma etching and deposition steps. Some transistors, called FinFETs, are themselves three-dimensional. Hundreds of chips are made at once on a 300-mm wafer of single-crystal silicon. The plasma must be uniform over the wafer. Chips near the boundary are usually not as good as those inside and could be sold as lower-frequency chips. The industry is pursuing the next step of 450-mm wafers, which will greatly increase the yield but would require an even larger uniform plasma.

The structure of a transistor is shown in Fig. 10.67. Its function is to control an electrical current with a gate that requires very little power, just as a dam can control a large amount of hydro power by opening and closing an actual gate. In

Fig. 10.67 electrons flow from the source S to the drain D, so that the positive current I_D goes from D to S. The current is controlled by the voltage V_G, which is capacitively coupled through an insulator to change the width of the current channel.

To make such transistors requires a series of steps of masking, etching, and deposition. Silicon can be etched chemically with chlorine or fluorine, or physically with energetic beams of atoms or ions. Winters and Coburn showed that silicon can be etched with XeF_2, a gas, but that the etch rate is greatly enhanced by Ar^+. Their famous graph is shown in Fig. 10.68. With only XeF_2 or with only an argon ion beam accelerated through the sheath, the etch rate is more than an order of magnitude lower than with both. This effect has been explained by molecular dynamics simulations by David Graves. By following the motion of each atom in a surface layer of hydrogenated amorphous carbon, he showed, in Fig. 10.69, that an argon ion beam sputters away the H atoms near the surface to a depth increasing with A^+ energy.

The patterns of Fig. 10.67 are built up layer by layer with photolithography, a process shown in Fig. 10.70. A photoresist layer is sensitive to UV light and areas exposed to light can be chemically dissolved or retained. The resist blocks the etching beams so that the desired pattern is etched in the Si layer underneath. The resist is then dissolved away. Figure 10.71 shows the etching process. A plasma containing Cl, F, and Ar ions and neutrals is at the top, and ions are accelerated by the sheath potential down to the substrate. At the top of each column is the mask protecting the layers below it.

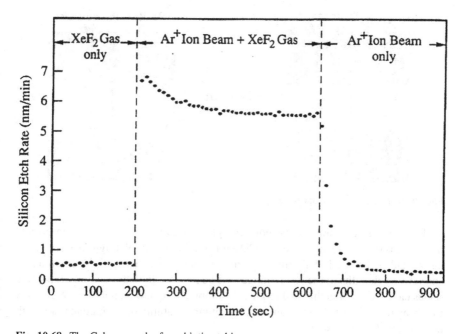

Fig. 10.68 The Coburn graph of symbiotic etching

Ar⁺ ion energy: 50 – 200 eV

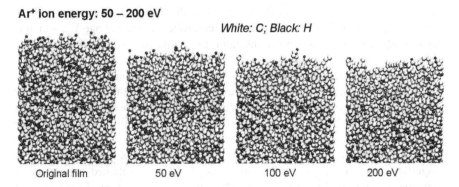

Original film   50 eV   100 eV   200 eV

Fig. 10.69 Molecular dynamics simulation of an aC-H (hydrogenated amorphous carbon) layer bombarded by argon ions at various energies [D.B. Graves, Gaseous Electronics Conference, 2011]

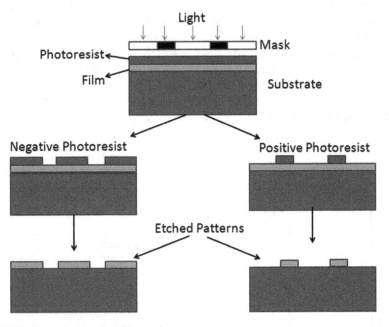

Fig. 10.70 Patterning by photolithography

The patterning of transistors requires light of wavelengths shorter than critical dimensions of their features. As the CDs got smaller, ultraviolet light from excimer lasers had to be used. This involved new optics that can pass UV. Eventually, shorter wavelengths in the X-ray range will be needed. In that case, only reflective optics can be used, and this will require development of new technology. We have a long way to go, however, to duplicate the computing ability of the human brain with an instrument of that size.

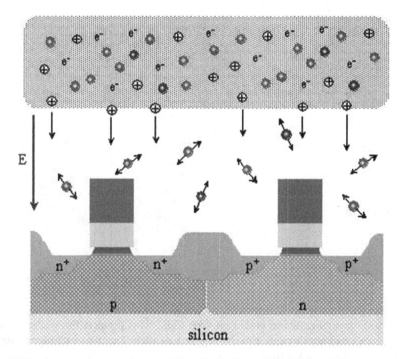

Fig. 10.71 Cartoon of semiconductor etching. The *top layer* of the columns is the mask

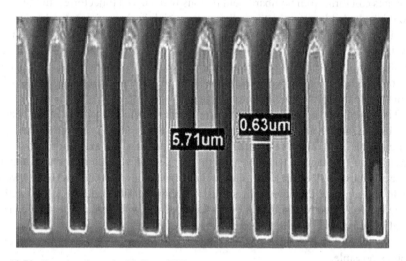

Fig. 10.72 Trenches formed with Deep RIE

When a deep trench or via is required, a process called DRIE (Deep Reactive Ion Etching, Fig. 10.72) is used. To keep the trench straight, etching and passivation are applied alternately. Etching is done with, say, SF_6; and, after a few seconds, an inert

Fig. 10.73 Notches at the
bottom of a trench

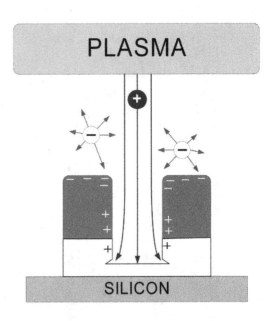

passivation layer like teflon is laid on with C_4F_8. This protects the walls, but the
etching ions can cut through it at the bottoms to allow chemical etching to occur.
This is sometimes called the Bosch process.

The bottoms of the trenches can have defects, such as notches or micro-trenches.
Notches can form from bombardment by ions with curved trajectories. In Fig. 10.73
the thick shaded layer at the top is the mask, which is charged negatively at the top
by electrons, which attract the ions, bending them towards the wall. The bottom of
the trench is charged positively by ions which have collected on the wall. The
resulting electric fields determine the ion trajectories which etch the notches at the
bottom. Figure 10.74 shows an SEM (Scanning Electron Microscope) image of an
actual trench deformed with microtrenches.

The mask shown in Fig. 10.73 is shown with rounded edges. Since the charges
that build up on these edges create E-fields, the ion orbits vary with these charges.
An example is shown in Fig. 10.75. This computation compares ion orbits with
sharp trench entrances with those with rounded ones, taking into account the
surface charges. A sharp edge shields ions from the sidewalls, whereas even a
slight rounding of the edge allows ions to deposit positive charge to the wall. A
more obvious electron shielding effect also occurs, with surface electron charges
repelling electrons from the nanometer-size trench.

This was a brief glimpse of the complex technology that makes modern electronics possible.

Fig. 10.74 Micro-trenches
at the bottom of a trench [R.
J. Hoekstra, M.J. Kushner,
V. Sukharev, and
P. Schoenborn, J. Vac. Sci.
Technol. B **16**, 2102 (1998)]

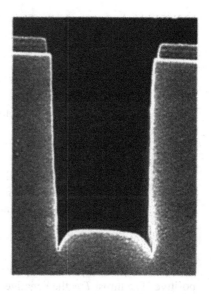

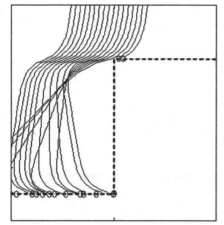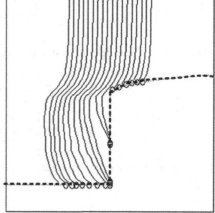

Fig. 10.75 Deflections of the orbits of ions from the plasma at the top caused by surface charges on the mask, for a square trench entrance (*left*), and a rounded one (*right*) [T.G. Madziwa-Nussinov, D. Arnush, and F.F. Chen, Phys. Plasmas **15**, 013503 (2008)]

10.6 Spacecraft Propulsion

10.6.1 General Principles

Once a satellite has been put into orbit, it needs a small amount of thrust once in a while to keep it there or to change its orbit. Chemical thrusters were formerly used but have been largely replaced by more efficient ion thrusters. These are of three kinds: gridded thrusters, Hall effect thrusters (HETs), and the new helicon thrusters, still under development. Ion thrusters are characterized by their *specific impulse* I_{sp}, defined by

$$I_{sp} \equiv v_{ex}/g \sec, \tag{10.22}$$

where v_{ex} is the exhaust velocity of the rocket, and g is the acceleration of gravity, of magnitude 9.8 m/s. Note that I_{sp} has the units of seconds. What this means physically is that if a rock is dropped from a height, it will reach the velocity v_{ex} in I_{sp} seconds.

The total amount of thrust is limited by the amount of mass available for ejection. Let M be the total mass of the load consisting of the intrinsic spacecraft mass m_s plus the decreasing propellant mass m_p:

$$M(t) = m_s + m_p(t). \tag{10.23}$$

In the frame of the spacecraft, its velocity v is zero, and its acceleration dv/dt is positive. The thrust T is the negative of the rate of change of exhaust momentum:

$$T = -\frac{d}{dt}\left(m_p v_{ex}\right) = -v_{ex}\frac{dM}{dt}. \tag{10.24}$$

The spacecraft's acceleration due to T is given by

$$T = \frac{d}{dt}(Mv) = M\frac{dv}{dt}, \tag{10.25}$$

$$\frac{dv}{dt} = -v_{ex}\frac{1}{M}\left(\frac{dM}{dt}\right) = -v_{ex}\left[\frac{d(\ln M)}{dt}\right]. \tag{10.26}$$

Integration from the initial velocity v_i to the final one v_f as m_p goes from m_{p0} to 0 gives

$$\int_{v_i}^{v_f} \frac{dv}{dt}dt \equiv \Delta v = -v_{ex}\int_{M_i}^{M_f} \frac{d(\ln M)}{dt}dt = -v_{ex}\ln\left(\frac{m_s}{m_s + m_{p0}}\right). \tag{10.27}$$

The mass of propellant needed for a given Δv is then

$$m_{p0} = m_s\left(e^{\Delta v/v_{ex}} - 1\right). \tag{10.28}$$

The amount of propellant needed to accelerate a given mass to a given velocity depends exponentially on the exhaust velocity v_{ex}.

10.6.2 Types of Thrusters

10.6.2.1 Gridded Thrusters

A schematic of a gridded thruster is shown in Fig. 10.76. A plasma is created by a voltage applied between a cathode and an anode grid. The cathode could be a good electron emitter like lanthanum hexaboride (LaB_6). Electrons are retained and ions are accelerated by grids. The ion beam will not detach from the spacecraft unless it is neutralized, so electron emitters have to be added at the sides to inject electrons into the accelerated ion beam and prevent the spacecraft from charging to a high negative potential. The neutralizers are usually hollow-cathode discharges. The ejected plume of plasma can then be formed by nozzles into an optimal shape. Sputtering limits the lifetime of the grids, but they have been designed to last at least 30,000 h.

10.6.2.2 Hall Thrusters

Elements of a Hall thruster are shown in Fig. 10.77, which is a cross section of a cylindrically symmetric device. A plasma is formed between the coaxial cylinders by applying a high voltage to the anode ring at the left. This voltage accelerates the

Fig. 10.76 A gridded ion thruster

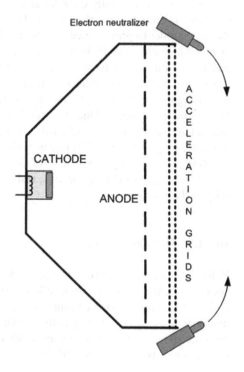

Fig. 10.77 A Hall effect
thruster

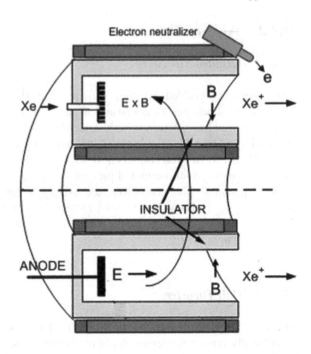

ions to v_{ex}. Holes in the ring also serve as the gas feed, as shown in the upper cross section. Xenon is normally used because of it is an inert, monatomic gas with high mass and low ionization potential. A radial magnetic field is created by coils (not as shown) in order to prevent electrons from following the ions. The electrons instead drift azimuthally with their $\mathbf{E} \times \mathbf{B}$ drift, forming the Hall current. Here again, electron neutralizers have to be added to neutralize the ejected ion beam. The function of the magnetic field is to prevent the electrons from moving axially and collapsing the anode voltage into a thin sheath at the anode. Hall thrusters have a few intrinsic problems. One is secondary emission of cold electrons from the internal surfaces which can upset the charge balance in the plasma. This can be minimized by coating the surfaces with carbon velvet. A second problem is instability, since many types of instabilities arise when there is a magnetic field (B-field). In spite of these problems, Hall thrusters have been engineered success-fully to fly in space.

10.6.2.3 Helicon Thrusters

Helicon plasmas, described in Sect. 9.5, could be advantageous for thrusters because of their high ionization efficiency and their thrust without an auxiliary E-field. Elements of a helicon thruster are shown in Fig. 10.78. The required DC magnetic field is generated by coils surrounding the tube. The RF is applied to an antenna, shown here as one side of a Boswell antenna, which is a version of the bidirectional Nagoya Type III antenna. Helical antennas can couple more

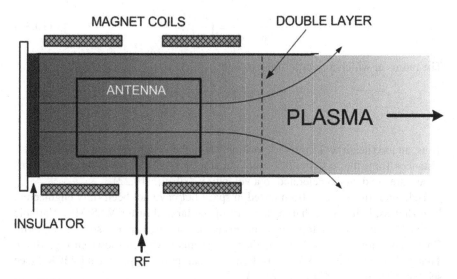

Fig. 10.78 Schematic of a helicon thruster [Drawn after C. Charles, J. Phys. D: Appl. Phys. **42** (2009) 163001]

efficiently to the dominant $m = +1$ azimuthal mode of the helicon wave, and ring antennas are used for the $m = 0$ mode. As the plasma leaves the uniform-field volume, the field lines diverge as the B-field expands. The electrons follow the field lines, and hence their density decreases. Since the electrons are essentially Maxwellian, the plasma potential must decrease with the density according to Eq. (3.73). There is therefore an electric field along the field lines shown in Fig. 10.78. This E-field accelerates ions; and, if $T_i \approx 0$, the ion velocity will eventually reach the acoustic velocity c_s. At that point, a sheath will form, according to Fig. 8.4, whether or not there is a wall there. This ion-rich sheath in "mid-air" will attract electrons which surround it, forming a *double layer* which, in 1D, shields the E-field from the rest of the plasma.

It is possible to calculate where this double layer will form. The ion energy at the sheath edge is $\frac{1}{2}Mc_s^2 = \frac{1}{2}KT_e$. To accelerate ions to this energy, the potential at the sheath edge, V_s, must be at least $-\frac{1}{2}KT_e/e$ if $V_s \equiv 0$ in the main plasma. Hence, the quasineutral density at the sheath edge, n_s, is given by Eq. (3.73) as

$$n_s = n_0 \exp(-\frac{1}{2}). \tag{10.29}$$

Since magnetic flux is conserved in the expansion from r_0 to r for each field line, and since electrons are constrained to follow the field lines, the field and density vary with r as

$$\frac{B}{B_0} = \frac{n}{n_0} = \left(\frac{r_0}{r}\right)^2. \tag{10.30}$$

The radius at which a sheath forms is then

$$\frac{r_s}{r_0} = \left(\frac{n_0}{n_s}\right)^{1/2} = e^{1/4} = 1.28. \tag{10.31}$$

Thus, an ion sheath will form at a position where the field lines have increased their distances from the axis by 28%. Ions passing through the potential drop of this sheath are suddenly accelerated to a velocity v_{ex} of Eq. (10.22).

Helicon thrusters have been tested in space but have not been fully engineered. Nonetheless, helicon discharges are part of the large thruster VASIMR (Variable Specific Impulse Magnetoplasma Rocket) being built by former astronaut Franklin Chang Díaz for travel to Mars. A diagram of this device is shown in Fig. 10.79. Though helicons are used for ionization, the main power is provided by ICRH (see section "Heating and Current Drive").

There will not be a more glamorous application of helicons!

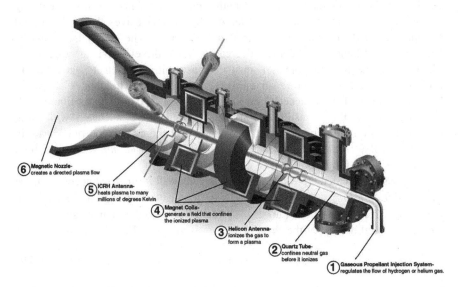

Fig. 10.79 The VASIMR rocket [Google images]

10.7 Plasmas in Everyday Life

Plasma physics may be an obscure and difficult science, but it is important because we see plasmas all the time. Every time we turn on the television, there are plasmas inside every pixel. Even before that, when we turn on the fluorescent lights, we are using a plasma. Older computer screens are also lit by fluorescents. The light of day is caused by the plasma in the sun's photosphere. Moonlight is a reflection of that. At night, the stars and nebulas can be seen by their plasma light. Without plasma, it would be very dark. Auroras are plasmas generated by the solar wind. When we get a shock when touching a doorknob after walking across a rug in winter, the spark is a plasma. Lightning is a larger form of that, coming from clouds. Ball lightning, though, is a slow-moving ball of plasma that no one understands.

We conclude with the topic that engendered plasma physics: nuclear fusion, which the general public has never heard of. By about 2050, there will be very little fossil fuel left; maybe dirty coal, but certainly not oil. Wind, solar, and hydro power are insufficient for the world's growing energy needs. Nuclear (fission) energy has its well known problems. To survive at least a few more centuries, mankind has to develop fusion power. This will require a workforce trained in plasma physics. The control of global warming starts in the classroom.

Errata to: Introduction to Plasma Physics and Controlled Fusion

Francis F. Chen

Erratum to:
F.F. Chen, *Introduction to Plasma Physics and Controlled Fusion*,
© Springer International Publishing Switzerland 2016
https://doi.org/10.1007/978-3-319-22309-4

The book was inadvertently published with the following errors and the same have been updated after publication.

1. The font of the symbol "nu" was changed to differentiate it from the symbol "vee".
2. Also, there were typos in vector and scalar notes throughout the book.

The updated online versions of these chapters can be found at
https://doi.org/10.1007/978-3-319-22309-4_2
https://doi.org/10.1007/978-3-319-22309-4_3
https://doi.org/10.1007/978-3-319-22309-4_4
https://doi.org/10.1007/978-3-319-22309-4_5
https://doi.org/10.1007/978-3-319-22309-4_6
https://doi.org/10.1007/978-3-319-22309-4_8
https://doi.org/10.1007/978-3-319-22309-4_10

The updated online versions of this book can be found at
https://doi.org/10.1007/978-3-319-22309-4

© Springer International Publishing Switzerland 2018   E1
F. Chen, *Introduction to Plasma Physics and Controlled Fusion*,
https://doi.org/10.1007/978-3-319-22309-4_11

Appendix A: Units, Constants and Formulas, Vector Relations

Units

The formulas in this book are written in the mks units of the International System (SI). In much of the research literature, however, the cgs-Gaussian system is still used. The following table compares the vacuum Maxwell equations, the fluid equation of motion, and the idealized Ohm's law in the two systems:

mks-SI	cgs-Gaussian
$\nabla \cdot \mathbf{D} = e(n_i - n_e)$	$\nabla \cdot \mathbf{E} = 4\pi e(n_i - n_e)$
$\nabla \times \mathbf{E} = -\dot{\mathbf{B}}$	$c\nabla \times \mathbf{E} = -\dot{\mathbf{B}}$
$\nabla \cdot \mathbf{B} = 0$	$\nabla \cdot \mathbf{B} = 0$
$\nabla \times \mathbf{H} = \mathbf{j} + \dot{\mathbf{D}}$	$c\nabla \times \mathbf{B} = 4\pi \mathbf{j} + \dot{\mathbf{E}}$
$\mathbf{D} = \epsilon_0 \mathbf{E} \quad \mathbf{B} = \mu_0 \mathbf{H}$	$\epsilon = \mu = 1$
$mn\frac{d\mathbf{v}}{dt} = qn(\mathbf{E} + \mathbf{v} \times \mathbf{B}) - \nabla p$	$mn\frac{d\mathbf{v}}{dt} = qn\left(\mathbf{E} + \frac{1}{c}\mathbf{v} \times \mathbf{B}\right) - \nabla p$
$\mathbf{E} + \mathbf{v} \times \mathbf{B} = 0$	$\mathbf{E} + \frac{1}{c}\mathbf{v} \times \mathbf{B} = 0$

The equation of continuity is the same in both systems.

In the Gaussian system, all electrical quantities are in electrostatic units (esu) except $\mathbf{B}$, which is in gauss (emu); the factors of c are written explicitly to accommodate this exception. In the mks system, $\mathbf{B}$ is measured in tesla (Wb/m^2), each of which is worth 10^4 gauss. Electric fields $\mathbf{E}$ are in esu/cm in cgs and V/m in mks. Since one esu of potential is 300 V, one esu/cm is the same as 3×10^4 V/m. The ratio of E to B is dimensionless in the Gaussian system, so that $v_E = cE/B$. In the mks system, E/B has the dimensions of a velocity, so that $v_E = E/B$. This fact is useful to keep in mind when checking the dimensions of various terms in an equation in looking for algebraic errors.

The original version of this chapter was revised. An erratum to this chapter can be found at https://doi.org/10.1007/978-3-319-22309-4_11

© Springer International Publishing Switzerland 2016
F.F. Chen, *Introduction to Plasma Physics and Controlled Fusion*,
DOI 10.1007/978-3-319-22309-4

The current density $\mathbf{j} = ne\mathbf{v}$ has the same form in both systems. In cgs, n and $\mathbf{v}$ are in cm^{-3} and cm/s, and e has the value $e = 4.8 \times 10^{-10}$ esu; then $\mathbf{j}$ comes out in esu/cm^2, where 1 esu of current equals c^{-1} emu or $10/c = 1/(3 \times 10^9)$ A. In mks, n and $\mathbf{v}$ are in m^{-3} and m/s, and e has the value $e = 1.6 \times 10^{-19}$ C; then $\mathbf{j}$ comes out in A/m^2.

Most cgs formulas can be converted to mks by replacing B/c by B and 4π by ϵ_0^{-1}, where $1/4\pi\epsilon_0 = 9 \times 10^9$. For instance, electric field energy density is $E^2/8\pi$ in cgs and $\epsilon_0 E^2/2$ in mks, and magnetic field energy density is $B^2/8\pi$ in cgs and $B^2/2\mu_0$ in mks. Here we have used the fact that $(\epsilon_0\mu_0)^{-1/2} = c = 3 \times 10^8$ m/s.

The energy KT is usually given in electron volts. In cgs, one must convert T_{eV} to ergs by multiplying by 1.6×10^{-12} erg/eV. In mks, one converts T_{eV} to joules by multiplying by 1.6×10^{-19} J/eV. This last number is, of course, just the charge e in mks, since that is how the electron volt is defined.

Useful Constants and Formulas

Constants

		mks	cgs
c	Velocity of light	3×10^8 m/s	3×10^{10} cm/s
e	Electron charge	1.6×10^{-19} C	4.8×10^{-10} esu
m	Electron mass	0.91×10^{-30} kg	0.91×10^{-27} g
M	Proton mass	1.67×10^{-27} kg	1.67×10^{-24} g
M/m		1837	1837
$(M/m)^{1/2}$		43	43
K	Boltzmann's constant	1.38×10^{-23} J/K	1.38×10^{-16} erg/K
eV	Electron volt	1.6×10^{-19} J	1.6×10^{-12} erg
1 eV	Of temperature KT	11,600 K	11,600 K
ϵ_0	Permittivity of free space	8.854×10^{-12} F/m	
μ_0	Permeability of free space	$4\pi \times 10^{-7}$ H/m	
πa_0^2	Cross section of H atom	0.88×10^{-20} m^2	0.88×10^{-16} cm^2
Density of neutral atoms at room temperature and 1 mTorr pressure		3.3×10^{19} m^{-3}	3.3×10^{13} cm^{-3}

Formulas (H) for hydrogen

		mks	cgs-Gaussian	Handy formula (n in cm^{-3})
ω_p	Plasma frequency	$\left(\dfrac{ne^2}{\epsilon_0 m}\right)^{1/2}$	$\left(\dfrac{4\pi ne^2}{m}\right)^{1/2}$	$f_p = 9000\sqrt{n}\,s^{-1}$
ω_c	Electron cyclotron frequency	$\dfrac{eB}{m}$	$\dfrac{eB}{mc}$	$f_c = 2.8$ GHz/kG
λ_D	Debye length	$\left(\dfrac{\epsilon_0 KT_e}{ne^2}\right)^{1/2}$	$\left(\dfrac{KT_e}{4\pi ne^2}\right)^{1/2}$	$740(T_{eV}/n)^{1/2}$ cm
r_L	Larmor radius	$\dfrac{mv_\perp}{eB}$	$\dfrac{mv_\perp c}{eB}$	$\dfrac{1.4 T_{eV}^{1/2}}{B_{kG}}$ mm(H)

(continued)

Formulas (H) for hydrogen

		mks	cgs-Gaussian	Handy formula (n in cm^{-3})
v_A	Alfvén speed	$\dfrac{B}{(\mu_0\rho)^{1/2}}$	$\dfrac{B}{(4\pi\rho)^{1/2}}$	$2.2 \times 10^{11}\dfrac{B}{\sqrt{n}}\dfrac{\text{cm}}{\text{s}}$ (H)
v_s	Acoustic speed ($T_i = 0$)	$\left(\dfrac{KT_e}{M}\right)^{1/2}$	$\left(\dfrac{KT_e}{M}\right)^{1/2}$	$10^6 T_{\text{eV}}^{1/2}\dfrac{\text{cm}}{\text{s}}$ (H)
v_E	$E \times B$ drift speed	$\dfrac{E}{B}$	$\dfrac{cE}{B}$	$10^8 \dfrac{E(\text{V/cm})}{B(\text{G})}\dfrac{\text{cm}}{\text{s}}$
v_D	Diamagnetic drift speed	$\dfrac{KT}{eB}\dfrac{n'}{n}$	$\dfrac{cKT}{eB}\dfrac{n'}{n}$	$10^8 \dfrac{T_{\text{eV}}}{B}\dfrac{1}{R}\dfrac{\text{cm}}{\text{s}}$
β	Magnetic/plasma pressure	$\dfrac{nKT}{B^2/2\mu_0}$	$\dfrac{nKT}{B^2/8\pi}$	
v_{the}	Electron thermal speed	$\left(\dfrac{2KT_e}{m}\right)^{1/2}$	$\left(\dfrac{2KT_e}{m}\right)^{1/2}$	$5.9 \times 10^7 T_{\text{eV}}^{1/2}\dfrac{\text{cm}}{\text{s}}$
ν_{ei}	Electron–ion collision frequency		$\approx \dfrac{\omega_p}{N_D}$	$\simeq 2 \times 10^{-6}\dfrac{Zn_e \ln\Lambda}{T_{\text{eV}}^{3/2}}s^{-1}$
ν_{ee}	Electron–electron collision frequency			$\simeq 5 \times 10^{-6}\dfrac{n\ln\Lambda}{T_{\text{eV}}^{3/2}}s^{-1}$
ν_{ii}	Ion–ion collision frequency		$Z^4\left(\dfrac{m}{M}\right)^{1/2}\left(\dfrac{T_e}{T_i}\right)^{3/2}\nu_{ee}$	
λ_{ei}	Collision mean free path		$\approx \lambda_{ee} \approx \lambda_{ii}$	$\simeq 3.4 \times 10^{13}\dfrac{T_{\text{eV}}^2}{n\ln\Lambda}\text{cm(H)}$
v_{osc}	Peak electron quiver velocity	$\dfrac{eE_0}{m\omega_0}$	$\dfrac{eE_0}{m\omega_0}$	$\dfrac{v_{\text{osc}}^2}{c^2} = 7.3 I_{19}\lambda_\mu^2$
				$\dfrac{v_{\text{osc}}^2}{v_e^2} = 3.7\dfrac{I_{13}\lambda_\mu^2}{T_{\text{eV}}}$

Useful Vector Relations

$$\mathbf{A} \cdot (\mathbf{B} \times \mathbf{C}) = \mathbf{B} \cdot (\mathbf{C} \times \mathbf{A}) = \mathbf{C} \cdot (\mathbf{A} \times \mathbf{B}) \equiv (\mathbf{ABC})$$

$$\mathbf{A} \times (\mathbf{B} \times \mathbf{C}) = \mathbf{B}(\mathbf{A} \cdot \mathbf{C}) - \mathbf{C}(\mathbf{A} \cdot \mathbf{B})$$

$$(\mathbf{A} \times \mathbf{B}) \cdot (\mathbf{C} \times \mathbf{D}) = (\mathbf{A} \cdot \mathbf{C})(\mathbf{B} \cdot \mathbf{D}) - (\mathbf{A} \cdot \mathbf{D})(\mathbf{B} \cdot \mathbf{C})$$

$$(\mathbf{A} \times \mathbf{B}) \times (\mathbf{C} \times \mathbf{D}) = (\mathbf{ABD})\mathbf{C} - (\mathbf{ABC})\mathbf{D} = (\mathbf{ACD})\mathbf{B} - (\mathbf{BCD})\mathbf{A}$$

$$\nabla \cdot (\phi\mathbf{A}) = \mathbf{A} \cdot \nabla\phi + \phi\nabla \cdot \mathbf{A}$$

$$\nabla \times (\phi\mathbf{A}) = \nabla\phi \times \mathbf{A} + \phi\nabla \times \mathbf{A}$$

$$\mathbf{A} \times (\nabla \times \mathbf{B}) = \nabla(\mathbf{A} \cdot \mathbf{B}) - (\mathbf{A} \cdot \nabla)\mathbf{B} - (\mathbf{B} \cdot \nabla)\mathbf{A} - \mathbf{B} \times (\nabla \times \mathbf{A})$$

$$(\mathbf{A} \cdot \nabla)\mathbf{A} = \nabla\left(\tfrac{1}{2}A^2\right) - \mathbf{A} \times (\nabla \times \mathbf{A})$$

$$\nabla \cdot (\mathbf{A} \times \mathbf{B}) = \mathbf{B} \cdot (\nabla \times \mathbf{A}) - \mathbf{A} \cdot (\nabla \times \mathbf{B})$$

$$\nabla \times (\mathbf{A} \times \mathbf{B}) = \mathbf{A}(\nabla \cdot \mathbf{B}) - \mathbf{B}\nabla \cdot \mathbf{A} + (\mathbf{B} \cdot \nabla)\mathbf{A} - (\mathbf{A} \cdot \nabla)\mathbf{B}$$

$$\nabla \times [(\mathbf{A} \cdot \nabla)\mathbf{A}] = (\mathbf{A} \cdot \nabla)(\nabla \times \mathbf{A}) + (\nabla \cdot \mathbf{A})(\nabla \times \mathbf{A}) - [(\nabla \times \mathbf{A}) \cdot \nabla]\mathbf{A}$$

$$\nabla \times \nabla \times \mathbf{A} = \nabla(\nabla \cdot \mathbf{A}) - (\nabla \cdot \nabla)\mathbf{A}$$

$$\nabla \times \nabla\phi = 0$$

$$\nabla \cdot (\nabla \times \mathbf{A}) = 0$$

Cylindrical Coordinates (r, θ, z)

$$\nabla^2\phi = \frac{1}{r}\frac{\partial}{\partial r}\left(r\frac{\partial\phi}{\partial r}\right) + \frac{1}{r^2}\frac{\partial^2\phi}{\partial\theta^2} + \frac{\partial^2\phi}{\partial z^2}$$

$$\nabla \cdot \mathbf{A} = \frac{1}{r}\frac{\partial}{\partial r}(rA_r) + \frac{1}{r}\frac{\partial}{\partial\theta}A_\theta + \frac{\partial}{\partial z}A_z$$

$$\nabla \times \mathbf{A} = \left(\frac{1}{r}\frac{\partial A_z}{\partial\theta} - \frac{\partial A_\theta}{\partial z}\right)\hat{\mathbf{r}} + \left(\frac{\partial A_r}{\partial z} - \frac{\partial A_z}{\partial r}\right)\hat{\mathbf{\theta}} + \left[\frac{1}{r}\frac{\partial}{\partial r}(rA_\theta) - \frac{1}{r}\frac{\partial A_r}{\partial\theta}\right]\hat{\mathbf{z}}$$

$$\nabla^2\mathbf{A} = (\nabla \cdot \nabla)\mathbf{A} = \left[\nabla^2 A_r - \frac{1}{r^2}\left(A_r + 2\frac{\partial A_\theta}{\partial\theta}\right)\right]\hat{\mathbf{r}}$$

$$+ \left[\nabla^2 A_\theta - \frac{1}{r^2}\left(A_\theta + 2\frac{\partial A_r}{\partial\theta}\right)\right]\hat{\mathbf{\theta}} + \nabla^2 A_z\hat{\mathbf{z}}$$

$$(\mathbf{A} \cdot \nabla)\mathbf{B} = \hat{\mathbf{r}}\left(A_r\frac{\partial B_r}{\partial r} + A_\theta\frac{1}{r}\frac{\partial B_r}{\partial\theta} + A_z\frac{\partial B_r}{\partial z} - \frac{1}{r}A_\theta B_\theta\right)$$

$$+ \hat{\mathbf{\theta}}\left(A_r\frac{\partial B_\theta}{\partial r} + A_\theta\frac{1}{r}\frac{\partial B_\theta}{\partial\theta} + A_z\frac{\partial B_\theta}{\partial z} - \frac{1}{r}A_\theta B_r\right)$$

$$+ \hat{\mathbf{z}}\left(A_r\frac{\partial B_z}{\partial r} + A_\theta\frac{1}{r}\frac{\partial B_z}{\partial\theta} + A_z\frac{\partial B_z}{\partial z}\right)$$

Appendix B: Theory of Waves in a Cold Uniform Plasma

As long as $T_e = T_i = 0$, the waves described in Chap. 4 can easily be generalized to an arbitrary number of charged particle species and an arbitrary angle of propagation θ relative to the magnetic field. Waves that depend on finite T, such as ion acoustic waves, are not included in this treatment.

First, we define the dielectric tensor of a plasma as follows. The fourth Maxwell equation is

$$\nabla \times \mathbf{B} = \mu_0 \left(\mathbf{j} + \epsilon_0 \dot{\mathbf{E}} \right) \tag{B.1}$$

where $\mathbf{j}$ is the plasma current due to the motion of the various charged particle species s, with density n_s, charge q_s, and velocity $\mathbf{v}_s$:

$$\mathbf{j} = \sum_s n_s q_s \mathbf{v}_s \tag{B.2}$$

Considering the plasma to be a dielectric with internal currents $\mathbf{j}$, we may write Eq. (B.1) as

$$\nabla \times \mathbf{B} = \mu_0 \dot{\mathbf{D}} \tag{B.3}$$

where

$$\mathbf{D} = \epsilon_0 \mathbf{E} + \frac{i}{\omega} \mathbf{j} \tag{B.4}$$

Here we have assumed an $\exp(-i\omega t)$ dependence for all plasma motions. Let the current $\mathbf{j}$ be proportional to $\mathbf{E}$ but not necessarily in the same direction (because of the magnetic field $B_0 \hat{\mathbf{z}}$); we may then define a conductivity tensor $\boldsymbol{\sigma}$ by the relation

© Springer International Publishing Switzerland 2016
F.F. Chen, *Introduction to Plasma Physics and Controlled Fusion*,
DOI 10.1007/978-3-319-22309-4

$$\mathbf{j} = \boldsymbol{\sigma} \cdot \mathbf{E} \tag{B.5}$$

Equation (B.4) becomes

$$\mathbf{D} = \epsilon_0 \left(\mathbf{I} + \frac{i}{\epsilon_0 \omega} \boldsymbol{\sigma} \right) \cdot \mathbf{E} = \boldsymbol{\epsilon} \cdot \mathbf{E} \tag{B.6}$$

Thus the effective dielectric constant of the plasma is the tensor

$$\boldsymbol{\epsilon} = \epsilon_0 (\mathbf{I} + i\boldsymbol{\sigma}/\epsilon_0 \omega) \tag{B.7}$$

where $\mathbf{I}$ is the unit tensor.

To evaluate $\boldsymbol{\sigma}$, we use the linearized fluid equation of motion for species s, neglecting the collision and pressure terms:

$$m_s \frac{\partial \mathbf{v}_s}{\partial t} = q_s (\mathbf{E} + \mathbf{v}_s \times \mathbf{B}_0) \tag{B.8}$$

Defining the cyclotron and plasma frequencies for each species as

$$\omega_{cs} \equiv \left| \frac{q_s B_0}{m_s} \right| \qquad \omega_{ps}^2 \equiv \left| \frac{n_0 q_s^2}{\epsilon_0 m_s} \right|, \tag{B.9}$$

we can separate Eq. (B.8) into x, y, and z components and solve for $\mathbf{v}_s$, obtaining

$$v_{xs} = \frac{iq_s}{m_s \omega} \frac{\left[E_x \pm i(\omega_{cs}/\omega)E_y \right]}{1 - (\omega_{cs}/\omega)^2} \tag{B.10a}$$

$$v_{ys} = \frac{iq_s}{m_s \omega} \frac{\left[E_y \mp i(\omega_{cs}/\omega)E_x \right]}{1 - (\omega_{cs}/\omega)^2} \tag{B.10b}$$

$$v_{zs} = \frac{iq_s}{m_s \omega} E_z \tag{B.10c}$$

where $\pm$ stands for the sign of q_s. The plasma current is

$$\mathbf{j} = \sum_s n_{0s} q_s \mathbf{v}_s \tag{B.11}$$

so that

$$\begin{aligned}
\frac{i}{\epsilon_0 \omega} j_x &= \sum_s \frac{i n_{0s}}{\epsilon_0 \omega} \frac{i q_s^2}{m_s \omega} \frac{E_x \pm i(\omega_{cs}/\omega)E_y}{1 - (\omega_{cs}/\omega)} \\
&= \sum_s -\frac{\omega_{ps}^2}{\omega^2} \frac{E_x \pm i(\omega_{cs}/\omega)E_y}{1 - (\omega_{cs}/\omega)}
\end{aligned} \tag{B.12}$$

Using the identities

$$\frac{1}{1-(\omega_{cs}/\omega)^2} = \frac{1}{2}\left[\frac{\omega}{\omega \mp \omega_{cs}} + \frac{\omega}{\omega \pm \omega_{cs}}\right]$$

$$\pm\frac{\omega_{cs}/\omega}{1-(\omega_{cs}/\omega)^2} = \frac{1}{2}\left[\frac{\omega}{\omega \mp \omega_{cs}} + \frac{\omega}{\omega \pm \omega_{cs}}\right],$$

$$\text{(B.13)}$$

we can write Eq. (B.12) as follows:

$$\frac{1}{\epsilon_0\omega}j_x = -\frac{1}{2}\sum_s \frac{\omega_{ps}^2}{\omega^2}\left[\left(\frac{\omega}{\omega \mp \omega_{cs}} + \frac{\omega}{\omega \pm \omega_{cs}}\right)E_x\right.$$

$$\left. + \left(\frac{\omega}{\omega \mp \omega_{cs}} + \frac{\omega}{\omega \pm \omega_{cs}}\right)iE_y\right].$$

$$\text{(B.14)}$$

Similarly, the y and z components are

$$\frac{1}{\epsilon_0\omega}j_y = -\frac{1}{2}\sum_s \frac{\omega_{ps}^2}{\omega^2}\left[\left(\frac{\omega}{\omega \pm \omega_{cs}} + \frac{\omega}{\omega \mp \omega_{cs}}\right)iE_x\right.$$

$$\left. + \left(\frac{\omega}{\omega \mp \omega_{cs}} + \frac{\omega}{\omega \pm \omega_{cs}}\right)E_y\right]$$

$$\text{(B.15)}$$

$$\frac{i}{\epsilon_0\omega}j_z = -\sum_s \frac{\omega_{ps}^2}{\omega^2}E_z$$

$$\text{(B.16)}$$

Use of Eq. (B.14) in Eq. (B.4) gives

$$\frac{1}{\epsilon_0}D_x = E_x - \frac{1}{2}\sum_s \left[\frac{\omega_{ps}^2}{\omega^2}\left(\frac{\omega}{\omega \mp \omega_{cs}} + \frac{\omega}{\omega \pm \omega_{cs}}\right)E_x\right.$$

$$\left. + \frac{\omega_{ps}^2}{\omega^2}\left(\frac{\omega}{\omega \mp \omega_{cs}} - \frac{\omega}{\omega \pm \omega_{cs}}\right)iE_y\right].$$

$$\text{(B.17)}$$

We define the convenient abbreviations

$$R \equiv 1 - \sum_s \frac{\omega_{ps}^2}{\omega^2} \left(\frac{\omega}{\omega \pm \omega_{cs}} \right)$$

$$L \equiv 1 - \sum_s \frac{\omega_{ps}^2}{\omega^2} \left(\frac{\omega}{\omega \mp \omega_{cs}} \right)$$

$$S \equiv \frac{1}{2}(R + L) \qquad D \equiv \frac{1}{2}(R - L)^* \tag{B.18}$$

$$P \equiv 1 - \sum_s \frac{\omega_{ps}^2}{\omega^2}$$

Using these in Eq. (B.17) and proceeding similarly with the y and z components, we obtain

$$\epsilon_0^{-1} D_x = S E_x - i D E_y$$
$$\epsilon_0^{-1} D_y = i D E_x + S E_y \tag{B.19}$$
$$\epsilon_0^{-1} D_z = P E_z$$

Comparing with Eq. (B.6), we see that

$$\boldsymbol{\epsilon} = \epsilon_0 \begin{pmatrix} S & -iD & 0 \\ iD & S & 0 \\ 0 & 0 & P \end{pmatrix} \equiv \epsilon_0 \boldsymbol{\epsilon}_R \tag{B.20}$$

We next derive the wave equation by taking the curl of the equation $\nabla \times \mathbf{E} = -\dot{\mathbf{B}}$ and substituting $\nabla \times \mathbf{B} = \mu_0 \boldsymbol{\epsilon} \cdot \dot{\mathbf{E}}$, obtaining

$$\nabla \times \nabla \times \mathbf{E} = -\mu_0 \epsilon_0 \left(\boldsymbol{\epsilon}_R \cdot \ddot{\mathbf{E}} \right) = -\frac{1}{c^2} \boldsymbol{\epsilon}_R \cdot \ddot{\mathbf{E}} \tag{B.21}$$

Assuming an exp $(i\mathbf{k} \cdot \mathbf{r})$ spatial dependence of $\mathbf{E}$ and defining a vector index of refraction

$$\boldsymbol{\mu} = \frac{c}{\omega} \mathbf{k}, \tag{B.22}$$

---

*Note that D here stands for "difference." It is not the displacement vector $\mathbf{D}$.

we can write Eq. (B.21) as

$$\boldsymbol{\mu} \times (\boldsymbol{\mu} \times \mathbf{E}) + \boldsymbol{\epsilon}_R \cdot \mathbf{E} = 0. \tag{B.23}$$

The uniform plasma is isotropic in the $x - y$ plane, so we may choose the y axis so that $k_y = 0$, without loss of generality. If θ is the angle between $\mathbf{k}$ and $\mathbf{B}_0$, we then have

$$\mu_x = \mu \sin \theta \qquad \mu_z = \mu \cos \theta \qquad \mu_y = 0 \tag{B.24}$$

The next step is to separate Eq. (B.23) into components, using the elements of $\boldsymbol{\epsilon}_R$ given in Eq. (B.20). This procedure readily yields

$$\mathbf{R} \cdot \mathbf{E} \equiv \begin{pmatrix} S - \mu^2 \cos^2\theta & -iD & \mu^2 \sin\theta \cos\theta \\ iD & S - \mu^2 & 0 \\ \mu^2 \sin\theta \cos\theta & 0 & P - \mu^2 \sin^2\theta \end{pmatrix} \begin{pmatrix} E_x \\ E_y \\ E_z \end{pmatrix} = 0. \tag{B.25}$$

From this it is clear that the E_x, E_y components are coupled to E_z only if one deviates from the principal angles $\theta = 0, 90°$.

Equation (B.25) is a set of three simultaneous, homogeneous equations; the condition for the existence of a solution is that the determinant of $\mathbf{R}$ vanish: $\|\mathbf{R}\| = 0$. Expanding in minors of the second column, we then obtain

$$\begin{aligned} (iD)^2 (P - \mu^2 \sin^2\theta) &+ (S - \mu^2) \\ &\times [(S - \mu^2 \cos^2\theta)(P - \mu^2 \sin^2\theta) - \mu^4 \sin^2\theta \cos^2\theta] = 0. \end{aligned} \tag{B.26}$$

By replacing $\cos^2\theta$ by $1 - \sin^2\theta$, we can solve for $\sin^2\theta$, obtaining

$$\sin^2\theta = \frac{-P(\mu^4 - 2S\mu^2 + RL)}{\mu^4(S - P) + \mu^2(PS - RL)}. \tag{B.27}$$

We have used the identity $S^2 - D^2 = RL$. Similarly,

$$\cos^2\theta = \frac{S\mu^4 - (PS + RL)\mu^2 + PRL}{\mu^4(S - P) + \mu^2(PS - RL)}. \tag{B.28}$$

Dividing the last two equations, we obtain

$$\tan^2\theta = \frac{P(\mu^4 - 2S\mu^2 + RL)}{S\mu^4 - (PS + RL)\mu^2 + PRL}.$$

Since $2S = R + L$, the numerator and denominator can be factored to give the cold-plasma dispersion relation

$$\tan^2\theta = \frac{P(\mu^2 - R)(\mu^2 - L)}{(S\mu^2 - RL)(\mu^2 - P)} \tag{B.29}$$

The principal modes of Chap. 4 can be recovered by setting $\theta = 0°$ and $90°$. When $\theta = 0°$, there are three roots: $P = 0$ (Langmuir wave), $\mu^2 = R$ (R wave), and $\mu^2 = L$ (L wave). When $\theta = 90°$, there are two roots: $\mu^2 = RL/S$ (extraordinary wave) and $\mu^2 = P$ (ordinary wave). By inserting the definitions of Eq. (B.18), one can verify that these are identical to the dispersion relations given in Chap. 4, with the addition of corrections due to ion motions.

The resonances can be found by letting μ go to ∞. We then have

$$\tan^2\theta_{\text{res}} = -P/S \tag{B.30}$$

This shows that the resonance frequencies depend on angle θ. If $\theta = 0°$, the possible solutions are $P = 0$ and $S = \infty$. The former is the plasma resonance $\omega = \omega_p$, while the latter occurs when either $R = \infty$ (electron cyclotron resonance) or $L = \infty$ (ion cyclotron resonance). If $\theta = 90°$, the possible solutions are $P = \infty$ or $S = 0$. The former cannot occur for finite ω_p and ω, and the latter yields the upper and lower hybrid frequencies, as well as the two-ion hybrid frequency when there is more than one ion species.

The cutoffs can be found by setting $\mu = 0$ in Eq. (B.26). Again using $S^2 - D^2 = RL$, we find that the condition for cutoff is independent of θ:

$$PRL = 0 \tag{B.31}$$

The conditions $R = 0$ and $L = 0$ yield the ω_R and ω_L cutoff frequencies of Chap. 4, with the addition of ion corrections. The condition $P = 0$ is seen to correspond to cutoff as well as to resonance. This degeneracy is due to our neglect of thermal motions. Actually, $P = 0$ (or $\omega = \omega_p$) is a resonance for longitudinal waves and a cutoff for transverse waves.

The information contained in Eq. (B.29) is summarized in the Clemmow–Mullaly–Allis diagram. One further result, not in the diagram, can be obtained easily from this formulation. The middle line of Eq. (B.25) reads

$$iDE_x + \left(S - \mu^2\right)E_y = 0 \tag{B.32}$$

Thus the polarization in the plane perpendicular to B_0 is given by

$$\frac{iE_x}{E_y} = \frac{\mu^2 - S}{D}. \tag{B.33}$$

From this it is easily seen that waves are linearly polarized at resonance ($\mu^2 = \infty$) and circularly polarized at cutoff ($\mu^2 = 0$, $R = 0$ or $L = 0$; thus $S = \pm D$).

Appendix C: Sample Three-Hour Final Exam

Part A (One Hour, Closed Book)

1. The number of electrons in a Debye sphere for $n = 10^{17}$ m^{-3}, $KT_e = 10$ eV is approximately

   (A) 135
   (B) 0.14
   (C) 7.4×10^3
   (D) 1.7×10^5
   (E) 3.5×10^{10}

2. The electron plasma frequency in a plasma of density $n = 10^{20}$ m^{-3} is

   (A) 90 MHz
   (B) 900 MHz
   (C) 9 GHz
   (D) 90 GHz
   (E) None of the above to within 10 %

3. A doubly charged helium nucleus of energy 3.5 MeV in a magnetic field of 8 T has a maximum Larmor radius of approximately

   (A) 2 mm
   (B) 2 cm
   (C) 20 cm
   (D) 2 m
   (E) 2 ft

4. A laboratory plasma with $n = 10^{16}$ m^{-3}, $KT_e = 2$ eV, $KT_i = 0.1$ eV, and $B = 0.3$ T has a beta (plasma pressure/magnetic field pressure) of approximately

   (A) 10^{-7}
   (B) 10^{-6}

© Springer International Publishing Switzerland 2016
F.F. Chen, *Introduction to Plasma Physics and Controlled Fusion*,
DOI 10.1007/978-3-319-22309-4

(C) 10^{-4}
(D) 10^{-2}
(E) 10^{-1}

5. The grad-B drift $\mathbf{v}_{\nabla B}$ is

   (A) always in the same direction as $\mathbf{v}_E$
   (B) always opposite to $\mathbf{v}_E$
   (C) sometimes parallel to $\mathbf{B}$
   (D) always opposite to the curvature drift $\mathbf{v}_R$
   (E) sometimes parallel to the diamagnetic drift $\mathbf{v}_D$

6. In the toroidal plasma shown, the diamagnetic current flows mainly in the direction

   (A) $+\hat{\boldsymbol{\phi}}$
   (B) $-\hat{\boldsymbol{\phi}}$
   (C) $+\hat{\boldsymbol{\theta}}$
   (D) $-\hat{\boldsymbol{\theta}}$
   (E) $+\hat{\mathbf{z}}$

7. In the torus shown above, torsional Alfvén waves can propagate in the directions

   (A) $\pm\hat{\mathbf{r}}$
   (B) $\pm\hat{\boldsymbol{\theta}}$
   (C) $\pm\hat{\boldsymbol{\phi}}$
   (D) $+\hat{\boldsymbol{\theta}}$ only
   (E) $-\hat{\boldsymbol{\theta}}$ only

8. Plasma A is ten times denser than plasma B but has the same temperature and composition. The resistivity of A relative to that of B is

   (A) 100 times smaller
   (B) 10 times smaller
   (C) approximately the same
   (D) 10 times larger
   (E) 100 times larger

9. The average electron velocity $\overline{|\mathbf{v}|}$ in a 10-keV Maxwellian plasma is

   (A) 7×10^2 m/s
   (B) 7×10^4 m/s
   (C) 7×10^5 m/s
   (D) 7×10^6 m/s
   (E) 7×10^7 m/s

10. Which of the following waves cannot propagate when $B_0 = 0$?

    (A) electron plasma wave
    (B) the ordinary wave

(C) Alfvén wave

(D) ion acoustic wave

(E) Bohm–Gross wave

11. A "backward wave" is one which has

(A) $\mathbf{k}$ opposite to $\mathbf{B}_0$

(B) $\omega/k < 0$

(C) $d\omega/dk < 0$

(D) $\mathbf{v}_i = -\mathbf{v}_e$

(E) $\mathbf{v}_\phi$ opposite to $\mathbf{v}_g$

12. "Cutoff" and "resonance," respectively, refer to conditions when the dielectric constant is

(A) 0 and ∞

(B) ∞ and 0

(C) 0 and 1

(D) 1 and 0

(E) not calculable from the plasma approximation

13. The lower and upper hybrid frequencies are, respectively,

(A) $(\Omega_p \Omega_c)^{1/2}$ and $(\omega_p \omega_c)^{1/2}$

(B) $\left(\Omega_p^2 + \Omega_c^2\right)^{1/2}$ and $\left(\omega_p^2 + \omega_c^2\right)^{1/2}$

(C) $(\omega_c \Omega_c)^{1/2}$ and $\left(\omega_p^2 + \omega_c^2\right)^{1/2}$

(D) $\left(\omega_p^2 - \omega_c^2\right)^{1/2}$ and $\left(\omega_p^2 + \omega_c^2\right)^{1/2}$

(E) $(\omega_R \omega_L)^{1/2}$ and $(\omega_p \omega_c)^{1/2}$

14. In a fully ionized plasma, diffusion across $\mathbf{B}$ is mainly due to

(A) ion–ion collisions

(B) electron–electron collisions

(C) electron–ion collisions

(D) three-body collisions

(E) plasma diamagnetism

15. An exponential density decay with time is characteristic of

(A) fully ionized plasmas under classical diffusion

(B) fully ionized plasmas under recombination

(C) weakly ionized plasmas under recombination

(D) weakly ionized plasmas under classical diffusion

(E) fully ionized plasmas with both diffusion and recombination

16. The whistler mode has a circular polarization which is

   (A) clockwise looking in the $+B_0$ direction
   (B) clockwise looking in the $-B_0$ direction
   (C) counterclockwise looking in the $+k$ direction
   (D) counterclockwise looking in the $-k$ direction
   (E) both, since the wave is plane polarized

17. The phase velocity of electromagnetic waves in a plasma

   (A) is always $>c$
   (B) is never $>c$
   (C) is sometimes $>c$
   (D) is always $<c$
   (E) is never $<c$

18. The following is *not* a possible way to heat a plasma:

   (A) Cyclotron resonance heating
   (B) Adiabatic compression
   (C) Ohmic heating
   (D) Transit time magnetic pumping
   (E) Neoclassical transport

19. The following is *not* a plasma confinement device:

   (A) Baseball coil
   (B) Diamagnetic loop
   (C) Figure-8 stellarator
   (D) Levitated octopole
   (E) Theta pinch

20. Landau damping

   (A) is caused by "resonant" particles
   (B) always occurs in a collisionless plasma
   (C) never occurs in a collisionless plasma
   (D) is a mathematical result which does not occur in experiment
   (E) is the residue of imaginary singularities lying on a semicircle

Part B (Two Hours, Open Book; Do 4 Out of 5)

1. Consider a cold plasma composed of n_0 hydrogen ions, $\frac{1}{2}n_0$ doubly ionized He ions, and $2n_0$ electrons. Show that there are two lower-hybrid frequencies and give an approximate expression for each. [Hint: You may use the plasma approximation, the assumption $m/M \ll 1$, and the formulas for v_1 given in the

text. (You need not solve the equations of motion again; just use the known solution.)]

2. Intelligent beings on a distant planet try to communicate with the earth by sending powerful radio waves swept in frequency from 10 to 50 MHz every minute. The linearly polarized emissions must pass through a radiation belt plasma in such a way that E and k are perpendicular to B_0. It is found that during solar flares (on their sun), frequencies between 24.25 and 28 MHz do not get through their radiation belt. From this deduce the plasma density and magnetic field there. (Hint: Do not round off numbers too early.)

3. When β is larger than m/M, there is a possibility of coupling between a drift wave and an Alfvén wave to produce an instability. A necessary condition for this to happen is that there be synchronism between the parallel wave velocities of the two waves (along B_0).

   (a) Show that the condition $\beta > m/M$ is equivalent to $v_A < v_{th}$.
   (b) If $KT_e = 10$ eV, $B = 0.2$ T, $k_y = 1$ cm^{-1}, and $n = 10^{21}$ m^{-3} find the required value of k_z for this interaction in a hydrogen plasma. You may assume $n_0'/n_0 = 1$ cm^{-1}, where $n_0' = dn_0/dr$.

4. When anomalous diffusion is caused by unstable oscillations, Fick's law of diffusion does not necessarily hold. For instance, the growth rate of drift waves depends on $\nabla n/n$, so that the diffusion coefficient $D_\perp$ can itself depend on ∇n. Taking a general form for $D_\perp$ in cylindrical geometry, namely,

$$D_\perp = Ar^s n^p \left(\frac{\partial n}{\partial r} \right)^q.$$

Show that the time behavior of a plasma decaying under diffusion follows the equation

$$\frac{\partial n}{\partial t} = f(r)n^{p+q+1}$$

Show also that the behavior of weakly and fully ionized plasmas is recovered in the proper limits.

5. In some semiconductors such as gallium arsenide, the current–voltage relation looks like this:

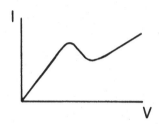

There is a region of negative resistance or mobility. Suppose you had a substance with negative mobility for all values of current. Using the equation of motion for weakly ionized plasmas with $KT = B = 0$, plus the electron continuity equation and Poisson's equation, perform the usual linearized wave analysis to show that there is instability for $\mu_e < 0$.

Appendix D: Answers to Some Problems

1.1 (a) At standard temperature and pressure, a mole of an ideal gas contains 6.022×10^{23} molecules (Avogadro's number) and occupies 22.4 L. Hence, the number per m^3 is $6.022 \times 10^{23}/2.24 \times 10^{-2} = 2.69 \times 10^{25}$ m^{-3}.

    (b) Since $PV = NRT$, $n = N/V = P/RT$. Hence $n_1/n_0 = P_1 T_0/P_0 T_1$. Taking n_0 to be the density in part (a) and n_1 to be that in part (b), we have

$$n_1 = (2.69 \times 10^{25}) \frac{10^{-3}}{760} \frac{273}{(273+20)} = 3.30 \times 10^{19} \text{m}^{-3}$$

Note that a diatomic gas such as H$_2$ will have twice as many *atoms* per torr as, say, He.

1.2

$$1 = \int_{-\infty}^{\infty} \hat{f}(u)du = Ah \int_{-\infty}^{\infty} e^{-u^2/h^2} d[u/h]$$

$$= Ah \int_{-\infty}^{\infty} e^{-x^2} dx = Ah\sqrt{\pi}$$

$$A = (2\pi KT/m)^{-1/2}$$

1.2a

$$\iint \hat{f}(u,v)dudv = 1 = A \iint e^{-(u^2+v^2)/h^2} dudv$$

$$1 = A \int_{-\infty}^{\infty} e^{-(u/h)^2} du \int_{-\infty}^{\infty} e^{-(v/h)^2} dv = Ah^2 \pi$$

$$A = (2\pi KT/m)^{-1}$$

© Springer International Publishing Switzerland 2016
F.F. Chen, *Introduction to Plasma Physics and Controlled Fusion*,
DOI 10.1007/978-3-319-22309-4

In cylindrical coordinates,

$$1 = A \iint_{-\infty}^{\infty} e^{-(u^2+v^2)/h^2}\,du\,dv = A \int_0^{\infty} \int_0^{2\pi} e^{-r^2/h^2} r\,dr\,d\phi = 2\pi A \int_0^{\infty} e^{-(r/h)^2} r\,dr$$

$$1 = 2\pi A h^2 \int_0^{\infty} e^{-x^2} x\,dx = \pi A h^2 \int_0^{\infty} e^{-y}\,dy = -\pi A h^2 e^{-y}\Big|_0^{\infty} = \pi A h^2$$

$$A = (2\pi KT/m)^{-1}$$

1.4

$$p = n(KT_e + KT_i) = 10^{21}\left(4 \times 10^4\right)\left(1.6 \times 10^{-19}\right)$$
$$= 6.4 \times 10^6\,\text{N/m}^2$$

$$1\,\text{atm} \simeq 10^5\,\text{N/m}^2 \qquad \therefore \quad p = 64\,\text{atm}$$
$$1\,\text{atm} \simeq 14.7\,\text{lb/in}^2. = (14.7)(144)/(2000)$$
$$= 1.06\,\text{tons/ft}^2$$

$$p \simeq 68\,\text{tons/ft}^2$$

1.5

$$\frac{d^2\phi}{dx^2} = -\frac{e(n_i - n_e)}{\varepsilon_0} = -\frac{1}{\varepsilon_0} n_\infty e\left(e^{-e\phi/KT_i} - e^{e\phi/KT_e}\right)$$

$$\simeq \frac{n_\infty e}{\varepsilon_0}\left(\frac{e\phi}{KT_i} + \frac{e\phi}{KT_e}\right)$$

$$\phi = \phi_0 e^{-|x|/\lambda_D}, \quad \text{where } \frac{1}{\lambda_D^2} = \frac{n_\infty e^2}{\varepsilon_0}\left(\frac{1}{KT_e} + \frac{1}{KT_i}\right)$$

$$\text{If } T_i \ll T_e \quad \lambda_D \simeq (\varepsilon_0 KT_i/n_\infty e^2)^{1/2}$$
$$\text{If } T_e \ll T_i \quad \lambda_D \simeq (\varepsilon_0 KT_e/n_\infty e^2)^{1/2}$$

However, this result is deceptive because in most experiments the ions move too slowly to shield charges. Electrons do the shielding, so λ_D depends on T_e, even when $T_e \gg T_i$, which is the usual case.

1.6 (a)

$$\frac{d^2\phi}{dx^2} = -\frac{nq}{\varepsilon_0}$$

Let $\phi = Ax^2 + Bx + C$; $\phi' = 2Ax + B$; $\phi'' = 2A$. At $x=0$, $\phi' = 0$ by symmetry
$\therefore$ $B = 0$. At $x = \pm d$, $\phi = 0$; therefore, $0 = Ad^2 + C$ and $C = -Ad^2$.

$$\phi'' = 2A = -\frac{nq}{\varepsilon_0} \quad \therefore \quad A = -\frac{1}{2\varepsilon_0}nq,$$

$$\phi = Ax^2 - Ad^2 = \frac{1}{2\varepsilon_0}nq\left(d^2 - x^2\right)$$

(b) Energy E to move a charge q from x_1 to x_2 is the change in potential energy
$\Delta(q\phi) = q(\phi_2 - \phi_1)$. Let $\phi_1 = 0$ at $x = \pm d$ and $\phi_2 = (\frac{1}{2}\varepsilon_0)nqd^2$ at $x = 0$. Then

$$E = \frac{1}{2\varepsilon_0}nq^2d^2.$$

Let $d = \lambda_D$; then

$$E = \frac{1}{2\varepsilon_0}nq^2\frac{KT\varepsilon_0}{nq^2} = \frac{1}{2}KT = E_{av}$$

for a one-dimensional Maxwellian distribution. Hence, if $d > \lambda_D$, $E > E_{av}$. If
the velocities are distributed in three dimensions, we have $E_{av} = \frac{3}{2}KT$ and
$E > \frac{1}{3}E_{av}$. The factor 3 is not important here. The point is that a thermal
particle would not have enough energy to go very far in a plasma ($d >> \lambda_D$)
if the charge of one species is not neutralized by another species.

1.7 (a) $\lambda_D = 7400(2/10^{16})^{1/2} = 10^{-4}$ m, $N_D = 4.8 \times 10^4$.

   (b) $\lambda_D = 7400(0.1/10^{12})^{1/2} = 2.3 \times 10^{-3}$ m, $N_D = 5.4 \times 10^4$.

   (c) $\lambda_D = 7400(800/10^{23})^{1/2} = 6.6 \times 10^{-7}$ m, $N_D = 1.2 \times 10^5$.

1.8

$$N_D = 1.38 \times 10^6 T^{3/2}/n^{1/2}$$
$$= \left(1.38 \times 10^6\right)\left(5 \times 10^7\right)^{3/2}/\left(10^{33}\right)^{1/2}$$
$$= 15.4$$

1.9 From Eq. (1.18), $\lambda_D = 69(T/n)^{1/2}$m, T in $^\circ$K

From Problem 1.5, $\lambda_D^{-2} = \frac{1}{69^2}\frac{n}{T} = \frac{10^6}{4760}\left(\frac{1}{100} + \frac{1}{100}\right) = 4.20$

Hence, $\lambda_D = 0.49$ m. The particle masses do not matter.

1.10 $\nabla^2\phi = \frac{1}{r^2}\frac{d}{dr}\left(r^2\frac{d\phi}{dr}\right) = \frac{e}{\varepsilon_0}(n_e - n_i) = \frac{e}{\varepsilon_0}n_0\left(e^{e\phi/KT_e} - 1\right) \approx \frac{e^2 n_0}{\varepsilon_0 KT_e}\phi = \phi/\lambda_D^2 = \kappa^2\phi$

where $\kappa \equiv 1/\lambda_D$.

Let $\phi = \Phi\frac{e^{-kr}}{r}$, where Φ is a constant in units of V-m.

$\frac{d\phi}{dr} = -\Phi\left(\frac{1}{r^2}e^{-kr} + \frac{k}{r}e^{-kr}\right) = -e^{-kr}\frac{\Phi}{r^2}(1+kr) \quad r^2\frac{d\phi}{dr} = \Phi(1+kr)e^{-kr}$

$\frac{d}{dr}\left(r^2\frac{d\phi}{dr}\right) = -\Phi\left[ke^{-kr} - k(1+kr)e^{-kr}\right] = -k\Phi e^{-kr}[1 - (1+kr)] = k^2 r^2\phi$

$\frac{1}{r^2}\frac{d}{dr}\left(r^2\frac{d\phi}{dr}\right) = k^2\phi \quad \therefore \quad k^2 = \kappa^2$

$\phi(r) = \Phi\frac{e^{-\kappa r}}{r} = \Phi\frac{e^{-r/\lambda_D}}{r}, \quad \phi_0 = \Phi\frac{e^{-\kappa a}}{a}, \quad \Phi = \frac{a\phi_0}{e^{-\kappa a}}$

$\phi = a\phi_0 e^{\kappa a}\frac{e^{-\kappa r}}{r} = \phi_0\frac{a}{r}e^{-\kappa(r-a)} = \phi_0\frac{a}{r}e^{-(r-a)/\lambda_D}$

1.11 Let $T_e = 300$ K. Then $\lambda_D = 69(T/n)^{1/2} = 69\left(300/(10^{22})\right)^{1/2} = 1.20 \times 10^{-8} =$ 12 nm

1.12 From Eq. (1.13), $f(u) = A\exp\left(-mu^2/2KT_e\right)$. From Eq. (1.6) and Problem 1.2,

$v_{th} \equiv (2KT_e/m)^{1/2}$ and $A = n(2\pi KT/m)^{-1/2} = n/v_{th}\sqrt{\pi}$

So $f(u) = (n/v_{th}\sqrt{\pi})\exp\left(-u^2/v_{th}^2\right)$. We wish to integrate $f(u)$ from u_{crit} to ∞ and from $-u_{crit}$ to $-\infty$. $\int_{u_{crit}}^{\infty} f(u)du = \frac{n}{v_{th}\sqrt{\pi}}\int_{u_{crit}}^{\infty} e^{-u^2/v_{th}^2}du =$ $\frac{n}{\sqrt{\pi}}\int_{y_{crit}}^{\infty} e^{-y^2}dy$, where $y \equiv u/v_{th}$.

Let $\frac{1}{2}mu_{crit}^2 = eV_{ioniz}$, $u_{crit} = (2eV_{ioniz}/m)^{1/2}$, where $V_{ioniz} = 15.8$ eV.

The error function erf(x) is defined as $\text{erf}(x) = \frac{2}{\sqrt{\pi}}\int_0^x e^{-t^2}dt$,

$\text{erfc}(x) = \frac{2}{\sqrt{\pi}}\int_x^{\infty} e^{-t^2}dt$

The integral from u_{crit} to ∞ is then $\frac{n}{\sqrt{\pi}}\int_{y_{crit}}^{\infty} e^{-y^2}dy = \frac{n}{2}\text{erfc}(y_{crit})$, where

$y_{crit} = u_{crit}/v_{th} = \left(\frac{2eV_{ioniz}}{m}\right)^{1/2}\left(\frac{m}{2KT_e}\right)^{1/2} = \left(\frac{eV_{ioniz}}{KT_e}\right)^{1/2}$. This density has to be doubled to account for negative velocities u. Finally, the fraction of electrons that can ionize is

$$\frac{\Delta n}{n} = \mathrm{erfc}\left(\frac{eV_{ioniz}}{KT_e}\right)^{1/2}.$$

2.1 (a) $E = \frac{1}{2}mv_\perp^2 \quad \therefore v_\perp = (2E/m)^{1/2}, r_L = mv_\perp/eB.$

$$v_\perp = \left[\frac{(2)(10^4)(1.6 \times 10^{-19})}{9.11 \times 10^{-31}}\right]^{1/2} = 5.93 \times 10^7 \mathrm{m/s}$$

$$r_L = \frac{(9.11 \times 10^{-31})(5.93 \times 10^7)}{(1.6 \times 10^{-19})(0.5 \times 10^{-4})} = 6.75\,\mathrm{m}$$

(b)

$$v_\perp = (300)(1000) = 3 \times 10^5 \mathrm{m/s}$$

$$r_L = \frac{(1.67 \times 10^{-27})(3 \times 10^5)}{(1.6 \times 10^{-19})(5 \times 10^{-9})} = 6.26 \times 10^5 \mathrm{m} = 626\,\mathrm{km}$$

(c)

$$v_\perp = \left[\frac{(2)(10^3)(1.6 \times 10^{-19})}{(4)(1.67 \times 10^{-27})}\right]^{1/2} = 2.19 \times 10^5 \mathrm{m/s}$$

$$r_L = \frac{(4)(1.67 \times 10^{-27})(2.19 \times 10^5)}{(1.6 \times 10^{-19})(5.00 \times 10^{-2})} = 0.183\,\mathrm{m}$$

(d)

$$r_L = \frac{2ME}{qB} = \frac{[(2)(4)(1.67 \times 10^{-27})(3.5 \times 10^6)(1.6 \times 10^{-19})]^{1/2}}{(2)(1.6 \times 10^{-19})(8)}$$

$$= 3.38 \times 10^{-2}\,\mathrm{m}$$

2.4 Let initial energy be $\mathcal{E}_0$, and Larmor radii r_1 and r_2, as shown. Energy at ① is $\mathcal{E}_1 = \mathcal{E}_0 + eEr_1$; energy at ② is $\mathcal{E}_2 = \mathcal{E}_0 - eEr_2$. (It would be acceptable to say: $\mathcal{E}_{1,2} = \mathcal{E}_0 \pm e\bar{E}r_L$ here.) Also $v_{\perp 1,2}^2 = 2\mathcal{E}_{1,2}/M$. We are asked to make the approximation

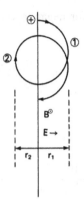

$$r_{1,2} = \frac{Mv_{\perp 1,2}}{eB} = \frac{M}{eB}\left(\frac{2\mathcal{E}_{1,2}}{M}\right)^{1/2}$$

$$= \frac{1}{\Omega_c}\left(\frac{2\mathcal{E}_0}{M}\right)^{1/2}\left(1 + \frac{eE}{\mathcal{E}_0}r_{1,2}\right)^{1/2}$$

For small E, expand the square root in a Taylor series:

$$r_{1,2} \simeq \frac{1}{\Omega_c}\left(\frac{2\mathcal{E}_0}{M}\right)^{1/2}\left(1 \pm \frac{1}{2}\frac{eE}{\mathcal{E}_0}r_{1,2}\right)$$

$$r_{1,2} = \frac{1}{\Omega_c}\left(\frac{2\mathcal{E}_0}{M}\right)^{1/2}\left[1 \pm \frac{1}{2}\frac{eE}{\mathcal{E}_0}\frac{1}{\Omega_c}\left(\frac{2\mathcal{E}_0}{M}\right)^{1/2}\right]^{-1}$$

$$\simeq \frac{1}{\Omega_c}\left(\frac{2\mathcal{E}_0}{M}\right)^{1/2}\left[1 \pm \frac{1}{2}\frac{eE}{\mathcal{E}_0}\frac{1}{\Omega_c}\left(\frac{2\mathcal{E}_0}{M}\right)^{1/2}\right]$$

Thus

$$r_1 - r_2 = \frac{eE}{\mathcal{E}_0}\frac{1}{\Omega_c^2}\left(\frac{2\mathcal{E}_0}{M}\right) = \frac{2eE}{M\Omega_c^2}$$

independent of $\mathcal{E}_0$. The guiding center moves a distance $2(r_1 - r_2)$ in a time $2\pi/\Omega_c$, so

$$v_{\text{gc}} = 2(r_1 - r_2)(\Omega_c/2\pi) = \frac{4eE}{M\Omega_c}\frac{1}{2\pi} = \frac{2}{\pi}\frac{E}{B} \simeq \frac{E}{B}$$

Thus the guiding center drift is independent of the ion energy $\mathcal{E}_0$. The factor $2/\pi$ would be 1 if we did not make the crude approximation.

2.5

(a)
$$n = n_0 e^{e\phi/KT_e} \quad \therefore \quad \phi = (KT_e/e)\ln(n/n_0)$$

$$\mathbf{E} = -\frac{\partial \phi}{\partial r}\hat{\mathbf{r}} = -\frac{KT_e}{e}\frac{1}{n}\frac{\partial n}{\partial r}\hat{\mathbf{r}} = \frac{KT_e}{e\lambda}\hat{\mathbf{r}}$$

(b) $\quad \mathbf{v}_E = -\dfrac{E_r}{B}\hat{\boldsymbol{\theta}} = -\dfrac{KT_e}{eB\lambda}\hat{\boldsymbol{\theta}}$

Consider electrons:

$$v_{\text{th}} = \left(\frac{2KT_e}{m}\right)^{1/2} \quad \therefore \quad |v_E| = \frac{KT_e}{m}\frac{m}{eB}\frac{1}{\lambda} = \frac{1}{2}\frac{v_{\text{th}}^2}{\omega_c}\frac{1}{\lambda}$$

Now, $r_L = mv_\perp/eB$, so for a distribution of velocities we must find an average r_L. Since $v_\perp$ contains two degrees of freedom, we have

$$\tfrac{1}{2}m\overline{v_\perp^2} = 2 \times \tfrac{1}{2}KT_e$$

The most convenient average is

$$\langle v_\perp \rangle_{\text{rms}} = (2KT_e/m)^{1/2} = v_{\text{th}}$$

Using this for $v_\perp$ in r_L, we have $\left|v_E\right| = \dfrac{1}{2}\dfrac{v_{\text{th}}}{\lambda}\dfrac{v_\perp}{\omega_c} = \dfrac{1}{2}\dfrac{v_{\text{th}}r_L}{\lambda}$

so that $|v_E| = v_{\text{th}}$ implies $r_L = 2\lambda$.

(c) If we take ions instead of electrons, we have $v_{\text{th}i} = (2KT_i/M)^{1/2} = v_{\perp i}$,

$r_{Li} = v_{\perp i}/\omega_{ci}$, and $\left|v_E\right| = \dfrac{1}{2\lambda}\left(\dfrac{2KT_e}{M}\right)\left(\dfrac{M}{eB}\right) = \dfrac{1}{2\lambda}\dfrac{T_e}{T_i}\dfrac{v_{\text{th}i}v_{\perp i}}{\omega_{ci}} = \dfrac{1}{2\lambda}\dfrac{T_e}{T_i}v_{\text{th}i}r_{Li}.$

If $|v_E| = v_{\text{th}i}$, it is still true that $r_{Li} = 2\lambda$ provided that $T_i = T_e$.

2.6 (a) $\quad n = n_0 \exp\left(e^{-r^2/a^2} - 1\right) = n_0\, e^{e\phi/KT_e} \quad \therefore \quad \dfrac{e}{KT_e}\phi(r) = e^{-r^2/a^2} - 1$

$$\mathbf{E} = -\nabla\phi = \frac{\partial\phi}{\partial r}\hat{\mathbf{r}} \qquad E_r(r) = -\frac{\partial\phi}{\partial r} = \frac{KT_e}{e}\frac{2r}{a^2}e^{-r^2/a^2}$$

$$\frac{dE_r}{dr} = \frac{2KT_e}{ea^2}\left(1 - \frac{2r^2}{a^2}\right)e^{-r^2/a^2} = 0 \qquad \frac{r^2}{a^2} = \frac{1}{2}$$

$$E_{\text{max}} = \frac{KT_e}{ea}\frac{2}{\sqrt{2}}e^{-1/2} = \frac{(0.2)\left(1.6 \times 10^{-19}\right)}{\left(1.6 \times 10^{-19}\right)(.01)}\sqrt{2}e^{-1/2} = 17\,\text{V/m}$$

$$\mathbf{v}_E = -\frac{E_r}{B}\hat{\boldsymbol{\theta}} \qquad V_{E\max} = \frac{E_{\max}}{B} = \frac{17}{0.2} = 8500 \text{ m/s}$$

(b) Compare the force Mg with the force eE for an ion. (mg for an electron would be 1836 times smaller.) $g = 9.80$ m/s^2. $Mg = (39)(1.67 \times 10^{-27})$ $(9.80) = 6.38 \times 10^{-25}$ N. $eE_{\max} = (1.6 \times 10^{-19})(17) = 2.75 \times 10^{-18}$ N $= 4 \times 10^6 Mg$. Hence gravitational drift is 4 million times smaller.

(c)

$$r_L = \frac{Mv_\perp}{eB} = 10^{-2}\text{m}$$

$$v_\perp = (2KT/M)^{1/2} = \left[\frac{(2)(0.2)(1.6 \times 10^{-19})}{(39)(1.67 \times 10^{-27})}\right]^{1/2}$$

$$= 9.9 \times 10^2\text{m/s}$$

$$B = \frac{(39)(1.67 \times 10^{-27})(9.9 \times 10^2)}{(10^{-2})(1.6 \times 10^{-19})} = 4.00 \times 10^{-2}\text{T}$$

2.8

$$B = \frac{c}{r^3} = \frac{0.3 \times 10^{-4}}{(r/R)^3}\text{T}$$

$$v_{\nabla B} = \frac{1}{2}v_\perp r_L \left|\frac{\mathbf{B} \times \nabla\mathbf{B}}{B^2}\right| = \frac{1}{2}v r_L \left|\frac{\nabla B}{B}\right|$$

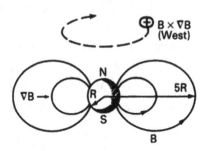

(a)

$$\nabla B = \frac{\partial B}{\partial r}\hat{\mathbf{r}} = -3\frac{c}{r^4}\hat{\mathbf{r}} = \frac{3}{r}B(-\hat{\mathbf{r}}) \qquad \left|\frac{\nabla B}{B}\right| = \frac{3}{r}$$

$$\frac{1}{2}v_\perp r_L = \frac{1}{2}\frac{v_\perp^2}{\omega_c} = \frac{1}{2}\frac{2KT/m}{eB/m} = \frac{KT}{eB} = \frac{(1.6 \times 10^{-19})(KT)_{eV}}{1.6 \times 10^{-19}}\frac{1}{B} = \frac{(KT)_{eV}}{B}$$

$$B(r = 5R) = \frac{0.3 \times 10^{-4}}{5^3} = 2.4 \times 10^{-7}\text{T}$$

$$5R = (5)(4000 \text{ miles})(1.6 \text{ km/mile})(10^3 \text{ m/km}) = 3.2 \times 10^7 \text{ m}$$

$$v_{\nabla B} = 10^8 \frac{(KT)_{eV}}{2.4 \times 10^{-7}} = 0.39(KT)_{eV} \text{m/s}$$

$$\text{Ions} : KT = 1\,\text{eV} \qquad v_{\nabla B} = 0.39\text{m/s}$$

$$\text{Electrons} : KT = 3 \times 10^4 \text{eV} \qquad v_{\nabla B} = 1.17 \times 10^4 \text{m/s}$$

(b) Ions: westward; electrons: eastward.
(c) $2\pi r = (6.28)(3.2 \times 10^7) = 2.0 \times 10^8$ m

$$t = \frac{2\pi r}{v_{\nabla B}} = \frac{(2.0 \times 10^8)}{(1.17 \times 10^4)} = 1.7 \times 10^4 s = 4.8\text{h}$$

(d)
$$j = nev_{\nabla B} \qquad \text{neglect ions}$$
$$= (10^7)(1.6 \times 10^{-19})(1.17 \times 10^4) = 1.87 \times 10^{-8} \text{A/m}^2$$

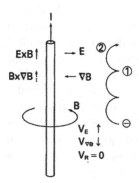

2.9 (a) $v_R = 0$, since the electron gains no energy in the parallel $\left(\hat{\theta} \right)$ direction. Since the electron starts at rest with no thermal energy, it will come back to rest after one cycle. Hence, the orbit has sharp cusps instead of loops. It is clear that the v_E drift must dominate, since the electron starts to the left, and the Lorentz force makes it move upwards.

(b) In cylindrical geometry, $\phi = A \ln r + B$. Since

$$\phi(10^{-3}) = 460V \quad \text{and} \quad \phi(0.1 \text{ m}) = 0,$$
$$460 = A \ln (10^{-3}) + B$$
$$0 = A \ln (0.1) + B \qquad B = -A \ln (0.1)$$
$$460 = A \ln (10^{-3}) - A \ln (0.1)$$
$$= A \ln (0.01) \qquad A = 460/\ln (0.01)$$
$$\phi(r) = \frac{460}{\ln(0.01)}[\ln r - \ln (0.1)] = 460\frac{\ln(0.1r)}{\ln 100} V$$
$$E = \frac{-\partial \phi}{\partial r} = \frac{-460}{\ln 100}\left(\frac{r}{0.1}\right)\left(\frac{-0.1}{r^2}\right) = \frac{460/r}{\ln 100} \frac{V}{m}$$
$$= \frac{460}{(4.6)(1)} = 10^4\frac{V}{m}\text{at} \quad r = 10^{-2}\text{m}$$
$$B = \frac{I(A)10^{-4}}{5r} = \frac{500 \times 10^{-4}}{(5)(1)} = 0.01 \text{ T}$$
$$|v_E| = |E/B| = 10^8 \frac{10^4 V/\text{cm}}{0.01 \text{ T}} = 10^6\text{m}/s$$

To estimate the ∇B drift, we must find $v_\perp$ in the frame moving with the guiding center. Remember that in deriving $v_{\nabla B}$, $v_\perp$ was taken as the velocity in the undisturbed circular orbit. Here, the latter is moving with velocity v_E, so that it does not look circular in the lab frame. Nonetheless, it can still be decomposed into a circular motion with velocity $v_\perp$ plus an $E \times B$ drift of the guiding center. Consider the z component of velocity (along the wire). At point ① on the orbit, $v_z = v_E + v \cos \omega_c t = 0$, where $\cos \omega_c t = -1$, its maximum negative value; hence, $v_E = v_\perp$. The same result can be obtained by considering that at point ② $v_z = v_E + v_\perp$ ($\cos \omega_c t = 1$). The energy there, $\frac{1}{2}(mv_z^2)$, must equal the energy gained in falling a distance $2r_L$ in an electric field. Thus

$$\frac{1}{2}m(v_E + v_\perp)^2 2r_L eE = 2eE\frac{mv_\perp}{eB} = 2mv_\perp\frac{E}{B} = 2mv_\perp v_E$$
$$v_E^2 + 2v_\perp v_E + v_\perp^2 = 4v_\perp v_E \quad (v_E - v_\perp)^2 = 0 \quad v_E = v_\perp$$

Now we can calculate $v_{\nabla B}$:

$$v_{\nabla B} = \frac{1}{2}\frac{v_\perp^2}{\omega_c}\left|\frac{\nabla B}{B}\right| \quad \omega_c = \frac{eB}{m} = \frac{(1.6 \times 10^{-19})(10^{-2})}{(9.11 \times 10^{-31})} = 1.76 \times 10^9 s^{-1}$$
$$\frac{dB}{dr} = \frac{I(-1)10^{-4}}{r^2} = -\frac{B}{r} \quad \left|\frac{\nabla B}{B}\right| = 10^2 \text{m}^{-1}$$
$$v_{\nabla B} = \frac{1}{2}\frac{v_E^2}{\omega_c}\frac{1}{210^{16}1.8 \times 10^9} = 2.8 \times 10^4 \text{m}/s$$

This amounts to a slowing down of the v_E drift due to a distortion of the orbit into a hairpin shape ⊱ because of the change in Larmor radius.

The *undisturbed* orbit is the path taken by the valve on a bicycle wheel as it rolls along:

Finally, we note that the finite Larmor radius correction to v_E is negligible:

$$\frac{1}{4}r_L^2 \nabla^2 \frac{E}{B} \simeq \frac{1}{4}\frac{r_L^2}{r^2}\frac{E}{B}$$

$$r_L = \frac{\left(9.11 \times 10^{-31}\right)\left(10^6\right)}{\left(1.6 \times 10^{-19}\right)(0.01)} = 5.7 \times 10^{-4}\text{m}$$

$$r \simeq 10^{-2}\text{m} \quad \therefore \frac{1}{4}\frac{r_L^2}{r^2} = 0.08\%$$

2.12 Let all velocities refer to the midplane, and let subscripts i and f refer to initial and final states (before and after acceleration).

(a) Given: $R_m = 5$, $v_{\perp i} = v_{\parallel i}$ since μ is conserved, $v_{\perp f} = v_{\perp i}$, and only $v_{\parallel}$ will increase. It will increase until the pitch angle θ reaches the loss cone:

$$\sin^2\theta_m = \frac{v_{\perp f}^2}{v_{\perp f}^2 + v_{\parallel f}^2} = \frac{1}{1 + v_{\parallel f}^2/v_{\perp i}^2} = \frac{1}{R_m} = \frac{1}{5}$$

Hence $v_{\parallel f}^2/v_{\perp i}^2 = 4$, $v_{\parallel f} = 2v_{\perp i}$. Energy is

$$E_f = \frac{1}{2}M\left(v_{\parallel f}^2 + v_{\perp f}^2\right) = \frac{1}{2}M(4 + 1)v_{\perp i}^2 = \frac{5}{2}mv_{\perp i}^2$$

$$E_i = \frac{1}{2}M\left(v_{\parallel i}^2 + v_{\perp i}^2\right) = \frac{1}{2}M(1 + 1)v_{\perp i}^2 = Mv_{\perp i}^2$$

$$\therefore E_f = 2.5E_i = (2.5)(1) = \underline{2.5\text{keV}}$$

(b) (1) Let particle have $v_0 > 0$ and hit piston moving at velocity $v_m < 0$. *In the frame of the piston, the particle bounces elastically and comes off with its initial velocity, but in the opposite direction. Let* ′ *refer to the frame of the piston. Initial and final velocities in this frame are*

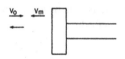

$$v_i' = v_0 - v_m \quad v_f' = -(v_0 - v_m)$$

(Note: v_m is negative.) Transforming back to lab frame,

$$v_f = v_f' + v_m = -v_0 + 2v_m$$

Since v_m is negative, the change in velocity is $2|v_m|$. QED

(2) At each bounce, the change in momentum is $\Delta p_{\parallel} = 2\,m|v_m|$. If N is the number of bounces, $p_{\parallel f} = p_{\parallel i} + N\Delta p$. Thus

$$N = \frac{p_{\parallel f} - p_{\parallel i}}{\Delta p} = \frac{v_{\parallel f} - v_{\parallel i}}{2v_m} = \frac{2v_{\perp i} - v_{\perp i}}{2v_m} = \frac{1 v_{\perp i}}{2 vm}$$

$$E_i = Mv_{\perp i}^2 = 1\,\text{keV} = (10^3)(1.6 \times 10^{-19}) = 1.6 \times 10^{-16}\text{J}$$

$$\therefore v_{\perp i} = \left(\frac{1.6 \times 10^{-16}}{1.67 \times 10^{-27}}\right)^{1/2} = 3.1 \times 10^5 \text{m/s}$$

$$v_m = 10^4 \text{m/s}$$

$$\therefore N = \frac{1}{2}\frac{(3 \times 10^5)}{10^4} = 15\,\text{bounces}$$

(3) Average $v_{\parallel}$ is

$$\bar{v} = \frac{1}{2}\left(v_{\parallel i} + v_{\parallel f}\right) = \frac{1}{2}(v_{\perp i} + 2v_{\perp i})$$

$$= \frac{3}{2}v_{\perp i} = 4.6 \times 10^5$$

$$L = 10^{13}\text{m}$$

$$\therefore t = \frac{NL}{\bar{v}} = \frac{(15)(10^{13})}{4.6 \times 10^5} = 3.2 \times 10^8 s$$

$$(= 10y)$$

However, L changes during this time by a distance

$$\Delta L = 2v_m t = (2)(10^4)(3.2 \times 10^8) = 6.4 \times 10^{12}\text{m}$$

so that actual time is more like 2.5×10^8 s. Since only factor-of-two accuracy is required, it is not necessary to sum the series—the above answer of 3.2×10^8 s will do.

2.13 (a)

$$\int v_{\parallel} ds \simeq v_{\parallel} L = \text{constant} \quad \therefore \dot{v}_{\parallel} L + v_{\parallel}\dot{L} = 0$$

(b)

$$\frac{\dot{v}_\parallel}{v_\parallel} = -\frac{\dot{L}}{L} \qquad \dot{v}_\parallel \simeq \frac{\Delta v_\parallel}{T} = \frac{v_\parallel}{L}(-\dot{L})$$

$$T \simeq \frac{\Delta v_\parallel}{v_\parallel}\frac{L}{-\dot{L}} = \frac{2v_{\perp i} - v_{\perp i}}{\frac{1}{2}(2v_{\perp i} + v_{\perp i})}\frac{L}{2v_m} = \frac{2}{3}\frac{10^{13}}{2 \times 10^4}$$

$$= \underline{3.3 \times 10^8 s}$$

2.14 As B increases, Maxwell's equation $\nabla \times \mathbf{E} = -\dot{\mathbf{B}}$ predicts an E-field. This induced E-field has a component along $\mathbf{v}$ and accelerates the particle. If B increases slowly and adiabatically, E will be small; but the integrated effect over many Larmor periods will be finite. The invariance of μ allows us to calculate the energy increases without doing this integration.

3.1 $\partial\sigma/\partial t + \nabla \cdot \mathbf{j} = 0$, where $\mathbf{j} = \mathbf{j}_P = (\rho/B^2)\dot{\mathbf{E}}$. Hence, $\dot{\sigma} = -\nabla \cdot [(\rho/B^2)\dot{\mathbf{E}}]$. The time derivative of Poisson's equation is $\nabla \cdot \dot{\mathbf{E}} = \dot{\sigma}/\epsilon_0$

$$\therefore \nabla \cdot \dot{\mathbf{E}} = -\left(\frac{1}{\epsilon_0}\right)\nabla \cdot \left(\frac{\rho}{B^2}\right)\dot{\mathbf{E}} \qquad \nabla \cdot \left(1 + \frac{\rho}{\epsilon_0 B^2}\right)\dot{\mathbf{E}} = 0$$

Assuming the dielectric constant $\in$ to be constant in time, we have $\nabla \cdot \dot{\mathbf{D}} = \nabla \cdot (\in\dot{\mathbf{E}}) = 0$. By comparison, $\in = 1 + \rho/\epsilon_0 B^2$.

3.2

$$\in \simeq 1 + \frac{nM}{\epsilon_0 B^2} \simeq \frac{\Omega_p^2}{\Omega_c^2} = \frac{ne^2}{\epsilon_0 M}\frac{M^2}{e^2 B^2} = \frac{nM}{\epsilon_0 B^2}$$

True if $\in \gg 1$.

3.3 Take divergence of Eqs. (3.56) and (3.58):

$$\nabla \cdot (\nabla \times \mathbf{E}) = -\nabla \cdot \dot{\mathbf{B}} = 0 \quad \therefore \frac{\partial}{\partial t}(\nabla \cdot \mathbf{B}) = 0$$

$\therefore \nabla \cdot \mathbf{B} = 0$ if it is initially zero. This is Eq. (3.57),

$$\nabla \cdot (\nabla \times \mathbf{B}) = 0 = \mu_0[q_i \nabla \cdot (n_i \mathbf{v}_i) + q_e \nabla \cdot (n_e \mathbf{v}_e)] + \frac{\nabla \cdot \dot{\mathbf{E}}}{c^2}$$

from Eq. (3.60), $\nabla \cdot (n_i \mathbf{v}_i) = -\dot{n}_i, \nabla \cdot (n_e \mathbf{v}_e) = -\dot{n}_e$

$$\therefore \mu_0(-q_i \dot{n}_i - q_e \dot{n}_e) + \frac{\nabla \cdot \dot{\mathbf{E}}}{c^2} = 0$$

$$\frac{\partial}{\partial t}\left[\nabla \cdot \mathbf{E} - \frac{1}{\epsilon_0}(n_i q_i + n_e q_e)\right] = 0$$

If $[] = 0$ initially, $\nabla \cdot \mathbf{E} = (1/\epsilon_0)(n_i q_i + n_e q_e)$. This is Eq. (3.55).

3.4

$$\mathbf{j}_D = (KT_i + KT_e)\frac{\mathbf{B} \times \nabla n}{B^2} \propto \frac{KT}{e}\frac{ne}{BL}$$

Since $KT \propto e\phi$ and $E \propto -\phi/L$, $KT/eL \propto E$ $\therefore$ $j_D \propto neE/B \propto nev$, since $E/B = v_E$.

3.5 Let j_D be constant in the box of width L. $\Delta n = n'L$, $|J_D| = |\Delta nev_y| = |n'Lev_y|$: from the difference between the currents on the two walls. This current J_D is over a box of width L, so the equivalent current density is

$$|j_D| = |J_D|/L = |n'ev_y|$$

Equation [3.69] gives $|j_D| \simeq |KT\nabla n/B| = |KTn'/B|$; hence, once v_y is chosen so the two formulas agree for one value of L, they agree for all L, since L cancels out.

3.6 (a)

$$\mathbf{v}_{De} = -\frac{\gamma KT_e}{eB}\frac{\hat{z} \times \nabla n}{n}$$

Isothermal means $\gamma = 1$.

$$\nabla n = \hat{x}\frac{\partial n}{\partial x} = -\frac{n_0 2x}{a^2}\hat{x}$$

$$\mathbf{v}_{De} = \hat{y}\frac{KT_e}{eB}\frac{2n_0}{a^2}\frac{x}{n_0}\left(1 - \frac{x^2}{a^2}\right)^{-1} = \hat{y}\frac{KT_e}{eB}\frac{2x}{a^2}\left(1 - \frac{x^2}{a^2}\right)^{-1}$$

(b)

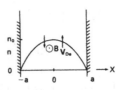

(c) $v_{De} = (2)/(0.2)\Lambda$

$$\Lambda^{-1} = \left|\frac{n'}{n}\right| = \frac{(2n_0/a^2)(a/2)}{n_0(1 - a^2/4a^2)} = \frac{1/0.04}{3/4} = 33.3\,\mathrm{m}^{-1}$$

$$\therefore v_{De} = (10)(33.3) = 333\,\mathrm{m/s}$$

3.7 $n = n_0 e^{-r^2/r_0^2} = n_0 e^{e\phi/KT_e}$

$$\phi = \frac{KT_e}{e}\ln\frac{n}{n_0} = \frac{KT_e}{e}\left(-\frac{r^2}{r_0^2}\right)$$

(a)

$$\mathbf{E} = -\frac{\partial \phi}{\partial r}\hat{\mathbf{r}} = \frac{KT_e}{e}\frac{2r}{r_0^2}\hat{\mathbf{r}}$$

$$\mathbf{v}_E = \frac{\mathbf{E} \times \mathbf{B}}{B^2} = -\frac{E_r}{B_z}\hat{\boldsymbol{\theta}} = -\hat{\boldsymbol{\theta}}\frac{KT_e}{eB}\frac{2r}{r_0^2}$$

$$\mathbf{v}_{De} = -\frac{\mathbf{B} \times \nabla p}{enB^2} = -\frac{KT_e}{eB}\frac{\partial n/\partial r}{n}\hat{\boldsymbol{\theta}} = -\hat{\boldsymbol{\theta}}\frac{KT_e}{eB}\frac{\partial}{\partial r}(\ln n)$$

$$= -\hat{\boldsymbol{\theta}}\frac{KT_e}{eB}\frac{\partial}{\partial r}\left(\frac{-r^2}{r_0^2}\right) = \hat{\boldsymbol{\theta}}\frac{KT_e}{eB}\frac{2r}{r_0^2} = -\mathbf{v}_E \quad \text{QED}$$

(b) From (a), the rotation frequency is constant whether we take $\mathbf{v}_E$, $\mathbf{v}_{De}$, $\mathbf{v}_{Di}$, or any combination thereof, since $\omega = v_\theta/r$ and $v_\theta \propto r$.

(c) In lab frame,

$$\mathbf{v} = \mathbf{v}_\phi + \mathbf{v}_E = 0.5\mathbf{v}_{De} + (-\mathbf{v}_{De})$$

$$= -\frac{1}{2}\mathbf{v}_{De}$$

3.8 (a)

$$j_D = ne(v_{Di} - v_{De}) = -\hat{\boldsymbol{\theta}}\frac{n_0(KT_e + KT_i)}{B} \cdot \frac{2r}{r_0^2}e^{-r^2/r_0^2}$$

(b)

$$j_D = \frac{(10^{16})(0.5)(1.6 \times 10^{-19})}{0.4(r_0^2/2r)(2.718)} = 0.147\,\text{A/m}^2$$

or:

$$j_D = ne(|v_{De}| + |v_{Di}|)$$

$$|v_{De}| = |v_{Di}| = \frac{(KT)_{\text{ev}}}{B}\frac{2r}{r_0^2} = \frac{(0.25)2r}{0.4r_0^2} = 1.25\frac{r}{r_0^2}\text{m/s}$$

Using $e = 1.6 \times 10^{-19}$ C, $\in = 2.718$,

$$j_D = (10^{16})(1.6 \times 10^{-19})(2)(1.25)\frac{r\in^{-1}}{r_0^2} = 0.147\frac{\text{A}}{\text{m}^2}$$

(c)

Since $\mathbf{v}_e = \mathbf{v}_E + \mathbf{v}_{De} = \mathbf{v}_E - \mathbf{v}_E = 0$ in the lab frame, the current is carried entirely by ions.

3.9

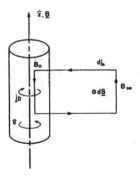

$$\nabla \times \mathbf{B} = \mu_0 \mathbf{j}_D$$

$$\int (\nabla \times \mathbf{B}) \cdot d\mathbf{S} = \mu_0 \int \mathbf{j}_D \cdot d\mathbf{S}$$

$$\oint \mathbf{B} \cdot d\mathbf{L} = \mu_0 \int \mathbf{j}_D \cdot d\mathbf{S}$$

Choose a loop with one leg along the axis ($B = B_0$) and one leg far away, where $B = B_\infty$. Since $\mathbf{j}_D$ is in the $-\hat{\boldsymbol{\theta}}$ direction, we can choose the direction of integration $d\mathbf{L}$ as shown, so that $\mathbf{j}_D \cdot d\mathbf{S}$ is positive. There is no B_r $\therefore$.

$$\oint \mathbf{B} \cdot d\mathbf{L} = (B_\infty - B_0)L$$

$$\mathbf{j}_D = -\hat{\boldsymbol{\theta}} \frac{n(KT_i + KT_e)}{B} \frac{2r}{r_0^2}$$

$$\int \mathbf{j}_D \cdot d\mathbf{S} = \frac{n_0(KT_i + KT_e)}{B_\infty r_0^2} \int_0^L \int_0^\infty e^{-r^2/r_0^2} 2r\, dr\, dz$$

$$= \frac{Ln_0(KT_i + KT_e)}{B_\infty} \left[-e^{-r^2/r_0} \right]_0^\infty = \frac{2Ln_0KT}{B_\infty}$$

where $T_e = T_i$. In this integral, we have approximated $B(r)$ by B_∞, since B is not greatly changed by such a small j_D. Thus,

$$\Delta B = B_\infty - B_0 = \mu_0 \frac{2n_0KT}{B_\infty}$$

$$= \frac{2(4\pi \times 10^{-7})(10^{16})(0.25)(1.6 \times 10^{-19})}{0.4}$$

$$= 2.5 \times 10^{-9}\,\text{T}$$

4.1 (a) Solve for ϕ_1:

$$\phi_1 = \frac{KT_e}{e} \frac{n_1}{n_0} \frac{\omega + ia}{\omega^* + ia} \times \frac{\omega^* - ia}{\omega^* - ia}$$

$$= \frac{KT_e}{e} \frac{\omega\omega^* + a^2 + ia(\omega^* - \omega)}{\omega^{*2} + a^2} \frac{n_1}{n_0}$$

If n_1 is real,

$$\frac{Im(\phi_1)}{Re(\phi_1)} = \frac{a(\omega^* - \omega)}{\omega\omega^* + a^2} = \tan \delta$$

Hence,

$$\delta = \tan^{-1} \left[\frac{a(\omega^* - \omega)}{\omega\omega^* + a^2} \right]$$

(b) $n_1 = \bar{n}_1 e^{i(kx-\omega t)}$, while $\phi_1 = An_1 e^{i(kx-\omega t+\delta)}$, where A is a positive constant. For $\omega < \omega^*$, we have $\delta > 0$. Let the phase of n_1 be 0 at (x_0, t_0): $kx_0 - \omega t_0 = 0$. If ω and k are positive and x_0 is fixed, then the phase of ϕ_1 is 0 at $kx_0 - \omega t + \delta = 0$ or $t > t_0$. Hence ϕ_1 lags n_1 in time. If t_0 is fixed, $kx - \omega t_0 + \delta = 0$ at $x < x_0$, so ϕ_1 lags n_1 in space also (since $\omega/k > 0$ and the wave moves to the right, the leading wave is at larger x). If $k < 0$ and $\omega > 0$, the phase of ϕ_1 would be 0 at $x > x_0$; but since the wave now moves to the left, ϕ_1 still lags n_1.

4.3

$$ikE_1 = \frac{1}{\epsilon_0} e(n_{i1} - n_{e1})$$

$$-i\omega m v_{e1} = -eE_1 \quad \text{(electrons)}$$

$$-i\omega M v_{i1} = eE_1 \quad \text{(ions)}$$

$$-i\omega n_{e1} = -ikn_0 v_{e1} \quad \text{(electrons)}$$

$$-i\omega n_{i1} = -ikn_0 v_{i1} \quad \text{(ions)}$$

$$n_{e1} = \frac{k}{\omega} n_0 \left(\frac{-ie}{m\omega} \right) E_1 \quad n_{i1} = \frac{k}{\omega} n_0 \left(\frac{ie}{M\omega} \right) E_1$$

$$ikE_1 = \frac{1}{\epsilon_0} \frac{k}{\omega} n_0 \frac{ie}{\omega} \left(\frac{1}{M} + \frac{1}{m} \right) E_1 = \frac{ikE_1}{\omega^2} \left(\Omega_p^2 + \omega_p^2 \right)$$

$$\underline{\omega^2 = \left(\omega_p^2 + \Omega_p^2 \right)}$$

4.4 Find ϕ_1, E_1, and v_1 in terms of n_1:

$$\text{Eq.[4-22]}: \quad v_1 = \frac{\omega}{k}\frac{n_1}{n_0}$$

$$\text{Eq.[4-23]}: \quad E_1 = \frac{ie}{\epsilon_0 k}n_1$$

But $E_1 = -ik\phi_1$,

$$\therefore \phi_1 = -\frac{e}{\epsilon_0 k^2}n_1$$

Hence, E_1 is 90° out of phase with n_1; ϕ_1 is 180° out of phase; and v_1 is either in phase or 180° out of phase, depending on the sign of ω/k. In (a), E_1 is found from the slope of the ϕ_1 curve, since $E_1 = -\partial\phi_1/\partial x$. In (b), $E_1/n_1 \propto i \times \text{sgn}\ (k)$ $\therefore \delta = \pm\ \pi/2$. If $\omega/k > 0$,

$$E_1 \propto \exp i(kx \pm |\omega|t \pm \pi/2)$$

the $\pm$ standing for the sign of k. Hence, E_1 leads n_1 by 90°. Opposite if $\omega/k < 0$.

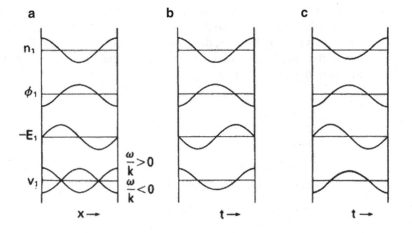

4.5

$$ikE_1 = -\frac{1}{\epsilon_0}en_1 = -\frac{1}{\epsilon_0}en_0\frac{k}{\omega}v_1 = -\frac{1}{\epsilon_0}en_0\frac{k}{\omega}\left(\frac{-ie}{m\omega}\right)E_1$$

$$ik\left(1 - \frac{n_0 e^2}{\epsilon_0 m\omega^2}\right)E_1 = 0 \quad\text{or}\quad \nabla\cdot\left(1 - \frac{\omega_p^2}{\omega^2}\right)E_1 = 0$$

$$\therefore \epsilon = 1 - \frac{\omega_p^2}{\omega^2}$$

4.7 (a)

$$mn_0(-i\omega)v_1 = -en_0E_1 - mn_0 v v_1$$

$$v_1\left(1 + \frac{iv}{\omega}\right) = \frac{ieE_1}{m\omega}$$

$$ikE_1 = -\frac{1}{\epsilon_0}en_1 \quad n_1 = \frac{k}{\omega}n_0 v_1 \quad \text{(continuity)}$$

$$ikE_1 = -\frac{1}{\epsilon_0}\frac{k}{\omega}e^{-n_0}\frac{ieE_1}{m\omega}\left(1 + \frac{iv}{\omega}\right)^{-1}$$

$$\omega^2\left(1 + \frac{iv}{\omega}\right) - \omega_p^2 \quad \underline{\omega^2 + iv\omega = \omega_p^2}$$

(b) Let $\omega = x + iy$. Then the dispersion relation is $x^2 - y^2 + 2ixy + ivx - vy = \omega_p^2$. We need the imaginary part: $2xy + vx = 0$, $y = (-1/2)v$ ∴ Im $(\omega) = -v/2$. Since $x = $ Re (ω), $v > 0$, and

$$E_1 \propto e^{-i\omega t} = e^{-i\omega t}e^{yt} = e^{-ixt}e^{-(1/2)vt}$$

the oscillation is damped in time.

4.8 $mn_0(-i\omega)v_1 = en_0E_1 - en_0(v_1 \times B_0)$. Take B_0 in the $\hat{z}$ direction and E_1 and k in the $\hat{x}$ direction. Then the y-component is

$$-i\omega m v_y = ev_x B_0 \quad \frac{v_x}{v_y} = -i\frac{\omega}{\omega_c}$$

Since $\omega = \omega_h > \omega_c$, $|v_x/v_y| > 1$; and the orbit is elongated in the $\hat{x}$ direction, which is the direction of k.

4.9 (a)

$$\nabla \cdot E_1 = -\frac{1}{\epsilon_0}en_1 \quad k = k_x\hat{x} + k_z\hat{z} \quad E_y = k_y = 0$$

$$i(k_xE_x + k_zE_z) = -\frac{1}{\epsilon_0}en_1$$

We need n_1:

$$\frac{\partial n_1}{\partial t} + n_0\nabla \cdot v_1 = 0 \quad -i\omega n_1 + n_0 i(k_xv_x + k_zv_z) = 0$$

We need v_x, v_z:

$$Mn_0(-i\omega)v_1 = -en_0E_1 - en_0(v_1 \times B_0)$$

$$x\text{-component}: \quad v_x = -\frac{ie}{m\omega}E_x - \frac{i\omega_c}{\omega}v_y$$

$$y\text{-component}: \quad v_y = 0 + \frac{i\omega_c}{\omega}v_x$$

$$v_x = -\frac{ie}{m\omega}E_x + \frac{\omega_c^2}{\omega^2}v_x = \frac{-ie}{m\omega}E_x\left(1 - \frac{\omega_c^2}{\omega^2}\right)^{-1}$$

$$z\text{-component}: \quad v_z = -\frac{ie}{m\omega}E_z$$

$$\text{Continuity}: \quad n_1 = \frac{n_0}{\omega}\left(\frac{-ie}{m\omega}\right)\left[k_x E_x\left(1 - \frac{\omega_c^2}{\omega^2}\right)^{-1} + k_z E_z\right]$$

$$k_x E_x + k_z E_z = i\frac{en_0}{\epsilon_0\omega}\left(\frac{-ie}{m\omega}\right)\left[k_x E_x\left(1 - \frac{\omega_c^2}{\omega^2}\right)^{-1} + k_z E_z\right]$$

$$k_x = k \sin\theta \qquad k_z = k \cos\theta$$

$$\therefore E_1\sin^2\theta + kE_1\cos^2\theta = \frac{\omega_p^2}{\omega^2}\left[kE_1\sin^2\theta\left(1 - \frac{\omega_c^2}{\omega^2}\right)^{-1} + kE_1\cos^2\theta\right]$$

$$1 = \frac{\omega_p^2}{\omega^2}\left[\sin^2\theta\left(1 - \frac{\omega_c^2}{\omega^2}\right)^{-1} + \cos^2\theta\right]$$

$$1 - \frac{\omega_c^2}{\omega^2} = \frac{\omega_p^2}{\omega^2}\left[1 - \cos^2\theta + \left(1 - \frac{\omega_c^2}{\omega^2}\right)\cos^2\theta\right]$$

$$\omega^2 - \omega_c^2 - \omega_p^2 = -\frac{\omega_p^2\omega_c^2}{\omega^2}\cos^2\theta$$

$$\omega^2\left(\omega^2 - \omega_h^2\right) + \omega_p^2\omega_c^2\cos^2\theta = 0 \quad \text{QED}$$

(b)

$$\omega^4 - \omega_h^2\omega^2 + \omega_p^2\omega_c^2\cos^2\theta = 0$$

$$2\omega^2 = \omega_h^2 \pm \left(\omega_h^4 - 4\omega_p^2\omega_c^2\cos^2\theta\right)^{1/2}$$

For $\theta \to 0$, $\cos^2\theta \to 1$,

$$2\omega^2 = \omega_h^2 \pm \left[\left(\omega_p^2 + \omega_c^2\right)^2 - 4\omega_p^2\omega_c^2\right]^{1/2}$$

$$= \omega_p^2 + \omega_c^2 \pm \left(\omega_p^2 - \omega_c^2\right)$$

$$\omega^2 = \omega_p^2, \omega_c^2$$

The $\omega = \omega_p$ root is the usual Langmuir oscillation. The $\omega = \omega_c$ root is spurious because at $\theta \to 0$, B_0 does not enter the problem. For $\theta \to \pi/2$, $\cos^2\theta \to 0$, $2\omega^2$

$= \omega_h^2 \pm \omega_h^2$, $\omega = 0, \omega_h$. The $\omega = \omega_h$ root is the usual upper hybrid oscillation. The $\omega = 0$ root has no physical meaning, since on oscillating perturbation was assumed.

(c)

$$\omega^4 - \omega_h^2\omega^2 + \frac{1}{4}\omega_h^4 = \frac{1}{4}\omega_h^4 - \omega_p^2\omega_c^2 \cos^2\theta$$

$$\left(\omega^2 - \frac{1}{2}\omega_h^2\right) + \left(\omega_p\omega_c \cos\theta\right)^2 = \left(\frac{1}{2}\omega_h\right)^2$$

$$(y-1)^2 + \frac{x^2}{a^2} = 1 \quad \text{QED}$$

(d)

ω_p/ω_c	$a = 1/2(\omega_c/\omega_p + \omega_p/\omega_c)$
1	1
2	5/4
∞	∞

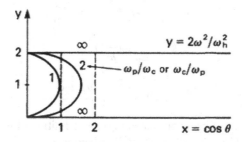

(e)

$$\omega^2 = \frac{1}{2}\left(\omega_p^2 + \omega_c^2\right) \pm \left[\left(\omega_p^2 + \omega_c^2\right)^2 - 4\omega_p^2\omega_c^2 \cos^2\theta\right]^{1/2}$$

Lower root: Take $(-)$ sign; ω is maximum when $\cos^2\theta$ is maximum $(=1)$. Thus

$$\omega_-^2 < \frac{1}{2}\left[\left(\omega_p^2 + \omega_c^2\right) - |\omega_p^2 - \omega_c^2|\right]$$
$$= \omega_c^2 \quad \text{if } \omega_p > \omega_c$$
$$= \omega_p^2 \quad \text{if } \omega_c > \omega_p$$

Upper root: Take $(+)$ sign; ω is maximum when $\cos^2\theta = 0$, $\omega^2 = \omega_h^2$. Thus $\omega_+^2 < \omega_h^2$. This root is minimum when $\cos^2\theta = 1$; thus

$$\omega_+^2 > \frac{1}{2}\left[\left(\omega_p^2 + \omega_c^2\right) + |\omega_p^2 - \omega_c^2|\right]$$
$$= \omega_p^2 \quad \text{if } \omega_p > \omega_c$$
$$= \omega_c^2 \quad \text{if } \omega_c > \omega_p$$

4.10 Use V_+, N_+ for proton velocity and density

V_-, N_- for antiprotons

v_-, n_- for electrons

v_+, n_+ for positrons

(a)

$$\nabla \times \mathbf{E} = -\dot{\mathbf{B}} \quad \nabla \times \mathbf{B} = \mu_0 \mathbf{j} + \frac{\dot{\mathbf{E}}}{c^2} \quad \nabla \times \nabla \times \mathbf{E} = -\left(\mu_0 \dot{\mathbf{j}} + \frac{\ddot{\mathbf{E}}}{c^2}\right)$$

$$-(\mathbf{k} \times \mathbf{k} \times \mathbf{E}) = -\left[\mu_0 n_0 e\,(\dot{\mathbf{v}}_+ - \dot{\mathbf{v}}_-) - \frac{\omega^2}{c^2}\mathbf{E}\right]$$

$$= k^2 \mathbf{E} - \mathbf{k}(\mathbf{k} \cdot \overset{0}{\mathbf{E}})$$

$$\left(\omega^2 - c^2 k^2\right)\mathbf{E} = \frac{1}{\epsilon_0} n_0 e(\dot{\mathbf{v}}_+ - \dot{\mathbf{v}}_-)$$

$$mn_0 \mathbf{v}_\pm = \pm e n_0 \mathbf{E} \quad \dot{\mathbf{v}}_\pm = \pm \frac{e}{m}\mathbf{E}$$

$$\omega^2 - c^2 k^2 = \frac{1}{\epsilon_0} n_0 e \frac{e}{m}(1+1) = 2\omega_p^2$$

$$\omega_p^2 = \frac{n_0 e^2}{\epsilon_0 m} \quad \underline{\omega^2 = 2\omega_p^2 + c^2 k^2}$$

(Or the 2 can be incorporated into the definition of ω_p.)

(b) $\nabla \cdot \mathbf{E}_1 = (1/\epsilon_0)(N_+ - N_- + n_+ - n_-)_1$, where $n_+ = n_0 e^{-e\phi/KT^+}$, $n_- = n_0 e^{e\phi/KT^-}$.
Let $T_+ = T_- = T_e$ $n_{1\pm} = \mp n_0 e\phi/KT_e$. Note: $N_{0\pm} = n_{0\pm} \equiv n_0$.

$$\frac{\partial N_\pm}{\partial t} + N_{0\pm}\nabla \cdot V_\pm = 0 \quad N_{1\pm} = N_{0\pm}\frac{k}{\omega}V_\pm = n_0 \frac{k}{\omega}V_\pm$$

$$M(-i\omega)V_\pm = \pm e\mathbf{E}_1 = \pm ike\phi \quad (M_+ = M_- = M)$$

$$V_\pm = \pm \frac{k}{\omega}\frac{e\phi}{M} \quad N_{1\pm} = \pm \frac{k^2}{\omega^2}\frac{n_0 e\phi}{M}$$

$$\nabla \cdot \mathbf{E}_1 = k^2 \phi = \frac{e}{\epsilon_0}\left(\frac{k^2}{\omega^2} + \frac{k^2}{\omega^2}\right)\frac{n_0 e\phi}{M} + \frac{e}{\epsilon_0}(-n_0 - n_0)\frac{e\phi}{KT_e}$$

$$= \frac{n_0 e^2}{\epsilon_0 M}\frac{2k^2}{\omega^2}\phi - \frac{n_0 e^2}{\epsilon_0 KT_e}2\phi = 2\phi\left(\Omega_p^2 \frac{k^2}{\omega^2} - \frac{1}{\lambda_D^2}\right)$$

$$k^2 \lambda_D^2 + 2 = \frac{2k^2}{\omega^2}\Omega_p^2 \lambda_D^2 = \frac{2k^2}{\omega^2}v_s^2 \quad v_s^2 \equiv \frac{kT_e}{M}$$

$$\frac{\omega^2}{k^2} = \frac{2v_s^2}{2 + k^2\lambda_D^2} = \frac{v_s^2}{1 + (1/2)k^2\lambda_D^2} \qquad \lambda_D \equiv \left(\frac{kT_e\epsilon_0}{n_0 e^2}\right)^{1/2}$$

4.11 $\tilde{n} = \dfrac{ck}{\omega} \qquad \omega^2 = \omega_p^2 + c^2 k^2 \qquad \dfrac{c^2 k^2}{\omega^2} = 1 - \dfrac{\omega_p^2}{\omega^2} = \epsilon$

$$\therefore \tilde{n} = \sqrt{\epsilon}$$

4.12 In $\nabla \times \mathbf{B} = \mu_0 \mathbf{j}_1$, $\mathbf{j}_1$ is the current carried by electrons only, since Cl^- ions are too heavy to move appreciably in response to a signal at microwave frequencies. Hence,

$$j_1 = -n_{0e} e v_e = -(1 - \kappa) n_0 e v_{e1}$$

If ω_p is defined with n_0 (i.e., $\omega_p^2 = n_0 e^2 / \epsilon_0 m$), the dispersion relation becomes

$$\frac{c^2 k^2}{\omega^2} = 1 - (1 - \kappa)\frac{\omega_p^2}{\omega^2}$$

Cutoff occurs for $f = (1-\kappa)^{1/2} f_p = (0.4)^{1/2}(9)(n_0)^{1/2}$, where

$$f = \frac{c}{\lambda} = \frac{3 \times 10^{10}}{3} = 10^{10}$$

Thus

$$n_0 = \left[\frac{10^{10}}{(0.63)(9)}\right]^2 = 3.1 \times 10^{18} \text{m}^{-3}$$

4.13 (a) Method 1: Let $N =$ No. of wavelengths in length $L = 0.08$ m, $N_0 =$ No. of wavelengths in absence of plasma.

$$N = \frac{L}{\lambda} \qquad N_0 = \frac{L}{\lambda_0} \qquad \lambda = \frac{2\pi}{k} \qquad \frac{ck}{\omega} = \left(1 - \frac{\omega_p^2}{\omega^2}\right)^{1/2}$$

$$\Delta N = N_0 - N = \frac{L}{\lambda_0} - \frac{Lk}{2\pi} = \frac{L}{\lambda_0} - \frac{L}{2\pi}\frac{\omega}{c}\left(1 - \frac{\omega_p^2}{\omega^2}\right)^{1/2}$$

$$\frac{\omega}{2\pi c} = \frac{1}{\lambda_0} \quad \therefore \Delta N = \frac{L}{\lambda_0}\left[1 - \left(1 - \frac{\omega_p^2}{\omega^2}\right)^{1/2}\right] = 0.1$$

$$\frac{L}{\lambda_0} = \frac{0.08}{0.008} = 10$$

$$\therefore \left(1 - \frac{\omega_p^2}{\omega^2}\right)^{1/2} = 1 - 10^{-2} \quad 1 - \frac{f_p^2}{f^2} = 1 - \left(2 \times 10^{-2}\right)$$

$$f_p^2 = f^2 \times 2 \times 10^{-2} = \left(\frac{c}{\lambda_0}\right)^2 2 \times 10^{-2} = 2.8 \times 10^{19}$$

$$n = \frac{2.8 \times 10^{19}}{(9)^2} = \underline{3.5 \times 10^{17} \mathrm{m}^{-3}}$$

Method 2: Let k_0 = free-space k. The phase shift is

$$\Delta\phi = \int_0^L \Delta k \, dx = (k_0 - k)L = (0.1)2\pi$$

This leads to the same answer.

(b) From above, ΔN is small if ω_p^2/ω^2 is small; hence expand square root:

$$\Delta N \approx \frac{L}{\lambda_0}\left[1 - \left(1 - \frac{1}{2}\frac{\omega_p^2}{\omega^2}\right)\right] = \frac{L}{\lambda_0}\frac{1}{2}\frac{\omega_p^2}{\omega^2} \propto n \quad \text{QED}$$

4.14 From Eq. (4.101a), we have for the X-wave

$$\left(\omega^2 - \omega_h^2\right)E_x + i\frac{\omega_p^2\omega_c}{\omega}E_y = 0$$

At resonance, $\omega = \omega_h)$ $E_y = 0$, $\mathbf{E} = E_x\hat{\mathbf{x}}$. Since $\mathbf{k} = k_x\hat{\mathbf{x}}$, $\mathbf{E}||\mathbf{k}$, and the wave is longitudinal and electrostatic.

4.15 Since $\omega_h^2 = \omega_c^2 + \omega_p^2$, clearly $\omega_p < \omega_h$. Further,

$$\omega_L = \frac{1}{2}\left[-\omega_c + \left(\omega_c^2 + 4\omega_p^2\right)^{1/2}\right]$$

$$< \frac{1}{2}\left[-\omega_c + \left(\omega_c^2 + 4\omega_c\omega_p + 4\omega_p^2\right)^{1/2}\right]$$

$$= \frac{1}{2}[-\omega_c + (\omega_c + 2\omega_p)] = \omega_p \quad \therefore \quad \omega_L < \omega_p$$

Also,

$$\omega_R = \frac{1}{2}\left[\omega_c + \left(\omega_c^2 + 4\omega_p^2\right)^{1/2}\right] > \omega_c$$

and

$$\omega_R^2 - \omega_R\omega_c - \omega_p^2 = 0 \quad \text{(Eq. [4-107])}$$
$$\therefore \quad \omega_R^2 = \omega_R\omega_c + \omega_p^2 > \omega_c^2 + \omega_p^2 = \omega_h^2$$

4.17 (a) Multiply Eq. (4.112b) by i and add to Eq. (4.112a):

$$\left(\omega^2 - c^2k^2 - \alpha\right)\left(E_x + iE_y\right) + \alpha\frac{\omega_c}{\omega}\left(E_x + iE_y\right) = 0$$

Now subtract from Eq. (4.112a):

$$\left(\omega^2 - c^2k^2 - \alpha\right)\left(E_x - iE_y\right) - \alpha\frac{\omega_c}{\omega}\left(E_x - iE_y\right) = 0$$

Thus

$$F(\omega) = \omega^2 - c^2k^2 - \alpha(1 + \omega_c/\omega)$$
$$G(\omega) = \omega^2 - c^2k^2 - \alpha(1 - \omega_c/\omega)$$

Since

$$\alpha \equiv \frac{\omega_p^2}{\left(1 - \omega_c^2/\omega^2\right)}$$

$$F(\omega) = \omega^2\left(1 - \frac{\omega_p^2/\omega^2}{1 - \omega_c/\omega} - \frac{c^2k^2}{\omega^2}\right)$$

$$G(\omega) = \omega^2\left(1 - \frac{\omega_p^2/\omega^2}{1 + \omega_c/\omega} - \frac{c^2k^2}{\omega^2}\right)$$

From Eqs. (4.116) and (4.117),

$$F(\omega) = 0 \quad \text{for the } R \text{ wave} \quad \text{and}$$
$$G(\omega) = 0 \quad \text{for the } L \text{ wave}$$

(b) $E_x = -iE_y)$ $E_y = iE_x$. Let $E_x = f(z)\,e^{-i\omega t}$. Then
$E_y = f(z)i\,e^{-i\omega t} = f(z)\,e^{-i\omega t + i(\pi/2)} = f(z)\,e^{-i[\omega t - (\pi/2)]}$

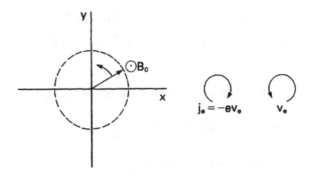

E_y lags E_x by 90°. Hence E rotates counterclockwise on this diagram. This is the same way electrons gyrate in order to create a clockwise current and generate a B-field opposite to $\mathbf{B}_0$. For the L wave, $E_y = -iE_x$ so that

$E_y = f(z)\, e^{-i(\omega t + \pi/2)}$ and E_y leads E_x by 90°.

(c) For an R-wave, $E_y = iE_x$. The space dependence is $E_x = f(t)\, e^{ikz}$, $E_y = f(t)i\, e^{ikz} = f(t)\, e^{i(kz+\pi/2)}$ For $k > 0$, E_y leads E_x (has the same phase at smaller z). For $k < 0$, E_y lags E_x (has the same phase at larger z).

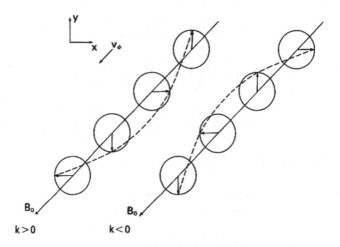

4.19

$$\frac{c^2 k^2}{\omega^2} = 1 - \frac{\omega_p^2/\omega^2}{1 - \omega_c/\omega} \qquad c^2 v_\phi^{-2} = 1 - \frac{\omega_p^2/\omega^2}{1 - \omega_c/\omega}$$

$$c^2(-2)v_\phi^{-3}\frac{dv_\phi}{d\omega} = -\omega_p^2 \frac{-1}{\left(\omega^2 - \omega\omega_c\right)^2}(2\omega - \omega_c) = 0$$

$$\therefore 2\omega - \omega_c = 0 \quad \omega = \frac{1}{2}\omega_c$$

At $\omega = 1/2\omega_c$,

$$\frac{c^2}{v_\phi^2} = 1 - \frac{\omega_p^2}{\frac{1}{4}\omega_c^2 - \frac{1}{2}\omega_c^2} = 1 + \frac{4\omega_p^2}{\omega_c^2} > 1$$

$$\therefore \quad v_\phi < c.$$

4.20

$$\frac{c^2 k^2}{\omega^2} = 1 - \frac{\omega_p^2/\omega^2}{1 - \omega_c/\omega} \quad c^2 k^2 = \omega^2 - \frac{\omega\omega_p^2}{\omega - \omega_c}$$

$$c^2 2k\,dk = 2\omega\,d\omega - \frac{(\omega - \omega_c) - \omega}{(\omega - \omega_c)^2}\omega_p^2\,d\omega$$

$$= \left[2\omega + \frac{\omega_c\omega_p^2}{(\omega - \omega_c)^2} \right] d\omega$$

$$\frac{d\omega}{dk} = \frac{kc^2}{\omega + \omega_c\omega_p^2/2(\omega - \omega_c)^2} \approx \frac{kc^2}{\omega + \omega_p^2/2\omega_c} \quad \text{if } \omega \ll \omega_c$$

But

$$ck = \left(\omega^2 - \frac{\omega_p^2}{1 - \omega_c/\omega} \right)^{1/2} \approx \left(\omega^2 + \frac{\omega\omega_p^2}{\omega_c} \right)^{1/2} \quad \text{if } \omega \ll \omega_c$$

$$\therefore \frac{d\omega}{dk} = c\frac{\left(\omega^2 + \omega\omega_p^2/\omega_c \right)^{1/2}}{\omega + \omega_p^2/2\omega_c} = c\frac{\left(1 + \omega_p^2/\omega\omega_c \right)^{1/2}}{1 + \omega_p^2/2\omega\omega_c}$$

To prove the required result, one must also assume $v_\phi^2 \ll c^2$, as is true for whistlers, so that $\omega_p^2/\omega\omega_c \ll 1$ (from line 1). Hence

$$\frac{d\omega}{dk} \approx 2c\left(\frac{\omega\omega_c}{\omega_p^2} \right)^{1/2} \propto \omega^{1/2}$$

4.21

$$(\omega^2 - c^2 k^2)\mathbf{E}_1 = \frac{1}{\epsilon_0}i\omega\mathbf{j}_1 \quad (\text{Eq.}[4\text{-}81])$$

$$\mathbf{j}_1 = n_0 e(\mathbf{v}_p - \mathbf{v}_e) \quad (v_p \text{ is the positron velocity})$$

From the equation of motion,

$$v_x = \frac{\pm ie}{m\omega}\left(E_x \pm \frac{i\omega_c}{\omega}E_y\right)\left(1 - \frac{\omega_c^2}{\omega^2}\right)^{-1}$$

$$v_y = \frac{\pm ie}{m\omega}\left(E_y \mp \frac{i\omega_c}{\omega}E_x\right)\left(1 - \frac{\omega_c^2}{\omega^2}\right)^{-1}$$

$$\therefore (\omega^2 - c^2 k^2)E_x = \left(-\frac{1}{\epsilon_0}i\omega\right)(n_0 e)\left(\frac{ie}{m\omega}\right)(1+1)E_x\left(1 - \frac{\omega_c^2}{\omega^2}\right)^{-1}$$

$$= \frac{2\omega_p^2}{1 - \omega_c^2/\omega^2}E_x$$

the E_y term canceling out. Similarly,

$$(\omega^2 - c^2 k^2)E_y = \frac{2\omega_p^2}{1 - \omega_c^2/\omega^2}E_y$$

the E_x term cancelling out. Both equations give

$$\frac{c^2 k^2}{\omega^2} = 1 - \frac{2\omega_p^2}{\omega^2 - \omega_c^2}$$

The R and L waves are degenerate and have the same phase velocities—hence, no Faraday rotation.

4.22 Since the phase difference between the R and L waves is twice the angle of rotation,

$$\int_0^L (k_L - k_R)\, dz = \pi$$

$$k_{R,L} = k_0\left(1 - \frac{\omega_p^2/\omega^2}{1 \pm \omega_c/\omega}\right)^{1/2}$$

To get a simple expression for $k_L - k_R$, we wish to expand the square root. Let us assume we can, and then check later for consistency:

$$k_{R,L} \approx k_0\left(1 - \frac{1}{2}\frac{\omega_p^2/\omega^2}{1 \pm \omega_c/\omega}\right)$$

$$k_L - k_R = \frac{1}{2}k_0\frac{\omega_p^2}{\omega^2}\left(\frac{1}{1-\omega_c/\omega}-\frac{1}{1+\omega_c/\omega}\right)$$

$$= \frac{1}{2}k_0\frac{\omega_p^2}{\omega^2}\frac{2\omega_c/\omega}{1-\omega_c^2/\omega^2}$$

$$\pi = L(k_L - k_R) = k_0 L\frac{\omega_p^2\omega_c}{\omega}\frac{1}{\omega^2-\omega_c^2} \quad k_0 = \frac{\omega}{c}$$

$$\omega_p^2 = \frac{\pi c}{L\omega_c}(\omega^2-\omega_c^2) \quad f_p^2 = \frac{c}{2L}\frac{f^2-f_c^2}{f_c}$$

$$f_c = 2.8 \times 10^{10}(0.1)\text{Hz}$$

$$f = \frac{c}{\lambda_0} = \frac{3 \times 10^8}{8 \times 10^{-3}} = 3.75 \times 10^{10}\text{Hz}$$

$$f_p^2 = \frac{(3 \times 10^8)}{(2)(1)}\frac{(1.41 \times 10^{21} - 7.8 \times 10^{18})}{2.8 \times 10^9}$$

$$= 7.5 \times 10^{19} = 9^2 n$$

$$n = 9.3 \times 10^{17}\text{m}^{-3}$$

To justify expansion, note that $f_c \ll f$, so that

$$\frac{\omega_p^2/\omega^2}{1\pm\omega_c/\omega} \approx \frac{f_p^2}{f^2} = \frac{7.5 \times 10^9}{(3.75 \times 10^{10})^2} = 0.05 \ll 1$$

4.24 12.7°.

4.25 (a) The X-wave cutoff frequencies are given by Eq. (4.107). Thus,

$$\omega_p^2 = \omega(\omega \pm \omega_c) = \frac{4\pi n e^2}{m}$$

$$n_{cx} = \frac{m\omega}{4\pi e^2}(\omega + \omega_c)$$

We choose the (+) sign, corresponding to the L cutoff, because that gives the higher density.

(b)

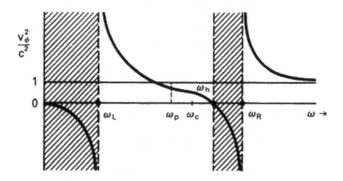

The left branch is the one that has a cutoff at $\omega = \omega_L$. One might worry that this branch is inaccessible if the wave is sent in from outside the plasma. However, if ω is kept less than ω_c, the stopband between ω_h and ω_R is avoided completely.

4.28 (a)

$$f_p = 9\sqrt{n} = (9)(10^{15})^{1/2} = 2.85 \times 10^8 \, \text{Hz}$$

$$f_c = 28\text{GHz/T} = (2.8 \times 10^{10}) \times (10^{-2}) = 2.8 \times 10^8 \, \text{Hz}$$

$$f = 1.6 \times 10^8 \text{Hz} \therefore \omega_p/\omega > 1 \quad \omega_c/\omega > 1$$

$$\omega_L = \frac{1}{2}\left[-\omega_c \pm \left(\omega_c^2 + 4\omega_p^2\right)^{1/2}\right] \approx \frac{1}{2}\left(-\omega_c + \sqrt{5}\omega_c\right)$$

$$= 0.62\omega_c \quad \text{for} \, \omega_c \approx \omega_p$$

$$f_L = (0.62)(2.8 \times 10^8) = 1.73 \times 10^8 > f$$

Also, $f >$ all ion frequencies.

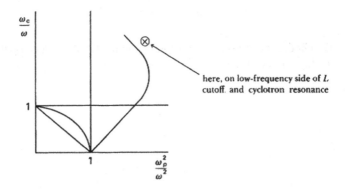

here, on low-frequency side of L cutoff. and cyclotron resonance

(b) The R-wave (whistler mode) is the only wave that propagates here.

4.29 (a)

$$v_A = \frac{B}{(\mu_0 nM)} = \frac{1}{\left[(1.26 \times 10^{-6})(10^{19})(1.67 \times 10^{-27})\right]^{1/2}}$$

$$= 6.9 \times 10^6 \text{m/s}$$

$$\Omega_c = \frac{eB}{M} = \frac{(1.6 \times 10^{-19})(1)}{(1.67 \times 10^{-27})} = 9.58 \times 10^7 \text{rad/s}$$

$$\omega = 0.1\Omega_c = 9.58 \times 10^6 \text{rad/s}$$

$$\omega = kv_A = 2\pi v_A/\lambda$$

If $\lambda = 2$ L,

$$L = \frac{\pi v_A}{\omega} = \frac{\pi(6.9 \times 10^6)}{9.58 \times 10^6} = 2.26 \text{m}$$

(b)

$$L \propto v_A/\omega \propto v_A/\Omega_c \propto B(nM)^{-1/2}B^{-1}M \propto (M/n)^{1/2}$$

$$\therefore L = (2.26)\left(\frac{133}{1}\right)^{1/2}\left(\frac{10^{19}}{10^{18}}\right)^{1/2} = 82 \text{m}$$

This is why Alfvén waves cannot be studied in Q-machines, regardless of B.

4.30 (a) $\qquad \omega^2 = \omega_p^2 + c^2k^2 \quad 2\omega\,d\omega = c^2 2k\,dk$

$$v_g = d\omega/dk = c^2 k/\omega$$

$$\frac{ck}{\omega} = \left(1 - \frac{\omega_p^2}{\omega}\right)^{1/2}$$

$$\therefore v_g = c\left(1 - \frac{\omega_p^2}{\omega}\right)^{1/2} \approx c\left(1 - \frac{1}{2}\frac{\omega_p^2}{\omega}\right) \quad \text{for } \omega^2 \gg \omega_p^2$$

$$v_g t = x \therefore t = x/v_g$$

$$\frac{dt}{d\omega} = \frac{x}{c}\left(1 - \frac{1}{2}\frac{\omega_p^2}{\omega^2}\right)^{-2}\left(-\frac{\omega_p^2}{\omega^2}\right) \approx -\frac{x\,\omega_p^2}{c\,\omega^3}$$

$$\therefore \frac{df}{dt} \approx -\frac{c\,f^3}{x\,f_p^2}$$

(b)

$$x = \frac{cf^3}{f_p^2}\left(-\frac{df}{dt}\right)^{-1} = \frac{(3 \times 10^8)(8 \times 10^7)^3}{(9)^2(2 \times 10^5)(5 \times 10^6)} = 1.9 \times 10^{18} \text{m}$$

$$= (1.9 \times 10^{18})(3 \times 10^{16})^{-1} = 63 \text{ parsec}$$

4.31 (a)   Let $n_0^{(1)} = (1 - \epsilon)n_0$, $n_0^{(2)} = \epsilon n_0$, $n_e = n_0 e\phi/kT_e$

Poisson: $ikE_1 = k^2\phi = \frac{1}{\epsilon_0}e\left(n_i^{(1)} + n_i^{(2)} - n_e\right)$

(Assume $z_{1,2} = 1$, since the ion charge is not explicitly specified.)

Continuity: $n_1^{(1)} = (1 - \epsilon)n_0\frac{k}{\omega}v_1^{(1)}$, $n_1^{(2)} = \epsilon n_0\frac{k}{\omega}v_1^{(2)}$

Equation of motion:

$$v_1^{(j)} = \frac{e}{M_j}\frac{k}{\omega}\phi\left(1 - \frac{\Omega_{cj}^2}{\omega^2}\right)^{-1} \tag{4.68}$$

$$\therefore k^2\phi = \frac{e}{\epsilon_0}\left[(1 - \epsilon)n_0\frac{k^2}{\omega^2}\frac{e}{M_1}\left(1 - \frac{\Omega_{c1}^2}{\omega^2}\right)^{-1}\right.$$

$$+\epsilon n_0\frac{k^2}{\omega^2}\frac{e}{M_2}\left(1 - \frac{\Omega_{c2}^2}{\omega^2}\right)^{-1} - n_0\frac{e}{kT_e}\right]\phi \approx 0 \,(\text{plasma approximation})$$

$$1 = (1 - \epsilon)\frac{k^2v_{s1}^2}{\omega^2 - \Omega_{c1}^2} + \epsilon\frac{k^2v_{s2}^2}{\omega^2 - \Omega_{c2}^2} \quad\Leftarrow$$

(b)   There are two roots, one near $\omega = \Omega_{c1}$ and one near $\omega = \Omega_{c2}$. If $\epsilon \to 0$, the root near Ω_{c2} approaches Ω_{c2} to keep the last term finite. The usual root, near Ω_{c1}, is shifted by the presence of the M_2 species:

$$\omega^2 - \Omega_{c1}^2 = k^2v_{s1}^2 - \epsilon\left[k^2v_{s1}^2 - k^2v_{s2}^2\frac{\omega^2 - \Omega_{c1}^2}{\omega^2 - \Omega_{c2}^2}\right]$$

In the last term, we may approximate ω^2 by $\Omega_{c1}^2 + k^2v_{s1}^2$. Thus,

$$\omega^2 \approx \Omega_{c1}^2 + k^2v_{s1}^2 + \epsilon\left[\frac{k^2v_{s2}^2}{\Omega_{c1}^2 - \Omega_{c2}^2} - 1\right]k^2v_{s1}^2$$

(c)

$$1 = \frac{1}{2} \frac{k^2 v_{sD}^2}{\omega^2 - \omega_{cD}^2} + \frac{1}{2} \frac{k^2 v_{sT}^2}{\omega^2 - \Omega_{cT}^2}$$

$$v_{sD}^2 = KT_e/M_D = \left(10^4\right)\left(1.6 \times 10^{-19}\right)/(2)\left(1.67 \times 10^{-27}\right) = 4.79 \times 10^{11}$$

$$v_{sT}^2 = \frac{2}{3} v_{sD}^2 = 3.19 \times 10^{11}$$

$$\Omega_{cD} = eB/M_D = \left(1.6 \times 10^{-19}\right)(5)/(2)\left(1.67 \times 10^{-27}\right) = 2.40 \times 10^8$$

$$\Omega_{cT} = \frac{2}{3}\Omega_{cD} = 1.60 \times 10^8 \quad k = 100\,\mathrm{m}^{-1}$$

$$\left(\omega^2 - \Omega_{cD}^2\right)\left(\omega^2 - \Omega_{cT}^2\right) = \frac{1}{2}k^2\left[v_{sD}^2\left(\omega^2 - \Omega_{cT}^2\right) + v_{sT}^2\left(\omega^2 - \Omega_{cD}^2\right)\right]$$

$$\omega^4 - \omega^2\left[\Omega_{cD}^2 + \Omega_{cT}^2 + \frac{1}{2}k^2\left(v_{sD}^2 + v_{sT}^2\right)\right]$$

$$+ \Omega_{cD}^2\Omega_{cT}^2 + \frac{1}{2}k^2\left(v_{sD}^2\Omega_{cT}^2 + v_{sT}^2\Omega_{cD}^2\right) = 0$$

$$\omega^4 - \omega^2\left[8.32 \times 10^{16} + 3.99 \times 10^{15}\right] + 1.47 \times 10^{33} + 1.53 \times 10^{32} = 0$$

$$\omega^4 - 8.72 \times 10^{16}\omega^2 + 1.63 \times 10^{33} = 0$$

$$\omega^2 = \frac{1}{2}\left[8.72 \times 10^{16} \pm \left(7.60 \times 10^{33} - 6.52 \times 10^{33}\right)^{1/2}\right]$$

$$= 6.0 \times 10^{16}, \quad 2.72 \times 10^{16}$$

$$\omega = 2.45, \ 1.65 \times 10^8 \, \sec^{-1} \quad f = 39 \, \text{and} \, 26.3 \, \text{MHz}$$

.32

$$E = n_0\left\langle \frac{1}{2}mv_e^2 \right\rangle \quad v_e = \frac{e}{im\omega}E$$

$$\therefore \langle v_e^2 \rangle = \frac{e^2}{m^2\omega^2}\langle E^2 \rangle$$

$$E = n_0\frac{1}{2}m\frac{e^2}{m^2\omega^2}\langle E^2 \rangle = \frac{\epsilon_0\omega_p^2}{\omega^2}\frac{\langle E^2 \rangle}{2}$$

But $\omega^2 = \omega_p^2 \ \therefore E = \frac{1}{2}\epsilon_0\langle E^2 \rangle$.

4.33

$$E = n_0\left\langle \frac{1}{2}Mv_i^2 \right\rangle \quad v_i \approx E_1/B_0$$

$$\therefore E = \frac{1}{2}Mn_0\langle E_1^2 \rangle/B_0. \ \text{But} \nabla \times \mathbf{E}_1 = -\dot{\mathbf{B}}_1 \therefore \langle E_1^2 \rangle = \left(\omega^2/k^2\right)\langle B_1^2 \rangle$$

$$E = \frac{Mn_0}{2B_0^2}\frac{\omega^2}{k^2}\langle B_1^2 \rangle.$$

For Alfvén wave,

$$\frac{\omega^2}{k^2} = \frac{B_0^2}{\mu_0 n_0 M} = \frac{\langle B_1^2 \rangle}{2\mu_0}$$

4.34 (a) With the L-wave, the cutoff occurs at $\omega = \omega_L$, so that one requires $\omega_L^2 < \epsilon\omega^2$. Since $\omega_L < \omega_p$ if n_0 is fixed (Problem 4.15), one can go to higher values of n_0 (for constant $\epsilon\omega^2$) with the L-wave than with the O-wave.

(b) For the L-cutoff,

$$\frac{\omega_p^2}{\omega^2} = 1 + \frac{\omega_c}{\omega} \therefore n_c = \frac{\epsilon_0 m\omega^2}{e^2}\left(1 + \frac{\omega_c}{\omega}\right)$$

Thus, to double the usual cutoff density of $\epsilon_0 m\omega^2/e^2$, one must have $f_c = f$

$$f = \frac{c}{\lambda} = \frac{3 \times 10^8}{337 \times 10^{-6}} = 8.9 \times 10^{11}\text{Hz}$$

$$f_c = 28 \times 10^9\text{Hz/T} \therefore B_0 = \frac{8.9 \times 10^{11}}{28 \times 10^9} = 31.8\,\text{T}$$

This would be unreasonably expensive.

(c) The plasma has a density maximum at the center, so it behaves like a convex lens. Such a lens focuses if $\tilde{n} > 1$ and defocuses if $\tilde{n} < 1$. The whistler wave always travels with $v_\phi < c$ (Problem 4.19), so $\tilde{n} = c/v_\phi > 1$, and the plasma focuses this wave.

(d) The question is one of accessibility. If $\omega < \omega_c$ everywhere, the whistler wave will propagate regardless of n_0. However, if $\omega > \omega_c$, the wave will be cut off in regions of low density. From (b) above, we see that a field of 31.8 T is required; this seems too large for the scheme to be practical.

4.35 The answer should come out the same as for cold plasma.

4.36 The linearized equation of motion for either species is

$$-i\omega m n_0 \mathbf{v}_1 = q n_0 (\mathbf{E} + \mathbf{v}_1 \times \mathbf{B}_0) - \gamma k T_i k n_1$$

Thus

$$-i\omega m n_0 \mathbf{k}.\mathbf{v}_1 = q n_0 (\mathbf{k} \cdot \mathbf{E} + \mathbf{k} \cdot \mathbf{v}_1 \times \mathbf{B}_0) - \gamma k T_i k^2 n_1.$$

But $\mathbf{k} \cdot \mathbf{E} = 0$ for transverse wave, and $\mathbf{k} \cdot (\mathbf{v}_1 \times \mathbf{B}_0) = -\mathbf{v}_1 \cdot (\mathbf{k} \times \mathbf{B}_0) = 0$ by assumption. The linearized equation of continuity is

$$-i\omega n_1 + n_0 i \mathbf{k} \cdot \mathbf{v}_1 = 0$$

Substituting for $\mathbf{k} \cdot \mathbf{v}_1$, we have

$$i\omega^2 m n_1 = i\gamma k T k^2 n_1$$

Thus n_1 is arbitrary, and we may take it to be 0. Then the ∇_p term vanishes for both ions and electrons.

4.44 For a given density, the highest cutoff frequency is ω_R. Thus the lowest bound for n is given by $\omega = \omega_R$.

$$\frac{\omega_p^2}{\omega^2} = \frac{f_p^2}{f^2} = 1 - \frac{\omega_c}{\omega} = 1 - \frac{\left(1.6 \times 10^{-19}\right)\left(36 \times 10^{-4}\right)}{\left(0.91 \times 10^{-30}\right)\left(2\pi\right)\left(1.2 \times 10^8\right)} = 0.16$$

$$n = f_p^2/q^2 = (0.16)\left(1.2 \times 10^8\right)^2 q^{-2} = 2.8 \times 10^{13} \text{m}^{-3}$$

4.46 Let $\omega = \omega_R$ at r_1 where $n = n_1$, $\omega_p = \omega_{p1}$; and $\omega = \omega_h$ at r_2, where $n = n_2$, $\omega_p = \omega_{p2}$ Then

$$\omega_{p2}^2 = \omega^2 - \omega_c^2 \tag{4.105}$$

$$\omega_{p1}^2 = \omega^2 - \omega\omega_c \tag{4.107}$$

Thus

$$\omega_{p2}^2 - \omega_{p1}^2 = \omega_c(\omega - \omega_c) = (n_2 - n_1)e^2/\epsilon_0 m$$

But

$$n_2 - n_1 \approx d|\partial n/\partial r| \approx n_1 d/r_0 = \left(\epsilon_0 m/e^2\right)(\omega)(\omega - \omega_c)(d/r_0)$$

So

$$d \approx (\omega_c/\omega)r_0$$

4.47 (a) The accessible resonance is on the far side, past the density maximum.

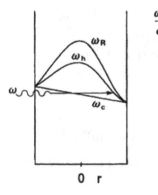
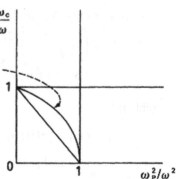

(b) Let ω_{c0} be ω_c at the left boundary, and ω_c be the value at the resonance layer, where $\omega = \omega_p$. Then we require

$$\omega_{c0} > \omega, \quad \text{where } \omega^2 = \omega_c^2 + \omega_p^2$$

Thus

$$\omega_{c0}^2 > \omega_c^2 + \omega_p^2 \quad \omega_{c0}^2 - \omega_c^2 > \omega_p^2$$
$$(\omega_{c0} + \omega_c)(\omega_{c0} - \omega_c) \approx 2\omega_c \Delta\omega_c > \omega_p^2$$

$$\frac{\Delta\omega_c}{\omega_c} = \frac{\Delta B_0}{B_0} > \frac{\omega_p^2}{2\omega_c^2}$$

4.48 These are the upper and lower hybrid frequencies and right- and left-hand cutoff frequencies with ion motions included. Note that $\omega_p^2/\omega_c = \Omega_p^2/\Omega_c$.

Resonance:

$$\omega^4 - \left(\omega_p^2 + \omega_c^2 + \Omega_p^2 + \Omega_c^2\right) + \omega_p^2\Omega_c^2 + \omega_c^2\Omega_p^2 + \omega_c^2\Omega_c^2 = 0$$
$$\omega_+^2 \approx \omega_h^2 + \Omega_p^2\left(1 - \omega_c^2/\omega_h^2\right) \quad \text{(upper hybrid)}$$

$$\omega_-^2 \approx \omega_c^2\Omega_p^2/\omega_h^2 \quad \text{or} \quad \frac{1}{\omega_-^2} = \frac{1}{\omega_c\Omega_c} + \frac{1}{\Omega_p^2} \quad \text{(lower hybrid)}$$

Cutoff:

$$\frac{\overline{\omega}_p^2}{\omega^2} = \left(1 \mp \frac{\omega_c}{\omega}\right)\left(1 \pm \frac{\Omega_c}{\omega}\right) \quad \left(\begin{matrix} R \\ L \end{matrix} \text{cutoff}\right)$$

This is more easily obtained, without approximation, from the form given in Problem 4.50.

5.1 (a) $D_e = KT_e/mv$

$$\sigma = (6\pi)\left(0.53 \times 10^{-10}\right)^2 = 5.29 \times 10^{-20}\text{m}^2$$
$$v = \left(\frac{2E}{m}\right)^{1/2} = \left[\frac{(2)(2)\left(1.6 \times 10^{-19}\right)}{\left(9.11 \times 10^{-31}\right)}\right]$$
$$= 8.39 \times 10^5\,\text{m/s}$$

From Problem 1.1b,

$$n_0 = \left(3.3 \times 10^{19}\right)\left(10^3\right) = 3.3 \times 10^{22}\text{m}^{-3}$$
$$v = n_0\overline{\sigma v} = n_0\sigma v = \left(3.3 \times 10^{22}\right)\left(5.29 \times 10^{-20}\right)\left(8.39 \times 10^5\right)$$
$$= 1.46 \times 10^9 s^{-1}$$

$$D_e = \frac{(2)(1.6 \times 10^{-19})}{(9.11 \times 10^{-31})(1.46 \times 10^9)} = 2.4 \times 10^2 \text{m}^2/\text{s}$$

(b) $j = \mu n e E$

$$\mu_e = eD_e/KT_e = \frac{(1.6 \times 10^{-19})(2.4 \times 10^2)}{(2)(1.6 \times 10^{-19})}$$
$$= 1.2 \times 10^2 \text{m}^2/\text{V s}$$

$$E = \frac{j}{\mu n e} = \frac{2 \times 10^3}{(1.2 \times 10^2)(10^{16})(1.6 \times 10^{-19})} = 1.04 \times 10^4 \text{V/m}$$

5.2

$$\frac{\partial n}{\partial t} = D\nabla^2 n - \alpha n^2$$

$$D\nabla^2 n = D \frac{\partial^2 n}{\partial x^2} = -Dn_0 \left(\frac{\pi}{2L}\right)^2 \cos \frac{\pi x}{2L} = -D \left(\frac{\pi}{2L}\right)^2 n = -\alpha n^2$$

$$\therefore n = \frac{D}{\alpha} \left(\frac{\pi}{2L}\right)^2 = \frac{0.4}{10^{-15}} \left(\frac{\pi}{0.06}\right)^2 = 1.1 \times 10^{18} \text{m}^{-3}$$

5.4 (a) From Problem 5.1a, $v_{en} = 1.46 \times 10^9 \text{ s}^{-1}$. We need to find whether $\mu_{e\perp}/\mu_{i\perp}$ is large or small:

$$\frac{\mu_e}{\mu_i} = \frac{M v_{in}}{m v_{en}} \qquad v_{jn} = n_n \sigma v_j \propto v_{thj} \propto m_j^{-1/2}$$

since σ is approximately the same for ion–neutral and electron–neutral collisions. Thus

$$\frac{\mu_e}{\mu_i} \approx \left(\frac{M}{m}\right)^{1/2} = (4 \times 1{,}836)^{1/2} = 85.7$$

$$\omega_c = \frac{eB}{m} = \frac{(1.6 \times 10^{-19})(0.2)}{9.11 \times 10^{-31}} = 3.52 \times 10^{10}$$

$$\omega_c \tau_{en} = \frac{3.52 \times 10^{10}}{1.46 \times 10^9} \times 24 \qquad 1 + \omega_c^2 \tau_{en}^2 = 580$$

$$\Omega_c \tau_{in} = \omega_c \tau_{en} \left(\frac{m}{M}\right) \left(\frac{M}{m}\right)^{1/2} = (24)(85.7)^{-1} = 0.28$$

$$\frac{\mu_{e\perp}}{\mu_{i\perp}} = \frac{\mu_e}{\mu_i} \frac{1 + \Omega_c^2 \tau_{in}^2}{1 + \omega_c^2 \tau_{en}^2} = (85.7)\frac{1.08}{580} = 0.16 \ll 1$$

$$\therefore D_{a\perp} = \frac{\mu_{i\perp} D_{e\perp} + \mu_{e\perp} D_{i\perp}}{\mu_{i\perp} + \mu_{e\perp}} \approx D_{e\perp} + \frac{\mu_{e\perp}}{m_{i\perp}} D_{i\perp}$$
$$= D_{e\perp} + 0.16 D_{i\perp}$$

But

$$D = \frac{KT}{e}\mu$$

$$\therefore \frac{D_{i\perp}}{D_{e\perp}} = \frac{\mu_{i\perp} T_i}{\mu_{e\perp} T_e} = \frac{1}{0.16}\frac{0.1}{2} = 0.3$$

$$\therefore D_{a\perp} = D_{e\perp}[1 + (0.16)(0.3)] = 1.05 D_{e\perp} \approx D_{e\perp}$$

(b)

$$\frac{a}{(D\tau)^{1/2}} = 2.4 \therefore \tau = \left(\frac{a}{2.4}\right)^2 \frac{1}{D_{a\perp}}$$

$$\tau = \frac{1}{(2.4 \times 10^{-2})^2} \frac{1}{D_{e\perp}}$$

$$D_{e\perp} = \frac{2.4 \times 10^2}{580} = 0.4140 \text{ (from Problem 5.1)}$$

$$\therefore \tau = 42 \ \mu s$$

5.5

$$\Gamma = -D \, dn/dx \quad n = n_0(1 - x/L)$$

$$\Gamma = Dn_0/L \quad (x > 0)$$

$$Q = 2\Gamma = 2Dn_0/L \therefore n_0 = QL/2D$$

5.7

$$\lambda_{ei} \approx v_{the}\tau_{ei} = v_{the}/\nu_{ei}$$

But $v_{the} \propto T_e^{1/2}$ and $\nu_{ei} \propto T_e^{-3/2}$

$$\therefore \lambda_{ei} \propto T_e^{1/2}/T_e^{-3/2} \propto T_e^2$$

5.8

$$\eta_\parallel = 5.2 \times 10^{-5} \frac{\ln\Lambda}{T_{ev}^{3/2}} \Omega\text{-m} \quad (\text{assume } Z = 1)$$

$$= \frac{(5.2 \times 10^{-5})(10)}{(500)^{3/2}} = 4.65 \times 10^{-8} \Omega\text{-m}$$

$$j = I/A = (2 \times 10^5)/(7.5 \times 10^{-3}) = 2.67 \times 10^7 \text{A/m}^2$$

$$E = \eta_\parallel j = (4.65 \times 10^{-8})(2.67 \times 10^7) = 1.2 \text{V/m}$$

5.9 (a)

$$KT_i = 20\,\text{keV}\quad KT_e = 10\,\text{keV}\quad n = 10^{12}\,\text{m}^{-3}$$

$$B = 5T\quad D_\perp = \frac{\eta n(KT_i + KT_e)}{B^2}$$

$$\eta_\perp = (2.0)(5.2 \times 10^{-5})\frac{\ln\Lambda}{T_{ev}^{3/2}} = \frac{(10^{-3})(10)}{(10^4)^{3/2}}$$

$$= 1.0 \times 10^{-9}\Omega\text{-m}$$

$$D_\perp = \frac{(1.0 \times 10^{-9})(10^{21})(3 \times 10^4)(1.6 \times 10^{-19})}{5^2}$$

$$= 3.0 \times 10^{-4}\text{m}^2/s$$

(b)

$$\frac{dN}{dt} = 2\pi rL\Gamma_r\quad \Gamma_r = -D_\perp\frac{\partial n}{\partial r}$$

$$\frac{\partial n}{\partial r} = \frac{n}{0.1}\quad r = 0.50\text{m}\ L = 100\text{m}$$

$$-\frac{dN}{dt} = (2\pi)(0.50)(10^2)(2.0 \times 10^{-4})(10^{21}/0.10) = 6 \times 10^{20}s^{-1}$$

(c)

$$\tau = \frac{N}{-dN/dt} = \frac{n\pi r^2 L}{-dN/dt}\quad r_{\text{effective}} = 0.55\,\text{m}$$

$$\tau = \frac{(10^{21})(\pi)(0.55)^2(10^2)}{6 \times 10^{20}} = 150s$$

5.13

$$\eta_\| = 5.2 \times 10^{-5}\frac{\ln\Lambda}{T_{ev}^{3/2}}\Omega\text{-m} = (5.2 \times 10^{-5})\frac{10}{10^{3/2}}$$

$$= 1.6 \times 10^{-5}\Omega\text{-m}$$

$$\eta j^2 = (1.6 \times 10^{-5})(10^5)^2 = 1.6 \times 10^5\,\text{W/m}^3$$

$$= 1.6 \times 10^5\,\text{J/(m}^3\text{-s)}$$

$$= (1.6 \times 10^5)/(1.6 \times 10^{-19}) = 10^{24}\text{eV/m}^3\text{-}s$$

$$= \frac{dE_{ev}}{dt}$$

$$E = \frac{3}{2}nKT_e \therefore \frac{dE_{ev}}{dt} = \frac{3}{2}n\frac{dT_{ev}}{dt}$$

$$\frac{dT_{ev}}{dt} = \frac{2}{3}\frac{1}{10^{19}}10^{24} = 0.67 \times 10^5 \text{eV}/s = 0.067 \text{eV}/\mu s$$

5.15 (a)

$$en(E\uparrow_\theta^0 -v_{ir}B) - \nabla\uparrow_\theta^0 p_i - e^2n^2\eta(v_{i\theta}- v_{e\theta}) = 0$$
$$-en(E\uparrow_\theta^0 -v_{er}B) - \nabla\uparrow_\theta^0 p_e + e^2n^2\eta(v_{i\theta}- v_{e\theta}) = 0$$

add:

$$-v_{ir}B + v_{er}B = 0. \therefore v_{ir} = v_{er}$$

(This shows ambipolar diffusion.)

(b)

$$en(E_r + v_{i\theta} B) - \frac{\partial p_i}{\partial r} - e^2n^2\eta(v_{ir}^{0} - v_{er}) = 0$$
$$-en(E_r + v_{e\theta} B) - \frac{\partial p_e}{\partial r} + e^2n^2\eta(v_{ir} - v_{er}) = 0$$
$$v_{i\theta} = -\frac{E_r}{B} + \frac{1}{enB}\frac{\partial p_i}{\partial r} = v_E + v_{Di}$$
$$v_{e\theta} = -\frac{E_r}{B} - \frac{1}{enB}\frac{\partial p_e}{\partial r} = v_E + v_{De}$$

(c) From the first equation in (a),

$$v_{ir} = -\frac{e^2n^2\eta}{enB}(v_{i\theta} - v_{e\theta})$$
$$= \frac{en\eta}{B}\frac{1}{enB}\left(\frac{\partial p_i}{\partial r} + \frac{\partial p_e}{\partial r}\right) = -\frac{\eta}{B^2}\frac{\partial p}{\partial r} = v_{er}$$

(This shows the absence of cross-field mobility.)

5.17 (a)

$$\rho_0\frac{\partial \mathbf{v}_1}{\partial t} = \mathbf{j}_1 \times \mathbf{B}_0 \qquad (1)$$

$$\mathbf{E}_1 + \mathbf{v}_1 \times \mathbf{B}_0 = \eta\mathbf{j}_1 \qquad (2)$$

$$\nabla \times \mathbf{E}_1 = -\dot{\mathbf{B}}_1 \quad \nabla \times \mathbf{B}_1 = \mu_0\mathbf{j}_1$$
$$\nabla \times \nabla \times \mathbf{E}_1 = -\nabla \times \dot{\mathbf{B}}_1 = -\mu_0\dot{\mathbf{j}}_1$$

$$-\mathbf{k}(\mathbf{k}^{0}\cdot \mathbf{E}) + k^2\mathbf{E}_1 = i\omega\mu_0\mathbf{j}_1 \qquad (3)$$

$$\mathbf{k} \cdot \mathbf{E} = 0 \quad \text{(transverse wave)}$$

Solve for $\mathbf{v}_1$ in Eq. (2):

$$\mathbf{E}_1 \times \mathbf{B}_0 + (\mathbf{v}_1 \times \mathbf{B}_0) \times \mathbf{B}_0 = \eta \mathbf{j}_1 \times \mathbf{B}_0$$

$$\underbrace{\qquad\qquad\qquad\qquad}_{-\mathbf{v}_{1\perp}B_0^2}$$

$$v_{1\perp} = \frac{\mathbf{E}_1 \times \mathbf{B}_0}{B_0^2} - \frac{\eta \mathbf{j}_1 \times \mathbf{B}_0}{B_0^2}$$

Substitute in Eq. (1), which has no parallel component anyway:

$$-i\omega\rho_0 \left(\frac{\mathbf{E}_1 \times \mathbf{B}_0}{B_0^2} - \frac{\eta \mathbf{j}_1 \times \mathbf{B}_0}{B_0^2} \right) = \mathbf{j}_1 \times \mathbf{B}_0$$

Since, by Eq. (3), $\mathbf{E}$ and $\mathbf{j}_1$ are in the same direction, take them both to be in the $\hat{\mathbf{x}}$-direction. Then the y-component is

$$\frac{E_1}{B_0} = \left(\frac{iB_0}{\omega\rho_0} + \frac{\eta}{B_0} \right) j_1$$

Equation (3) becomes

$$k^2 E_1 = \mu_0 i\omega \frac{E_1}{B_0} \left(\frac{iB_0}{\omega\rho_0} + \frac{\eta}{B_0} \right)^{-1}$$

$$= \mu_0 \omega^2 \left(\frac{B_0^2}{\rho_0} - i\eta\omega \right)^{-1} E_1$$

$$\frac{\omega^2}{k^2} = \mu_0 \left(\frac{B_0^2}{\rho_0} - i\omega\eta \right)$$

(b)

$$k = \left(\mu_0 \omega^2 \right)^{1/2} \left(\frac{B_0^2}{\rho_0} - i\omega\eta \right)^{-1/2}$$

$$= \omega \left(\frac{\mu_0 \rho_0}{B_0^2} \right)^{1/2} \left(1 - \frac{i\omega\eta\rho_0}{B_0^2} \right)^{-1/2}$$

$$\mathrm{Im}(k) = \omega \frac{\omega\eta\rho_0}{2B_0^2} \left(\frac{\mu_0 \rho_0}{B_0^2} \right)^{1/2} = \frac{\omega^2 \eta}{2} \frac{1}{v_A^3}$$

But for small η, $\omega \approx k v_A$, where $k = \mathrm{Re}\ (k)$

$$\therefore \mathrm{Im}(k) \approx \frac{(\eta)(k^2)}{2v_A}$$

6.4 (a)
$$\mathbf{j} \times \mathbf{B} = \nabla p = KT\nabla n \quad (KT = KT_e + KT_i \text{ here})$$
$$(\mathbf{j} \times \mathbf{B}) \times \mathbf{B} = KT\nabla n \times \mathbf{B} = \mathbf{B}(\mathbf{j} \cdot \mathbf{B}) - \mathbf{j}B^2$$

The parallel component is $0 = j_\parallel B^2 - j_\parallel B^2 \therefore j_\parallel$ is arbitrary. The perpendicular component is

$$\mathbf{j}_\perp = \frac{KT}{B^2}\mathbf{B} \times \nabla n = \frac{KT}{B}\frac{\partial n}{\partial r}\hat{\theta}$$

(b)
$$\int \nabla \times \mathbf{B} \cdot d\mathbf{S} = \mu_0 \int \mathbf{j} \cdot d\mathbf{S}$$
$$\oint \mathbf{B} \cdot d\mathbf{L} = \mu_0 \int \mathbf{j} \cdot d\mathbf{S} = \mu_0 L \int_0^\infty j_\theta dr$$

since $\mathbf{j}$ and $d\mathbf{S}$ are both in the $\hat{\theta}$ direction, and L is the width of the loop in the $\hat{\mathbf{z}}$ direction. By symmetry, there can be no B_r, so only the two z-legs of the loop contribute to the line integral. Substituting for j_θ, we have

$$(B_{ax} - B_0)L = \mu_0 L KT \int_0^\infty \frac{\partial n/\partial r}{B(r)} dr$$

(c) $\partial n/\partial r = -n_0\delta(r - a)$, since $\partial n/\partial r$ is a function that is zero everywhere except at $r = a$, is $-\infty$ there, and has an integral equal to $-n_0$. Thus

$$B_{ax} - B_0 = \mu_0 KT \int_0^\infty -n_0 \frac{\delta(r - a)}{B(r)} dr$$

Since all the diamagnetic current is concentrated at $r = a$, B takes a jump from a constant value B_{ax} inside the plasma to another constant value B_0 outside. (Remember that the field inside an infinite solenoid is uniform.) Upon integrating across the jump, one obtains the average value of B on the two sides, i.e., $B(a) = \frac{1}{2}(B_{ax} + B_0)$. Thus

$$B_{ax} - B_0 = \mu_0 KT n_0 \frac{-1}{\frac{1}{2}(B_{ax} + B_0)}$$

$$B_{ax}^2 - B_0^2 = -2\mu_0 n_0 KT$$

$$1 - \frac{B_{ax}^2}{B_0^2} = \frac{2\mu_0 n_0 KT}{B^2} \equiv \beta = 1 \therefore B_{ax} = 0$$

6.5 (a) By Faraday's law, $V = -d\Phi/dt$

$$\therefore \int V \, dt = -N \int \frac{d\Phi}{dt} dt = -N\Delta\Phi$$

Since $\Delta\Phi$ is the flux change due to the diamagnetic decrease in B,

$$-N\Delta\Phi = -N \int (\mathbf{B} - \mathbf{B}_0) \cdot d\mathbf{S}$$

The sign depends on which side of V is considered positive. In practice, this is of no consequence because the oscilloscope trace can easily be inverted by using the polarity switch.

(b) In Problem 6.4b, we can draw the loop so that its inner leg lies at an arbitrary radius r rather than on the axis. We then have

$$B(r) - B_0 = \mu_0 KT \int_r^\infty \frac{\partial n/\partial r}{B(r')} dr' \approx \mu_0 KT \int_r^\infty \frac{\partial n/\partial r'}{B_0} dr'$$

where again KT is short for $\Sigma \, KT$

$$\frac{\partial n}{\partial r} = n_0 \left(\frac{-2r}{r_0^2} \right) e^{-r^2/r_0^2}$$

$$B(r) - B_0 = \frac{\mu_0 KT \, n_0}{B_0 \, r_0^2} \int_\infty^r e^{-r'^2/r_0^2} 2r' \, dr'$$

$$= \frac{\mu_0 n_0 KT}{B_0} \left[e^{-r'^2/r_0^2} \right]_r^\infty = \frac{-\mu_0 n_0 KT}{B_0} e^{-r^2/r_0^2}$$

This is the diamagnetic change in B at any r. To get the loop signal, we must integrate over the plasma cross section.

$$\int V \, dt = -N \int (\mathbf{B} - \mathbf{B}_0) \cdot d\mathbf{S} = -N \int\int [B(r) - B_0] r \, dr \, d\theta$$

where both $\mathbf{B}$ and $d\mathbf{S}$ are in the $\hat{z}$ direction. Substituting for $B(r) - B_0$ and assuming the coil lies well outside the plasma, we have

$$\int V \, dt = N \frac{\mu_0 n_0 KT}{B_0} 2\pi \int_0^\infty e^{-r^2/r_0^2} r \, dr$$

$$= N\pi \frac{\mu_0 n_0 KT}{B_0} r_0^2 \left[e^{-r^2/r_0^2} \right]_\infty^0 = \frac{1}{2} N\pi r_0^2 \left(\frac{2\mu_0 n_0 KT}{B_0^2} \right) B_0$$

(c) The quantity in parentheses is β by definition; hence,

$$\int V \, dt = \frac{1}{2} N \pi r_0^2 \beta B_0$$

Both sides of this equation have units of flux,

6.6 (a)   For each stream, we have

$$m\left(\frac{\partial \mathbf{v}_1}{\partial t} + \mathbf{v}_0 \cdot \nabla \mathbf{v}_1\right) = -e\mathbf{E}_1 = (-i\omega + ikv_0)\mathbf{v}_1$$

$$\mathbf{v}_1 = \frac{-ie\mathbf{E}_1}{m(\omega - kv_0)}$$

$$\frac{\partial n_1}{\partial t} + n_0(\nabla \cdot \mathbf{v}_1) + (\mathbf{v}_0 \cdot \nabla)n_1 = 0$$

$$(-i\omega + ikv_0)n_1 + ikn_0v_1 = 0 \qquad n_1 = n_0 \frac{kv_1}{\omega - kv_0}$$

$$\therefore n_{1j} = n_{0j} \frac{-ikE_1 e}{m(\omega - kv_{0j})^2}$$

Poisson: $ikE_1 = (e/\epsilon_0)(n_{1a} + n_{1b})$, where stream a has $v_{0a} = v_0\hat{\mathbf{x}}$, $n_{0a} = \frac{1}{2}n_0$; stream b has $v_{0b} = -v_0\hat{\mathbf{x}}$, $n_{0b} = \frac{1}{2}n_0$. Thus

$$ikE_1 = -\left(\frac{e}{\epsilon_0}\right)\left(\frac{-ikeE_1}{m}\right)\left[\frac{\frac{1}{2}n_0}{(\omega - kv_0)^2} + \frac{\frac{1}{2}n_0}{(\omega + kv_0)^2}\right]$$

$$1 = \frac{n_0 e^2}{\epsilon_0 m} \cdot \frac{1}{2}\left[\frac{1}{(\omega - kv_0)^2} + \frac{1}{(\omega + kv_0^2)}\right]$$

$$1 = \frac{1}{2}\omega_p^2\left[\frac{1}{(\omega - kv_0)^2} + \frac{1}{(\omega + kv_0)^2}\right]$$

(b)

$$1 = \omega_p^2 \frac{\omega^2 + k^2 v_0^2}{(\omega^2 - k^2 v_0^2)^2}$$

$$\omega^4 - \left(\omega_p^2 + 2k^2 v_0^2\right)\omega^2 + k^2 v_0^2\left(k^2 v_0^2 - \omega_p^2\right) = 0$$

$$\omega^2 = \frac{1}{2}\left(\omega_p^2 + 2k^2 v_0^2\right) \pm \frac{1}{2}\left(\omega_p^4 + 8\omega_p^2 k^2 v_0^2\right)^{1/2}$$

Let

$$x = \frac{2k^2 v_0^2}{\omega_p^2} \qquad y^2 = \frac{2\omega^2}{\omega_p^2}$$

Then

$$y^2 = 1 + x \pm (1 + 4x)^{1/2}$$

y can be complex only if the $(-)$ sign is taken. This y is pure imaginary, and we can let $y = i\gamma$:

$$\gamma^2 = (1 + 4x)^{1/2} - (1 + x)$$
$$\frac{d}{dx}(\gamma^2) = 2(1 + 4x)^{-1/2} - 1 = 0 \quad x = \frac{3}{4}$$

Thus

$$\gamma^2 = (1 + 3)^{1/2} - \frac{7}{4} = \frac{1}{4}$$
$$\gamma = \frac{1}{2} = \frac{\sqrt{2}\,\text{Im}(\omega)}{\omega_p} \quad \text{Im}(\omega) = \frac{\omega_p}{2^{3/2}}$$

6.8 (a)

$$1 = \omega_p^2 \left[\frac{1}{\omega^2} + \frac{\delta}{(\omega - ku)^2} \right]$$

where $\omega_p^2 \equiv n_0 e^2 / \epsilon_0 m$.

(b) This equation is the same as Eq. (6.30) except that m/M is replaced by δ, which is also small, and that the rest frame has changed to one moving with velocity u. The maximum growth rate does not depend on frame, as can be seen from Fig. 6.11 by imagining γ to be plotted in the z direction vs. x and y; a shift in the origin of x will not affect the peak. Analogy with Eq. (6.35) then gives

$$\gamma_{\text{max}} \approx \delta^{1/3} \omega_p$$

(The exact constant that should appear here is $3^{1/2} 2^{-4/3} = 0.69$. The derivation of γ_{max}, which is difficult because the dispersion relation is cubic, and the proof that it is independent of frame for real k are left as exercises for the advanced student.)

6.9 (a) Since only the y component of $\mathbf{v}_j$ and $\mathbf{E}$ are involved, the given relation is easily found from Eqs. (4.98b) and (6.23), plus continuity and Poisson's equation. Note that Ω_p is defined with n_0, not $(1/2)n_0$.

(b) Let $\alpha \equiv \frac{1}{2}\Omega_p^2 \left(1 + \omega_p^2/\omega_c^2\right)^{-1}, \beta \equiv k^2 v_0^2$. Then the dispersion relation reduces to

$$\omega^4 - 2(\alpha + \beta)\omega^2 + \beta^2 - 2\alpha\beta = 0$$

The dispersion $\omega(k)$ is given by

$$\omega^2 = \alpha + \beta \pm \left(\alpha^2 + 4\alpha\beta\right)^{1/2}$$

Instability occurs if $(\alpha^2 + 4\alpha\beta)^{1/2} > \alpha + \beta$, or $\beta < 2\alpha$, i.e.,

$$k^2 < \left(\Omega_p^2/v_0^2\right)\left(1 + \omega_p^2/\omega_c^2\right)^{-1}$$

Where this is satisfied, the growth rate is given by

$$\gamma = \left[\left(\alpha^2 + 4\alpha\beta\right)^{1/2} - (\alpha + \beta)\right]^{1/2}$$

7.3 (a)

$$f_p(v) = \frac{n_p}{a\pi^{1/2}}e^{-v^2/a^2}$$
$$f_b(v) = \frac{n_b}{b\pi^{1/2}}e^{-(v-V)^2/b^2}$$

(b)

$$f_b'(v) = \frac{n_b}{b\pi^{1/2}}\frac{-2(v - V)}{b^2}e^{-(v-V)^2/b^2}$$
$$f_b''(v) = \frac{-2n_b}{b^3\pi^{1/2}}\left[1 - \frac{2(v - V)^2}{b^2}\right]e^{-(v-V)^2/b^2} = 0$$
$$v - V = \pm b/\sqrt{2} \quad v_\phi = V - b/\sqrt{2}$$
$$f_b'(v_\phi) = \left(\tfrac{2}{\pi}\right)^{1/2}\frac{n_b}{b^2}e^{-1/2}$$

(c)

$$f_b'(v_\phi) = \frac{n_p}{a\pi^{1/2}}\left(\frac{-2}{a^2}\right)\left(V - \frac{b}{2^{1/2}}\right)e^{-(V-b/\sqrt{2})^2/a^2}$$
$$\approx -\frac{2n_pV}{a^3\pi^{1/2}}e^{-V^2/a^2} \quad V \gg b$$

(d)

$$\left(\frac{2}{\pi}\right)^{1/2}\frac{n_b}{b^2}e^{-1/2} = \frac{2n_pV}{a^3\pi^{1/2}}e^{-V^2/a^2}$$

$$\frac{n_b}{n_p} = (2e)^{1/2}\frac{b^2}{a^3}Ve^{-V^2/a^2} \qquad \frac{b^2}{a^2} = \frac{T_b}{T_p}$$

$$\therefore \frac{n_b}{n} = \left(2e^{1/2}\right)\frac{T_bV}{T_pa}e^{-V^2/a^2}$$

7.8 From Eq. (7.127), we obtain $\sum\alpha_jZ'(\zeta_j) = 2T_i/T_e$, where $\alpha_j = n_{0j}/n_{0e}$, $\zeta_j = \omega/kv_{\mathrm{th}j}$. Assume at first that α_H is small, so that $\alpha_A \approx 1$, $\alpha_H = \alpha$; furthermore, small α means that v_ϕ will be nearly unchanged from v_s of argon. Then doubling the Landau damping rate means Im $Z'(\zeta_H) = $ Im $Z'(\zeta_A)$, where Im $Z'(\zeta_j) = -2i\sqrt{\pi}\zeta_je^{-\zeta_j^2}$. = Thus

$$\zeta_Ae^{-\zeta_A^2} = \alpha\zeta_He^{-\zeta_H^2} \qquad \alpha = \frac{\zeta_A}{\zeta_H}e^{-(\zeta_A^2-\zeta_H^2)}$$

$$\frac{\zeta_A}{\zeta_H} = \left(\frac{M_A}{M_H}\right)^{1/2} \qquad \alpha = (40)^{1/2}e^{-\zeta_A^2(1-1/40)}$$

$$\zeta_A^2 = \frac{KT_e + 3KT_i}{M_A}\cdot\frac{M_A}{2KT_i} = \frac{13}{2}$$

$$\alpha = \sqrt{40}e^{-6.5(0.975)} = 1.12\times10^{-2} \approx 1\%$$

Thus α is so small that our initial assumptions are justified.

7.9 (a)

$$\frac{2k^2}{k_{Di}^2} = Z'(\zeta_i) + \frac{1-\alpha}{\theta_e}Z'(\zeta_e) + \frac{\alpha}{\theta_h}Z'(\zeta_h)$$

(b)

$$Z'(\zeta_i) \approx -2 - 2i\sqrt{\pi}\zeta e^{-\zeta^2}$$

Since $\zeta_h \ll \zeta_e \ll 1$,

$$\left|\mathrm{Im}Z'(\zeta_h)\right| \ll \left|\mathrm{Im}Z'(\zeta_e)\right|$$

(c) Since $Z'(\zeta_h) \approx Z'(\zeta_e) \approx -2$), the ζ_h term in (a) is negligible compared with the ζ_e term if $\theta_h \gg \theta_e$ and $\alpha < 1/2$. Now the dispersion relation is

$$Z'(\zeta_i)| = \frac{2k^2}{k_{Di}^2} + \frac{2(1-\alpha)}{\theta_e} = \frac{2T_i}{T_e}\left(1-\alpha+\frac{T_ek^2}{T_ik_{Di}^2}\right)$$

The last term is $\approx k^2\lambda_D^2$ and is negligible when quasineutrality holds. Thus the ion wave dispersion relation is the same as usual, except that T_i/T_e has been replaced by $(1-\alpha)\,T_i/T_e$. Since small T_i/T_e means less Landau damping, the hot electrons have decreased ion Landau damping.

8.3 Refer to Fig. 8.4. Take a number of ions with $v=u_0$ and split them into two groups, one with $v=u_0+\Delta$ and one with $v=u_0-\Delta$. After acceleration in a potential ϕ, the faster half will have less fractional energy gain (because it started with more energy) and, hence, will have less fractional density decrease. The opposite is true for the slower half, and to first order the total density decrease is the same as if all ions had $v=u_0$. However, there is a second-order effect which makes the slower group dominate. This can be seen by making Δ so large that $v\approx0$ for the slower half, which clearly must then suffer a huge density decrease. To compensate for this, u_0 must be *increased* to higher than the Bohm value.

8.4 The maximum current occurs when the space charge of decelerated ions near grid 3 decreases the electric field to zero. Thus we can apply the Child-Langmuir law to the region between grids 2 and 3.

$$J = \frac{4}{9}\left[\frac{(2)(1.6\times10^{-19})}{(4)(1.67\times10^{-27})}\right]^{1/2}\frac{(8.85\times10^{-12})(100)^{3/2}}{(10^{-3})^2} = 27.2\frac{A}{m^2}$$

$$A = \frac{\pi}{4}(4\times10^{-3})^2 = 1.26\times10^{-5}\,m^2$$

$$I = JA = \underline{0.34mA}$$

8.6 (a) At $\omega_p=\omega$,

$$F_{NL} = -\frac{\epsilon_0\langle E^2\rangle}{2L} = -\nabla p_{eff} = \frac{p_{eff}}{L}$$

$\therefore p_{eff} = \frac{1}{2}\epsilon_0\langle E^2\rangle$. But $I_0 = c\epsilon_0\langle E^2\rangle = P/A$, where $P=10^{12}$ and $A = (\pi/4)\,(50\times10^{-6})^2 = 1.96\times10^{-9}\,m^2$

$$p_{eff} = \frac{P}{2cA} = \frac{10^{12}}{(2)(3\times10^8)(1.96\times10^{-9})} = 8.50\times10^{11}\,\frac{N}{m^2}$$

$$= \frac{(8.50\times10^{11})(0.2248)}{(39.37)^2} = 1.23\times10^8\,\frac{lb}{in.^2}$$

(b)
$$F = pA \quad P/2c = 10^{12}/(2)(3\times10^8) = 1667\,N$$

$$F = Mg \quad M = F/g = 1667/9.8 = 170\,kg = 0.17\,tonnes$$

(c)

$$2nKT = p_{\text{eff}}$$

$$\therefore n = \frac{8.5 \times 10^{11}}{(2)(10^3)(1.6 \times 10^{-19})} = 2.66 \times 10^{27} \text{m}^{-3}$$

8.7

$$F_{\text{NL}} = \nabla p \therefore \frac{\partial}{\partial r}(nKT) = -\frac{n}{n_c}\frac{\partial}{\partial r}\left(\frac{\epsilon_0\langle E^2\rangle}{2}\right)$$

$$\frac{1}{n}\frac{\partial n}{\partial r} - \frac{\epsilon_0}{2n_cKT}\frac{\partial}{\partial r}\langle E^2\rangle \quad \ln n = -\frac{\epsilon_0\langle E^2\rangle}{2n_cKT} + \ln n_0$$

$$n = n_0 e^{-\epsilon_0\langle E^2\rangle/2n_cKT}$$

At r = 0,

$$n_{\min} = n_0 e^{-\epsilon_0\langle E^2\rangle\max/2n_cKT} = n_0 e^{-\alpha}$$

$$\therefore \alpha = \frac{\epsilon_0\langle E^2\rangle_{\max}}{2n_cKT}$$

8.9

$$k_0 = 2\pi/\lambda_0 = 2\pi/1.06 \times 10^{-6} = 5.93 \times 10^6 \text{m}^{-1}$$

$$k_i \approx 2k_0 = 1.19 \times 10^7 \text{m}^{-1}$$

$$v_s = \left(\frac{KT_e + 3KT_i}{M}\right)^{1/2} = \left[\frac{(10^3)(1.6 \times 10^{-19})}{(2)(1.67 \times 10^{-27})}\right]^{1/2}\left(1 + \frac{3}{\theta}\right)^{1/2}$$

$$\omega_i = \Delta\omega = k_i v_s = (1.19 \times 10^7)(2.19 \times 10^5)\left(1 + \frac{3}{\theta}\right)^{1/2}$$

$$= 2.61 \times 10^{12}\left(1 + \frac{3}{\theta}\right)^{1/2}$$

$$\frac{\Delta\omega}{\omega_0} = -\frac{\Delta\lambda}{\lambda_0} \therefore \Delta\omega = -\frac{\omega_0}{\lambda_0}\Delta\lambda = -\frac{2\pi c}{\lambda_0^2}\Delta\lambda$$

$$= -\frac{(2\pi)(3 \times 10^8)}{(1.06 \times 10^{-6})^2}(21.9 \times 10^{-10})$$

$$= 3.67 \times 10^{12}$$

$$1 + \frac{3}{\theta} = \left(\frac{3.67 \times 10^{12}}{2.61 \times 10^{12}}\right)^2 = 2 \quad \theta = \frac{T_e}{T_i} = 3 \therefore T_i = \frac{1}{3}\text{keV}$$

8.10 (a)

$$\langle E_0^2 \rangle = \frac{1}{2}\bar{E}^2 = \frac{8\omega_1\omega_2\Gamma_1\Gamma_2}{c_1 c_2}$$

$$c_1 c_2 = \frac{\epsilon_0 k_1^2 \omega_p^4}{n_0 \omega_0^2 M} \qquad \Gamma_2 = \frac{\omega_p^2}{\omega_2^2}\frac{\nu}{2}$$

$$\langle E_0^2 \rangle = \frac{4\omega_1\Gamma_1\omega_0^2\nu}{\omega_2 k_1^2}\frac{n_0 M}{\epsilon_0 \omega_p^2} = \frac{4\omega_1\Gamma_1\omega_0^2\nu M m}{\omega_2 k_1^2 e^2}$$

$$\langle v_0^2 \rangle = \frac{e^2\langle E_0^2 \rangle}{m^2 \omega_0^2} = \frac{4\omega_1\Gamma_1\nu M}{\omega_2 k_1^2 m}$$

$$k_1^2 = \frac{\omega_1^2}{v_s^2} = \frac{\omega_1^2 M}{KT_e} = \frac{\omega_1^2 v_e^2 M}{M} \quad \therefore \frac{\langle v_0^2 \rangle}{v_e^2} = \frac{4\Gamma_1\nu}{\omega_1\omega_2}$$

(b)

$$\frac{\langle v_0^2 \rangle}{v_e^2} = \frac{4\Gamma_1\nu_{ei}}{\omega_1\omega_0}$$

since $\omega_2 \approx \omega_0$ when $n \ll n_c$.

$$\omega_0 = \frac{2\pi c}{\lambda_0} = \frac{(2\pi)(3 \times 10^8)}{10.6 \times 10^{-6}} = 1.78 \times 10^{14}s^{-1}$$

$$v_e^2 = \frac{KT_e}{m} = \frac{(10^2)(1.6 \times 10^{-19})}{(0.91 \times 10^{-30})} = 1.76 \times 10^{13}\frac{m^2}{s^2}$$

$$\frac{\Gamma_1}{\omega_1} = \left(\frac{\pi}{8}\right)^{1/2}\theta(3+\theta)^{1/2}e^{-(3+\theta)/2} \qquad \theta = \frac{T_e}{T_i} = 10$$
$$= 3.40 \times 10^{-2}$$

$$\eta = 5.2 \times 10^{-5}\frac{\ln\Lambda}{T_{eV}^{3/2}} = \frac{(5.2 \times 10^{-5})(10)}{(100)^{3/2}} = 5.2 \times 10^{-7}\Omega\text{-m}$$

$$v_{ei} = \frac{ne^2\eta}{m} = \frac{(10^{23})(1.6 \times 10^{-19})^2(5.2 \times 10^{-7})}{(0.91 \times 10^{-30})} = 1.46 \times 10^9 s^{-1}$$

$$\langle v_0^2 \rangle = \frac{(4)(3.4 \times 10^{-2})(1.46 \times 10^9)}{1.78 \times 10^{14}}(1.76 \times 10^{13}) = 1.96 \times 10^7\frac{m^2}{s^2}$$

From Problem 8.6(a):

$$I_0 = c\epsilon_0\langle E^2 \rangle = c\epsilon_0\frac{m^2\omega_0^2}{e^2}\langle v_0^2 \rangle$$

$$I_0 = (3 \times 10^8)(8.854 \times 10^{-12}) \frac{(0.91 \times 10^{-30})^2 (1.78 \times 10^{14})^2 (1.96 \times 10^7)}{(1.6 \times 10^{-19})^2}$$

$$= 5.34 \times 10^{10} \frac{W}{m^2} = 5.34 \times 10^6 \frac{W}{cm^2}$$

8.11
$$\left(\omega_s^2 + 2i\gamma\omega_s - \omega_1^2\right)\left[(\omega_s + i\gamma - \omega_0)^2 - \omega_2^2\right] = \tfrac{1}{4}c_1 c_2 \overline{E}_0^2.$$

If $\omega_s^2 = \omega_1^2$, $(\omega_s - \omega_0)^2 = \omega_2^2$, and $\gamma/\omega_s \ll 1$, then

$$(2i\gamma\omega_s)[2i\gamma(\omega_s - \omega_0)] = \frac{1}{4}c_1 c_2 \overline{E}_0^2 = 4\gamma^2 \omega_s \omega_2$$

From Problem 8.10,

$$c_1 c_2 = \frac{\epsilon_0 k_1^2 \omega_p^4}{n_0 \omega_0^2 M} = \frac{k_1^2 \omega_p^2 e^2}{\omega_0^2 mM}$$

$$\gamma^2 = \frac{k_1^2 \omega_p^2 e^2 \overline{E}_0^2}{16\omega_s \omega_2 \omega_0^2 mM} = \frac{k_1^2 \omega_p^2 \overline{v}_0^2 m}{16\omega_s \omega_2 M} \approx \frac{(2k_0)^2 \Omega_p^2 \overline{v}_0^2}{16\omega_0 \omega_s}$$

$$= \frac{\omega_0^2 \Omega_p^2 \overline{v}_0^2}{4c^2 \omega_0 \omega_s} \quad \therefore \gamma = \frac{\overline{v}_0}{2}\left(\frac{\omega_0}{\omega_s}\right)^{1/2} \Omega_p$$

8.13 (a)
$$Mn_0 \frac{\partial v}{\partial t} = en_0 E - \gamma_i KT_i \nabla_n - Mn_0 \nu v + F_{NL}$$

$$Mn_0(-i\omega + \nu)v = en_0(-ik\phi) - \gamma_i KT_i ikn_1 + F_{NL}$$

with $e\phi/KT_e = n_1/n_0$, this becomes

$$(\omega + i\nu)v = kv_s^2 \frac{n_1}{n_0} + \frac{iF_{NL}}{Mn_0}$$

Continuity:

$$-i\omega n_1 + ikn_0 v = -i\omega n_1 + ikn_0(\omega + i\nu)^{-1}\left[kv_s^2 \frac{n_1}{n_0} + \frac{iF_{NL}}{Mn_0}\right] = 0$$

$$(\omega^2 + i\nu\omega - k^2 v_s^2)n_1 = ikF_{NL}/M$$

When $F_{NL} = 0$,

$$\omega^2\left(1+i\frac{\nu}{\omega}\right)=k^2 v_s^2 \therefore \omega \approx k v_s\left(1-\frac{1}{2}i\frac{\nu}{\omega}\right)=k v_s-\frac{i}{2}\nu$$

Hence $-\operatorname{Im}\omega\equiv\Gamma=\nu/2.$ So$\left(\omega^2+2i\Gamma\omega-k^2 v_s^2\right)n_1=ikF_{NL}/M$

(b)

$$F_{NL}=-\frac{\omega_p^2}{\omega_0\omega_2}\nabla\epsilon_0\langle E_0 E_2\rangle=-\frac{\omega_p^2}{\omega_0\omega_2}ik\epsilon_0\langle E_0 E_2\rangle$$

Thus,

$$c_1=\frac{ikF_{NL}}{M}\frac{1}{\langle E_0 E_2\rangle}=\frac{ik}{M}\left(\frac{-\omega_p^2}{\omega_0\omega_2}ik\ \epsilon_0\right)=\frac{\omega_p^2}{\omega_0\omega_2}\frac{k^2\epsilon_0}{M}$$

8.14 The upper sideband has $\hbar\omega_2=\hbar\omega_0+\hbar\omega_1$, so that the outgoing photon has *more* energy than the original photon $\hbar\omega_0$. The lower sideband would be expected to be more favorable energetically, since it is an exothermic reaction, with $\hbar\omega_2=\hbar\omega_0-\hbar\omega_1$.

8.18 $U(\xi-c\tau)=3c\ \operatorname{sech}^2\ [(c/2)^{1/2}(\xi-c\tau)]$, where $\xi=\delta^{1/2}(x'-t')$, $\tau=\delta^{3/2}t'$, $x'=x/\lambda_D$, $t'=\Omega_p t$, $\delta=M-1$

$$\zeta=\xi-ct=\delta^{1/2}\left(\frac{x-v_s t}{\lambda_D}-\delta c\frac{v_s}{\lambda_D}t\right)$$

since $\lambda_D\Omega_p=v_s$

$$\zeta=\frac{\delta^{1/2}}{\lambda_D}[x-(1+\delta c)v_s t]$$

The soliton has a peak at $\zeta=0$. The velocity of the peak is $dx/dt=(1+\delta c)v_s$. By definition,

$$\frac{dx}{dt}=Mv_s=(1+\delta)v_s$$

$$\therefore c=1\quad\therefore U_{max}=3c=3$$

From Eq. (8.111),

$$x_{\max} \equiv \frac{e\phi_{\max}}{KT_e} \approx \delta x_{1\max} = \delta U_{\max}$$

$$\therefore \delta \frac{e}{KT_e} \frac{\phi_{\max}}{U_{\max}} = \frac{121}{103} = 0.4$$

$$v_\phi = (1+\delta)v_s = 1.4 v_s$$

$$v_s = \left(\frac{KT_e}{M}\right)^{1/2} = \left[\frac{(10)(1.6 \times 10^{-19})}{1.67 \times 10^{-27}}\right] = 3.10 \times 10^4$$

$$\underline{v_\phi = 4.33 \times 10^4 \text{m/s}}$$

At half maximum, $\text{sech}^2 a = \frac{1}{2} \therefore a = 0.8814 = \sqrt{\frac{1}{2}\zeta} \therefore \zeta = 1.25 = \delta^{1/2} x/\lambda_D$ at $t = 0$, say.

$$\delta^{1/2} = \sqrt{0.4} = 0.632$$

$$\lambda_D = \left(\frac{\epsilon_0 KT_e}{n_0 e^2}\right)^{1/2} = 2.35 \times 10^{-4}\text{m} = 0.235\,\text{mm}$$

$$x = \frac{1.25\lambda_D}{0.632} = 0.46\,\text{mm} \quad \text{FWHM} = 2x = \underline{0.93\text{mm}}$$

8.21 $$|u| = 4A^{1/2}|\sec hx| \therefore |u|^2 = 16A|\sec hx|^2$$

$$\delta n = \frac{1}{4}|u|^2 \left(\frac{V^2}{\mathcal{E}^2} - 1\right)^{-1} \approx -\frac{1}{4}|u|^2 = -4A|\sec hx|^2$$

$$\overline{\delta n} = -4A\overline{|sechx|^2} \approx -2A$$

$$\frac{\delta\omega_p}{\omega_p} = \frac{1}{2}\frac{\delta n}{n} = -\frac{1}{2}(2A) = -A$$

$\therefore A$ is frequency shifted due to δn.

8.22 In real units,

$$v = \frac{v}{v_e} = 4A^{1/2}\text{sech}\left[\left(\frac{2A}{3}\right)^{1/2}\left(\frac{x}{\lambda_D} - \frac{V}{v_e}\omega_p t\right)\right]\exp\left\{-i\left[\left(\frac{\omega_0}{\omega_p} + \frac{1}{6}\frac{V^2}{v_e} - A\right)\right]\omega_p t\right.$$

$$\left. -\frac{V}{3v_e}\frac{x}{\lambda_D}\right\}$$

$$v_e = \left(\frac{KT_e}{m}\right)^{1/2} = 5.93 \times 10^5 \text{m/s} \quad \omega_p = \left(\frac{ne^2}{\epsilon_0 m}\right)^{1/2} = 1.78 \times 10^9 \frac{\text{rad}}{\text{s}}$$

$$\lambda_D = \frac{v_e}{\omega_p} = 3.33 \times 10^{-4} \text{m} \quad k = \frac{(k\lambda_D)}{\lambda_D} = \frac{0.3}{\lambda_D} = 9.02 \times 10^2 \text{m}^{-1}$$

$$u_{p-p} = 4A^{1/2} \qquad -i\omega mv = -eE = -e(-ik\phi) \therefore \phi = -\frac{m\omega v}{ek}$$

$$\phi_{p-p} \approx \frac{m\omega}{ek} 4A^{1/2} v_e \quad \omega \approx \left(\omega_p^2 + 3k^2 v_e^2\right)^{1/2} 2.01 \times 10^9$$

$$A^{1/2} = \frac{ke\phi_{p-p}}{4m\omega v_e} = \frac{k}{4\omega}\frac{e\phi_{p-p}}{KT_e}\frac{1}{v_e} = \frac{kv_e}{4\omega}\frac{e\phi_{p-p}}{KT_e}$$

$$= \frac{kv_e}{4\omega}\frac{3.2}{2} = 0.106$$

$$A = 1.13 \times 10^{-2}$$

(a)

$$\text{sec h} X = \frac{1}{2} \quad X = 1.315 = \left(\frac{2A}{3}\right)^{1/2}\frac{x}{\lambda_D}$$

$$x = \left(\frac{3}{2}\right)^{1/2}\frac{(1.315)(3.33 \times 10^{-4})}{0.106} = 5.04 \times 10^{-3}$$

$$\text{FWHM} = 2x = 1.01 \times 10^{-2} = 10.1 \text{ mm}$$

(b)

$$N = \frac{1.01 \times 10^{-2}}{2\pi/k} = 1.45$$

(c)

$$\delta\omega = A\omega_p = \left(1.13 \times 10^{-2}\right)\left(1.78 \times 10^9\right) = 2 \times 10^7 \text{rad/s}$$

$$\delta f = \delta\omega/2\pi = 3.2 \times 10^6 = 3.2 \text{MHz}$$

8.23

$$3v_e^2 = \frac{(3)(3)(1.6 \times 10^{-19})}{0.91 \times 10^{-30}} = 1.58 \times 10^{12} \text{m}^2/\text{s}^2$$

$$\omega_p^2(\text{out}) = \frac{(10^{16})(1.6 \times 10^{-19})^2}{(8.824 \times 10^{-12})(0.91 \times 10^{-30})} = 3.18 \times 10^{19} \frac{\text{rad}^2}{\text{s}^2}$$

$$\omega_p^2(\text{in}) = 0.4\omega_p^2(\text{out})$$

$$k_{max}^2 = \frac{\omega_p^2(\text{out}) - \omega_p^2(\text{in})}{3v_e^2} = \frac{3.18 \times 10^{19}}{1.58 \times 10^{12}}(1 - 0.4)$$

$$= 1.21 \times 10^7 \text{m}^{-2}$$

$$\lambda_{min} = \frac{2\pi}{k_{max}} = 1.81 \times 10^{-3}\text{m} = \underline{1.81 \text{ mm}}$$

Index

© Springer International Publishing Switzerland 2016
F.F. Chen, *Introduction to Plasma Physics and Controlled Fusion*,
DOI 10.1007/978-3-319-22309-4

Printed in the United States
By Bookmasters